Science in India

A Historical Perspective

Science in India

A Historical Perspective

B. V. Subbarayappa

GUTENBERG
INDIA

First Edition, 2018
ISBN 978-93-86240-41-5

Published by
Gutenberg Inc.
237/2 BT, Victoria Street, Chowk,
Lucknow 226003 INDIA
call. 9005090070, 9005005005

Science in India

A Historical Perspective

B. V. Subbarayappa

First Edition, 2018
ISBN 978-93-86240-41-5

Published by
Gutenberg Inc.
237/2 BT, Victoria Street, Chowk,
Lucknow 226003 INDIA
call. 9005090070, 9005005005

Dedicated to my family
Lalitha, Sheshu, Satya and Sarvamangala

CONTENTS

PREFACE

The history of science, like history, owes its authenticity to the preserved records or visible, knowable sources. A historian of science is governed by them, as it were, in his analysis, interpretations and perspectives. There are mainly two kinds of sources for the history of science and technology in India—literary and archaeological. The wealth of scientific manuscripts (by and large in Sanskrit) and preserved in a number of repositories in and outside of India, is indeed impressive. They shed light on the scientific tradition in astronomy, mathematics, medicine, plants and the like, which have been fostered in India for over two millennia. However, there are not many manuscripts on technological aspects; practically none on metallurgy in which India had attained great heights, particularly in iron, steel and zinc, as evidenced by extensive archaeological data relating to them. It is amazing that, even though astronomers, mathematicians and medical men lived separately in far off places and in different centuries, they adopted the same scientific terminology and wrote their treatises more or less in the same style; likewise, the metallurgists and other technicians functioned and displayed their skills in the ancient and medieval periods.

In the eighteenth and nineteenth centuries, when the methodology-centred modern science was introduced into India by the British colonial administration, Indians welcomed it because of their scientific tradition and contributed to its progress, specially in the twentieth century. The records of the East India

Company and of the colonial government are reputed to be the best preserved records in the world, and they are valuable sources for evaluating the advent and growth of modern science in India.

The dedicated endeavours of several European scholars in the seventeenth and eighteenth centuries, and later of Indian savants, have led to valuable publications for understanding traditional sciences, besides the archival documents, for the evaluation of the historical development of modern science in India.

A historian of science in India now has a large repertory of source-materials—both primary and secondary including books and articles—and has the benefit of the wisdom of a great many scholars of yore. The author of this book is enormously indebted to all of them. The preparation of this book, intended for university students and enlightened general readers, would not have been possible but for their valuable publications bequeathed to us.

An attempt has been made in this book to present a historical perspective on science in India, from early times to Independence. Its Introductory Overview, the first chapter, is to provide the reader a bird's-eye view of the whole canvas of the book. The second chapter tries to delineate briefly the early scientific ideas and practices, especially those of the physical and plant world. Chapters 3-7 deal with traditional astronomy, mathematics, chemical practices and medicine. The eighth and last chapter discusses the development of modern science in India before Independence, with a brief account of its growth in the West so that it could serve as a background for understanding its nature and structure.

Science in India, whether traditional or modern, needs to be understood in the context of the developments in the contemporaneous cultures or regions, as well as its educational dimensions. This book also attempts to discuss briefly these aspects, besides the advent and growth of Islamic astronomy and medicine (Unāni) in India.

It is my fond hope that this book will be found useful by all those who are interested in understanding our scientific tradition over the ages. My humble efforts will be amply rewarded if the book appeals to young students and induces them to undertake researches in this fascinating field.

B.V. Subbarayappa

FOREWORD

The Indus Valley Civilization or the Harappa Culture as the archaeologists prefer to call it, was by far the largest of the three ancient civilizations, the other two being the Egyptian and the Mesopotamian. It flourished for well over 700 years (2600-1700 BCE) and lasted in a decadent phase for a couple of centuries more. This civilization was especially noted for its functional town planning, drainage system, public water bath, standardized bricks, copper-bronze technology, agricultural production and management. However, we are still in the dark about its thoughts on astronomy, mathematics, medicine and the like, since the so-called Indus script has not been deciphered so far.

In its historical matrix, science in India has two facets: one, traditional astronomy, mathematics, medicine and the physical worldview, the origins of which can be traced to the Vedic and the post-Vedic period; and two, the advent and growth of modern science in India. In the ancient and medieval periods up to the nineteenth century, India witnessed remarkable achievements in computational astronomy, mathematics, medicinal as well as surgical practices and metallurgy. Some of them, like the adoption of the decimal place-value system using only nine digits and zero, certain surgical practices like rhinoplasty, zinc metallurgy, iron macro-technology, and the production of high quality steel, had no parallels elsewhere then. Besides, Indian mathematical achievements were of high order.

The *Ṛgveda* had set the tone: 'Let noble thoughts come

from everywhere unhindered and overflowing'—and thus open-mindedness characterized our scientific heritage. In the ancient and medieval periods, Indian scientific and technical achievements were due to both endogenous intellectual efforts and exogenous influences, but much more of the former. Among the exogenous influences, special mention needs to be made of the Hellenistic geocentric planetary astronomy—along with its epicyclic and eccentric methods, which are still a debatable issue—and its concept of, and naming of the seven days of the week after planets; Graeco-Arabic or Unāni medicine; Chinese Taoist ideas and practices which provided seed ideas for the *Rasaśāstra* and for Siddha medicine; paper production and pyrotechnics.

Although the recording tradition was not the forte of India, which preferred oral transmissions for a long time, the large number of scientific and related manuscripts, especially from the thirteenth century onwards, preserved in various repositories in and outside of India, speak volumes about the scientific activities fostered by Indian savants over the centuries—and these are of great value to the understanding of our scientific tradition.

Such a tradition knew no barriers of religion or region. Thus, when modern science (it was called Western science then) was introduced into India, largely by the British, in the nineteenth century and in the colonial period, Indian intellectuals and even the common people did not think of it as an intangible intruder; instead, they welcomed it. In the twentieth century, several Indian scientific pioneers emerged, including a Nobel Laureate in physics, who won international acclaim. In addition, the freedom movement was also a source of inspiration to them. The charismatic leader Jawaharlal Nehru was a votary of science, and the scientific attitude or 'temper' as he called it, even before India attained Independence. Under his leadership, modern science and technology took rapid strides for the betterment of the poverty-stricken masses of the country.

It is significant that, even in the modern scientific milieu, traditional sciences have not taken a back seat. For traditional astronomy coexists with modern astronomy in India; likewise, traditional medicine and technology.

In this book, Professor B.V. Subbarayappa has brought out in a lucid fashion the various facets of the developments in science and technology in an exceedingly comprehensive manner. He has brought to bear his extraordinary scholarship in delving into the details of Indian contributions to astronomy, mathematics, material sciences and the various aspects of the medical systems over millennia. A striking and unique aspect of his presentation relates to the coherent structuring of the evolution of ideas and knowledge in different epochs in each of the topics—which demands the highest level of versatility and familiarity with the related details of these fundamental and applied sciences. The way Prof. Subbarayappa has, for example, dealt with the presentation and discussions in the chapter on mathematics—starting from the Vedic period through succeeding centuries, going through Āryabhaṭa I, to contributions in mathematics, geometry, trigonometry, followed by the growth of the Kerala School in the fourteenth to nineteenth century—really highlights this versatility. Further, he goes on to critically appraise the Indian contributions in these fields, in the context of the influences from Greece, Islamic traditions, along with the links with Chinese contributions. A similar trait can be seen in the discussions related to astronomy, chemical techniques and alchemy, Ayurveda and other medical systems like Siddha and Unāni.

In the last chapter, 'The New Entrant: Modern Science in India', Prof. Subbarayappa has broadly outlined the developments in science during the Renaissance in Europe, going through a chronological sequence of developments and covering the works of such titans as Galileo, Johannes Kepler and Isaac Newton. He further traces the developments in nineteenth-century sciences including atomic theory, the emergence of the periodic table, the seminal contributions made in electromagnetics, developments in biology including evolutionary theories followed by the more modern theories such as quantum mechanics and the general theory of relativity. The corresponding Indian developments, starting from the seventeenth century to the initiation of the major scientific endeavours of Jawaharlal Nehru also help to update the reader on the continuing march of science and its relationship to society.

On the whole, it is the most comprehensive account of the historic march of science in our country that I have come across. Prof. Subbarayappa's narration emphasizes the pride of position that India enjoyed in developing and contributing ideas and enlightenment for the welfare of humankind over millennia. I am sure this book—which has a host of unique information that Prof. Subbarayappa has carefully included—will not only serve the purpose of providing a comprehensive account of the great contributions that have emerged from this continent since time immemorial, but will also trigger fresh ideas and research in new directions for those who are interested in pursuing scientific endeavours. As expected, this work will certainly be an outstanding contribution to the scientific heritage of this country for all times to come.

K. Kasturirangan
Former Chairman, ISRO and
Member, Planning Commission

New Delhi
July 2013

ACKNOWLEDGEMENTS

At the outset, I offer my grateful thanks to the Governing Council of the National Centre for History of Science (est. 2004 in the campus of Mysore University) for asking me to take up the preparation of this book, offering the necessary financial support.

I am beholden to Dr K. Kasturirangan, former Chairman, ISRO and now Member, Planning Commission, New Delhi, for writing an encouraging Foreword amid his onerous responsibilities. My special thanks are due to Prof. V.V. Raman, Rochester University, US and Prof. M.D. Srinivas, Chairman, Centre for Policy Studies, Chennai, who are well known scholars in the history of science in India, for going through the book and giving their valuable opinion (extracts reproduced on the back cover).

To Shri D.R. Ananda Kumar I owe a great deal for his devoted efforts in going through the manuscript as well as its proofs, leading to the refinement of the language and clarity of expression, wherever it was necessary. I offer my grateful thanks to him. I am also thankful to Prof. K. Ramasubramanian (IIT, Powai), Prof. M.D. Srinivas and Prof. B.G. Gopinath (Art of Living Foundation, Bangalore) for their valuable suggestions for improving the chapters on astronomy, mathematics and medicine respectively.

To Prof. Bhuvan Chandel, Member-Secretary, PHISPC, New Delhi, my thanks are due, for giving permission to reproduce several illustrations from the related volumes published by PHISPC.

1

AN INTRODUCTORY OVERVIEW

The word *science* owes its origin to the Latin word *scientia* or *scrire*, which means *knowledge*, similar to the word *Veda* (literally, 'to know' or 'knowledge' itself). The word *science*, as we understand and use it now, does not represent a specific academic discipline or subject, but is in the nature of an over-arching nomenclature, including under its umbrella subjects like physics, chemistry, botany, zoology, geology, astronomy and their subsidiaries, as well as their inter-links, such as bio-physics, geo-chemistry, molecular biology and astrophysics. In all of these, the underlying commonality is the scientific methodology—induction, deduction, mathematical modelling, hypothesis, theories and laws. Till the time of the Renaissance in science in Europe (fifteenth to sixteenth centuries), *scientia* encompassed the acquisition of knowledge as a whole, and such knowledge was by and large based on the sensorial perceptions, mental processes and the intellection associated with them. Since the time of the Renaissance, however, the strength of all scientific pursuits, whether in Europe or elsewhere in the world, has been its own methodology, which has empowered modern science as the greatest exploratory and innovative intellectual enterprise of mankind today, leading to progressive enrichment of human material life through its beneficial applications.

A simple definition of science about two hundred years ago was and continues to be in general terms, that science is a body of systematized rational knowledge gained through sensorial observations, and inferences based on them. Since then, science has traversed a long way and its canvas has become so wide in both the physical and biological realms that any type of capsuled or oversimplified definition would be inadequate and less meaningful. Logical methodology, verifiability, reproducibility and even falsifiability constitute the innate rational strength of modern science.

The foundational ideas of science have undergone changes from time to time, since science has been an incessant search leading to a coherent knowledge of man and *nature*, man and the *universe*. This incisive search, however, is characterized by the duality of the *subjective* and the *objective*, the knower and the known, through the senses, which are the gateways of human knowledge. Nevertheless, the search is punctuated by certain synthetic endeavours which have led to a kind of non-duality in duality, like those of matter and energy, organic and inorganic, space and time. These have not only provided extraordinary insights into the phenomenal world, but have also revolutionized the very process of our thinking as well as knowing the physical and biological world through our senses.

In recent times science, through its rational method and a wide variety of instrumentation that is in the nature of extending the capabilities of our senses, has been attempting to understand the origin of the universe, of life and consciousness. In these pursuits it has to encounter and overcome its own limitations in order to bring them onto the anvil of its method. But scientists have been continuing their investigations of the primordial origins as well as the relationship between the apparent dualities of the physical and biological world. Such an approach was also the concern of thinkers of early civilizations, but in their own way.

The saga of early human knowledge goes back to the ancient civilizations—the Mesopotamian, the Egyptian, the Indian and the Chinese. The notable characteristic of that knowledge was what might be called the *man-spirit-cosmos view*—a holistic vision that was the fountainhead of all forms of knowledge and experience. It was a way of both thinking and experiencing concordance or consonance of man *in* the universe, but not man *and* the universe. Several religious and mythological ideas, too, were woven into this triple-stranded vision of the macrocosm and microcosm. Of all the ancient cultures, the Indian culture-area developed some pristine ideas and practices leading to such a vision, the beginnings of which could be seen among the Vedic people. The approach and attainments of Vedic people in such fields as astronomy, mathematics, medicine and plant science were an inseparable part of their vision of *man-spirit-cosmos*. These were also culture-specific, in contradistinction to modern science, which, as noted earlier, has been methodology-specific and has no geographical label. Thus in the ancient and medieval periods, one can think in terms of Indian science, Greek science, Chinese science and the like, whereas these distinctions have been overcome in respect of modern science.

SOURCES OF INDIAN SCIENCE

The extensive database of Indian science is both literary and archaeological, but far more of the former. The earliest source is the extensive Vedic literature, which comprises the four Vedas (*Ṛgveda*, *Yajurveda*, *Sāmaveda* and *Atharvaveda*), their liturgical texts, which go by the name of the Brāhmaṇas, and the Āraṇyakas (forest treatises that have philosophical ideas). Also significant are the Upaniṣads, which are in the nature of discourses on metaphysical concepts leading to a progressive understanding of the seminal Upaniṣadic monistic idea of the universal self (Brahman), and the non-duality of the self (ātman) and the universal self. The main *Brāhmaṇas* and the *Upaniṣads* associated with each of the four Vedas are as follows:

The *Ṛgveda*: *Aitareya Brāhmaṇa; Aitareya Upaniṣad* and *Kauśītakī Brāhmaṇa* and *Upaniṣad.*

The *Yajurveda*: in two śākhās (schools)—*Kṛṣṇa* (Black) *Yajurveda: Taittirīya, Maitrāyaṇī, Kāṭhaka,* and *Kāṭhaka-kāpiṣthala*; Upaniṣads: *Taittirīya, Kāṭhaka, Śvetāśvatara; Maitrāyaṇīya* and *Mahānārayaṇa*; *Śukla* (White) *Yajurveda: Vājasaneyī*; *Śatapatha Brāhmaṇa*; and Upaniṣads: *the Bṛhadāraṇyaka* and *Īśa.*

The *Sāmaveda: Pañcaviṃśa Brāhmaṇa* or *Tāṇḍya Mahābrāhmaṇa, Ṣaḍviṃśa Jaiminīya Brāhmaṇa;* Upaniṣads: *Chāndogya* and *Kena* (*Talavakāra*).

The *Atharvaveda: Gopatha Brāhmaṇa;* Upaniṣads: *Māṇḍūkya, Praśna,* and *Muṇḍaka.* There are also several other Upaniṣads.

It needs to be borne in mind that the relationship between the *Brāhmaṇas* with their inclusive orientation towards various liturgical aspects, and the Upaniṣads with their exalted metaphysical or philosophical speculations, is not perceptibly direct, but is found through the *Āraṇyakas* in several cases. In fact, one of the major Upaniṣads is titled *Bṛhadāraṇyakopaniṣad.* Certain types of allegorical explanations and metaphorical nuances of the *Brāhmaṇas* could also have paved the way for the Upaniṣadic way of thinking. It would appear that the dividing line between the *Brāhmaṇas* and the Āraṇyakas is somewhat arbitrary. The Upaniṣads, which connote in a general sense the relationship or communication between a preceptor and his disciples, constitute *vedānta* (literally 'end of the Vedas').

Regarded as the foremost among the Vedas is the *Ṛgveda*, the earliest part of which is dated to 1500 CE. It comprises 1028 hymns or *sūktas* (each śukta includes several *mantras*) under ten *maṇḍalas* or segments as follows:

Ṛgvedic segments

Maṇḍala	Sūktas	Mantras	Ṛṣis
I	191	2006	several *ṛṣis*
II	43	429	Gṛtsnamada
III	62	617	Viśvāmitra
IV	58	589	Vāmadeva and Gautama
V	87	727	Atri
VI	75	765	Bharadvāja
VII	104	841	Vasiṣṭha
VIII	103	1716	Kaṇva
IX	114	1108	several *ṛṣis*
X	191	1754	several *ṛṣis*

The *Ṛgveda* is also divided into eight parts called *aṣṭakāṣ*, which have more or less equal numbers of *sūktas* each. This sublime compendium has set the tone for a holistic understanding of the visible universe in a three-fold format: celestial, atmospheric, and terrestrial. It makes some references to a few constellations, and also attempts to reflect upon the origin of the universe, which it expounds as an ordered whole, governed by cosmic order (*ṛtam*). Its cosmogonic hymn (X.129) as well as its concept of *one* in *many* or *one* becoming *many* is indeed a lofty ideal, which culminated in the later Upaniṣadic concept of *Brahman*.

The *Yajurveda* is not only a compendium of certain sacrificial ideas and practices, but also has several *mantras* borrowed from the *Ṛgveda*. The *Śatapatha Brāhmaṇa,* which belongs to the *Śukla Yajurveda,* provides some information on the twenty-eight *nakṣatras* (asterisms) including Abhijit; so does the *Kṛṣṇa Yajurveda,* or the *Taittirīya Saṃhitā*. The *Sāmaveda* contains a large portion of the *Ṛgvedic* hymns, but chanted in a musical manner. Hence it is stated to be the primary source of Indian music. The *Atharvaveda* also lists twenty-eight *nakṣatras* and, like the *Taittirīya Saṃhitā* and the *Śatapatha Brāhmaṇa,* its list begins with *Kṛttikā* and ends with *Bharaṇī,* pointing to the fact that the vernal equinoctial point was then related to *Kṛttikā*. The *Atharvaveda*, strangely, recognizes time (*kāla*) as the first cause of all existence. It also presents some details about magico-religious healing and extols some plants as having an extraordinary curative potential, which could be used as healing talismans.

Ayurveda is believed to have its origin in the *Atharvaveda*. But the rational approach of Ayurveda to the diagnosis as well as the cure of ailments is entirely different from that of the religious, esoteric medical cures of the *Atharvaveda*. The doctrine of five elements (*pañcabhūtas)* is a dominant and viable base of the former, while it is totally absent in the latter. However, one of the eight branches of Ayurveda, called *Bhūtavidyā* (demonology and psychiatry), is concerned with the cure of such diseases (especially the mental ones), the causes for which cannot be easily diagnosed. Nevertheless, for curing mental afflictions and chronic diseases, Ayurveda, too, recommends the use of methods like incantation, performance of fire-worship (*havana, hoama*) or 'religious rites' more or less similar to those found in the *Atharvaveda*. Even now such methods are being practised by people of all walks of life, urban or rural, rich or poor.

The Upaniṣads, with a spectrum of incisive discourses on the ultimate reality, are also important from the point of view of the early foundational idea of the five elements: *pṛthivī* (earth); *ap* (water); *tejas* (fire); *vāyu* (air); and *ākāśa* (an ubiquitous principle). There are several Upaniṣads. Among them, the *Chāndogya* has dealt with a group of three elements, placing the water element between the two elements, fire and earth. The postulate of five elements was expounded first perhaps in the *Taittirīya* and the *Aitareya* Upaniṣads, which also relate these elements to the *Self* and *Brahman*.

The Sūtra Literature

The Upaniṣadic metaphysical or ontological five elements underwent a transformation into physical entities from about the sixth century CE, with the emergence of the *Sāṃkhya* and the *Vaiśeṣika* systems. The authors of these and other orthodox (owing allegiance to the Vedas) thought-structures adopted in general the form of *sūtra* or the condensed aphoristic style, thus setting the kernel of knowledge in short forms, avoiding the use of sentences as far as possible. The main reason for preferring this style was to facilitate easy and lasting memorization of a vast body of knowledge. There is probably no such attempt at precision and brevity in the entire literature of the world. The six orthodox systems of Indian thought are: *Sāṃkhya, Vaiśeṣika, Nyāya, Yoga, Pūrva-Mīmāṃsā* and *Uttara-mīmāṃsā*. These along with the six *Vedāṅgas* are the torchbearers of this style of presentation.

The six *Vedāṅgas* are: Śikṣā (phonetics); *Kalpa* (ritual); *Vyākaraṇa* (grammar); *Nirukta* (etymology); *Chandaḥ* (metrics) and *Jyotiṣa* (astronomy). Among these the *Kalpa* (its Śulba-sūtras), *Chandaḥ* and *Jyotiṣa* are important for understanding the beginnings of Indian mathematics and astronomy. The Śulba-sūtras (*Āpastamba, Baudhāyana, Kātyāyana* and *Mānava*) deal with Vedic geometry and algebra; the *Chandaḥ*, with permutations and combinations; and *Jyotiṣa*, also called the *Vedāṅga-jyotiṣa*, with its *nakṣatras*, establishes a five-year lunar-solar cycle (*yuga*) and the determination of the length of the day between the two solstices—the shortest day being twelve *muhūrtas* at the winter solstice and the longest day being eighteen *muhūrtas* (one *muhūrta*=forty-eight min.) at the summer solstice. The *Vedāṅga-jyotiṣa* also adopted the stellar zodiac for understanding positions of the sun and the moon. Planetary observations are absent in this text.

The sūtra-type of presentation scaled new heights in the religio-philosophical schools. Of special interest to the history of Indian science in general, and the concepts concerning the physical world in particular, are the *Sāṃkhya*, the author of which is believed to be sage Kapila, the *Vaiśeṣika* sūtras of *Kaṇāda* and the *Nyāya* sūtras of Gautama. The *Sāṃkhya*, in an evolutionary manner, expounds the manifestation of *prakṛti* (an intertwined ensemble of the three qualities (*guṇas):* *sattva* (righteous), *rajas* (energetic), and *tamas (*inertia) in equilibrium. This manifestation is presented in the form of twenty-four principles, with the interaction of *Puruṣa* (the sentient principle). Of them, the five elements, as noted before, constitute the physical realm. The classical *Sāṃkhya*, as expounded by Īśvarakṛṣṇa (second century CE), includes what are called the five *tanmātras* (or subtle essences), each for one element, perhaps sharing both the non-material and material characteristics. The *Sāṃkhya*, however, does not have much to offer by way of its exposition on the properties of matter, motion, space and time. The *Vaiśeṣika* sūtras, on the other hand, throw adequate light on them with an epistemology strengthened by the *Nyāya* sūtras.

Since the sūtras were generally enigmatic, they needed commentaries or *bhāṣyas*. Therefore, several commentaries were produced on the *Sāṃkhya*, the *Vaiśeṣika*, the *Nyāya*, and the syncretic *Nyāya-Vaiśeṣika*, from about the fourth century CE onwards. These are valuable sources for understanding the Hindu or orthodox views on the physical world. During this period or a few centuries earlier, the two major heterodox systems of thinking, the Jaina and different Buddhist schools, produced several texts, a study of which would also be necessary, especially for understanding the nuances of Indian perspectives on the physical world.

Nyāya-Vaiśeṣika Texts

Among the important *Nyāya-Vaiśeṣika* textual sources which deserve special mention are the following: the *Padārthadharamasaṅgraha* by Praśastapāda (fourth century CE); *Nyāya-bhāṣya* by Vātsyāyana (fourth century); *Nyāya-*

vārttika by Udyotakara (seventh century); *Nyāya-sāra* by Bhāsvarajña (eighth century); *Nyāya-vārttika-tātparyaṭīkā* by Vācaspati Miśra (ninth century); *Nyāyamañjarī* by Jayanta Bhaṭṭa (ninth century); *Kiraṇāvalī* by Udayana (tenth century); *Nyāya-kandalī* by Śrīdhara (tenth century); *Saptapadārthī* by Śivāditya (eleventh century); *Nyāya-līlāvatī* by Vallabhācārya (twelfth century); *Nyāya-siddhāntadīpikā* by Śaśidhara (twelfth century); *Tattvacintāmaṇi* by Gaṅgeśa (thirteenth century); *Upaskāra* and *Kaṇādarahasya* by Śaṅkara Miśra; *Padārthatattva nirūpaṇa* by Raghunātha Śiromaṇi (sixteenth century); *Bhāṣāpariccheda* and *Nyāya-Muktāvalī* by Viśvanātha (seventeenth century).

Jaina and the Bauddha Texts

Of the Jaina texts, Umāsvāti's *Tattvārthādhigamasūtra* (second century CE), *Bhagvatīsūtra* (fourth century), *Nyāyāvātāra* by Siddhasena (sixth century), *Dravyasaṅgraham* by Nemicandra (twelfth century), and the *Syādvādamañjarī* by Mallisena (thirteenth century) deserve special mention. Among the noted early Buddhist sources are the *Saṃyuttanikāya* and the *Dhammasaṅgaṇi,* Buddhaghoṣa's *Atthaśālinī* and *Viśuddhimagga* (fifth century), Vasubandhu's *Abbidhammakośa* (sixth century), Diṅnāga's *Pramāṇasamuccaya* and *Nyāya-praveśa* (sixth century), and Dharmakīrti's *Nyāyabindu* (seventh century), all of which are of special value for studying the Buddhist approach to the physical world, and the logic associated with it.

The *Nyāya-Vaiśeṣika* texts deal extensively with the concept of substance in terms of the five elements (four material and the non-material *ākāśa*), space, time, the self and the mind. This is an integrated concept of substance, which is inclusive of the material and the non-material, the knower and the knowing agency (mind). The five elements are also correlated with the five senses (*pṛthivī*: nose, smell; *ap*: tongue, taste; *tejas*: eyes, vision; *vāyu*: skin, tactile; and *ākāśa*: ears, sound). The doctrine of five elements as expounded in the *Vaiśeṣika* texts is in the nature of an attempt to explain the maximum phenomena (observed through senses) with the minimum postulates.

These texts also discuss various aspects of atomism, space, and time. The epistemological frame of their exposition is known as *ārambhavāda*, or discontinuous causation, with the basic premise that the cause produces the effect, and the effect is a new event. This is in sharp contrast with the epistemic position of the *Sāṃkhya*, which posits the idea of a continuous causation (*parīṇāmavāda*) inasmuch as the effect is an evolutionary one in which the cause is already present. The *Vaiśeṣika* developed an atomism of the four elements, excluding the fifth element, *ākāśa*, and the *Nyāya* fortified it with its epistemology.

The Jainas and the Sautrāntikas as well as the Vaibhāṣikas (the two schools of the Hīnayāna form of Buddhism) also developed atomism, but within their philosophical framework. Indian atomism, however, was more a speculative exercise than a rational elucidation of the problem of matter. In several ways, Indian atomism was at variance with the Greek atomism that was propounded by Leucippus (c. 475 BCE) and expounded by Democritus (470-400 BCE). It was later elaborated upon by the Epicurean philosopher, Lucretius (first century BCE). However, Greek atomism lay dormant till the seventeenth century, when it was resurrected by Robert Boyle (1627-91) and Isaac Newton (1642-1727) in Britain, and Pierre Gassendi in France (1592-1655). In the next century, it underwent reappraisal and reformulation, strengthened by several chemical experiments with gases. Early in the nineteenth century it took the shape of an atomic theory enunciated by John Dalton (1766-1844). The wide variety of chemical reactions began to be studied for an understanding of their quantitative relationship and, with the law of conservation of mass expounded by A.L. Lavoisier (1743-94), chemistry emerged as a branch of science in the nineteenth century. On the other hand, Indian atomism continued as a vibrant polemical thought till about the thirteenth century, especially among the followers of the *Nyāya-Vaiśeṣika*. It did not come out of this groove; nor did it enter the arena of chemical experiments of the *Rasavādins,* whose thoughts were of a different nature, being more alchemical and iatrochemical than chemical. The *Nyāya-Vaiśeṣika* concepts of space and time, too, did not become components of the explanation of motion in a quantitative and verifiable manner. Such an approach was witnessed in the West (from the sixteenth century), which brought about a new understanding, as well as a measurable explanation of motion in terms of the laws of motion in relation to space and time, besides the concepts of mass and force, including gravitational force.

Ayurveda: Its Main Texts

During the six centuries before the Christian Era, there was a gradual expansion of what some historians call the second urbanization, the first one being much earlier—the Indus Valley Civilization, also called the Harappa culture (2350-1700 BCE) was its mature phase, and its decadent phase continued for a couple of centuries more. This period was noted for the systematization of six orthodox religio-philosophies. Jainism and Buddhism, too, were experiencing a sort of consolidation or crystallization of their positions. The religio-philosophies of the *Sāṃkhya*, the *Vaiśeṣika* and the *Nyāya* provided a foundation for medicinal ideas and healing practices which, since the Vedic times, were often associated with magico-religious rites and other superstitious methods. The result was that quackery had to recede to the background, yielding place to new concepts of man and nature, man and matter, the relationship of microcosm and macrocosm. The doctrine of five elements, which had taken shape by then, provided a new conceptual framework. These led to a rational understanding of the body and mind as well as the physiological processes within the body in terms of the harmony or equilibrium of the inner processes of the body and mind on the one hand and, on the other, the harmony of life as a whole with the macrocosm. Medicine could no longer be considered an empirical healing art, but came to be understood as a rational curative system sustained by sound foundational ideas and standardized diagnostic as well as curative methods. This new mind-set towards medicine engendered a systematic classification or categorization of diseases, their causes and remedial procedures, which took on the name of *Ayurveda* (literally, the knowledge of a long, healthy life). Its main objectives are: prolongation of a healthy, spiritual life and the prevention of the onset of diseases as well as curing diseases in a holistic manner.

The two main classics of Ayurveda are the *Caraka Saṃhitā* (second century CE) and the *Suśruta Saṃhitā* (fourth century). Nevertheless they were redactions of the Ayurvedic knowledge and practices of the preceding centuries even before the Christian Era. In the introductory passages of these texts are found expositions of the divine origin of the Ayurveda, tracing it to *Brahmā,* the Creator, in the *Caraka Saṃhitā,* and *Dhanvantari* in the *Suśruta Saṃhitā*. The former text deals with several medicinal aspects and the latter, with surgery in addition. Among the most striking and fundamental concepts of Ayurveda are the three-humoural *(triodoṣa-tridhātu)* theory, the various types of diagnostics and metabolism in terms of digestive agencies and elements as well as metabolic products *(dhātvagnis, rasa-guṇa-vīrya-vipāka)*, and *Kriyā kāla*, which are based on the versatile doctrine of the five elements. Surgery of various types, including plastic surgery, laparotomy and lithotomy, was an accomplished skill with effective pre- and post-operative treatment. In fact, Suśruta has been hailed as the Father of Indian Surgery.

The literature on Ayurveda, both primary and secondary sources, is quite considerable. Among the important primary sources are the three seminal works namely, *Caraka Saṃhitā* and the *Suśruta Saṃhitā,* which were redacted by Dṛḍhabala (fourth-fifth century CE) and Nāgārjuna (eighth century CE), later by Ḍalhaṇa (eleventh century CE), respectively, and the *Aṣṭāṅgahṛdaya* of Vāgbhaṭa (seventh century); *Aṣṭāṅgasaṅgraham* by Vṛddha Vāgbhaṭa (sixth century); *Bhela Saṃhitā, Hārīta Saṃhitā; Kāśyapa Saṃhitā; Nīdāna* by Mādhava (ninth century); *Dravyaguṇa saṃgraha* of Cakrapāṇi Datta (tenth-eleventh centuries); *Cikitsāsaṃgraham* of Vaṅgasena (eleventh century); *Bhaiṣajya Ratnāvalī* of Govinda, *Dhanavantari Nighaṇṭu; Nāḍīparīkṣā* of Rāvaṇa; *Śataślokī* of Vopadeva; *Śoḍhala Nighaṇṭu*; *Cikitsākalikā* of Tiṣatācārya (fifteenth century); *Śāṅgadhara saṃhitā* by Śāṅgadhara (thirteenth century); *Dravyaguṇa Śataślokī* by Trimalla Bhaṭṭa; *Āyurveda saukhya* of Todarānanda (sixteenth century); *Bhāvaprakāśa* by Bhāvamiśra (sixteenth century); *Vaidyacintāmaṇi* of Vallabhācārya; and the *Vaidyājīvana* and *Vaidyāvataṃśa* of Lolimbarāja (sixteenth-seventeenth centuries). This list is not exhaustive and the dates of works are also debatable.

An interesting source on medicine, which was discovered in Central Asia, deserves special mention. Titled the *Bower Manuscript,* it was discovered in 1890 by A. Bower, a British army Lieutenant. It was studied by the philologist A.F.R. Hoernle, and published in 1912. It is a collection of seven texts authored by some scholars who were perhaps the residents of a Kuchean monastery in Chinese Turkestan. Out of the seven texts, three are purely medical. Several therapeutic formulae described in this manuscript are also found in the *Caraka Saṃhitā*. The first medical tract deals with the medicinal value of garlic as well as some recipes for the cure of eye diseases. The second, which is also extensive, is called *Navanītakam* (or quintessence) and, in its sixteen chapters, describes various recipes and procedures for healing several common diseases. The third tract has fourteen medical formulae, some of them similar to those given in the *Caraka* and *Suśruta Saṃhitās*. But the language associated with the contents of the manuscript is older or pre-classical. The date of the manuscript is about the fourth century CE.

Ayurveda's rational approach is governed by its means

of acquiring knowledge, which comprises four methods: (i) Scriptural testimony (*āptopadeśa*); (ii) Direct perception (*pratyakṣa*); (iii) Inference (*anumāna*); and (iv) Experimental verification (*yukti*). In fact, the last one occupies a distinct position in Ayurvedic therapeutic practices. To Caraka belongs the credit of being the first to have adopted *yukti*. The concept of *yukti* found able exponents later in Varāhamihira (*Gandhayukti:* techniques of perfumery) and others.

Siddha Medicine: Sources

In its origin and growth, Ayurveda has been an indigenous system of medicine. The vast Indian culture-area is also known for its medical pluralism, the other systems being the Siddha, Unāni, and modern medicine apart from homeopathy. The name *Siddha* owes its inspiration to medicinal ideas and practices of a class of Tamil sages called *Siddhas* (literally, holy immortals; 'perfected beings') who were believed to have acquired superhuman powers, and who had firm faith in the 'deathlessness' of the human body being in tune with the spiritual immortal 'soul'. According to Tamil tradition, Siddha medicine is associated with eighteen Siddhas, among whom Agastyar, Tirumūlar, and Bhogar had a distinct position. The dates of the Siddhas are uncertain. The three well-known Siddhas might have lived during the fifth to eight centuries, and the other Siddhas were of a later date.

Siddha medicine is a belief system to the core, showing immense faith in the 'miracles' of mercurial drugs and in the prolongation of life through rejuvenation techniques and intense yogic practices, among which regulated breathing has pride of place. It has alchemical undertones too. Its seed-ideas were those of Taoism, which percolated into India through Buddhist itinerants in the milieu of tāntrism during the fourth-fifth centuries.

The literary sources of Siddha medicine are extensive, numbering more than seven hundred texts, all in Tamil. Of them, special mention needs to be made of the following: Agastyar's *Amṛtukalaijñānam*, *Auṣadasāram*; *Ayurvedam Chenturam sūtram; Vaidyaciñtāmaṇi* and *Kāyasiddhi Valalai;* Nandīsar's *Yeḷāriyam Kāṇḍam* and *Vaidyam;* Bhogar's *Yamakaveṇba; Yeḷāiram-Iraṇḍāiyaram* and *Kappiaṃ;* Tirumūlar's *Tirumantiram;* Theraier's *Cigiccaisārasaṅgraham; Yamaka venba;* Machhamuni's *Karukkādal;* Ramadevar's *Vaidyam* and *Perunūl; Vaidya Cintāmaṇi* of Thanvantari; and *Vaidyam* of Pulippāṇi. Siddha medical treatises have identity marks, as it were, in terms of the numbers of verses contained in each of them and included in their titles.

The range of Siddha drugs (metallic and mineral ones which are processed elaborately with a wide variety of herbs) is not extensive. In its diagnostic methods, Siddha medicine attaches considerable importance to the examination of the pulse and urine although, like Ayurveda, it accepts the three humoural concepts. But, unlike Ayurveda, its anatomy is bizarre, with tāntric speculation of veins and nerves as well as six vital centres or *cakras*. Siddha medicine is known especially for its intricate rejuvenation method called the *Kāya-kalpa*. A combination of three salts (natural exudation from the earth called *pūnīru*, processed sodium and potassium salts) known by the name of *muppu* has an important position in the dispensation of Siddha medicine and in the *Kāya-kalpa* treatment. Astrology and incantations are also part of Siddha therapy. Confined to Tamil Nadu and Tamils in the adjoining areas, Siddha medicine continues to be a healing tradition in these regions, encouraged by both the central and Tamil Nadu state governments.

Unāni Medicine

Though the foundational ideas of the Siddha system were influenced by the exogenous Chinese Taoist thoughts and practices, it is not in the nature of a foreign system, for it grew in the Indian tāntric environment and assimilated certain aspects of the indigenous Ayurvedic ideas. Unāni, on the other hand, was an alien system which entered India in the thirteenth-fourteenth centuries along with Muslim rulers, and began to flourish during Mughal rule. Its fundamental concepts, too, are of a different character, their formulation being derived from Greek or Hippocratic medicine, and modified by Arabic medical men. It is, therefore, called 'Graeco-Arabic medicine'. The greatest authority of this medical system is Ibn Sīna (980-1037), a Central Asian, who wrote (in Arabic) his magnum opus, *Al-Qānūn* (The Canon of Medicine), which is regarded as the most authoritative text of Unāni medicine.

Unāni medicine believes in the Greek four elements: earth, water, fire, and air with their four basic but opposite pairs of qualities, namely, *hot* and *cold*, *dry* and *moist*. It posits four humours—blood, phlegm, yellow bile and black bile—as integrated components of the bodily fluids. However, like Ayurveda, this system explains that their equilibrium is health while their imbalance is disease, the difference being that Ayurveda believes in three humours, based on five elements. The humoural physiology as well as pathology is also somewhat different in Unāni from those in Ayurveda. However, like Ayurveda, it takes into consideration the nature of the body called *tabi'yat* (physical nature), but regards seven components as being responsible for the existence of human body. They are, *Al-Arkān* (the elements); *Al-Mizāj* (temperament); *Al-Akhlāt* (humours); *Al-A'ḍa* (organs); *Al-Arwah* (pneuma or vital spirit); *Al-Quwa* (faculties); and *Al-Afāl* (functions). The diagnostic methods of Unāni include the pulse and urine examination.

The literature on Unāni authored by celebrated hakims (practitioners of medicine), both Persian and Indian, is quite considerable, spanning over five hundred years: the Sultanate period; Deccan period; Mughal period; and the post-Mughal period. A number of reputed hakims made independent contributions to therapeutics and pharmaceutical techniques, and used several drugs that were unknown to the followers of Greek medicine.

Unāni medicine began to register progress from the time of Alauddin Khalji's reign (1296-1316) and also through the Muslim rulers in the Deccan. It was greatly patronized by the Mughal monarchs. Several noted hakims produced elegant treatises, some of them being voluminous, and it is not possible to mention all of them. During the reign of Muhammad bin Tughluq (1325-51) and even later, hakims also had knowledge of Ayurvedic practices. The Muslim rulers of Gujarat encouraged several hakims to write medical treatises in Persian. In the Deccan, under the Qutb Shahi dynasty, there were reputed Unāni physicians.

The Mughal period (beginning from the sixteenth century) witnessed several renowned hakims coming to India, particularly from Iran. They wrote a substantial number of authoritative treatises on Unāni medicine, ophthalmology, and tested prescriptions. In the post-Mughal period, there were two illustrious families—the Sharifi family of Delhi and the Azizi family of Lucknow. Hakims of these families played a key role in the production of valuable works on Unāni *tibb*. Hakim Ajmal Khan, who wrote several medical works, belonged to the Sharifi family and established educational institutions as well as Unāni research units. Among them the Tibbiyah College in Delhi (inaugurated by Gandhiji in 1921) and the Hindustan Dawakhana are still holding the Unāni tradition aloft. The Azizi family also produced many famous hakims who, in turn, were the authors of authentic medical works. Though during the British colonial rule, Unāni did not receive its due encouragement, it continued to be an important component of medical care during that time. Since Independence, Unāni, like the Ayurveda and Siddha systems, has been receiving adequate support from the Government of India.

Astronomy: Important Texts

The beginning and exposition of planetary or mathematical astronomy in India could be found in what is known as Siddhāntic astronomy, led by the most illustrious astronomer-cum-mathematician, Āryabhaṭa I (b. 476 CE). Before him, however, there were several texts under the generic name of *Siddhānta*. Eighteen such *Siddhānta*s have been recognized, but most of the texts alluded to have been lost. However, Varāhamihira (early sixth century) in his *Pañcasiddhāntikā*, redacted five of them, namely, *Saura, Paitāmaha, Vaśiṣṭha, Pauliśa* and *Romaka* as known to him then. In any case, they deal with planetary astronomy and some mathematical applications for arriving at accurate astronomical parameters, including planetary positions and the occurrence of eclipses.

The *Sūryasiddhānta* (author unknown) is an important text of Indian astronomy. The other major *Siddhāntic* texts and the *Karaṇa* works (which take a particular epoch for computations) are: the *Āryabhaṭīya* (fifth century CE); *Mahābhāskarīya, Laghubhāskarīya*, and *Āryabhaṭīya bhāṣya* of Bhāskara I (seventh century); *Brāhmasphuṭa Siddhānta* and *Khaṇḍakhādyaka* of Brahmagupta (b. 598; seventh century CE); *Karaṇaratna* of Devācārya (later half of the seventh century); *Śiṣyadhīvṛddhidatantra* of Lallācārya (eighth century); *Vaṭeśvara Siddhānta* and *Karaṇasāra* of Vaṭeśvara (ninth century); *Laghumānasa* of Muñjāla (tenth century); *Ārya* or *Āryabhaṭa Siddhānta* (also called *Mahāsiddhānta*) of Āryabhaṭa II (late tenth century); *Siddhāntaśekhara* and *Dhīkoṭī* by Śrīpati (eleventh century); *Bhāsvatīkaraṇa* of Śatānanda (latter half of the eleventh century); *Siddhāntaśīromaṇi* and *Karaṇakutūhala* by Bhāskarācārya II (twelfth century). In the succeeding centuries after Bhāskarācārya II, several commentaries appeared generally on the *Āryabhaṭīya, Khaṇḍakhādyaka* of Brahmagupta, *Sūryasiddhānta, Bhāsvatīkaraṇa*, and the *Mahābhāskarīya* as well as the *Laghubhāskarīya*. These commentaries carried on the tradition of astronomy in different parts of India, each in its own way, but in the overall framework of geocentric astronomy, the kernel of which was the *Āryabhaṭīya*.

The tradition of mathematical astronomy began to

scale new heights with what is generally known as the Kerala School of Astronomy and Mathematics (fourteenth-nineteenth centuries). The doyen of the savants of this School was Mādhava, the author of *Sphuṭa Candrapti*, *Veṇvāroha*, *Agaṇitagrahacāra*, and *Candravākyāṁ*. Parameśvara, who formulated the *Dṛk* system of calculation based on observed data to replace the earlier *Parahita* system introduced by Haridatta (seventh century), produced a large number of treatises: *Goladīpikā*, *Grahānayanadīpikā*, *Grahanamaṇḍala*, *Dṛggaṇita*, *Candrachhāyāgaṇita*, and *Vākyakaraṇa*. Alongside, he wrote commentaries on the *Āryabhaṭīya*, *Mahābhāskarīya*, and *Laghubhāskarīya* as well as on the *Sūrya Siddhānta*. He was hailed as *Paramaguru* by the most noted Kerala astronomer, Nīlakaṇṭha Somayājī. Nīlakaṇṭha surpassed his preceptor, Damodara, the son of Parameśvara, and wrote several texts on astronomy. His *Golasāra*, *Jyotirmīmāṃsā*, *Grahaṇanirṇaya*, and *Siddhantadarpaṇa*, as well as his *Taṅtrasaṁgraha* and commentary on the *Āryabhaṭīya*, became landmarks of Indian astronomy in general and the Kerala School of Astronomy in particular. The credit of presenting a new planetary model (1500 CE), which was based on accurate computations, belongs to Nīlakaṇṭha. In this new model he showed how the inner as well as the outer planets revolve round the sun, and the sun, along with them, revolves round the earth. Such a model was conceived by Tycho Brahe in the Copenhagen observatory nearly ninety years later. The *Yuktibhāṣā* of Jyeṣṭhadeva, *Karaṇasāra*, and the commentary on Nīlakaṇṭha's *Taṅtrasaṅgraha* by Śaṅkara Vāriyar, *Karaṇottama* and commentaries by Acyuta Piṣāraṭī, *Karaṇapaddhati* and *Veṇvārohāṣṭaka* by Putumana Somayājī, and *Sadratnamālā* by Śaṅkaravarman—these erudite treatises added a veneer of excellence to Indian astronomy.

From the sixteenth century onwards, a centre of astronomy that came to be recognized as the Maharashtra School arose in western India. Gaṇeśa Daivajña was well known for his elegant treatise called *Grahalāghavam*, which is by and large a *karaṇa* (computational) text, published around 1520 CE. He did not use the usual trigonometric functions found in the other Siddhāntic texts; instead, he developed methods of easy calculations with the result that the *Grahalāghavam* was, and continues to be, a popular source-book for makers of the *pañcāṅga* (traditional almanac) in several parts of India. The other works of Gaṇeśa Daivajña comprise the *Laghu-cintāmaṇi* and *Bṛhat-cintāmaṇi*, *Buddhivilāsinī* (a commentary on the *Līlāvatī* of Bhāskara II), besides a commentary on his own work, a text on astronomical instruments titled *Pratodayantra*, and some astronomical tables.

Divākara was a devoted disciple of Gaṇeśa Daivajña and his sons, grandsons, and great-grandsons carried on in an exemplary manner the family tradition that added new dimensions to Indian astronomy in the succeeding centuries. Of them, Viṣṇu Daivajña, Kamalākara, author of the well known work *Siddhāntatattvaviveka*, Nṛsimha, Viśvanātha, and Mallāri produced several commentaries on the *Sūryasiddhānta*, *Siddhāntaśiromaṇi*, and others. Contemporaneous with them were other noted astronomers like Munīśvara, author of the *Siddhāntasārvabhauma*, and Kṛṣṇa Daivajña.

More importantly, the astronomer-ruler Mahārāja Sawai Jai Singh II (1686-1734) bequeathed to the nation the five astronomical observatories at Benaras, Mathura, Ujjain, Delhi and Jaipur. Under his direction, astronomical tables were compiled and these were dedicated to his Mughal emperor and patron under the title *Zīj-Muḥammad Shāhī* in Persian, and an incomplete one in Sanskrit. The Mahārāja was also the author of *Yantrarājaracana*, a valuable text that describes the instruments in his observatories in detail. Jai Singh's court astronomer-preceptor was the erudite Jagannātha Paṇḍita who translated Ptolemy's *Almagest* into Sanskrit from its Arabic version, besides Euclid's *Elements*, also from its Arabic version, under the titles *Samrāṭ-siddhānta* and *Rekhāgaṇita*, respectively. In the middle of the nineteenth century, there was an astronomer of exceptional ability in Orissa (now Odisha)—

Sāmanta Chandrasekhara, who was also from a royal family but lived in poverty. He produced a voluminous work called the *Siddhāntadarpaṇa,* written on palm leaves in Oriya script. In his own way he made observations through the instruments devised by him, and developed a planetary model similar that of Nīlakaṇṭha or Tycho Brahe.

The wide range of texts in Sanskrit on astronomy received considerable attention from Western scholars, especially during the nineteenth and early twentieth centuries. Some Indian scholars, too, were involved in the translation of a few texts into English, and produced their critical editions. David Pingree, an American, has catalogued extensively the available manuscript-sources on Indian astronomy. However, his studies on Indian astronomy are biased, and his historiography leaves much to be desired. A critical understanding of mathematical astronomy in India reveals its own development, particularly in computational astronomy, for which several innovations were made, including the word-numerals and alphabetical numeration of the decimal place-value system.

The practical facet of Indian astronomy is the *pañcāṅga* or the traditional almanac with its five components: *tithi* (lunar day); *vāra* (week-day); *nakṣatra* (asterism); *yoga* (related to the longitudes of the sun and the moon), and *karaṇa* (related to *tithi*). There are several *pañcāṅgas* that are produced annually in different parts of India. They can be broadly classified under two heads: (i) Luni-solar and (ii) Solar. In the former, which belongs to the *amānta* system, the lunar month begins with the first day of the bright half or *Śukla pakṣa* and ends with the last day of the dark half or *Kṛṣṇa pakṣa*. The luni-solar calendar is generally followed in the Karnataka, Andhra, Maharashtra, and Gujarat regions. In the other system, called *pūrṇimānta,* the lunar month begins with the full moon and ends with the next full moon. In a purely solar *pañcāṅga,* the solar year commences when the sun enters the first zodiacal sign, *Meṣa,* and it is known as *Meṣa-saṅkramaṇa* (solar ingress). Both the lunar and solar years have twelve months each, which are reckoned by different names. Since the lunar month is shorter than a solar month, an extra or intercalary lunar month is added to the lunar year, approximately once in three years. A *pañcāṅga* deals with several aspects, including the auspicious and inauspicious times, festivals, *nakṣatra*s, and is used extensively in predictive planetary astrology.

Planetary astrology, however, is not indigenous. Originally from the Chaldean/Graeco-Roman area, it entered India in the fifth or sixth century CE. The earliest astrological text is *Bṛhajjātaka* (CE sixth century) by Varāhamihira who, in no uncertain terms, hails the Mlechhas or Yavanas as experts in this type of astrology. During the Vedic times and later, there was the *nakṣatra vidyā*, according to which the *nakṣatras* were categorized into the auspicious or inauspicious ones, and their propitiation was also prescribed. With the advent of planetary astrology, the *nakṣatra vidyā* too was absorbed into planetary astrology and thus Indian astrology is a combination of both the *nakṣatra vidyā* and planetary predictive astrology. It is also a mixture of irrational and superstitious beliefs. Nevertheless, it continues to be a part of Indian ways of living even in the modern scientific milieu.

Inscriptions

One of the present superstitious beliefs is concerned with the occurrence of eclipses. A *pañcāṅga* refers to the date of the occurrence of an eclipse and prescribes some propitiatory measures to those in whose birth *nakṣatra* and the zodiacal sign the eclipse is stated to occur. There are some propitiatory injunctions vis-à-vis the initial obscuration as well as during and after the eclipse. It is a common experience even now that, during a total solar eclipse, the entire movement of people in the place where it can be seen comes almost to a standstill, out of fear, and often even highly educated people do not come out of their homes to observe the spectacle through any contrivance.

Such has been the impact of an eclipse on people since the Vedic times. However, Indian astronomers from Āryabhaṭa I onwards, except Brahmagupta (who continued to believe in the *Rāhu-Ketu* seizing either the sun or the moon thus causing an eclipse), have given in their texts a scientific explanation of the occurrences of eclipses. Besides, they have presented the method of calculations involved in determining the time of the occurrence, duration, and the ending of an eclipse. However, certain corrections are needed in view of the changes in planetary positions over a period of time. At present, *Pañcānga*-makers do not undertake this task of applying corrections; instead, they reproduce in the *pañcāṅga* the timings of eclipses given (in hours and minutes) in the modern ephemeris prepared by the Positional Astronomy Centre, Kolkata.

Eclipses have been recorded in several inscriptions, which provide the *pañcāṅga* dates on which grants of land, donations and the like were made by kings or chieftains. More than 90,000 inscriptions on stone and copper plates have been maintained by the Directorate of Epigraphy, Archaeological Survey of India, located in the city of Mysore. It is estimated that there may be an equal number of inscriptions in Sanskrit and other languages, and the survey is continuing.

Of over 90,000 known inscriptions, only about 25 per cent have been published in several volumes: *Epigraphia Indica* (42 volumes), *South Indian Inscriptions* (27), *Corpus Inscriptionarum Indicarum* (7), *Epigraphia Carnatica* (17), and some volumes in other languages like Bengali, Gujarati, Malayalam, and Telugu, besides those that are known as Jaina inscriptions.

The early inscriptions are Aśokan (third century BCE); they are in the Brāhmī, Kharoṣṭi, and Aramaic scripts, but in Prākṛt languages, and there is also one in the Greek language. But none of them contains any reference to eclipses or presents any other astronomical data. Even details of the exact dates of the eclipses are lacking in several of them, since they mention only the regnal years. The later inscriptions like those of Śuṅga (first century BCE), Kuśānas (first-second century BCE), Śātavāhanas (second century BCE to CE second century), and Kṣatrapas of western India (second century CE), also do not contain any information on eclipses. In general, the inscriptions in Prākṛt dialects do not make a mention of eclipses.

It is only from the Gupta period, the Classical Age of India, even so from the fifth century CE onwards, that the inscriptions provide some details of the dates of eclipses according to the Hindu calendar. The recordings in several cases indicate the actual dates on which the eclipses occurred. In some other cases, the inscriptions provide the dates on which grants or gifts were given on account of an eclipse (*grahaṇanimitta*) and such a date could be set even after the eclipse had occurred. While the inscriptions are by and large a source of information on the eclipses, some of them do shed light on the genealogy of an astronomer or the prediction of an eclipse, and also record the solar ingress (*saṁkrāntis*) and new moon days (*amāvāsyās*) with dates when land, villages, and other gifts were presented. They also deal with different types of eras, which is of value for understanding Indian chronology.

Mathematics

Indian astronomers were able mathematicians too. The origins of mathematical activity can be traced to the Vedic period. Like the *Vedāṅga-jyotiṣa* which deals with astronomy, a section of the Śulba-sūtras of another *Vedāṅga* called the *Kalpa,* deal with geometrical aspects of a sacrificial altar. The number-reckoning on the ascending decimal scale up to 10^{12} with appropriate terms had already been accomplished by the Vedic sacerdotal class in the context of quantifying the merits that a sacrificer would attain and in the calculation of a sacrificial fee. The measured chantings of the Vedic *mantras* for certain rites involved a progressive arithmetical series employing the odd and the even numbers. Significantly, the Śulba-sūtras—the *Baudhāyana, Āpastamba, Kātyāyana,* and *Mānava*—raised the level of Vedic geometry and some aspects of algebra, as well. Their inspirational source was the Vedic injunction for the accurate construction of sacrificial altars of different shapes for attaining different desires. Their mathematical spin-offs were indeed impressive: combinations and transformations of rectilinear figures, especially the squares and rectangles, squaring the circle, surds, and values of $\sqrt{2}$, $\sqrt{3}$ etc., the so-called Pythagorean theorem, methods of constructing various forms of sacrificial altars in shapes like those of a falcon, tortoise and chariot wheel, along with the related mensuration. The altar geometry provides also a glimpse of the beginnings of algebraic notions in Vedic mathematics.

Another *Vedāṅga*, the *Chandaḥ*, in pursuance of the Vedic concern for exactitude in meters and their variations, had also laid the foundation of combinations called the *meruprastāra,* or a methodical triangular array of short and long sounds. This was elaborated by Halāyudha in the tenth century CE in order to find the number of combinations of n syllables taken 1, 2, 3...n at a time, an accomplishment of great merit at that time. It was only six centuries later that such a triangular array of that type appeared in Europe—and it was known as Pascal's triangle in the seventeenth century.

From these pristine beginnings in a religious matrix, mathematics in India began to attain new heights, particularly from the time of Āryabhaṭa I (fifth century CE). Two or three centuries before his time, the now universal decimal place-value system had been developed, the exact authorship of which is unknown. Using nine digits and zero, giving a name and place-value for zero, this versatile system energized certain aspects of mathematical thinking and practice. More importantly, it acted as a catalytic medium of computational astronomy in which Indian astronomers excelled, unlike their compeers in other cultures. In the succeeding centuries not only Āryabhaṭa I, but Bhāskara I, Brahmagupta, Vaṭeśvara, Āryabhaṭa II, Śrīdhara, Śrīpati and Bhāskarācārya II, made significant contributions in arithmetic, geometry and algebra. The solutions of indeterminate equations of the second degree conceived by Brahmagupta Jayadeva and Bhāskara II were far ahead of similar exercises in Europe. In trigonometry, too, Indian mathematicians had registered substantial achievements. It is significant to note that the modern words, zero, sine, and cosine owe their origin to the Sanskrit words, *śūnya*, *jyā* and *ko-jyā*, respectively, through their translations and transformations in Arabic and Latin. Such has been the seminal influence of Indian mathematical accomplishments that attracted the attention of the West from time to time.

The mathematical endeavours of Āryabhaṭa I, Brahmagupta and others, including Bhāskara II, were in tandem with their works on astronomy. But Mahāvīra (ninth century), a Jaina by birth and native of Karnataka, produced a large treatise comprising about 1100 verses, divided into nine chapters dealing exclusively with mathematics—arithmetic, algebra, geometry, and mensuration—under the title *Gaṇitasārasaṅgraha*, the manuscript of which is housed in the Oriental Research Institute, Mysore. Apart from its mathematical contents, which are in the nature of a discussion or explanation of the topics dealt with by his predecessors, Āryabhaṭa I, Bhāskara I and Brahmagupta among others, the *Gaṇitasārasaṅgraha* has to its credit some notable contributions. A perceptible aspect of this treatise is that several problems have been presented in an exquisitely poetic manner. Jaina scholars were also interested in large numbers and used the term *sthāna* for the decimal place. They closely followed the Vedic śulba experts in their approach to geometry and mensuration, and arrived at methods of permutations and combinations.

The Kerala School of Mathematics (fourteenth-nineteenth centuries CE) led by Mādhava, was an intellectual hub that blazed new trails in formulating the power-series—the trigonometrical sine, co-sine, and $\tan^{-1}x$ as well as *pi* series, which antedated similar mathematical formulations in Europe by a couple of centuries. Geometrical algebra was also developed, as evidenced by the sixteenth-century work, the *Kriyākramakarī*.

In North India, there were also notable mathematicians who made important contributions. Nārāyaṇa Paṇḍita (fourteenth century CE) was one of the leading mathematicians, and produced two works: the *Gaṇitakaumudī* and the *Bījagaṇitāvataṃśa*. The former is modelled on the *Līlāvatī* of Bhāskara II, but is more extensive. The latter is a work on algebra like that of Bhāskara II, but only its first few chapters are availble. Nārāyaṇa also worked on magic squares. Muṇīśvara and Kamalākara were among the other noted mathematicians.

It is sometimes said that arithmetic and algebra were the forte of Indian mathematics, while geometry was the strong pursuit of Greek mathematicians. However, Brahmagupta's geometrical formulations were of a high order. It is true that the work on geometry by Euclid, the *Elements,* was translated into Sanskrit from its Arabic version only in the early eighteenth century. It is also true that al-Bīrūnī, who was in India in the first half of the eleventh century CE, tried to introduce Euclid's works along with Ptolemy's *Almagest* from their Arabic versions, but his efforts did not meet with the desired success. However, some Indian Muslim mathematicians had evinced an interest in Euclidian geometry and produced its contents in Persian. Ghulam Ḥusain Jaunpūrī, well known for his astronomical *zījes* (tables), threw light on as many as 142 theorems along with the proofs given by Euclid dealing with the propositions selected from the first six books of Euclid's *Elements*. A significant effort of Jaunpūrī was that, while providing only the statements of Euclid, he preferred to give his own proofs using the Euclidean techniques, thus revealing his originality.

In Sanskrit tradition, there was no axiomatic geometry as such, but geometry was by and large a constructional one. It would appear that in the early stages of the development of mathematics in Central Asia, the situation was somewhat similar, and trigonometry and geometry were not thought of as specific branches of mathematics till the time of Umar b. Ibrahim al-Khayyāmi (twelfth century CE), whose commentaries on the third and fourth books of *Euclid* are well known. Umar b. Ibrahim al-Khayyāmi, however, was famous for his work on algebra. His treatise, *Maqālah fi-l-Jabr-i* (twelfth century), is regarded as one of the greatest medieval texts on algebra. In this work he has elegantly classified the equations and dealt with geometric and algebraic solutions of quadratic equation, besides some aspects of binomial series. This work is also known for its arrangement of groups of cubic equations and their solutions.

In India from the sixteenth century onwards, some Muslim mathematicians began to produce books on mathematics. In 1555, Abū Isḥāq b. Abdu'llah, a mathematician who lived in Golconda (near Hyderabad), produced an arithmetical work in Persian, which was a commentary on the text written by a pupil of Naṣir al-Din-al-Tusi. In 1587, Abū'l-Faiz Faizī, who was a poet-laureate in the court of Akbar, translated Bhāskara II's *Līlāvatī* into Persian. About fifty years later, the well-known scholar of the times, Atau'llāh Rashīdi, brought out a Persian translation of Bhāskara II's *Bījagaṇita*. Based on *Līlāvatī*, an arithmetical treatise in Persian under the title *Badāi'ul-kanūn* by Mednī Mal also came out. There were also other works which showed the interaction, though not as strong as it should have been, between the Islamic and Sanskritic traditions in mathematics.

In the westward transmission of the Indian decimal place-system and the associated Brāhmī numerals (one to nine and zero), Islamic savants played a seminal role, diligently encouraged by the Caliphate at Baghdad (eighth-tenth centuries CE), which was then a renowned academic centre. Al-Khwārizmī (ninth century), a versatile Central Asian mathematician who had moved to Baghdad, became the key figure in this transmission to the Islamic West Asia. In a century or so, the Indian decimal place-value method along with numerals for one to nine and zero, spread even to far-off parts of Europe, which, like Spain, were under Islamic dispensation. Indian arithmetic, based on this decimal place-value numeration, passed first into Latin Europe, which became possible through translations of Arabic texts. There was intense activity of such translations from the twelfth century CE. Adelard of Bath, John of Serville, Robert of Chester, Villedieu, Leonardo Pisano, and Sacrobosco were among the noted scholars of this transmission into Latin Europe. The erudite Adelard of Bath translated an arithmetical work attributed to Al-Khwārizmī, under the title *Liber ysagogarum Alchorismi*; Robert of Chester rendered into Latin Al-Khwārizmī's *Ḥisāb al-jabr wal-muqābala*, which contained several Indian algebraic elements, and John of Seville wrote *Liber algorismī* based on Al-Khwārizmī's and other works. Early in the thirteenth century an arithmetical classic appeared; it was *Liber abaci* by Leonardo Pisano, which expounded in Latin the Indian arithmetic along with related contributions by Arab mathematicians. More importantly, it became a foundational text for Indian decimal place-value numeration as well as aiding its spread to different parts of Europe. All of these introduced a new arithmetic called *algorism* based on the decimal place-value system, and the new arithmetic became one of the catalytic elements of the Renaissance itself.

Chemical Practices

Chemical practices have been prevalent since ancient times, as is evidenced by the technical skills of antiquity possessed by the practical chemists: potters, brewers, dyers, glass-makers, perfumers and, more importantly, metal smiths. They contributed a great deal to the enrichment of the material culture of their societies, despite the fact that they did not have the relevant theoretical knowledge of the chemical transformations that would take place in the processes they dealt with. But they had sound empirical knowledge acquired over centuries of practices, innovating and improving them from time to time by responding to the challenges posed by chemical techniques, particularly in metallurgy.

In India, the beginnings of chemical practices are noticeable in remote antiquity, in the Neolithic period as well as in the advanced settlements of the Indus Valley Civilization or Harappa culture. Over the millennia, diverse chemical practices—ceramic and glass; metallurgy and metal working; dyes and pigments; cosmetics and perfumery; paper-making and preparation of metallic compounds—have been fostered assiduously, as evidenced by extensive archaeological explorations as well as by some literary sources. The archaeological data concerning various types of pottery are valuable for an incisive understanding of the types and the techniques of making them.

It was, however, in the metallurgical and metal-working techniques that Indian metal-smiths excelled. Of the three most important metals, which have served man in various ways since early times—namely copper, iron and zinc—the metallurgy of copper was developed first, around 4000 BCE, in Central Asia. The method of extraction of iron from its ores was first known perhaps to West Asians, particularly the Hittites around the eighteenth to seventeenth centuries BCE. Meanwhile, the extraction of zinc from its ores, by a method called downward distillation, was first developed in India around 400 BCE, as evidenced by archaeological excavations in the mining region of Rajasthan. It was a difficult process, which required extraordinary metallurgical skills, and the Indian zinc-smiths displayed them in an innovative manner.

Metallurgical Skills

The archaeological explorations of what is called the Megalithic Culture, which spread extensively, especially south of the Vindhyas up to the tip of the southern peninsula, have brought to light a wide variety of iron objects as grave-goods of several megalithic burials, some of which have been dated to about 1300 BCE. Non-megalithic iron objects, too, have been considerable and date to about 1000 BCE in northern

India. Indian metallurgists had notably attained excellence in the macro-technology of both iron and copper, which was unparalleled elsewhere in ancient times. The Delhi Iron Pillar of the early fifth century CE (over 24 feet in height and 6 tonnes in weight) and the Copper Statue of the Buddha of the fifth century CE (Sultanganj in Bihar; over 7 feet in height and 1 tonne in weight) are vivid examples of macro-metallurgy. The forge-welding technique of the the Iron Pillar is indeed fascinating, mostly for its long-lasting corrosion-resistance. The reason for its being corrosion-resistant over sixteen centuries has been scientifically known. Nevertheless, it is still a mystery as to how the ancient iron-smiths accomplished the forge-welding of such a massive pillar with an admirable uniformity of its chemical constituents. Indian iron-smiths were unable to produce cast iron, as their Chinese counterparts did, which required the generation of higher temperatures of about 1400^0C. Their furnaces could go up to a temperature of about 1100^0C, which was sufficient for the production of red-hot spongy iron. But it was their technique of selecting the right type of ore, use of the reducing agent in right proportions, and the forging techniques that enabled them to produce iron of high purity close to cent per cent—a technique unmatched elsewhere at the time.

In the macro-technology of copper, too, ancient Indian copper-smiths showed admirable accomplishments. The Chinese pilgrim Hiuen Tsang in seventh century CE referred to a huge copper image of the Buddha (24.2m in height) and a brass temple being built by Emperor Harsha. While there is no archaeological evidence that confirms the statement of the Chinese pilgrim, a remarkable and imposing huge statue of the Buddha (2.25m in height and nearly a tonne in weight) was unearthed in the ruins of a Buddhist monastery at Sultanganj (Bihar) in 1864 by an East India Company engineer. This statue was perhaps a reminder of the importance of Buddhism at a time when it was showing signs of decline in India, while Hinduism, through its *Purāṇas*, Āgamic manuals and tāntric practices, had begun to worship a plethora of gods and goddesses in temples and homes. In the next two to three centuries iconometry as well as iconography began to receive extensive attention.

The South Indian bronze idols of the Chola period (eighth century CE), and even those of one or two centuries earlier, have been shining examples of not only artistic sensibilities in human anatomy but also of the expertise in metal-working or icon-casting by the *cire-purdue* or lost-wax process. The origin of this process in India goes back to the Harappan civilization. Beeswax was generally used for shaping the desired bronze icon and the iconography coupled with iconometry was accomplished according to certain *Āgamic* manuals. Some of the South Indian divine metal idols are also made of five metals (*pañcaloha*), comprising, besides the main components (copper and tin), some quantities of silver, gold, zinc or iron.

Steel: India was well known for the micro-technology of steel, an alloy of iron and carbon in the right proportions. Even as early as the fourth century BCE, Indian steel had become a prized commodity fit enough to be presented to Alexander the Great, who had invaded the northwestern parts of India then. But it was from about the sixth century CE that South Indian iron-smiths began to produce high-quality steel that was sought after in West Asia. The famous damascened swords made of pellets of Indian steel imported through Persia (Iran), were in great demand because of their enduring edges and sharpness. In the British colonial period, the steel produced by indigenous iron-smiths in the Deccan was known as the 'Wootz steel' (about 1.3 to 1.8 per cent carbon), the word, *Wootz*, being a corrupt derivative of the Kannada word, *ukku,* or the Tamil word, *urukku*. The process involved in the production of such high-quality steel has been vividly described authoritatively by Francis Buchanan-Hamilton in his travelogue: *A Journey from Madras through the Countries of Mysore, Canara and Malabar* (1834). Also called the crucible steel, since the crucible plays an important role in the process of making this steel by carburization as well as de-carbonization to obtain the iron-carbon alloy in the desired proportion, 'Wootz' steel has attracted the attention of even modern metallurgists.

Zinc Metallurgy: The available archaeological evidence indicates that ancient Indian metalsmiths could take legitimate pride in their metallurgical skills, especially in the extraction of zinc from its ores. They showed a rare expertise in overcoming the difficulties in the process of the downward distillation of zinc in a reducing atmosphere and at the optimum temperature so that the metal would not get oxidized, which would have defeated the very purpose of obtaining the metal. For the first time in the history of extracted metals, zinc metallurgy was successfully achieved in India around BCE 400. It was only in the eighteenth century that zinc metallurgy was practiced in Europe more or less on similar lines to those adopted in India.

There is, nevertheless, practically no literature either in Sanskrit or other languages that provides details of the metallurgy of either copper, zinc, or iron. But the archaeological data and a wide spectrum of metal objects unearthed in different parts of India, and their chemical analysis, enable us to understand the metallurgical practices of yore. However, there are literary sources, in addition to archaeological data, pertaining to the bronze icon-casting method, iconography, and iconometry: Āgamic manuals like the *Vaikhānasa Āgama, Vaiṣṇava Pāñcarātra, Bṛhatsaṃhitā, Kāśyapa Saṃhitā*, and *Mānasollasa* or *Abhilaṣitārtha-cintāmaṇi*. In general, the literature on technology is not as extensive as that on astronomy, mathematics or medicine, except for an appreciable number of texts on cosmetics and perfumery.

Cosmetics and Perfumery: The tradition of cosmetics and perfumery has been a vibrant one from its roots in the Harappan culture to the preparation of a wide variety of perfumes even to this day. The second urbanization, which began to expand from about the sixth century CE into the Indo-Gangetic plains and other regions in the west, east and the south, leading to an elevated material life, provided a fillip to the emergence of a wide variety of cosmetics and perfumery. New techniques emerged under the appellation of *Gandhayukti,* in which permutations and combinations of a chosen number of aromatic ingredients were processed to obtain a large number of pleasantly fragrant perfumes.

In the Classical Age of India and later (from about the fourth century CE onwards), it was mainly royal patronage and the extensive use of perfumes by the royal household as well as the wealthy class that stimulated the growth of the perfume industry. Added to this were the prescribed uses of incenses and other perfumes as part of the worship of diverse divine icons, both at home and in temples. The religious undertone of perfumery is revealed particularly in two texts—the *Viṣṇudharmottara Purāṇa* and the *Kālikā Purāṇa*. Around this time (fifth-sixth centuries CE), appeared the encyclopaedic work of Varāhamihira, the *Bṛhat Saṃhitā,* in which the art of perfumery and the preparation of several enchanting fragrances by permutations and combinations of some basic aromatic ingredients have been presented. The *Mānasollāsa* (twelfth century) of King Someśvara III, another encyclopaedic work, is a valuable source for understanding the types of fragrant incenses (*dhūpabhoga*) which were in use then. The *Purāṇas* as well as the *Mahābhārata* hail incense as the most popular form of perfume for offering to the divine, and as a purifier of the temple environment, leading to an ennobling experience of the divinity by the devout.

The *Gandhasāra* of Gaṅgādhara, probably belonging to the twelfth or thirteenth century, and an earlier text called *Gandhavāda* by an unknown author, with a commentary in Marāṭhi, are valuable sources for understanding not only the nature and extent of a wide variety of cosmetics and perfumes being produced then but also the techniques associated with their production. The *Gandhasāra* has presented a classification of aromatic substances in terms of eight *vargas* or categories. These reveal the extensive empirical knowledge of aromatic plants and their parts possessed by the perfume-makers. In the medieval period, the Muslim monarchs, who were fond of perfumes, encouraged the perfume industry to a great extent. The now familiar *attar* of roses or rose-water made its debut during the time of Akbar and Jahangir. The Āin-i-Akbarī (CE 1590) of Abul Fazl deals elaborately with perfumes and the Perfume Office of Akbar. It may be noted that there is no reference to roses in the earlier Sanskrit literature—the cultivation of the rose plant which was introduced from Persia was greatly encouraged in the Mughal period.

Pyrotechnics: The art and technique of preparing devices containing combustible materials which, when ignited or exploded, produce light, sound, and a variegated display of colour had been perfected in India by about the fifteenth century. However, two or three centuries earlier, pyrotechnics, perhaps derived from China, was already in practice. In the course of time, the display of fireworks became part of celebrations, such as festivals, marriages, victories in battles, and the procession of temple deities, a tradition that continues even today. One of the festivals is called *Dīpāvalī*, the festival of lights, which lasts for two or three days as determined by the traditional lunar *pañcāṅga*, and celebrated with éclat all over India with the display of fireworks of different types.

There are special texts in Tamil, Malayalam, Marāthi and Bengali, as well as in Sanskrit, with details about different firework devices. The *Bānasāttiram* (*Bāṇaśāstra*) in Tamil, ascribed to one Bogar (tenth century CE) who, according to Tamil tradition, was actually Chinese, has described as many as ninety-eight formulae for the production of twenty types of fireworks, including rockets. In Odisha, it would appear that Brahmins were also engaged in the production of such pyrotechnic devices as rockets, and they were even referred to as *bānuya* or those who were making *bāṇa* or rockets. At present, the production of pyrotechnics is extensively carried out in south Tamil Nadu in pursuance of its long tradition.

Rasaśāstra Texts

A wide variety of chemical practices have been described in the *Rasaśāstra* texts. These involve several processes: distillation of three types (upward, downward, and sideways), evaporation, sublimation, fumigation, steaming, prolonged incineration, and extraction. The apparatus and contrivances of the *rasavādins* were simple but purposeful, and the source of heat was either cow dung cake, wood or charcoal. The ingredients included mercury, sulphur, and several metals and minerals. An appreciable number of plants, too, were employed in their processes. Though they sometimes indulged in esoteric practices of the tāntric milieu, their experimental skills were of high order. Both the Sanskritic and Tamilian alchemical practices were soon transformed into medicinal iatrochemistry, more so in Tamilian Siddha medicine.

The dates of most of the *Rasaśāstra* texts are uncertain, but they were possibly redacted into the textual forms between the ninth and the eighteenth centuries, the period between the eleventh and the fifteenth centuries being perhaps the prolific one. However, chemical practices and the associated ideas found in them were a few centuries older. Among the noted *Rasaśāstra* texts are: *Rasārṇavakalpa* (part of the *Rudrayāmaḷa tantra*); *Rasaratnākara* and *Rasakacchapuṭa* by Siddha Nāgārjuna; *Rasārṇava* by an unknown author; *Rasaḥrdaya* by Govinda Bhāgavat; *Rasaprakāśasudhākara* by Yaśodhara; *Rasaratnasamuccaya* by Vāgbhaṭa; *Rasendra Cūḍāmaṇi* by Soma Deva; *Rasasāra* by Govindācārya; *Rasarājalakṣmī* by Rameśvara Bhaṭṭa; *Rasakaumudī* by Jñana Candra; *Rasasaṅketakalikā* by Cāmuṇḍa; *Rasapaddhati* by Bindu Paṇḍita; *Rasamaṅgala* by Gahanānanda Muni; *Rasacandrikā* by Mādhava Kavindra; *Rasakaṅkali* by Kaṅkali; *Rasaratnapradīpa* by Rāmarāja;

Rasasārasaṅgraham by Gangādhara Paṇḍita; *Rasasudhānidhi* by Vājarāya Śukla; and *Rasayogamuktāvalī* by Narahari Bhaṭṭa. There are several other works, the authorships of which have not been established satisfactorily, and some of them appear to be fragments of earlier major works. The *Rasaśāstra* texts are a repository of extensive technical vocabulary in Sanskrit, not only of the chemical ingredients used but even of the various processes and apparatus employed. There are also some secretive types of words, generally called *sandhyābhāṣā* (intentionally used language), which were intelligible only to those who were initiated into the secret lore of alchemy. Such a method was perhaps intended to keep away charlatans.

Tamilian alchemy was not far removed from the Sanskritic *Rasaśāstra* in its basic ideas and even terminology. The exponents of Tamilian alchemy were the celebrated Siddhas, of whom Bogar, Rāmadevar, Agastyar, Kanganavar, and Nandiśar, dealt with some aspects of alchemy. There are about 700 works on Siddha medicine and not a few of them have alchemical undertones. They deal with mercury, sulphur, arsenic substances, and other minerals more or less in the same way as in the *Rasaśāstra* texts. However, the concept of the 'pill of immortality' (*rasaguḷikai*), a mercurial preparation, is more pronounced in Tamil texts with extraordinary claims of its miraculous powers.

THE EMERGENT PICTURE

The extensive archaeological data on metallurgy and metal-working, ceramics and glass-making, wall-paintings, and temple architecture (some of which have an astronomical orientation) on the one hand and, on the other hand, the large body of scientific manuscripts on astronomy, mathematics, medicine, and plant science, definitively indicate the technological and scientific tradition which has been fostered in India over three or four millennia. Said al-Andalūsi, an Islamic thinker and historian of science of the eleventh century CE, in his treatise *Kitāb Ṭabakāt al-Umam* (The Book of Category of Nations), tried to evaluate the scientific and other attainments of the people of eight nations, including the Greeks, Persians, Egyptians, and the Chaldeans. He accorded the first place to Hindus (the people of India) for their scientific achievements coupled with their wisdom. The reason was not far to seek. India by then had made notable progress in astronomy, mathematics, medicine, and metallurgy. Indian astronomical and medical texts had been rendered into Arabic, and the Indian decimal place-value system and numeration had won appreciation in the Islamic world. In the succeeding centuries, contrary to the general view that Indian scientific and technological vigour was on the decline from the twelfth century, there were new developments and attainments in astronomy and mathematics in Kerala and elsewhere, as well as certain aspects of technology like pyrotechnics, paper production, and perfumery. Muslim monarchs also encouraged activities in these fields. The manuscript tradition continued unabated and reached its pinnacle in the seventeenth-eighteenth centuries. However, it remained, by and large, the treasure and preserve of the literati and the orthodoxy.

Between the fifteenth and seventeenth centuries, the period of the scientific Renaissance in Europe, there ensued a new movement leading to what is called the 'Scientific Revolution', which heralded the dawn of a new way of looking at nature and the universe, matter and motion, space and time. The Copernican heliocentric theory, Galileo's questioning of the Aristotelian dogmatic views, Kepler's laws regarding the motion of planets in elliptical orbits, and the concept of mass and the measurable laws of motion as well as of universal gravitation enunciated by Newton, changed the entire complexion of traditional views on matter and motion, space and time. The word, *science*, began to take on a new definition in terms of measurement, verifiability, and reproducibility. Motion, too, was understood quantitatively in the context of the mass of a body, force, space and time, all measurable and verifiable. The trinity of experimentation, observation, and inference became the bedrock of all investigations—physical, chemical, and biological. By the middle of the nineteenth century, science became synonymous with its methodology of induction and deduction, of verifiability, reproducibility, and even falsifiability. Its divorce from philosophy and religion on methodological grounds was complete while the 'traditional science', with its moorings in some kind of religio-philosophy, began to pass into the pages of history, except in India.

The question is: why did India, China and the Islamic world—which, over the centuries, had made their own contributions in the fields of astronomy, medicine, and the physical world of matter and motion, space and time—not develop a new attitude towards them in the same way as their contemporaries in Europe did? A cognate question is: why did the savants of these areas continue to move along the already charted grooves, some of which were governed by the respective religio-philosophical ideas? Questions such as these are indeed too difficult to answer. However, as far as India is concerned, some explanation can be found in: (a) the absence of a dominant philosophy of nature in contradistinction to the incisive philosophy of man and his emancipation from suffering, which was pursued in India over the centuries; (b) the absence of centres of learning like the medieval universities which came up in Europe from the thirteenth century; (c) the preceptor-disciple relationship for the communication of knowledge generally by oral methods; (d) the pursuit of knowledge in the context of the four *puruṣārthas*—*dharma, artha, kāma*, and *mokṣa*; and (e) the social rigidity that was not permissive of any type of interaction between the thinkers and the artisans or technicians governed by caste considerations and stratified intellectual endeavours, of which the priestly class or Brāhmins were the privileged caste. There could well have been other reasons like the general ethos of the people preferring the traditional ideas and milieu to new modes of thinking. The result was that India remained with its own tradition, but had displayed its potentiality for selective assimilation of ideas from elsewhere.

In the history of the dynamics of science and technology, India, by virtue of its geographical position and attainments, has been an active participant in the cross-cultural transmissions right from the time of the Harappan culture (third millennium BCE), when it had perceptible contacts with the Mesopotamian region, and later with Central Asian cultures. The *Ṛgveda* had enjoined: *Let noble thoughts come from anywhere unhindered and overflowing* (I. 161), and this had set the tone for the exchange of ideas and practices, including those in astronomy, mathematics, medicine, and metallurgy. In the early centuries of the Christian era, perhaps some Hellenistic astronomical concepts (still a debatable issue) might have been assimilated by Indian astronomers. Likewise, Chinese alchemical seed-ideas could have made their way into the Indian tāntric milieu and became the forerunner of medicinal chemistry involving metallic and mineral medicines. But later, during the eighth and tenth centuries, Indian astronomy and mathematics, especially the decimal place-value numeration as well as Āyurvedic ideas and practices, had reached West Asia and even beyond. During the thirteenth-fourteenth centuries, the Graeco-Arabic or Unāni medicine entered India and registered substantial progress, particularly during Mughal rule. Open-mindedness is a characteristic of scientific attitude, and Indian thinkers as well as the general public, despite their traditional compulsions, were generally known for their impartial, tolerant understanding and for being open to ideas. For this reason, when Western science made its debut and began to take firm roots in India in the nineteenth century, largely through the efforts of the British colonial masters, Indians did not think of the new entrant as alien to their own ethos; instead, they welcomed it, and Indian intellectuals began to make notable contributions towards its progress, particularly in the twentieth century.

WESTERN SCIENCE IN INDIA

The diffusion of Western (European) science into India, perhaps also into other non-European countries during the eighteenth and nineteenth centuries, reveals certain trends. First, the European naturalists, engineers and medical men who were employed by the East India Company became deeply involved in the investigations of the new colonial land's flora, fauna, minerals, weather, and other aspects, on which they brought to bear their training in, and knowledge of, systematic observations. Thus, in the beginning, the botanical, zoological, geological and meteorological explorations and surveys appealed to them. The fresh knowledge gained by them through these explorations was methodically presented, preserved, and evaluated in the context of its global importance. Second, alongside, there sprang up in the new land several institutions to impart the Western type of liberal education, including scientific education, besides some scientific establishments or organizations. These were noticeably instrumental in expanding the range of investigations. Third, several intelligent Indians, who were well trained in the new educational institutions, joined the colonial masters in their scientific pursuits. The last quarter of the nineteenth century witnessed the emergence of Indian scientific pioneers who provided a lead either by establishing research institutions or by undertaking scientific researches on their own, imbued with national spirit and in furtherance of national regeneration.

It has been rightly stated that the nineteenth century was a dividing line. For, these hundred years changed the face of India far more than did the preceding one thousand years. One of the important catalytic instruments or agents of this change was that in the nineteenth century, in the colonial dispensation, there was a gradual expansion of scientific investigations through field surveys in botany, geology, trigonometrical survey, meteorology, zoology, and even archaeology. These led to the establishment of survey organizations like the Trigonometrical Survey, Geological Survey, Botanical Survey, and Meteorological Department. Education in general and specifically scientific and technical or professional education, received some support from the colonial government, and several colleges came up. The first three universities were established only as affiliating and degree-awarding apex bodies—in Calcutta, Bombay, and Madras in 1857—which were followed later by two other universities at Lahore (1882) and Allahabad (1887). Telegraph and the railways, too, made their home in India in the middle of the nineteenth century, and soon developed a network of their own. Some research institutions like the Imperial Bacteriological Laboratory at Poona, the Plague Research Laboratory (later named as the Haffkine Institute) at Bombay, and the Solar Physics Laboratory at Kodaikanal in the Madras Presidency, were also established. Raja Ram Mohun Roy and other Indian intellectuals and social reformers played a leading role in the introduction of Western liberal education with English as the medium of instruction, which was strongly recommended by Lord Macaulay in 1835 and fully supported by Lord William Bentinck, the then Governor-General-in-Council.

Among the early Indian scientific pioneers was Mahendra Lal Sircar (1833-1904) who, in cooperation with certain wealthy citizens of Calcutta, set up the Indian Association for the Cultivation of Science in 1876. J.N. Tata (1839-1904), a philanthropic industrialist, came forward to start a university, which eventually resulted in the establishment of the Indian Institute of Science in Bangalore (1909). In the meantime, in Calcutta, Ashutosh Mukherjee (1864-1924), J.C. Bose (1858-1937) and P.C. Ray (1861-1944) blazed new trails in mathematics, physics and chemistry, respectively. The Indian National Congress came into being in 1885 and began to spearhead the freedom movement. An intense nationalism transcending the barriers of region, religion, caste, and sex pervaded all parts of India. In this ambience and in the first three decades of the twentieth century, there were outstanding scientists, such as C.V. Raman (1888-1970), M.N. Saha (1893-1955), and S.N. Bose (1894-1974), besides the mathematical prodigy, S. Ramanujan (1887-1920). Their renowned researches in the then frontier areas of physics won for them international acclaim, including the Nobel Prize for Raman in 1930. On the eve of Independence, the scientific spectrum encompassing the physical, chemical, biological, geological, and related sciences was rather impressive, with nineteen universities, some with postgraduate research departments. Several scientific research institutions and coordinating bodies like the Indian Science Congress Association (1914), the Indian National Science Academy (1935), the Council of Scientific and Industrial Research (1942); and the Tata Institute of Fundamental Research in Bombay (1945), were established.

Even in the new climate of modern science, it is significant to note that traditional astronomy, mathematics, and medicine too received their due attention by some Western scholars, who called them Hindu sciences.

Hindu Science: The alien government, as a policy, did not interfere with Indian religious thoughts, nor with their diverse obnoxious practices. The British government in London had issued strict instructions to the Governors-General in India to maintain neutrality and non-interference in religious matters. Nevertheless, some European scholars began to study religious texts and translated them into English, German, or French. The nature of the religio-philosophical thoughts of Hinduism, Jainism, and Buddhism began to be understood by Western savants. More importantly, the missionaries and other Europeans in the colonial dispensation soon observed that a wide variety of religious rites, periodical festivals, and marriages were being performed according to Hindu calendars that were being prepared annually based on the traditional texts of astronomy. The colonial administration, too, depended on the Hindu calendars for declaring religious-festival holidays.

There was a recrudescence of interest in Hindu astronomy. In the late eighteenth century, the Hindu methods of calculating the duration of lunar and solar days and planetary positions, were studied and published by French astronomers M. Le Gentil (1772) and M. Jean Sylvain Bailly (1787). In 1784, the Asiatick Society (later, the Asiatic Society) was founded in Calcutta; this institution encouraged oriental studies, astronomy being one of them, and published these studies in its journal, *Asiatick Researches.* Its first president, William Jones, himself led the researches, and threw light on the antiquity of the Hindu zodiac (1790). He was ably followed by several other British and continental scholars. It was through the sustained, methodical studies of several European scholars and their publications in *Asiatick Researches* and books throughout the nineteenth century, that what they called 'Hindu astronomy' came to light; likewise, the associated 'Hindu mathematics'.

Notable among the studies on Hindu astronomy were those of Samuel Davis (*Sūryasiddhānta;* Jupiter's cycle, 1792); John Bentley (*Sūryasiddhānta* and the fixation of its date): H.T. Colebrooke, who also brought to light Indian mathematical elements (1809-17); Charles M. Whish, who also unearthed the mathematical-astronomical texts, and the trigonometrical acumen of the Kerala School of Astronomy (1827-35); L. Wilkinson (1834; (tr) *Siddhāntaśiromaṇi)*; Reverend E. Burgess (1860; (tr) *Sūryasiddhānta*); W. Brennand (1886; *Hindu Astronomy*); and George Thibaut (1894-95; *Pañcasiddhāntikā*).

Some Westerners also became interested in indigenous medical practices and studied the classical texts on medicine. J. Forbes Royle wrote on the *Antiquity of Hindu Medicine* (1838) and T.A. Wise on the *Hindu System of Medicine* (1860). Several Europeans, in the course of their field surveys or official administrative work, began to collect manuscripts (in Sanskrit) on astronomy, mathematics, medicine, and biological aspects, which were traditionally preserved by Hindu, Jaina, and

Buddhist scholars in different parts of India, and these served as the foundation for the emergence of oriental manuscripts' libraries in several places. The pioneering efforts of foreign scholars were a source of inspiration for some Indian scholars to pursue studies in 'Hindu science' in the closing decade of the nineteenth century and later.

A question that naturally arises is: did the new religious movements like those of the Brahmo Samaj and Ārya Samaj, and the emergence of 'Hindu science' pose a challenge to the advent and growth of Western science in India?. Or, was there any serious confrontation between them? The new religious reformatory movements sought to uplift the vast mass of people from the morass of superstitions, drawing upon the enlightening ancient wisdom. At the same time they recognized the importance of Western scientific education.

Did Hindu astronomy come in the way of modern astronomical pursuits in India? It is significant to note that, while the *Asiatick Researches* was publishing articles on Hindu astronomy (from 1790 onwards), the East India Company established the Madras Observatory (1792) with telescopes and other instruments. Throughout the nineteenth century, this observatory was engaged in systematic observations of the sun's transit every day at noon, Jupiter's satellites and the like, besides the compilation of a *Star-Catalogue* (1844). Six minor planets and seven variable stars were also discovered (1861-67). An orthodox Hindu, C. Raghoonathachary, was involved in some of these observations. There were two other observatories set up by princely endowments in Trivandrum (1836) and Poona (1890). In the latter, an Indian (K.D. Naegamvala) made important observations on the solar corona during the eclipse that occurred in 1898. Side by side, the traditional almanacs based on some Hindu astronomical texts continued to be prepared. However, the almanac-makers began to use the modern ephemeris to study the occurrences of eclipses.

P.C. Ray (1861-1944), who had obtained his doctorate in

Chemistry (1887) from the University of Edinburgh, produced two volumes under the title *A History of Hindu Chemistry* (1902 and 1907). In the preface to his first volume, he has recorded: 'The progress of chemical knowledge among ancient nations had a fascination for me and, after studying such knowledge in the other cultures, I was naturally led to an inquiry into the exact position which India occupies therein.' In this endeavour he was influenced by the French chemist, Marcelin Bertholet, with whom he was in correspondence. With the help of traditional scholars he studied chemical practices found in Ayurvedic and alchemical texts. He was also drawn to the mercury-centred alchemy since he had investigated mercurous compounds and successfully synthesized mercurous nitrite. The nationalist that he was, Ray tried to show the superiority of some Hindu chemical practices to those of other ancient cultures. But the modern chemist in him asserted itself when he wrote: '...I paused in my antiquarian studies and put aside my completion of the second volume of Hindu chemistry as I had now to catch up and be in current with modern chemical literature.' His humanitarian concern for the medical care of the poor motivated him toward the production of low-cost drugs from indigenous herbals. But he adopted scientific methods, and his efforts led to the establishment of the first pharmaceutical company in colonial India with an in-house research laboratory. Ray created a spirit of inquiry among young students of modern chemistry and, in the early decades of the twentieth century, schools of modern chemical research sprang up in Calcutta and elsewhere.

Ayurveda, Unāni, Siddha and folk medicine being indigenous medical systems were at the service of practically the entire population of India, even in the eighteenth century. It was only in 1792 that the British East India Company opened its first hospital to civilians in Calcutta. The next ten years saw just sixteen general hospitals—in Bombay (3), Calcutta (6), and Madras (7)—a woefully inadequate number for the large population, even of urban areas. As for the education in

European medicine, the whole of the nineteenth century had witnessed the establishment of only three medical colleges (one in each of the three presidencies) and some medical schools here and there. The colonial government, though it initially tried to encourage education in Ayurveda and Unāni, was by and large lukewarm to the indigenous systems of medicine. The result was that these systems had to plough their weary way; but the vast masses of people continued to depend upon them to a very great extent.

'Hindu medicine' was not an obstacle to the introduction of Western medical education and medical investigations. For, the dissection of the human body was undertaken in medical colleges by Hindu students, including those from the orthodox Brāhmin caste. The proficiency of native medical students was such that the medical degree of Calcutta University was recognized by the General Medical Council of Great Britain (1892); and the medical colleges in Bombay and Madras had received recognition of the Royal College of Surgeons of London in 1854 and 1855, respectively. In the early decades of the twentieth century, several Hindu doctors were also involved in clinical trials and medical researches on such terrible diseases as cholera, malaria, filariasis, tuberculosis, and several types of fever. At the same time, efforts were also made for the revival of both Ayurveda and Unāni. The Benaras Hindu University had as its constituent an Ayurvedic College (1920), and the liberal Muslim, Hakim Ajmal Khan established an Ayurvedic and Unani Tibbia College in Delhi (1921). A significant aspect was that certain modern medical courses, too, were included in the curricula of these institutions.

Darwin's theory of evolution did not raise a dissonant debate in India as it did in the West. The main reason for this was that Hindu religio-philosophies did not think in terms of the duality of the creator and the created, but postulated that the creator was immanent in all manifestations. Swami Vivekānanda was of the view that the yogic spiritual or inner evolution of man to the divine state was superior to the outward one, and he did not approve of the idea of the struggle for existence in this context. Swami Dayananda Saraswati, the founder of Ārya Samaj, and Aurobindo Ghosh, however, were not inclined to accept it. Despite such demurrers, Darwinism found its exponents in Indian biologists and other men of science, who continued to teach it in the universities.

In 1885, the Indian National Congress was founded and it went on to spearhead the subsequent freedom movement against British rule. It was a source of inspiration, too, for scientists to pursue scientific investigations which, as noted earlier, were undertaken by outstanding scientists. But, during the colonial rule, science and its application could not be channelized toward uplifting the nation from its poor socio-economic milieu. Nevertheless there was one great leader, Jawaharlal Nehru who, even when he was in jail between 1942 and 1944, was thinking of science and technology as instruments of social betterment. In his *Discovery of India*, which he penned during his imprisonment, he emphasized: 'It is the scientific approach, the adventure and the critical temper of science, its search for truth and new knowledge, its refusal to accept anything...its reliance on facts and not pre-conceived notions—all of these are necessary not merely for the application of science but for life itself and the solution of its many problems'. (*Discovery of India*, pp. 512-14).

As Independent India's first prime minister, Jawaharlal Nehru lent his prestige and support to the planned socio-economic development of the nation and, as an integral component of it, promoted science and technology, enlisting the cooperation of scientific pioneers like H.J. Bhabha, S.S. Bhatnagar and M.N. Saha. On the eve of Independence, there were about 170 colleges, many of which offered courses in physics, chemistry, mathematics, botany, zoology and geology. As for technical education, four colleges each for medicine and engineering, several medical, engineering and agricultural schools as well as some industrial ones were active in different parts of the country. Among the universities, of which there were seventeen, Calcutta University had emerged as the top most in offering postgraduate courses in science and promoting scientific investigations. All of these provided a reasonably well-laid take-off ground for the progress of science and technology soon after Independence.

2

EARLY SCIENTIFIC IDEAS AND PRACTICES

AN ANCIENT CIVILIZATION COMES TO LIGHT

In 1856, two British engineering contractors were involved in laying a railway track between Lahore and Multan. Soon, another railway track between Karachi and Lahore was taken up, and it was while quarrying the surface bricks, which lay buried in the mounds at Harappa (Montgomery district of the Punjab) as well as in Mohenjo-daro (Sind), that they encountered a number of antiquities. These included steatite seals bearing what the excavators thought were some 'pictographic symbols'. Alexander Cunningham, who was then heading the Archaeological Survey of India, collected some more seals at Harappa in 1875. In the next three decades, more and more artefacts came to light. Around 1920, Dayaram Sahni began his excavations at Harappa, and in 1922, R.D. Banerjee took up excavations at Mohenjo-daro. John Marshall, who was then the Director-General of the Archaeological Survey of India, published in 1924 an account of these excavations in the *Illustrated London News*. He thus announced for the first time the discovery of an ancient civilization in the Indus Valley. This civilization found a new but distinct place along with the other two ancient civilizations—the Egyptian and the Mesopotamian.

Of the three, the Indus Valley Civilization (archaeologists choose to call it the Harappa culture) was the youngest, but by far the largest, embracing an extensive area of over one million sq. km. even beyond the confines of the Indus Valley in all directions. It had several hundreds of settlements, both urban and rural, including the metropolitan cities like Mohenjo-daro and Harappa and several townships, such as Chanhu-daro, Kalibangan, Rupar, Lothal, Rangpur, Dholavira and Daimabad.

Mohenjo-daro: Literally meaning the mound of the dead, Mohenjo-daro comprises two mounds—the high mound called the 'Stupa Mound' since a Buddhist stupa of the second century CE existed there, and a lower one called the 'Lower City'. Mohenjo-daro, like the other urban settlements of the Harappa culture, was noted for its town-planning. Its acropolis (upper city) had large buildings measuring 200 × 400 m, while in the lower city, houses were packed together.

In addition, Mohenjo-daro had what the archaeologists call the Great Bath, which is rectangular in shape and measures about 12 m (north to south) and about 7 m (east to west), with a depth of about 2.4 m. This hydropathic system had several bathrooms and effective functional arrangements for the supply of water. Possibly, as some archaeologists opine, this system might have been related to the religious life of the city or its rulers.

To the west of the Great Bath lay a large granary (some archaeologists have disputed its existence) which was constituted originally of as many as twenty-seven blocks of brickwork. It had criss-crossed and well-laid-out passages between the blocks to ensure air circulation. It appears to have had a timber stair or ramp and along its northern side was a brick platform with a brick-floored alcove near its western end. The granary was indeed a massive construction at that point of time, which throws light on the engineering skills, and the attention bestowed on ventilation and techniques of preservation of foodgrains.

The Great Bath, Mohenjo-daro

The engineering craft that shaped the metropolitan city could also be seen in what the archaeologists have designated the 'College of Priests' that lay towards the northeast of the Great Bath. This was in the form of a square (10 m each side) courtyard and a series of rooms paved with bricks along with a verandah on three sides. There was also a large building, which was nearly 27.5 m square and divided into five aisles. A much larger building, over 76 m long, which is regarded by the excavators as a palace, existed in the southern sector. In addition to its inner structural arrangements, this building had a bread-oven. Also in Mohenjo-daro were several other structures for different purposes, like premises for workers, attendants and labourers. A striking feature was the inter-locked drainage system, which was well planned and executed with great care with a view to maintaining a high standard of sanitation. Triangular terracotta forms and alabaster lattice were used for cross-ventilation in the houses. To counter the onslaught of riverine floods, houses in the 'Lower City' were built on high platforms, reinforced by either mud or baked-brick walls. It would appear that the goal ever in sight of the people in Mohenjo-daro, Harappa and other advanced settlements, was to lead a comfortable, civilized life.

Harappa: The city of Harappa (upstream the river Ravi, a tributary of the river Indus) was nearly 5 km in circuit with neatly built fortifications and a citadel. The parallelogram-shaped citadel (415 m north to south and 193 m east to west) had a huge platform constructed out of mud-bricks with a thick wall also of mud-bricks on three sides (east, south, and north), but of burnt bricks on the western side. The 'Lower City' which lay to the west of the citadel had many structures: two ranges of barrack-like buildings, five rows of circular platforms, two rows of houses, courtyard, and others. There were eighteen circular platforms, each created in the form of five concentric rings of bricks with a hole at the centre, and situated in the vicinity of the granary, possibly for pounding or dehusking grains.

The Harappa granaries were located not far away from the riverbed. They were well-planned and executed constructions, each unit measuring about 15 m in length and 6 m in width. There were two rows of six units built symmetrically with a central passage of about 7 m. The combined floor-space of the Harappa group of granaries was more or less the same as that of a single Mohenjo-daro granary. The granaries both at Mohenjo-daro and Harappa were storehouses of the main agricultural produce—barley, wheat and cotton.

Other towns: About 130 km south of Mohenjo-daro, on the other side of the river Indus, lay Chanhu-daro, which, like the other two metropolitan cities, had a well-laid-out drainage system and streets and lanes with rows of houses built of mud-bricks. It was more or less a town of craftsmanship, as evidenced by the archaeological data relating to bead-making, shell-working, and copper-smithy.

Lothal was connected at that time with the navigable estuaries of the Sabarmati and Bhogavo rivers flowing into the Gulf of Cambay of the Arabian Sea. According to the excavator, it had emerged as a port-town of commercial importance. The acropolis of Lothal was in the shape of a trapezoid and had within it several types of constructions. It would appear that on a 3.5-metre-high platform stood what is designated as the residence of a ruler or chieftain, which measured about 126 × 30 m. Like Mohenjo-daro and Harappa, Lothal had inter-connected streets and lanes, blocks of houses, an efficient drainage system and wells for the supply of potable water. An important aspect of Lothal was the distinct warehouse, which was erected on a 4-metre-high platform with an extensive floor area of over 1,900 sq. m. Lothal was obviously the busiest port of the Indus Civilization and could well have handled a large volume of trade. The much larger 'Lower Town' at Lothal had houses, besides workshops and a bead-production unit. It did not lag behind in drainage and other sanitary arrangements.

The dockyard of Lothal was supposed to be the largest structure (as a dockyard) ever built by the Bronze Age community. It had water-locking devices as well as a water-regulatory mechanism by taking advantage of high tides of the river. Trapezoid in shape, the embankment walls of the dockyard were about 212 (west), 209 (east), 35 (south) and 36 (north) m, and were built with burnt bricks. The dockyard was also capable of harbouring sea-borne ships of considerable size.

Kalibangan was situated on the left bank of the now dried-up river Ghaggar in the present Ganganagar district of Rajasthan, and lay about 160 km southeast of Harappa. Though smaller in area than Mohenjo-daro, Harappa, or Lothal, it was also a well-planned town with a drainage system as well as residential houses, some with flooring of ornamental baked-tiles. The drainage system, however, left much to be desired. There is also some archaeological evidence to show that certain ritualistic practices were in vogue at Kalibangan, including the performance of sacrifices in fire-altars. The furrowed field which came to light during excavations has revealed the *modus operandi* of agricultural operations about 4,500 years ago.

Dholavira was another important Harappan settlement, which had standardized town-planning and artefacts. It is known for a unique inscription (Indus script) nearly three metres long with ten signs, each over 35 cm high. Dholavira also had a notable water management system. Daimabad (in Maharashtra), a late Harappan site, displayed its skills in copper-bronze technology.

The economy of the Harappa culture was noticeably agriculture-based. The fertile lands of the riverine beds and the tropical climate were conducive to growing several types of crops. Although no agricultural implements of any significance have been found, the use of a type of plough and operations like de-husking, threshing, and pounding can be inferred. Charred or decomposed seeds of several grains have been found at Mohenjo-daro, Harappa and some other places. Bread wheat (*Triticum compactum* and *Triticum sphaerococcum*), barley *Hordeum vulgare* (six-, four- and two-rowed), and cotton (*Gossypium arboreum*) were grown extensively. Barley appeared to be the most important foodgrain, followed by wheat. There is evidence to indicate that paddy was being grown to some extent in the region of Saurashtra at a later time. Sesame, peas, melon, dates and *rai* (*Brassica* variety) were also among the agricultural products. It would appear that the entire agricultural production and management were under the control of an efficient administration. Harappans were familiar with several animals like the elephant, bull, tiger, spotted deer, and rhinoceros, all of which have been depicted on several seals and sealings. The domesticated ones included sheep, humped and humpless cattle, buffalo, dogs and pigs. Perhaps horses of a superior breed like the ones with which the Ṛgvedic people were familiar, were not in use then.

The technology of Harappans encompassed glazing and painting on exquisite pottery, statues, terracotta craft, spinning and weaving, boat-building, bead-making, production of standardized bricks (generally 1:2:4) and, more importantly, copper-bronze-casting by the *cire purdue* or lost-wax process. However, we are still in the dark about their ideas on astronomy,

Some seals with animal motifs: Mohenjo-daro

mathematics, and medicine; the so-called Indus script found on seals, sealings and other inscribed objects, remains yet undeciphered. Nevertheless, it would be reasonable to assume that the Harappans had evolved several scientific ideas, developed their own calendar-making based on astronomical observations, practised the healing art and developed curative medicines. They were also concerned with ecological and environmental aspects in the context of their sustainable agricultural production and management. This civilization, but a silent zone of early history in India, lasted for nearly eight hundred years (2600-1900 BCE was its mature phase, followed by its decadent phase up to 1600 BCE) till the emergence of what came to be regarded as Vedic culture (around 1500 BCE).

THE VEDIC PERIOD

The earliest sources that deal with certain scientific ideas amid their hymnal prayers and rites are the four Vedas, their liturgical texts called the *Brāhmaṇas*, and the metaphysical treatises known as the *Āraṇyaka-Upaniṣads*. Such ideas were by and large concerned with the knowable world with a few speculations or reflections on the universe as a whole.

The *Ṛgveda* has expounded some ideas relating to the cosmos or the universe, but presented them within a divine framework. A mantra of a hymn (II, 12.2) says: '(the Divinity) solidified the molten earth, established the mountains, created the heavenly regions and firmly fixed the skies.' There are over a hundred references in this compendium to what is called *ṛtam*, or the natural order or law (which had moral dimensions too), and it was extolled as governing the cosmos, including all types of movements and occurrences within it. The concept of *ṛtam* was integral to the Vedic vision of the triple-stranded *man-spirit-cosmos,* which envisaged the synergy of the 'outer man', his inner spirit or soul, and the two together in consonance with cosmos.

That the cosmos is an ordered whole and not an ensemble of disjointed entities was recognized by the Vedic seers. It is indeed significant to note that modern science, from its beginnings in the Renaissance period to the present, has proceeded on the fundamental idea that the universe is not capricious, but that there is an order underlying it, and the aim of scientific thinking is not only to take cognizance of this but also to discover it by its own rational methodology. In the middle of the last century, a mathematician-philosopher, A. N. Whitehead, rightly observed in his thoughtful book, *Science and the Modern World*: 'In the first place there can be no *living science* unless there is a widespread intuitive conviction in the existence of an order of things, and in particular of an order of Nature.' (p. 4). What goes by the appellation *scientific laws* are really the laws of nature, which have been, or are being, discovered by man through the methodology of science.

Natural Law

Ṛtam, the natural law, is extolled in many ways in the *Ṛgveda*.

The Vedic gods were enjoined to follow and also uphold it, and they were expected to chant its rhythm as well as cherish its value, besides rejoicing in the fulfilment of it. The sun was eulogized as being a dweller in eternal law since it was recognized that its movements and periodicity take place on the stipulated course with precise timing because of *ṛtam*. The morning goddess, Uṣas, was deemed to wake up from the stratum of this cosmic law, follow its path and illumine the world with the reddish effulgence that appears with dawn. Agni, the ruler of *yajña* (sacrifice), was invoked as the guardian of this eternal law. Another Vedic divinity, Bṛhaspati, was regarded as the destroyer of guilt and evil in order to uphold the all-powerful law. Of other celestial entities, Mitra-Varuṇa had pride of place as the custodians of *ṛtam*. Yet another important god, Indra, according to the *Ṛgveda,* would light up and energize the entire world as a result of this law. According to Vedic seers, all phenomena—the cyclic changes of seasons, the occurrence of day and night, the flow of rivers, birth, growth and decay of life-forms—happen in accordance with *ṛtam*. They perceived in *ṛtam* a pristine unitary principle, ever-active and present everywhere as an immanent governing agent.

The Vedic priests had envisioned an interrelatedness of the celestial and the terrestrial in their own way. The Vedic primal performance of *yajña* when they offered *soma* juice as an oblation, was intimately related to *ṛtam* which, in turn, was believed to be the binding thread of *yajña* itself, both of them being inseparably associated with the sacrificial performer's inner soul. The idea of the natural or moral order seems to be much older than the *Ṛgveda,* since one notices a similar concept called *aśa* in the expression of the earlier integrated group called Indo-Iranians, which finds succinct exposition especially in the Avestan literature. It would also appear that, just as Varuṇa, among all the Vedic gods, was intimately associated with *ṛtam*, and regarded as the upholder of natural and moral order, so also was the Avestan Ahura Mazda. The

Indian *ṛtam* and the Avestan *aśa* or natural law have a much wider connotation and significance than the later Chinese *Tao,* and the Greek *moira*, which also admit that the universe is an ordered entity.

Yajña: The Vedic religious kernel was *yajña*, or the performance of sacrifices. The word 'sacrifice' as an equivalent in English for *yajña* is woefully inadequate because the Vedic conception of *yajña* had a broader significance. *Yajña* was regarded not only as the navel (*nābhiḥ*) of the world but also extolled as the abode of *ṛtam*. The performance of *yajña* was, therefore, an inseparable part of the Vedic way of living. Moreover, the act of sacrifice itself was believed to protect *ṛtam* by which all occurrences, both in the terrestrial world and the celestial sphere, would continue to be orderly and harmonious. There was also the Vedic view that the performance of *yajña* was an efficacious pathway leading to divine experience and eventually to the abode of the Creator.

Early Mathematics and Astronomy

Vedic geometry: As noted before, one of the six auxiliaries of the Vedas is known as *Kalpa,* in the form of several *sūtras* or cryptic formulae or prescriptions. An integral part of the *Kalpa* or Śrauta-sūtras goes by the name, the Śulba-sūtras. The word *śulba*, means a 'cord' or a rope or string, and its root *śulb* has the connotation of measurement. Thus the Śulba-sūtras literally mean a compendium or collection of rules of measurements (with the help of a cord) of various linear or three-dimensional figures. These represent the liturgical or Brāhmaṇa geometry or mensuration. The methods employed by them are referred to as *Śulba-vijñāna,* and an expert in this type of geometry was called *śulbavid*, who enjoyed an exalted position among the learned priestly class.

Of the several *Śrauta*-sūtras, only those of the different schools of the *Yajurveda*—the *Kṛṣna* (Black) and the *Śukla*

(White)—to which the Śulba sūtras are attached have come down to us. The most comprehensive Śulbas were produced by the perceptibly dynamic *Taittirīya* school of the *Kṛṣna* Yajurvedins. Among these are *Baudhāyana*, *Vādhula*, *Āpastamba* and *Hiraṇyakeśin*. The Śulbas, *Mānava* and *Vārāha*, belong to the *Maitrāyaṇī*, and the one by Laugākṣi is of *Kāṭhaka-Kapiṣthala* school. The *Kātyāyana* śulba of white *Yajurveda*, is somewhat more accurate, but is comparatively small, with only 102 sūtras. The *Baudhāyana* is the oldest (600 BCE) and the most voluminous (525 sūtras). The *Āpastamba* comprises 223 sūtras. Both the *Mānava* and *Kātyāyana* provide specifications of the measuring device and discuss the gnomon as well as its application in determining the cardinal points. The *Baudhāyana* and *Āpastamba*, on the other hand, deal with geometrical propositions concerned with the construction of sacrificial altars of different designs, like those of the falcon, tortoise and chariot shapes, to attain the desired goals (*kāmyāgni*).

The Śulba-sūtras, which generally deal with the rules for measurement and construction of sacrificial altars, are involved *inter alia* in the elucidation of certain geometrical propositions and problems. These relate to rectilinear figures and their combinations as well as transformations, circling the square, squaring the circle, and similar other solutions. Although the exact rules are not specifically stated, it would appear that the *śulbavid* (the geometer) knew that (i) a circle can be divided into a number of equal parts by drawing diameters; (ii) each diagonal bisects a rectangle, and both diagonals bisect each other; (iii) a rectangle could be divided into four equal parts with the vertically opposite ones being equal in all respects; (iv) a straight line joining the vertex with the mid-point of the opposite side would divide an isosceles triangle into two equal halves; (v) squaring of the circle and its converse, circling of the square and such other ideas.

The Śulba texts also provide various rules like those for constructing a straight line which is perpendicular to another straight line, drawing a square, rectangle, an isosceles triangle or a trapezium with given sides or altitude. The rules for the combinations and transformations of squares and rectangles are indeed fascinating. The so-called Pythagorean theorem (the square on the hypotenuse of a right-angled triangle is equal to the sum of the squares on the other two sides) was already known to the Vedic geometers. A *Baudhāyana* Śulba-sūtra (i.48) says: 'The diagonal of a rectangle produces by itself both (the areas) produced separately by its sides.' The *Āpastamba* also has a similar definition. These geometrical relationships were employed for finding the perpendicular directions (north-south) at the east or the west point of the east-west line for the desired positioning of a sacrificial altar. For measurements, a cord or rope was generally used. It is interesting to note that Egyptians used ropes for measuring lands, which is why the Egyptian geometers were called 'rope-stretchers'.

Vedic geometers were experts in circling the square or squaring the circle. They had also arrived at a value of π (the circumference of a circle divided by its diameter) as being equal to 3.0883 (approximately). Their algebraic notions included indeterminate and quadratic equations, the determination of the value of irrational numbers, and so on. The forms of *Mahāvedi* (shape of an isosceles trapezium) and *Śyena-cit* (falcon-shaped altar) and their constructions with prescribed dimensions speak of the algebraic knowledge even at that early period (about the fifth century BCE). The prescriptions for the construction of altars were rigorous; for example, the *Baudhāyana* says that a *gārhapatya vedi* (the altar at home) should be constructed with five layers of bricks, each layer containing twenty-one bricks so arranged that their rift in two consecutive layers do not coincide. For constructing a square altar in which square bricks are used, the rule envisages the use of three kinds of bricks with their sides $\frac{1}{6}$, $\frac{1}{4}$, and $\frac{1}{3}$ of the side. Surds such as $\sqrt{2}$ and $\sqrt{3}$ as well as the irrationality of their quantities were also known.

Vedic mathematics was thus developed in the religious context of performing sacrifices.

Vedic astronomy: Keen observers of the sky and of the motion of the sun and the moon, the Vedic priestly-astronomers in a way mapped the sky in terms of twenty-seven or twenty-eight *nakṣatras* or asterisms in relation to the motion of the moon, and thus developed a lunar zodiac. They were also well aware of the solstices, both winter and summer, and generally the year-long sacrifice would start on the day of winter solstice. The universe was regarded as divided into three well-defined regions: the earth (*pṛthivī*); the firmament (*antarikṣa*); and heaven or the celestial (*dyaus*). The Vedic month was the natural unit of time between the consecutive new moons or full moons, which was further divided into two halves (the bright and the dark), depending upon the phases of the moon. The names of lunar months were derived from the *nakṣatras* in which or near which the full moon occurred. Solar months were also known as well as the intercalation at regular intervals for luni-solar adjustments. Though the length of the year was taken to be 360 days, there were different types of years with their varying duration.

Number-reckoning

Even number-reckoning on the ascending decimal scale was part of religious performances. For example, the *Yajūrveda Saṃhitā* enumerates numbers up to 10^{12}: *eka*; *daśa* (10^1); *śata* (10^2); *sahasra* (10^3); *ayuta* (10^4); *niyuta* (10^5); *prayuta* (10^6); *arbuda* (10^7); *nyarbuda* (10^8); *samudra* (10^9); *madhya* (10^{10}); *anta* (10^{11}), and *parārdha* (10^{12}). This type of enumeration is in the context of the benefit of merit that a person obtains through the performances of sacrifices and offering cows as sacrificial fees. The system of number-words was facile enough for addition, subtraction, and multiplication. It was ingeniously developed in three tiers: (i) the first nine digits: *eka, dvi, tri, catur, pañca, saṣṭa, sapta, aṣṭa* and *nava*; (ii) the second group was obtained by multiplying each of the above digits by 10, *daśa, viṃśat, triṃśat, catvāriṃśat, pañcāśat, saṣṭi, saptati, aśīti* and *navatī*; and (iii) the third group would begin with 100, followed by multiples of 10. Fundamental arithmetical operations with elementary fractions were known even in the Ṛgvedic times.

Among the other mathematical attainments during the Vedic period were the use of progressive series and geometrical and algebraic exercises involving even indeterminate equations.

The Vedic oral tradition encompassed number-reckonings, number-words, and several mathematical attainments without the use of numerals or number forms. It is, however, conceivable that numerical forms would have been in vogue for commercial or trade transactions although no archaeological evidence of that period is available.

Healing Art

The healing art in the Vedic times was a triple combination of gods, plants, and diseases. Vedic religious healing was in the nature of an appeasement, through hymnal prayers and certain rituals, of several gods or divinities whose imagery was symbolic of different natural ailments. In this process of thinking, plants acquired a place of reverence and the dreaded diseases were sought to be cured through propitiation. The two principal cults that emerged were the Agni and the Soma. Agni, the fire-god, was considered as Aṅgiras, and Aṅgirasas were fire-worshippers; later, the Atharvans and Bhṛgus represented the fire-cult. Significantly, these were regarded as the authors of the *Atharvaveda*, a repository of several remedial measures or healing powers through magico-religious rites. There was, on the other hand, what was known as the Soma cult, which centred round the plant *soma*, the juice of which was offered as a libation to the gods during the performance of sacrifices, and also partaken by the priests. Soma juice, besides being an energizing and exhilarating drink, was supposed to confer immortality and thus soma was raised to the divine level. The concept of attaining a long life associated with soma found its way into the Ayurvedic medical practices in the form of a special therapy called the *Rasāyana*, one of the eight components of *Ayurveda*.

The cult of Soma was essentially that of medicinal plants. *Oṣadhi* and *bheṣaja* are the two words used in Vedic literature for medicinal herbs, and the latter was also indicative of medicine. Though Bṛhaspati was the sage of the *oṣadhis*, there were other divinities, too, who were supposed to have had proficiency or expertise in herbal cures. A large number of plants or herbs that were supposed to have medicinal or medicinal-cum-magical powers have been dealt with mainly in the *Atharvaveda*.

Vedic medicine was a combination of both curative practice and magical rites in which several chosen plants were used not only as medicines but also as amulets (*maṇi*). The Vedic concept of disease was a queer blend of both irrational beliefs and a rational approach. The irrational aspect comprised the performance of certain types of sacrificial rites, incantations, and the use of amulets and talismans. The curative practices too were in line with them. The treatment of diseases through amulets and spells was always accompanied by one or the other type of ceremonial magic. Some diseases were regarded as demoniac or demon-based, and 'expulsion' was performed by propitiation and exorcism, including drawing a so-called 'ring of magic powder' around the house of an afflicted person in order to prevent the return of the 'demon of disease'. There was also a ritual which was intended to 'trap' the demon by offerings made to fire surrounded by a ditch filled with hot water, and also chanting of specific mantras. The inhalation of fumes generated by igniting dried parts of certain plants was also a practice to 'drive out demons' and worms. Some of the Vedic gods like Agni, Āditya (or Sūrya), Indra, Varuṇa, Rudra and Vāyu were extolled as god-physicians for the cure of specific diseases. For example, Agni was supposed to be a reliever from insanity, fever, snake-bite and other poisons, besides his capability for preventing ageing; Indra, too, was hailed as the divinity for the bestowal of a long life and the cure of consumption and other wasting diseases; and Varuṇa, for the cure of water-borne diseases. Rudra was eulogized as the lord of medicine for combating a wide variety of diseases, while Aśvinī-kumāras were venerated as the twin divine physicians.

Amid the irrationality of ascribing demoniac and similar origins of several diseases and their magico-religious cures, Vedic texts have also some details about anatomical, physiological, and pathological ideas that are rational. The derangement of bodily humours, seasonal changes, germs or worms, heredity factors and the effect of unwholesome food were also regarded as the causes of disease. The *Atharvaveda* describes diseases as being *abhraja* (born of clouds, that is, phlegm), *vātaja* (born of wind); and *suṣmaja* (of heat, that is, bile). Nearly one hundred diseases of different types have been named in the Vedic literature and some of them are under the general category of fever.

The Worldview

The Vedic seers envisioned the unmanifest primordial energetic ensemble as cosmic energy, and the cosmic light (*viśvajyoti*) as its manifestation. Significantly the generic nomenclature, *deva*, of the Vedic gods, means cosmic light, and Vedic gods were deemed to be the natural forms of cosmic light or cosmic energy. One of the Vedic goddesses, Uṣas, was regarded as heaven's daughter in the eastern region (sunrise) and attired in garments of light. Savitṛ was praised as the light immortal for all human beings. More importantly, Vedic seers conceived of cosmic light as the very principle of life and even longevity, the 'soul of all gods', and the 'womb' of all created things. In the Vedic view, Aditi, the Divine Mother, encompassed all creation, including gods who were supposed to be an inseparable part of *infinite light*. The Vedic vision of infinite light was in the nature of an energetic principle which attained great height in the

Upaniṣads, in some of which Brahman, the Supreme Principle, is referred to as 'self-luminous light'.

In the *Ṛgveda,* there are certain symbolic images concerning the origin of the universe in a speculative manner. One of them is the idea of a cosmic or primordial man, an idea which had also found mystical expression in other ancient mythologies, such as the Scandinavian and the Chinese ones. The personification articulated in the *Ṛgveda* (*Puruṣa sūkta*; *RV*. X. 90) is presented as a Supreme Being from whom emanated the sun and the moon, gods like Indra, Agni, and Vāyu, the sky, air, earth, animals, and humans belonging to the four castes. But the *Ṛgveda*, in tandem with its seminal concept of *yajña*, tells us that all of these came forth when sacrifice was performed with *Puruṣa* as the oblation. This personalized approach was soon overshadowed by an impersonal exposition in the *Nāsadīyasūkta* (*RV. X.*129) of the *Ṛgveda* which runs as follows:

(1)

Non-being, then existed not, nor being,
There was no air, nor sky that is beyond it,
What was concealed? Wherein? In whose protection?
And was there deep unfathomable water?

(2)

Death then existed not, nor life immortal,
Of neither night nor day was any token,
By its inherent force, the one breathed windless
No other thing than that beyond existed.

(3)

Darkness there was at first by darkness hidden,
Without distinctive marks, this all was water,
That which becoming, by the void, was covered,
That One by force of heat came into being.

(4)

Desire entered the One in the beginning,
It was the earliest seed of thought, the product,

The sages searching in their hearts with wisdom,
Found out the bond of being in non-being.

(5)

Their ray extended light across the darkness,
But was the One above or was it under?
Creative force was there and fertile power,
Below was energy, above was impulse.

(6)

Who knows for certain? Who shall here declare it?
Whence was it born, and whence came this creation?
The gods were born after the world's creation,
Then who can know from whence it has arisen.

(7)

None knoweth whence creation has arisen,
And whether he has or has not produced it,
He who surveys it in the highest heaven,
He only knows or haply, he may know not.

—(Macdonell's translation)

It will be observed that in this hymn the primordial or the First Cause has been articulated without any mythological or supernatural association. It speaks of an inward fervour or warmth and thereafter, the desire, the primal seed. It ends with a question about the exact knowledge of the origin of the world or universe, but projects the ultimate principle as *Tad Ekam* or *That One,* formless and primordial, beyond observable matter and motion, space and time. The concept of *one* becoming *many* has been a fundamental one in many cultures, but in no other culture has this been so well structured and articulated as in India from the early times, metaphysical though it was in its approach.

There is an interesting speculation by the *Śatapatha Brāhmaṇa* on the origin and the main components of the universe. It thinks of primeval *water,* an inorganic and impersonal entity, as the only one that existed in the beginning, and the *waters* (plural) desired to propagate their kind. Through their churning or mortification, a golden egg originated in them and swam in the waters for about a year. After another year, the cosmic person, Prajāpati, arose out of it and uttered the word *bhūḥ*, which became Earth; his word *bhuvaḥ* became the Atmosphere; his utterance, *svar*, became the Sky; and finally with these five syllabi he created five seasons. Such a conception of the primeval water as in the *Śatapatha Brāhmaṇa*, which is under a question mark in the *Nāsadīyasūkta*, was also noticeable among the Mesopotamians and later among the Greeks, with Thales of Miletus (sixth century BCE), the pre-Socratic thinker, being the ablest exponent of *water* as the first principle of the universe, but without any mythological association as has been postulated in the *Śatapatha Brāhmaṇa*.

Five Elements

Although the monistic conceptions like those of *one* becoming *many*, the cosmic person, and the primeval water represented an exalted thinking, they could not explain the apparently diverse dimensions of the physical or the knowable world through the five senses. A new approach was therefore necessary, which would lead to a newer formulation, pluralistic in its structure but holistic in nature. Such a formulation was known as the *pañcabhūtas* consisting of five elements: *pṛthivī* (earth); *ap* (water); *tejas* (fire); *vāyu* (air); and *ākāśa* (a non-material ubiquitous substance). It should, however, be noted that the Indian five elements expressed in Sanskrit words have a wider connotation and significance than the familiar material earth, water, fire, and air. The non-material and all-pervading *ākāśa* is totally different from what sometimes is understood as *ether* of the Greeks.

The origin of the postulate of five elements can be found in some of the Upaniṣads. The *Chāndogya* speaks about a group of three wherein the *water* element is placed between two elements, *fire* and *earth*, each supposed to be in succession from its predecessor. The *five elements*, appear in an orderly manner perhaps first in the *Taittirīya Upaniṣad* (2.1), which has an exposition as follows: From this ātman (also identified with *Brahman*) in reality has arisen *ākāśa*; from *ākāśa*, *vāyu* (air); from *vāyu*, agni (fire); from agni, *ap* (water); from *ap*, *pṛthivī* (earth); from earth, herbs and so on.

Another Upaniṣad, the *Aitereya*, in its exposition of ātman or the self identified with *Brahman*, enumerates the five elements in a different order, namely, *pṛthivī*, *vāyu*, *ākāśa*, *āp* and, instead of *tejas*, it preferred a synonym, *jyotiṃṣi*. The *Chāndogya Upaniṣad* speaks of *Brahman* involving water (*abbrahma*), *tejas* (*tejobrahma*) and *ākāśa* (*ākāśabrahma*). It needs to be emphasized that the elements (*bhūtas*) as conceived and explained in the Upaniṣads are more metaphysical than physical, and in the nature of certain thought processes towards a progressive understanding of, and experiencing *Brahman*, the ultimate cosmic principle. Thus the Indian doctrine of five elements, though pluralistic apparently, was developed on the canvas of monism. Moreover, each element was not regarded or considered as having its own individuality; for, in each element the other four in some proportions would also be present, according to some exponents of this doctrine. The metaphysical nuances of the five elements, however, began to recede to the background and, as they did, their physical characteristics came to the fore, not on their own or independently but as integral components of the emerging religio-philosophical systems: the *Sāṃkhya*, the *Vaiśeṣika* as well as its associate the *Nyāya*, and the *Pūrva-Mīmāṃsā* of the orthodox ones which owed allegiance to the Vedas. The ideas of four elements (not including *ākāśa*) also found its way to some schools of Buddhism and Jainism, the Jaina concept of *ākāśa* being different from that of the orthodox systems.

The Indian doctrine of five elements had attained its full stature (by the sixth century BCE) so as to offer a coherent explanation of the knowable phenomena through the five senses. Each of the five elements was associated with a sense organ: *pṛthivī* (smell, nose); *ap* (taste, tongue); *tejas* (vision, eyes); *vāyu* (tactile, skin); and *ākāśa* (sound, ears). Pre-Socratic Greek thinkers, too, had evolved a theory of elements gradually, perhaps a century later (fifth century BCE). As noted before, Thales of Miletus (sixth century BCE) thought that *water* was the primal matter or the mobile essence of the world. His pupil, Anaximander, expounded a limitless or infinite substance as the first principle of the knowable world. Another Miletian, Anaximenes, attempted to specify the concept of Anaximander and enunciated that the first principle was *air* or *pneuma*, speculating that the processes of rarefaction and condensation would be causative of the varied and observed transformations of this primary substance. An Ionian philosopher, Heracletus of Ephesus, emphasized that change alone was real, and *fire,* the ever-changing element, was the primordial substance. Since each of the foregoing three elements had its inherent limitations for explaining all of the diverse phenomena, a Sicilian thinker, Empedocles (fifth century BCE) attempted to provide a pluralistic but an integrated presentation of the four elements by including *earth* as one more element. Calling them the 'roots' of the world, he introduced two sets of opposite qualities: *heat* and *cold*; *dryness* and *moistness.* These, he thought, through their affinity and opposition as well as their association with the elements, would be able to explain the multiplicity of all terrestrial phenomena. Over a century later, Aristotle, who was a pupil of Plato, introduced the idea of *ether* (quintessence) as the fifth element that would be a constituent of all celestial bodies. Thus in the Greek theory of elements, there was a divide between the celestial and the terrestrial bodies, unlike the Indian holistic view of the universe. More importantly, the Indian five elements were associated with the five senses, the gateways of human knowledge, and such an approach was totally absent in the Greek theory of elements.

The Chinese, too, had developed a theory of the five elements in terms of wood, fire, earth, metal and water. In both the Indian and the Greek postulates, air (*vāyu* or *pneuma*) played an important role not only in the explanation of the motive phenomena of the physical world but also in medicine, but this element was totally absent in the Chinese theory of elements. They were generally classified according to basic properties, such as solidity, liquidity or fluidity, heat and combustion. In the Chinese scheme, wood was accorded the first place, and water the last, with a fanciful explanation by a Chinese scholar (second century CE) that: 'Wood produces fire; fire produces earth; earth produces metal; metal produces water; and water produces wood.'

POST-VEDIC PERIOD: PHILOSOPHICAL SYSTEMS AND THE FIVE ELEMENTS

The *Sāṃkhya* enumerates the twenty-four entities as follows:

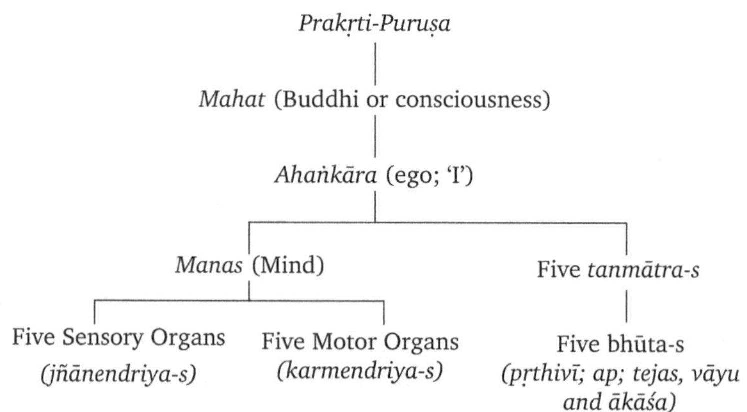

It needs to be understood that *prakṛti* is the unmanifested one, while sentient *puruṣa* (in plural, *puruṣas*) is intimately related to it. The former is regarded as the 'rootless root' of the universe and also intertwined, as in a three-stranded rope, with the three *guṇas* (qualities), namely, *sattva* (bright, righteous), *rajas* (energetic), and *tamas* (dark and inert). This is known as classical *Sāṃkhya*

and, since it does not project any type of divinity, it is also called the *Nirīśvara Sāṃkhya*. Much later, however, a godhead was also included in its enunciation and it was, therefore, called *Śeśvara Sāṃkhya*. However, it is the *Nirīśvara Sāṃkhya* that has a distinct position in Indian philosophy.

Of relevance to us is its presentation of five elements as belonging to the physical world as well as the physical body. Each of the elements is also believed to be associated with or emanated from its subtle state called the *tanmātra*. The five *tanmātras*, according to the *Sāṃkhya* exposition, are also the constituents of what is called *liṅgam*, or the mark by which the individual *puruṣa* is distinguished. It is also deemed to be the *sūkṣma śarīra* (finest body) that would pass on from one birth to the other. The *liṅgam* is said to comprise, besides the five *tanmātras*, the *mahat* (or the consciousness), *ahaṅkāra* (ego), *manas* (mind), the five sensory and the five motor organs. Thus each *puruṣa* has his own *liṅgam* and the different characteristics of individuals are stated to be due to the quantitative (and even qualitative) variations in respect of the three *guṇas—sattva, rajas* and *tamas*—since the *tanmātras* are also part of the *liṅgam.* The exponents of the *Sāṃkhya* believe that at each unfolding of the self, *puruṣa,* the five elements which are the off-shoots of the *tanmātras,* would also be in conjunction with the individual. Each individual thus would have his own body, mindset, perceptions and experience of the physical, material world. These ideas influenced the foundational concepts of Ayurveda as will be seen later.

The *Sāṃkhya* thought-structure is evolutionary in its details. In tandem with it is also a kind of evolutionary approach to the five elements vis-à-vis the related *tanmātras*. The important classic text on the *Sāṃkhya* (second century CE) by Iśvara Kṛṣṇa, called the *Saṃkhya Kārikā,* as well as the commentary on it by Vijñānabhikṣu titled *Sāṃkhya Pravacana-bhāṣya* (eighth century CE) states the following:

From the *śabda* (sound) *tanmātra* emerges *ākāśa*; from the *śabda* and *sparśa* (tactile) *tanmātras* emerges *vāyu* (air);

from the *śabda, sparśa,* and *rūpa* (colour, vision) *tanmātras* comes *tejas* (light or fire); from the *śabda, sparśa, rūpa* and *rasa* (taste) *tanmātras, ap* (water) comes into being; and from the *śabda, sparśa, rūpa, rasa* and *gandha* (smell) *tanmātras pṛthivī* (earth) emerges.

It follows from the above that *ākāśa* can only be heard; *vāyu,* heard and felt (touch); *tejas,* heard, felt and seen; *ap,* heard, felt, seen and tasted; and *pṛthivī,* heard, felt, seen tasted and smelt. Each element, as noted before, is stated to have one specific quality and is associated with one of the five sensory organs—*ākāśa*: sound and ears; *vāyu*: tactile and skin; *tejas*: vision and eyes; *ap*: taste and tongue; and *pṛthivī*: smell and nose. This type of delineation notwithstanding, the *Sāṃkhya* does not throw any further light whatsoever on the nature of the sensorial world of matter, qualities, and motion, nor on space and time in the context of the knowable world. It would appear that the *Sāṃkhya* was more concerned with the nature of an individual and his interaction with the phenomenal world.

The *Vaiśeṣika* System

The *Vaiśeṣika,* in its approach towards an explanation or understanding of the physical world through the senses, the mind, or the knowing agency and the knower or the self, has provided a detailed account of the five elements under one of the six generic categories of this system called *dravya,* or substance. Nine entities, which are also called *dravyas,* comprise the five elements along with space, time, self and mind. The significant aspect of this categorization is the enunciation of the qualities or attributes of each of the nine entities. The *Vaiśeṣika* is a realistic system as it believes in the reality of the physical world.

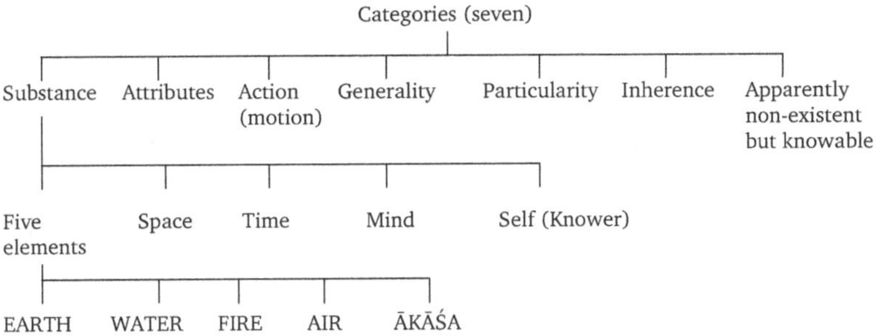

Since four of the five elements—earth, water, fire, and air—are regarded as being material, and their gross products can undergo division and changes, this realistic system, of necessity, conceived of their indivisible, permanent or finest atomic states. It has postulated that each of these four elements would be of two kinds—namely, eternal in its atomic form and non-eternal or evanescent in its produced state. Continuing its coherent way of thinking, the *Vaiśeṣika* also thought of the products of each of them having three facets: (i) (a) solidity (*pṛthivī*); (b) liquidity (*ap*), (c) energy or brilliance (*tejas*); and (d) movement (*vāyu*); (ii) association of each element with a sense-organ; and (iii) their possessing certain attributes or qualities. Thus, according to the *Vaiśeṣika, pṛthivī* includes all types of earthy material, such as rocks, minerals, gems and even vegetables, flowers and fruits. *Ap* encompasses all kinds of water—rivers, oceans, rain and other liquid substances. *Tejas* includes luminous heavenly bodies and all forms of fire (agni), including the Ayurvedic concept of stomachic fire (*jaṭharāgni*). *Vāyu* is representative of atmospheric air, breathing and the Ayurvedic idea of *vāyu* or *vāta* that moves various fluids in the body.

The *Vaiśeṣika* has also listed the various attributes of each of the five elements. The original *Vaiśeṣika* sūtras of Kaṇāda (sixth century BCE), had presented in all, seventeen qualities under the category of *guṇa*. His scholiast, Praśastapāda (fourth century CE) added seven more, one of them being *gurutva* (heaviness). But he appears to have associated this quality only with *pṛthivī* and *ap*. He also included sound as a quality that could be produced by either in conjunction or disjunction. The five elements differ from one another in respect of some attributes; but five of the attributes—conjunction, disjunction, dimension, distinctness and number—are common to all of them as well as the other substances.

Jaina View

That the universe is uncreated, beginningless, and endless is the fundamental postulate of Jainism. Expounding what is called the *Anekantavāda* or the Doctrine of Many (that is, reality is multiple and not one), Jaina metaphysics regards matter (*pudgala*) and spirit (*jīva*) as independent realities, categorizing the universe broadly into *jīva* (living) and *ajīva* (non-living). The former is believed to comprise five qualities, namely, *dharma* (motion), *adharma*, *ākāśa* (space), *pudgala* (matter), and *kāla* (time). Each of them, excepting *kāla*, is known as *astikāya* or that which exists, has pervasiveness, and occupies space (*pradeśa*), space being regarded as an integral part of *ākāśa*. The Jaina concept of *ākāśa* is altogether different from that of either the *Vaiśeṣika* or the *Sāṃkhya* inasmuch as the former is not associated with the auditory sense organ. The word, *pudgala*, for inanimate matter, means that which can fill up (*pur*) and dissolve (*gala*). In the Jaina view, all atoms are in the nature of *pudgala*.

The Bauddha Approach

The four well-known Buddhist Schools are: the Sautrāntika and the Vaibhāṣika of the Hīnayāna, the Yogācāra or Vijñānavāda, and the Mādhyamika of the Mahāyāna; the latter has also postulated the *Śūnyavāda* or nihilism.

The Bauddha view about the reality of the physical world, in general, is that there is no absolute reality—everything is momentary and is in a state of flux (*santāna*), a series of similar states or a fleeting, ordered succession. The objects are regarded as having no stability or permanency, but continue in a state of incessant, but ordered momentariness. Nevertheless, according to Buddhists, such an order does not owe its origin to any type of supernatural agency, but is governed by necessity and natural conditions. Buddhism admits that each phenomenon can be known by its cause, but emphasizes that it undergoes changes continuously. This postulate of Buddhists (Sarvāstivādins) is called *Kṣaṇabhaṅgavāda* or the doctrine of momentariness. Also known as *Santānavāda* or a theory of ceaseless flow or flux, it emphasizes that everything is momentary, and the change, the very process associated with objects, is alone real, but not the objects by themselves.

According to Bauddhas, the manifold phenomena experienced through the senses are relative, not real. Reality is above all kinds of categories, and is to be realized through spiritual experience—a state of enlightenment that transcends the duality of subject and object as well as the plurality of phenomena. The *Śūnyavāda* of the Mahāyana school posits that matter is not the ultimate reality, the knower as well as the known is not at all substantial, and the world of our experience is but unreal or void. The *Vijñānavāda* followers, while being in general agreement with the Śūnyavādins, do not accept the latter's concept of void. According to the Vijñānavāda followers, void has no meaning, since it is contradicted by the conscious perception which comes up every moment in an undeniable manner.

An important aspect of one school of Buddhism (*Sarvāstivādin*), is its admission of the atomic state of matter, but within its general thought-structure on momentariness. The Jaina as well as the *Vaiśeṣika* (including the syncretic *Nyāya-Vaiśeṣika*) also developed atomistic postulates, each on its own metaphysical canvas.

ATOMISM

That the external world is real and exists independently of the knower, and that it is capable of being experienced through the five senses, which are the instruments of human knowledge, has been integral to Indian realism. Likewise, the concept of phenomenal matter which can undergo division but will reach a point at which further division ceases, has also been a part of the thought-structures of some Indian religio-philosophies. Of them, the *Vaiśeṣika*, the Jaina texts, and the Hīnayāna school of Buddhism, have expounded atomism in considerable detail.

Nyāya-Vaiśeṣika Atomism

The atomic views of the *Vaiśeṣika* received the epistemic support of the *Nyāya* theory of knowledge. Thus the followers of both the *Nyāya* and the *Vaiśeṣika* promoted what came to be known later as the *Nyāya-Vaiśeṣika* atomism. In the *Vaiśeṣika* sūtras, the word *aṇu* is consciously used, meaning an atom, which is indestructible, indivisible, and partless. All atoms are regarded as being globular (*pārimāṇḍalya*). Later texts like the *Nyāya-bhāṣya* of Vātsyāyana, also use the word *paramāṇu* for an atom.

The main features of the *Nyāya-Vaiśeṣika* atomism are briefly as follows: of the five elements, it is only *pṛthivī* (earth), *ap* (water), *tejas* (fire) and *vāyu* (air) that are atomic in their eternal states, and non-eternal in their produced states. Each element has its own class of atoms with specific attributes—earth atom: *odour*; water atom: *taste*; fire atom: *colour* and air atom: *touch*. As noted before, these specific attributes are also of the gross elements themselves. Thus the *Nyāya-Vaiśeṣika* atomism does not make a distinction between the atomic and gross states of a particular element in relation to its special quality, which is also associated with the concerned sense-organ, a coherent way of explaining different states of matter. The atoms can, however, be differentiated by their specific attributes, that is, qualitatively; otherwise, the four classes of atoms have similar characteristics.

An interesting dimension of the *Nyāya-Vaiśeṣika* atomism is its logical exposition (apart from its inconsistencies here and there) of the formation of gross bodies from the minutest atoms. This exposition envisages that in the beginning of the universe, the atoms were all in motion and there was thus a possibility of the combination of some of them. However, to start with, only two atoms of the *same* element entered into combination, although during such a combination the atoms of the other elements could also be present. Two like atoms (those of the same element) are regarded as the material cause, while those of the other elements could play a supportive role. Such an approach was in tune with the *Nyāya-Vaiśeṣika* holistic view of the doctrine of five elements. In any case it is pointed out that two unlike atoms, for example, an atom of earth and that of air, would not enter into combination.

If two like atoms are the material cause, their effect is the formation of a dyad (*dvyaṇuka*). Even though a *dvyaṇuka* is formed out of two atoms, it is regarded as supra-sensible and infinitesimal. For, it is explained that an atom is so minute that it cannot be a causative entity contributing any magnitude to the product. It is also argued that atoms have no magnitude of their own, but are minute, and their minuteness cannot cause grossness. Hence, a dyad is not superior in magnitude to the two atoms that give rise to it, and the magnitude of a dyad must be derived from the *number* of atoms that constitute it—namely, two.

The *Nyāya-Vaiśeṣika* postulate of the formulation of a triad follows the same trend of thinking. It explains that *three* dyads are the cause, and the effect is a triad (*trasareṇu* or *tryaṇuka*), which is stated to be the size of a mote in a sunbeam. The question as to why three dyads and not *six* atoms enter into combination is discussed as follows: Since atoms are eternal, if six of them give rise to a triad directly, the product would also be eternal and cannot undergo division. But the experience is to the contrary since such products are destructible. It is, therefore, necessary that there is an intermediary stage in the form of dyad, which is supra-sensible, between the atoms and the minimum gross matter. Here again the minimum standard of plurality is argued in terms of number, that is, number *three* is thought of, and hence three dyads constitute a triad; likewise four triads give rise to a tetrad (*caturaṇuka*), and so on. There were, also, differing views that a triad could be produced by the combination of three atoms in which case the former would not remain atomic but would possess magnitude, unlike the atoms. There was even an extreme view that a triad could be constituted of several atoms. However, the generally accepted position of the *Nyāya-Vaiśeṣika* was that three dyads alone would form a triad, four triads would give rise to a tetrad, and the sequence would continue.

According to some exponents of the *Nyāya-Vaiśeṣika*, triads of different elements could also enter into combination while forming a tetrad and the like. Since the atoms of different elements would have their own specific qualities, and there would be a structural arrangement in the triads, tetrads and so on, it is possible to explain different qualities of substances on the basis of the structural arrangement (*vyūha*), a refreshing concept even from the modern scientific point of view. Alongside, the *Nyāya-Vaiśeṣika* strengthened its concept with the principle of causality known as *ārambhavāda* or *asatkāryavāda*, which emphasizes that the effect, a new event, always has a cause. Atoms are the material cause of a dyad, and the dyads are the material cause of a triad. The cause in each case brings about an effect, and is immediately absorbed or inheres in the effect which, in turn, performs the causal functions to continue the sequence, thus pointing out the prior non-existence of the effect.

Two questions—why primordial atoms enter into combination and what causes the union of two atoms to begin with—engaged the incisive minds of the Nyāya-Vaiśeṣikas. Their explanation for the former is no doubt in terms of the motion of atoms, but the initial motion of atoms is attributed to an unseen force or power (*adṛṣta*) that would set off the process of combination. Thus the idea of a prime mover or divine dispensation is projected as the efficient cause, while atoms are held to be the material cause. It would, however, appear that neither the *Nyāya* nor the *Vaiśeṣika* system in its early stages, thought of *adṛṣta* in a theistic framework. Nowhere in the *Vaiśeṣika* sūtras does the idea of God or Īśvara occur specifically, nor is there a clear exposition of the recognition of the authority of the Vedas as a *pramāṇa* or means of valid knowledge. It is only in the fourth century CE that Prasatapāda opens his commentary on the *Vaiśeṣika sūtras* of Kaṇāda with a homage to Iśvara and also expresses a belief in the authority of the Vedic scriptures. However, he does emphasize the importance of the precise knowledge of the six or seven categories (*padārthas*) of the original *Vaiśeṣika sūtras* that would eventually lead to liberation. According to him, such knowledge would result from *dharma* enjoined by Iśvara. He also points out that the periodic processes of creation (*sṛṣti*) and dissolution (*saṃhāra*) are primarily caused by Iśvara's design. He even goes to the extent of stressing that the duality (as in dyads) and plurality (as in triads) are governed by divine intent.

A significant aspect of the *Nyāya-Vaiśeṣika* atomism is that it regards mind (*manas*), an internal senses-related organ (*antaḥkaraṇa*), not only as active participant in the cognitive process along with the self, but also as atomic. It is also stated to be endowed with enormous or immeasurable motion, since it enables the cognizance of objects in the quickest possible time. One of the texts, *Yoga-Vāsiṣṭha* (eighth century CE), though not belonging to the *Nyāya-Vaiśeṣika* genre, presents the view that atoms are an essential part of living systems, including mind and consciousness. However, unlike Greek atomism, the *Nyāya-Vaiśeṣika* does not accept the premise that the soul is composed of atoms.

Jaina Atomism

The atomic postulate of the Jainas is somewhat circumscribed. It does not accept the atomic pluralism like the *Nyāya-Vaiśesika* which, as stated before, thinks of four kinds of atoms. One of the noted Jaina thinkers, Kundakundācārya (first century CE) has dealt with atomism in great detail. He enumerates, however, four forms of material bodies (*pudgala-kāya*) namely, *skandha* (aggregate), *skandhadeśa, skandha-pradeśa,* and *paramāṇu.* The first three have been regarded as aggregates of *paramāṇus* or atoms, which are held to be indivisible. Several aggregates constitute *pudgala* (matter) in its gross state. Since Jainas firmly believe in the doctrine of *karma* and its effects on birth after birth, they regard what is known as *kārmic* matter also as being atomic in a subtle state (*sūkṣma paramāṇu*). In the Jaina view, atoms are the prime cause of the four elements: *pṛthivī, ap, tejas* and *vāyu*, and also of their specific qualities. Even though in Jaina thinking *ākāśa* does not figure as the fifth element, the Jaina savant, Umāsvāti (second century CE) suggests that the quality of sound (which is associated with *ākāśa* in the *Nyāya-Vaiśeṣika* scheme) can accrue during the formation of aggregates from the atoms. A particular characteristic of Jaina atomism is its recognition of the positional relationship between an atom and the space it occupies.

The aggregates are stated to be formed, according to Umāsvāti, either by the union of atoms or by the division of larger aggregates, while an atom is produced only by division. He explains the combination of atoms as being due to two mutually antagonistic qualities—namely, attraction or greasiness (*snigdhatva*) and repulsion or dryness (*rūkṣatva*). He emphasizes that no such combination could take place if the atoms have the lowest or the minimum (one) degree of these two qualities, or of equal degrees of the same qualities. However, there are no quantitative details in respect of the degrees.

Jaina atomism postulates, unlike the *Nyāya-Vaiśeṣika*,

that the combination between the atoms of the same kind, and those of the other kind, occurs provided that there is a difference of two degrees—all in a speculative manner. There are some strange ideas, too, like the distinction between what is called the *sūkṣma paramāṇu* (the minutest or the finest) and the *vyāvahārika paramāṇu* (literally, the practical ones), stating that it is only the former ones that are indivisible, impassable, indestructible, and incombustible. Another speculation of Jaina atomism is that it regards an atom as being endowed with one colour, one taste, one smell, but two types of touches, namely, either cold and cohesive, warm and cohesive, cold and dry, or warm and dry. The general Jaina view is that these qualities exist equally in each of the infinite number of atoms, but their varied degrees of transformation would give rise to different materials of differing qualities.

Jaina atomism does not think of the motion of atoms as being due to *adṛṣta*, a postulate presented by the *Nyāya-Vaiśeṣika*. It accepts that there is an active motion which ensues as a sequel to what has been termed as *pudgalagati,* or the motion of matter by itself. It is hard to understand the Jaina view that such a motion of an atom can be from one space to another, regular or irregular and also revolving about itself. Even the rate or speed of motion of an atom has been speculated upon by the Jaina thinkers who thought that an atom would have a maximum speed when it travels from one border of the universe to another, and a minimum speed when it moves from one space-point to the adjacent space-point. The motion can be either in straight line or curvi-linear if, during its motion, an atom is acted upon by others all in a fanciful manner.

An interesting Jaina postulate, however, is the reckoning of the minutest part or division of time as being atomic or instants of time (*kālāṇus*), distinctly separate from one another. Jaina thinkers thus do not accept the view that time is one or absolute. Since time-units are separate, the practical (*vyavahāra*) dimensions of time—simultaneity, the present, the past and the future—could be understood as variations of time-units without any reference to the *present* itself. The *Nyāya-Vaiśeṣika*, on the other hand, subscribes to the relationship of the *present* with both the *past* and the *future*, but it does not accept the atomic nature of time.

Bauddha Atomism

Of the different Buddhist schools, the Vaibhāṣika and the Sautrāntika, which are regarded as somewhat realistic to a limited extent, admit the atomic state of matter, but within their overall philosophical frame and as part of their concept of *rūpa*. Their view is that matter is in the nature of a collection of a four-fold substratum of colour, taste, odour and touch. They regard an atom as the minutest unit of *rūpa* (that which has the potentiality of affecting our senses), thus relating it in a way indirectly to an observational entity. An atom is explained in terms of its indivisibility, invisibility, inaudibility, unassailability, and even intangibility. In tune with the philosophical position of these two schools of Buddhism, atoms are considered to be momentary, in the sense that they continuously undergo phase-changes.

Envisaging two types of atoms, that is, simple and compounded, the followers of these two schools postulate that compound atoms could be a product of direct combination or in the form of just an aggregation of several atoms. Some Buddhists view gross matter as a conglomeration or a cluster of atoms, with one atom at the centre and the others around it, without any type of interaction or interpenetration, all in a speculative way. In general, according to these schools, there are eight types of atoms, the four fundamental ones of earth (solid); water (fluid); fire (hot); and of air (moving) and four secondary atoms (of odour, taste, colour and touch), obviously relating them to the four specific qualities which are also held to be atomic. There is also an attempt in Buddhist atomism to quantify an aggregate in terms of four fundamental atoms and the four secondary ones, that is, $4 \times 4 + 4 = 20$ atoms of a minimum aggregate, if it does not have the quality of sound; if it is to sound, the minimum aggregate would be $5 \times 4 + 5 = 25$ atoms. In any case, Buddhists of the Hīnayāna school do not regard atoms as indivisible material particles like the *Nyāya-Vaiśeṣika* or the Jainas; instead they think of them as force or energy, and even try to explain them in terms of their presence in all things and in the same proportion. They argue that different states of bodies like solid, liquid or gas, are perceived because of the variation in what they call the 'force-contents' of different elements, even though the proportionality of the elements themselves could remain unchanged. The structural details of Buddhist atomism are somewhat limited and hard to understand when compared with those of either the *Nyāya-Vaiśeṣika* or the Jainas. The atomism of all of them, of course, is essentially conjectural.

In the history of human adventure in ideas, atomism has a distinct place. The Greek thinkers Leucippus (c. 475 BCE) and Democritus (470-400 BCE) put forward the idea that the primordial stuff of the world consisted of atoms along with the *void*. In their coherent way of thinking, everything, including the soul, was composed of atoms. Greek atomism found able exponents in Epicurus (342-270 BCE), and Lucretius (first century BCE), an Epicurean, elaborated upon it. Thereafter, it lay dormant till about the seventeenth century, when it was revived particularly by Pierre Gassendi, Robert Boyle and Isaac Newton. Early in the nineteenth century, John Dalton put forward an atomic theory on scientific lines. Even so, the fundamental idea that an atom was an indestructible, indivisible entity still held the fort. Towards the end of the century, however, with the discovery of the electron by J.J. Thomson in 1897, the indivisibility of the atom was laid to rest, and in the twentieth century, the discovery of protons by Ernest Rutherford (1920) and neutrons by James Chadwick (1935) opened up fresh vistas, and extensively enlivened the physics of fundamental particles.

There is some evidence which indicates that the *Nyāya-Vaiśeṣika* atomism was known to some Arabic savants in Baghdad through the works of al-Ashari and al-Rāzi in the ninth-tenth centuries. It is well known that Indian astronomical, medical, and probably other texts were rendered into Arabic in the Caliphate at Baghdad between the eighth and tenth centuries, and they found their way into several works in Arabic. The latter were translated into Latin in the twelfth-thirteenth centuries by European scholars who were selective in their efforts and gave preference to Indian astronomical, mathematical and medicinal ideas and practices. The *Nyāya-Vaiśeṣika* atomism did not receive its due attention because of their preference of Greek atomism with which they were familiar, but speculative atomism continued to be a vibrant thought-structure in India till about the thirteenth century.

European thinkers viewed Greek atomism favourably because it had an acceptable dimension inasmuch as it did not make distinctions like allotting different atoms to different elements. It considered them different only in form and size, and thus they could be the building blocks of any substance. Jaina atomism, too, projected the idea that atoms were of one class, but also thought of four specific sensorial qualities—smell, taste, colour, and touch—in relation to the four elements: earth, water, fire and air. Even Buddhists (the Hiṇayāna school) viewed atoms in terms of these qualities, but in the form of force or energy. This type of circumscribed, speculative thinking was not permissive of a broader concept of atoms as the possible building units of the phenomenal world. Moreover, the other dimension of Greek atomism—namely, the void or vacuum—became important for later physical investigations in Europe. The idea of the void (*śūnya*) was never associated with Indian atomism, nor did it come down from its philosophical pedestal to the physical plane. However, it gave rise to the concept of zero in the decimal place-value number system. Eventually, Indian atomism remained within philosophical or metaphysical confines and did not stimulate any type of new thinking about the knowable physical world.

CONCEPT OF SUBSTANCE

There was, nevertheless, an extraordinary concept developed by the *Nyāya-Vaiśeṣika* relating to what could be called the *world-stuff* or *substance.* What is substance? This question has remained unresolved in modern science despite the amazing advances towards understanding the basic stuff of the physical and biological world. The Cartesian duality of mind and matter, the observer and the observed, the subject and the object, still rule the scientific approach to the world of matter and motion, space and time, even though there have been such synthetic strands as matter and energy, the space-time continuum, field theory, and the like. It needs to be recognized that scientific methodology takes cognizance of the 'observables' or the sensorial data and the associated causal mechanism. The unobservable at some stage or the other also needs to be understood in one form or another. In the name of objectivity, the scientific approach excludes the observer or the participant from the whole process. The exclusion of the *self* or the observer in scientific methodology seems to have come in its way of arriving at a definite, precise conclusion regarding *substance* or *world-stuff*.

On the other hand, the ancient *Vaiśeṣika* system conceived of substance in an *inclusive* manner. It comprised the four material elements; the knowable but non-material (*ākāśa*); the absolute, also relativistic space and time; the cognizer (self) and the cognizing agency (mind). These nine entities were so integrated that as a whole they would constitute what might be called the *substance*. The *Vaiśeṣika* concept was not, however, based on any experimental or mathematical model. It was essentially in the nature of a thought-model consistent with the *Nyāya-Vaiśeśika* approach to the reality of a knowable world.

SPACE AND TIME

Space and time are the two entities that are intricately associated with our perceptions of every event or occurrence, and yet they remain enigmatic, defying succinct understanding of their true nature. Each of them has, no doubt, been recognized in its twin aspects—the absolute and the practical—by philosophers and men of science alike, the latter even admitting and explaining a space-time continuum. In the formulation of the Special Theory of Relativity, Einstein stated that (i) there is no absolute time, and space and time have to be fused into a four-dimensional space-time continuum; (ii) space contracts and time dilates in a moving medium, its extent depending upon velocity; and (iii) the idea of simultaneity no longer has any meaning. The Big Bang cosmology envisages the beginnings of time and different dimensions or aspects of space.

Is *time* absolute or relative? Are the divisions of time that are in vogue merely pragmatic or is there a unitary principle underlying them? Is space bound or boundless? Questions such as these have been consistently addressed by Indian thinkers of different shades. The Indian concept of time has not evolved independently, but in the matrix of ontology and some sort of epistemology. Even in the Vedic period, there were savants of time called *kālaciṅtakas* (thinkers of time) and *kālavādins* or *kālakāraṇavādins* (advocates of the causality of time). The *Ṛgveda* does not appear to have dealt with time in any meaningful way, but the *Atharvaveda* has a hymn called *Kālasūkta* which is entirely devoted to the glorification of time as the creator and the sustainer of the universe, regarded as *Brahman* itself. The Upaniṣads, however, describe *Brahman* as transcending time, although one of them (the *Śvetāśvatara*) hails time as the originator of the universe. Two forms of *Brahman* have also been conceived: *time* and *not-time*; that which existed before the sun being *non-time*, and that which began with the sun, as time that is divisible. In general the Upaniṣadic *Brahman* has been projected as being beyond both space and time.

A clear categorization and exposition of space and time can be found in the *Vaiśeṣika* concept of substance (*dravya*). Among the nine components of the *Vaiśeṣika*, substance, space, and time have distinct positions. Time is recognized as one, all-pervading and eternal, like space and *ākāśa*. But, unlike *ākāśa* (which, as noted before, is one of the five elements associated with a specific quality of sound and the auditory organ), time has no specific sensorial attribute. According to Kaṇāda, the author of the *Vaiśeṣika* sūtras, the knowledge of time is inferential and is related to priority, posteriority, succession, and simultaneity as well as quickness and slowness. Time is also regarded as an instrumental cause (*nimittakāraṇa*) of every produced object.

Priority and posteriority, young and old, are words of practical import, and are supposed to be in relation to a larger or smaller number of solar revolutions. For, as some Nyāya-Vaiśeṣikas argue, if one person is prior to another person, or the other person is posterior to the first person, the other person obviously has been connected with a larger number of solar revolutions, thus regarding the number of solar days as being the yardstick of the arrow of time. The existence of time is sought to be established by *reductio ad absurdum* of the other causes, time having been considered as an indirect relation between an individual entity and solar motion. Implicit in this approach is the recognition of the fact that solar motion and time are two-in-one and are real, but apparently divisible for purposes of practicality and knowability.

The three major divisions of time, namely, the *past*, the *present*, and the *future*, have been contentious issues, especially among the Nyāya-Vaiśeṣikas and the Buddhists. Such issues are more polemical, centring round the basic idea of the *present*. The Buddhists do not even subscribe to the notion of the present time, while the followers of the *Nyāya*, like Vātsyāyana, aver that, if there is no *present*, the other two—namely, the *past* and the *future*—would be inconceivable, as these are related to the *present*.

Buddhists also think of time in relation to space. But Vātsyāyana denies this relation and asserts that time is manifested by action (*kriya*). If the action is continuing, it is *present*; if it has happened, it is *past*; and, if it is to happen, it is *future*, according to him. The 'tensed time' and the 'tenseless space' have also been explained in terms of the action in the case of time, and no action but conjunctions between points in the case of space. However, the Nyāya-Vaiśeṣikas view both space and time as reals. They do not regard time as merely a summation of momentary reals, nor atomic, as the Jainas believed. Buddhists, while discarding the oneness and all-pervasiveness of time, accept that time is instantaneous but impermanent, and this view fits well with their doctrine of momentariness. Likewise, they also posit that space is impermanent and mutable through epithets like *here* and *there*. In contrast, the Jainas thought of time and space in a different way. They recognized a persistent aspect of time as well as practical time, calling the former the *paramārtha kāla* (noumenal), being persistent through all the changes and acting as the accompanying cause of the modifications of material objects. Practical time was called *vyāvahārika kāla*.

Ākāśa in the Jaina thought-structure does not mean the same substance as in the *Nyāya-Vaiśeṣika*. It stands for space, which in relation to matter (*pudgala*), is considered to be infinite, formless, and even inactive. Jainas think of two forms of *ākāśa*: the *lokākāśa*, and the *alokākāśa*. The former is regarded as mundane space that provides room for the other entities like *dharma, adharma,* and even *kāla*. It is stated to be a limited one, while beyond it lies the *alokākāśa*, which is devoid of any type of objects, itself infinite and indestructible. The view of time and space as being infinite and eternal features in one form or the other in various shades of Indian thinking.

The problem of space and time had seized some Western

thinkers, too, for a long time. Among the Greeks, Plato thought of time as belonging to the world of appearances, not a real one. On the other hand, his illustrious pupil, Aristotle, viewed the universe as being limited in space (finite) but contained within an outer sphere that was unlimited in time. He sought to explain space and time with assertions, such as that the motion of heavenly bodies was perfect and they move with uniform circular velocity in crystalline spheres, while terrestrial motion was imperfect. He regarded time as an attribute of motion, but offered no definitive explanation of time. In the fifth century CE, the great Father of the Church, St Augustine said: 'We understand when we speak of time; we understand also when we hear it spoken of by another. What then is Time? If one asks me, I know; if I wish to explain it to one that asked, I know not.'

Early in the seventeenth century, when Galileo laid the scientific foundation for dynamics or mechanics, the concept of measurable time was built into it. Newton's laws of motion comprised mass, space (measurable), and time (also measurable). However, Newton had articulated that real time was absolute—that it flows on its own and was distinct from practical time like hours, months, and so on. He held that space was a three-dimensional substratum in which material objects were located and events like motion took place, while space by itself remained unchanged. In his magnum opus, the *Mathematical Principles of Natural Philosophy* (briefly called the *Principia*), Newton expressed: 'Space is an absolute, infinite, three-dimensional and eternally fixed, uniform receptacle into which God placed the Universe at the moment of creation.' Likewise, he viewed time as a one dimensional frame and eternal. Newton's contemporary and opponent, Gottfried Leibniz, advanced the relational aspects of space and time.

There were other views too in the West. Immanuel Kant, the celebrated German philosopher of the eighteenth century, regarded time as a form of our sensational intuition, but not as an object of perception, being in *a priori* form. In his view, time was real only empirically as a condition of subjective experience. Bishop Berkeley rejected the notions of absolute space and time since, according to him, they had no corroboration in our sense experience. Nearly two centuries later, Ernst Mach of the Positivistic School supported this argument and added that the relativistic aspects of space and time would provide the needed simplicity for understanding the physical world. There were other interpretations of time like those of Joseph-Louis Lagrange, Edmund Husserl, and Martin Heidegger in the nineteenth and twentieth centuries. Still, the concept of space and time continues to be problematic, while their practical, measurable aspects have been scientifically understood and presented.

TRADITIONAL EDUCATION

In India, from the Vedic times and over the millennia, oral education, which was largely religious in character, was in vogue. Liberal education on the pattern of Western education and, as part of it, scientific and technical education was introduced into India by the British colonial government in the beginning of the nineteenth century. A bird's-eye view of the system or the nature and extent of education that was prevalent in India in the ancient and medieval periods is presented below.

Education in the Ancient and Medieval Periods

In the Vedic period, more than three thousand years ago, the way of living was interwoven with ritualistic practices. This necessitated a system of instruction for the continuous training of Vedic scholars, leading to their attainment of high proficiency not only in the accurate phonetic articulation of the Vedic *mantras* but also in the ritualistic procedures associated with them, particularly in the performance of sacrifices of different types. There is no tangible evidence of a written tradition at that time. Oral communication, memory, and mnemonic exercises were the preferred modes of training in the preceptor-disciple tradition. Of the four castes, those belonging to the Brāhmin caste were a privileged lot. The disciple, in his young age, had to undergo what is known as *upanayana* (thread ceremony) which literally means, 'taking near' or the act of handing over one's son as a student to a teacher. This type of initiation was also available to those of the Kṣatriya and Vaiśya castes, but the Śudras who formed a majority of the population even then were not permitted to have this type of initiation, and thus were deprived of any type of education.

Vedic literature, including the *Brāhmaṇas,* or the liturgical texts, and the *Āraṇyakopaniṣads*, the metaphysical treatises, is a repository of extensive knowledge. Of the six auxillaries or the *Vedāṅgas*, as noted before, the *Kalpa* and the *Jyotiṣa* dealt with scientific aspects of mathematics and astronomy. A disciple or a student in the Vedic environment also received training in these traditional scientific branches. This type of training and the knowledge gained through it was called *aparāvidyā*, while that of the *Āraṇyakopaniṣad's* quest of the Supreme Being *(Brahman)* was called the *parā vidyā* or higher knowledge. Both these types of education went hand-in-hand in the Vedic and the post-Vedic periods. Though education and the educated individual enjoyed a distinct position, the latter was enjoined to be beneficial to the society at large. The *Ṛgveda* stressed that 'if one human being was superior to the other it was not because the former had an extra eye or hand, but because education had sharpened his mind and intellect, and thus rendered him more efficient' (RV. X 117). At the same time, the preceptor or the teacher was expected to have high standard of scholarship and lead a life of exemplary character in order to earn respect and veneration not only from his pupils but also from the society in which he lived. He was to be a competitive teacher with communication skills, and his reputation rested upon the number of students he would attract into his fold. The *Mahābhārata* (V.33) and *Māḷavikāgnimitra* (Act I) of Kālidāsa have presented these aspects in a succinct manner.

The institutionalization of teaching, which was effectively personalized in practice, revolved round the teacher-student relationship. Thus the concept of a school was mainly associated with the household of the teacher or his hermitage in the forest, if the sage-teacher lived there. The *Rāmāyaṇa* and the *Māhabhārata* speak of such hermitages as educational institutions, like Vālmīki's Citrakūṭa hill, Viśvāmitra's hermitage on the banks of the river Kauśikā, Bharadvāja's at the confluence of the river Gangā and the Yamunā. Even later this tradition continued in Atri's (*Raghuvaṃśa* of Kālidāsa) and Jābali's hermitages (*Kādambarī* of Bāṇa). The principal reason behind this approach was that education was regarded as the pathway for strengthening oneself spiritually, in contradistinction to leading just a material life. Though the course and intensity of this oral education appeared to be religious ephemerally, its ultimate aim was one of elevating the disciple into the spiritual realm. However, there were concerted efforts from time to time for the exchange or promotion of scholarship through collective assemblies, as evidenced by the *saṃgha* or *pariṣad* mentioned in the Vedic texts.

Ancient Universities: The early attempts at the institutionalization of education were by Buddhists and Jainas. In the *vihāras* or the serene places of meditation, communities of Buddhist monks were engaged in canonical studies. Senior monks would educate the junior ones as part of their religious duty, and also examine from time to time the students and their level of acquisition of knowledge. In the course of time, even those who did not belong to the order of monks and who were only householders but religious, were enrolled as students and offered instructions in such subjects as grammar, philosophy, medicine, astronomy, arts and crafts. It was not long before such a body of dedicated teachers and students living and working together became the nucleus for the emergence of a large educational institution or university, such as those at Takṣaśīla and Nalanda.

Takṣaśīla, in northwest India (now near Islamabad in Pakistan) was a reputed place of learning. Several Buddhists spoke of its excellent teachers, and students from various parts of India, including Ujjain, Mithila, Benaras and Kośala sought admission to the educational institutions there. However, not much information is available on its origin, its possible development first as a Brāhmiṇical centre of learning, and later its growth and functioning under the influence of Buddhism. The archaeological data reveal the existence of three separate advanced settlements or towns in that region: (i) the Bhir Mound (150-180 BCE); (ii) the Sirkap area probably founded by the Greeks and occupied in succession by the Scythians and Parthians (up to the first century CE), and (iii) Sirsukh, founded by the Kuśānas. Takṣaśīla became the capital of several rulers and hence its culture was a composite one. During the Achaemanian reign, the Kharoṣtī script was in vogue and was employed by teachers and students in Takṣaśīla. The eminent scholar in the Sautrāntika (Hīnāyanā) form of Buddhism, Kumāralabdha, is stated to have worked in this university. Later, the invasions of the Huns resulted in its destruction.

Nalanda, situated about 60 km south of Patna (now capital of the state of Bihar) was originally a seat of Brāhmaṇical learning till about the fifth century CE. Āryabhaṭa I, the famous astronomer (fifth century) lived in Patna, then called Kusumapura (identified as Pataliputra) and was a *kulapa* (the status of vice-chancellor) of an institution of higher learning here. Perhaps around the middle of the sixth century, it became a centre of Buddhist learning.

The Gupta emperors encouraged the construction of *vihāras* extensively in Nalanda. The archaeological evidence indicates that Nalanda University covered an area of about 1.5 km long and 0.75 km wide. The college of instruction in the centre had seven large halls and about 300 classrooms. Besides, there were other monastery buildings, a large stūpa, and what were known as *caityas*, embedded with precious stones as well as a water-clock which kept time. It has been estimated that the number of students at any time in the Nalanda University was of the order of a few thousands. According to Hiuen Tsang (the Chinese traveller), the population of both monks and students was about 10,000, while I-Tsing, another Chinese itinerant, has stated that the number of students was more than 3,000. The university had a large library comprising three buildings called Ratnasāgara, Ratnodadhi, and Ratnarañjaka. True to its learning tradition, Nalanda University was a scholarly abode for the study of not only the Māhāyana and the Hinayana schools of Buddhism but also the Vedic texts, the *Hetuvidyā (Nyāya)*, and Ayurveda. The works of Nagārjuna, Vasubandhu, Asaṅga, and Dharmakīrṭi received special attention because of their logical expositions of the Buddishist philosophy. Students not only from different places in India but even from far off lands like China, Korea, Tibet, and Mongolia sought admission to this university. The notable erudite teachers included Śāntarakṣita, Padmasambhava, Kamalaśīla and Sthiramati, who were also involved in preaching Buddhism and translating the Buddhist works in Sanskrit into Pāli and Tibetan. It was much later, in about the twelfth century that the decline of Nalanda University began.

In western India, Valabhi in Kathiawar (now in Gujarat) was a seat of another Buddhist university, which was as famous as Nalanda. This higher educational institution was patronized by the royal family of the Maitraka kings during the fifth to the eighth centuries. Hiuen Tsang who also visited this centre of learning, states that it had about a hundred *saṁghārāmas* (monasteries) and 6,000 monks. I-Tsing too admired Valabhi University for its students who came from all parts of India. However, this university was somewhat different from Nalanda inasmuch as secular subjects like *nīti* (morals), *vārtta* (commerce), and *cikitsāvidyā* (medicine) were accorded more importance, and its inclination was more towards the Hināyāna school of Buddhism.

Another university, Srī Vikramaśīla *Vihāra* or Vikramaśīla University, which came into being around the eighth century in Bihar was founded by King Dharmapāla. The exact place of this centre of learning has not been established beyond doubt; possibly it was not far away from Nalanda. However, scholars trained in Vikramaśīla played a pivotal role in the dissemination and consolidation of Buddhism, particularly in Tibet. This university had halls for six colleges connected with a central hall, and the entrance of each college from the central hall was guarded by a *dvāra-paṇḍita* (scholar gate-keeper) with the names: (i) East gate: Ratnākara Śānti; (ii) West gate: Vāgīśvarakīrti of Benaras; (iii) North gate: Naropa; (iv) South gate: Prajñā-karāmati; (v) First central gate: Ratnavajra of Kāśmīra; and (vi) Second gate: Jñānaśrīmitra of Gauḍa. These names implied that any student who sought admission to this university had to interact with scholar gate-keepers and win their approval. Although no definite information is available on the exact courses of study, it would appear that logic, metaphysics, grammar, tāntra, and ritualistic literature were among the subjects taught in this university. Successful students were awarded degrees or certificates as recorded in Tibetan sources. The well-known Buddhist scholar, Dīpaṅkara Śrījñāna Atiśa, who was the president of Vikramaśīla University (1034-38 CE) went to Tibet at the invitation of its king, stayed there for nearly thirteen years, produced several works on Tibetan Buddhism (*Vajrayāna* form) and also translated certain Sanskrit works into Tibetan. The revered monks, Dīpankara, Vairocana Rakṣita and others had produced their erudite impact in Tibet leading to the emergence of Tibet as an important centre of Buddhism.

Odantapuri monastery turned out to be another university which owed its foundation to King Gopāla of the Pāla dynasty in the middle of the eighth century in the Bengal region. It is interesting to note that Ratnākaraśānti, the scholar whose name was associated with the east gate of Vikramaśīla, was educated and ordained in the *Sarvāstivāda* (Hīnayāna) school of Buddhism in Odantapuri. This university was also noted for imparting instruction in Tāntric Buddhism. It flourished especially in the eleventh-twelfth centuries and had about a thousand monks at one time, and a well-stocked library to its credit. However, it crumbled and perished as a result of the Turkish onslaught.

Another Pāla king of Bengal had established a *vihāra* in Jaggadāla on the banks of the river Ganga in the beginning of the twelfth century, but this centre of learning, too, perished when Mohammed Bukhtyar's army invaded Bengal in 1203. Though this institution existed for less than a hundred years, it had produced eminent scholars like Vibhūticandra, Danaśīla, and Mokṣakaragupta.

Gurukula System and Temples: The Hindu method of education was largely carried out through the Gurukula system, generally in the ambience of temples and holy places. There were, however, small educational schools called *tols* in Bengal, Bihar, and the United Provinces (Gangetic region), each of which had around twenty-five students at a time. They were taught in thatched mud huts in a range of subjects: literature (poetry), philosophy (six orthodox systems), astronomy (*jyotiṣa*), grammar, logic, *Śrutis*, and *tāntra*. Small schools of this type were also active even in the colonial period.

In South India particularly, the temples played a significant role in the promotion of education in the Vedas, Upaniṣads, and Śāstras, Sanskrit literature, epics and the *Purāṇas* as well as *jyotiṣa* (astronomy and astrology). There are authentic inscriptional records which reveal land grants as well as other forms of endowments provided for traditional education. An inscription of the Rāṣtrakūta King Krishna III (945 CE) has interesting details of a temple-college established by his minister, Nārāyaṇa. According to this inscription, the temple-college had a spacious hall for education, which could accommodate a large number of students from different parts of the country. As many as twenty-seven houses were attached to it to provide free board and lodging for about 200 students. The temple, for this purpose, had received endowments of adequate lands, including some cultivable land, for the uninterrupted maintenance of a lit lamp before the deity. This inscription also says that the village community participated in funding this institution by contributing five gold coins on the occasion of each marriage, two and a half at each *upanayana* (thread ceremony), and one and a quarter at each *cūḍākarma* (head-shaving) ceremony, thus revealing the importance attached to education (*Epigraphia Indica*, IV, 66).

Another inscription refers to the Vedic college which functioned at Ennayiram (South Arcot district) which was established by the King Rajendra Cola I (1023 CE). It had a land grant of about 300 acres made by the villagers themselves, offered free lodging, board and tuition to about 350 students and maintained sixteen teachers. The inscription also gives (i) details of the number of students for each subject including the four Vedas, grammar, *dharmaśāstra, mimāṃsā,* and *vedānta;* (ii) remuneration paid to teachers; and (iii) allowance for students (*Annual report, South Indian Epigraphy,* 1918, p.145). Another temple-college called Venkateśa Perumal at Triumakkūdal (eleventh century), though a smaller institution providing instruction to about sixty students at a time, had an Ayurvedic hospital attached to it, in addition to a hostel for students. There was even a temple-college in Tiruvoriyūr (Chingelput district near Chennai), which was a centre of linguistic studies, dedicated especially to the study of Pāṇinī's grammar. The foregoing examples amply illustrate not only the role of temples in education but also the abiding interest in education evinced by kings and common villagers alike.

In the Karnataka region, as elsewhere in South India, three types of educational institutionalization existed under the names of *agrahāra, ghaṭikā* and *Brahmapuri,* all of which comprised an assemblage of erudite Brāhmins. A special feature of these schools was that they were supported by land grants. According to an inscription (Altekar, p. 140), a place called Kādiyur (in the district of Dharwar), became an *agrahāra* during the reign of the Raṣṭrakūṭas (tenth century). In it resided about 200 Brāhmanas who were scholars in the Vedas and Upaniṣads, *Purāṇas*, and logic. The kings had also provided some endowments for paying remuneration to teachers, free board and lodging for students. The concept of the *agrahāra* goes back to the time of the Śātavāhanas (first to second centuries CE), as evidenced by an *agrahāra* centre called Belgame, which gradually expanded, owning as many as 144 villages and running seven *Brahmapuris* and five *maṭhas*. A *ghaṭikā* was smaller than an *agrahāra* both in its ambit and educational activities, which more or less conformed to the preceptor-disciple relationship. According to an inscription (*Epigraphia Carnatica*; VII, SK. 176), Kāncīpuram had a number of *ghaṭikās*. Different nomenclatures apart, they bear testimony to the fact that education in India received its due attention and patronage from the kings as well as the public throughout the ancient and medieval periods.

Even during the colonial regime in the late seventeenth to early nineteenth century, there was an active indigenous educational system, according to the available records of Europeans like William Adam and Thomas Munro (*The Beautiful Tree: Indigenous Indian Education in the Eighteenth Century* by Dharmpal, Goa, 2000, pp. 20-35). The Bengal and Bihar regions alone had over one lakh schools, each village having a school supported by the revenue contributed by the local community. Elementary education was being imparted in Sanskrit, Persian, Arabic and local languages. A notable aspect of this spectrum was that a large number of students belonged to Sūdra and other lower castes. Indian schooling was more broad based and in no way inferior to that in Britain then. The education of girls, however, left much to be desired.

Education in Scientific Subjects: Both in the Hindu *gurukula* system and Buddhist centres of learning, astronomy, mathematics, and medicine were among the subjects taught. Though the exact syllabus or the structure of teaching these subjects is not available, the Vedic texts like the *Śulba Sūtras* (mathematics), *Vedāṇga-Jyotiṣa,* as well as Siddhāntic texts like the *Āryabhaṭīya, Mahābhāskarīya, Brāhmasphuṭa Siddhānta,* Lalla's *Śiṣyadhīvṛddhida* tantra, Bhāskarācārya II's *Siddhāntaśiromaṇi* (astronomy), and medical texts like *Caraka* and *Sūsruta Saṃhitās* could well have been prescribed and studied.

It is significant to note that the astronomer, Lalla, in the beginning of the first chapter of his work says: 'Lalla now depicts the motion, etc., of the heavenly bodies, a subject which is rather difficult. Therefore, Lalla is composing the work, *Dhīvṛddhida,* the enlighter of intellect towards enhancing the astronomical knowledge of the students.' His inclusion of certain astronomical problems at the end of his treatise reveals its practical dimension intended for students. Likewise Vaṭeśvara, another noted astronomer, in his *Siddhānta,* has given examples at the ends of the chapters on longitudinal corrections, elements of *pancāñga,* graphical representation of shadow, eclipses and lunar horns, with a view to helping a student understand the concerned subject matter clearly. It would, however, appear that education in astronomy and mathematics was, by and large, in the nature of the preceptor-disciple relationship or sometimes, a family concern. In the medieval period (from the fourteenth century onwards), there were schools of astronomy and mathematics in Kerala, Mahārāṣhtra, and Benaras centering around relationships of this type, and such schools produced well educated and astute astronomers-mathematicians.

Medical Education: Similarly, there were noted teachers in the field of medical education, which was essential for training competent physicians and surgeons. For, surgery was also an accomplished healing practice. The Ayurvedic classics, the *Caraka* and *Suśruta Saṃhitās* have laid stress on the characteristics of a good physician as well as the service he is expected to render to the society. They speak of the harm caused by an untrained person or a quack. According to Caraka (*CS*. Sū. 29, 6-9), a physician should have a deep knowledge of the anatomy and physiology of the entire body, its growth characteristics as well as the origin and evolution of the universe. He must have a thorough understanding of the eight branches of Ayurveda and should be capable of applying his knowledge for the treatment of diseases. Suśruta even equated the untrained quacks with 'killers of men out of stupidity'. His prescriptions for a physician were that he should be clean, courageous, fully equipped with adequate quantities of medicine and surgical instruments and appliances; and he should also be intelligent and well-read, a person with readily available resources who commands decent practice, and endowed with all moral virtues (*SS*. Sū 34.9).

To achieve these levels of excellence, proper education of select students by competent teachers was needed. Caste was no bar for selecting a student. Caraka, though he stipulates that Ayurveda can be mastered by anyone of the first three castes, clarifies that it can be studied by all for the attainment of virtues, wealth and pleasure (*CS*. Sū. 30.29). Suśruta, too, accepts that the right to study medicine belonged to anyone, regardless of caste—but he strongly qualifies his approach by stating that at the time of initiation, the related *mantra* should not be recited if the student is a Śudra. However, where imparting medical education was concerned, the intrinsic merit and knowledge of the philosophical systems mattered most in his selection of a student. Moreover, it was the teacher who was to decide about the worthiness of the candidate for imparting medical training to him. In turn, the pupil also sought a good physician-teacher, even if he lived far away. Caraka has laid down the qualifications of a good teacher as follows: 'He should be well-versed in scriptures, equipped with practical knowledge, intelligent and skilful, pious and infallible in his perceptions for curing a patient. He should also be free from anger, envy and vanity, studious, affectionately disposed towards his pupils and capable of expressing his views with clarity'. The simile employed by Caraka is: 'A preceptor possessed of such qualities infuses medical knowledge to a good disciple just as the seasonal clouds (rain) help bring about good crop in a fertile land.' (*CS. Vi*.8.4)

Selected pupils were initiated into training by a consecration ceremony, featuring oath-taking and elaborate advice by the teacher on medical ethics to be meticulously followed by each pupil during and after his studentship. In accordance with the tradition of ceremonial functions, the teacher would initiate his pupil in the presence of fire (*agni*), Brāhmaṇas and other senior physicians, by saying that the pupil should observe celibacy, be truthful always and follow a vegetarian diet to sharpen his intellect. A student was enjoined to obey his teacher's instructions, respect him, be devoted to him like his son and avoid egoism, jealousy, and other evils inherent in a human being.

The advice tendered to students for their success in the medical profession was indeed extensive and humane (summarized below):

> '...you must never bear any ill-will towards your patients even at the cost of your life... Do not commit adultery, nor aspire for any property belonging to others... Do not undertake to treat women in the absence of their husbands or guardians... In entering the house of a patient make sure that you are accompanied by a person who knows the house and has secured the permission to enter premises... Your speech, intellect and sense should be entirely on the patient and his welfare. You

> should treat as confidential all information about the patient and his family. If you apprehend impending death to the patient, do not disclose it to the patient or to his relatives as it may cause a shock; though you are well-grounded in professional knowledge, do not brag it to others...and try all the time to learn something more of this medical science.' (*CS. Vi.*8.13-14)

Ayurveda, (see ch. 6), has been divided into eight branches and this division was valuable for its teaching. But the texts like the *Caraka Saṃhitā* and *Suśruta Saṃhitā* have their own arrangement, such as *Sūtrasthāna* (fundamental principles), *Śārīrasthāna* (anatomy and physiology), *Nidānasthāna* (pathology), *Cikitsāsthāna* (treatment of diseases), and others. They contain easily understandable verses in Sanskrit. The teacher, then as now, would explain a set of verses at a time, and the pupils were expected to memorize and understand them. According to Caraka, the pupil should repeat his recitation of verses with a view to overcoming his own defects in their rendering, and this practice would continue in the forenoon, afternoon, and night continually—such was the rigour of training then. Caraka also advocated the organization of seminars and professional discussions from time to time for clarification of doubts, which led to the acquisition of precise knowledge. However, he added that such seminars and discussions must be friendly and free from hostility (*CS. Vi.* 8.16-18).

Suśruta stressed the importance of practical training for the development of surgical skills. He emphatically stated that only a physician who is well versed in the principles of surgery as well as experienced in the practice of medicine, was capable of curing disease. Both Caraka and Suśruta have also stressed the need for proper and affectionate nursing in the successful treatment of a patient, and even specified the qualifications of nursing attendant. The Buddhist text, *Mahāvagga*, states that a good nurse should possess five qualities: (i) be capable of dispensing medicines as directed by the physician; (ii) be knowledgeable in the appropriate dietary regimen for the patient; (iii) have love and affection for the patient; (iv) should willingly and without disgust, clean up faecal and vomited matter; and (v) be capable of teaching, arousing and gladdening the patient with religious discourses. All of these needed proper education and training, constituting, as they did, high medical ethics and professional conduct.

ISLAMIC EDUCATIONAL SYSTEM IN INDIA

Following the precept of Islam enjoining the acquisition of knowledge, the Muslim rulers were patrons of learning. They established in India a large number of Islamic educational institutions called *madrasas*. Among the pre-Mughal rulers, Firoz Shah Tughluq, who was associated with men of learning, such as Maulana Jalal ud-din Rumi, Zia-ud-Din Barni, Siraj Arif, and Qazi Abdul Qadir, founded as many as fifty *madrasas*, including the famous one in Delhi called the Qudam Sharif (near Lahore Gate). Sikandar Lodhi encouraged the study of Persian and Islamic literature by Hindus and also the translation of Ayurvedic works into Persian. During the Mughal rule, beginning with Akbar, education received more attention—Abul Fazl wrote in *Āin-i-Akbarī* that Hindustan was particularly famous for its seminaries. Akbar had founded a substantial number of *madrasas,* and even a big college at Fathehpur Sikri. A remarkable feature was that both Hindu and Muslim students studied together in these institutions. Scientific subjects like arithmetic, geometry, astronomy, and natural philosophy were also taught along with agriculture, accounts, divinity and history.

Medical education received special attention. As noted before, Muslim rulers (the Delhi Sultans, provincial rulers and, more importantly, the enlightened Mughal kings) patronized medical training both in Unāni and Ayurveda, and established hospitals. According to Abul Fazl, the capital city of Akbar had thirty-nine hakims, but their number added up to more than a thousand in other parts of India. The reputed hakims were involved in imparting medical education in Unāni, and teaching medicine as expounded in Ibn Sīna's *Al-Qānūn*, an encyclopaedia of medicine in five books which received special attention. Several hospitals also had medical schools attached to them. Though the *madrasas* and the *gurukula* systems of education continued in the medieval period, the fact remained that the vast mass of people, including even women of the upper castes, were still illiterate, superstitious and mentally bound by traditional ideas and practices.

PLANTS: EARLY IDEAS AND NEW APPROACHES

The realm of plants as a whole had a fascinating appeal for people of the Vedic period, as evidenced by a large number of allusions to different types of plants, especially in the *Ṛgveda* and the *Atharvaveda*. In the Vedic idea of creation, plants were believed to have appeared even before the gods, which is why several of them were deified. In the course of time, a cult of vegetation appeared, with the *soma* plant earning the appellation of the Lord of Herbs as well as the Lord of Forests (*vanarāja)*. In the ninth *maṇḍala* of the *Ṛgveda*, there is hymnal praise for the divine powers of *soma* as well as its exhilarating juice. The plant, however, has not been identified botanically yet, mainly because of the absence of its specific descriptions in the *Ṛgveda* and other texts. The intoxicating power of its juice is indicative that either it could be an *Ephedra* (a gymnosperm), or a type of flowering plant (*Sarcostemma*) or a mushroom (a fungus).

Among the other venerated plants and trees were the peepal tree (*Aśvattha*; *Ficus religiosa*) and banyan (*Nyagrodha*; *Ficus bengalensis*). The wood of the banyan tree was preferred for making bowls used in sacrificial performances, although the *Ṛgveda* also speaks of utensils or vessels made from the wood of the peepal tree. Vedic people knew a large number of plants including the flower-bearing ones, and studied their characteristics in their own way. It is interesting to note that the *Ṛgveda* mentions two varieties of lotus *(Nelambo nucifera)*: white (*puṇḍarīka*) and blue (*puṣkara*). The *Atharvaveda* refers to the white lily, too (*kumuda*). Among the trees mentioned in the Vedic texts are *khadira (Acacia catechu), vibhītaka (Terminalia bellerica), śami (Prosopis* spp.*)* and śimśupa (*Dalbergia sisso*). Sugarcane (*ikṣu*) was a cultivated plant, even in the Vedic times.

Morphological Ideas

The *Atharvaveda* provides probably the earliest glimpses of plant morphology by way of categorizing the growth habits of trees into eight types: (i) with spreading branches; (ii) with leaves in long clusters; (iii) bushy plants; (iv) with monopodial growth; (v) expanding ones; (vi) creeping ones; (vii) with many stalks; and (viii) with knotty joints. The *Yajurveda* delineates different parts of a plant right from the roots to the fruits. Medicinal herbs received careful attention and nurture, and the *Atharvaveda* has a substantial list of plants which include apāmārga (*Achyrenthus aspera*); asikinī or asurī (*Indigofera tinctoria*); guggulu (Bdellium); jaṅgiḍa (*Terminalia arjuna*); pippalī (*Piper longum*); haridrā (*Curcuma longa*); and tila or sesame (*Sesamum indicum*). Some plants, like sadaṃpuṣpa (*Calotropis gigantea*) and palāśa or *parṇa* (*Butea frondosa*), were also used as amulets after the associated incantations.

An impressive image, comparing a human being with a tree, can be found in the *Bṛahadāraṇyakopaniṣad* (iii, 9.28): 'A man is indeed like a mighty tree; his hairs are its leaves and his skin is its outer bark. The blood flows from the skin of man; so does the sap from the skin of a tree that is struck (wounded). Man's flesh corresponds to that which is within the inner bark of a tree; his nerves are as strong as the inner fibres of a tree. His bones lie behind the flesh as the wood lies behind the soft tissue. The marrow of human bones resembles the pith of a tree.'

The *Amarakośa*, a lexicon compiled by Amarasiṃha (fifth

century CE), deals extensively with plants as well as their parts: the stalk, tubular petiole and rhizome; androecium; leaf primordium; inflorescences; pollen; bark; leaves and flowers. A later text, the *Vṛksāyurveda,* of Paraśara, has categorized leaves according to their shape as well as their resemblance to other organisms. It also describes the parts of plants in terms of leaves, bark, flowers, fruits, stem, root, seed and even sap and exudation, thus revealing the careful observations of plants made at that time. This text also speaks of two types of leaves, parallel-veined and net-veined, pointing out that the first one is characteristic of monocotyledons (*ekamātrika*), and the latter, of dicotyledons (*dvimātrika*). It is indeed refreshing to observe that in these texts, precise Sanskrit words of a technical nature have been used, which indicates that the plant kingdom was studied meticulously and the characteristics of different parts of a plant as well as of different plants themselves, were diligently and precisely recorded, which led to their systematic classification.

Classification

The *Ṛgveda* has categorized plants under three heads: (i) trees (*vṛkṣa*); (ii) herbs or medicinal plants (*oṣadhi*); and (iii) creepers (*vīrūdh*), besides the flowering (*puṣpavatī*) and fruit-bearing ones (*phalavatī*). The *Atharvaveda* has classified plants based on their morphological details as well as their special properties. Manu, in his *Dḥarmaśāstra,* has provided a classification of plants under eight heads as follows: (i) *oṣadhi* (bearing fruits and flowers; but they wither away after yielding fruits); (ii) *vanaspati* (bearing fruits); (iii) *vṛkṣa* (which bear flowers and fruits); (iv) *guccha* (bushy herbs); (v) *gulma* (shrubs which are succulent); (vi) *tṛṇa* (grasses); (vii) *pratāna* (creepers that spread); and (viii) *vallī* (climbers and entwiners).

The Ayurvedic classic, *Caraka Saṃhitā,* in its chapter titled *Vibhāga vidyā* (classification of those which are used as medicine), has categorized medicinal substances of plant origin (*udbidha*) into four kinds: (i) *vanaspati* (plants which bear fruits and have no flowers); (ii) *vṛkṣa* (those bearing both fruits and flowers); (iii) *vīrūdh* (creepers); and (iv) *oṣadhi* (herbs which wither away after fruition). Ḍalhana (tenth-eleventh CE), in his commentary on the *Suśruta Saṃhitā* gives *plakṣa* (*Ficus infectoria*) and *udumbara* (*Ficus racemosa*) as examples of *vanaspati;* mango and *jambu* (*Eugenia jambalana*), of flower-bearing and fruit-yielding trees. He has divided *vīrūdhs* also into two groups: (i) creepers which spread their stems on the ground; and (ii) succulent herbs. His examples of *oṣadhi* are wheat and barley. Cakrapāṇi, the commentator of the *Caraka Saṃhitā,* divides *oṣadhis* into annuals or perennials bearing fruits, and grasses that wither away without fruits. These are referred to as *udbhida,* further divided into fifty groups based on their ability to cure certain diseases and pathological disorders. Suśruta has provided a list of thirty-five categories of plants in the ascending order of their medicinal properties.

The *Caraka Saṃhitā* has also classified flowering plants under seven heads based on their dietetic qualities: cereals, pulses, pot-herbs, fruits, vegetables, oils, and sugarcane. The identification or classification of some plants according to the number of leaves has been attempted, and this is in the nature of an accurate description of a plant taking the leaves into account—for example, one leaflet; two leaflets; three, four, five, six , seven and multiple ones (*bahupatrikā*). The medical lexicon *Dhanvantari Nighaṇṭu* emphasizes that a physician should master *Bheṣaja vidyā* (knowledge specially of medicinal plants) by studying the various names and characteristics of plants in Sanskrit and languages of common people, and this was in the nature of a separate discipline of study.

The *Padārthadharmasaṃgraha* of Praśastapāda (an erudite commentary on the *Vaiśeṣika* sūtras) has classified plants under seven categories: *vṛksa, tṛna; oṣadhi, gulma, latā, avatāna,* and *vanaspati*. Their characteristics have been well defined by Udayana, an exponent of the *Nyāya-Vaiśeṣika* in his *Kiraṇāvalī,* giving examples for each of the categories.

A later text probably belonging to the eleventh or twelfth century, the *Vṛkṣāyurveda* of Paraśara has dealt with, in considerable detail, the floral characteristics, their similarities and differences—a categorization based on the relationship between thalamus (*śalāka*) and gynoecium (*bījadhara*). Significantly, the chapter that describes these is titled *Puṣpānga sūtrādhyāya* (principles of describing the organs of the flower). The four categories presented in this chapter are: (i) hypogynous (*tuṇḍapuṣpa maṇḍala*); (ii) epigynous (*kumbhapuṣpamaṇḍala*); (iii) perigynous (*tuṅgapuṣpamaṇḍala*); and (iv) flowers in which stamens are above the gynoeceium (*vatyapuṣpamaṇḍala*). The words in Sanskrit are clearly suggestive of the methodology involved in the classification, which throws light on the consonance between the technical words in Sanskrit and their precise connotation. Such a concordance was also noticeable in the classification of plants based on certain characteristics and similarities, as well as differences.

The *Vṛksāyurveda* of Paraśara has attempted to classify plants into families too (*gaṇavibhāga*). For example: (i) *Leguminosae* (*samīgaṇīya*): represented by *śamīvṛkṣa,* a plant bearing a legume or pod, with leaves on a common stalk and leaflets arranged like a feather; flowers are hypogynous and five-petalled. (ii) *Rutaceae* (*pupilikagaṇīya*): these plants have spines, leaves which have odour, winged petioles and hypogynous flowers that have free petals and stamens. (iii) *Cruciferae* (*svastikāgaṇīya*): this name is derived from the shape of the calyx which looks like a *svastikā*; it has four sepals, four petals and six stamens (free) and superior ovary *(tuṇḍa maṇḍala)*; the flowers are arranged in rows in inflorescence. (iv) *Compositae* (*kūrcagaṇiya*): flowers are directly attached and borne on a common axis surrounded by common calyx, and the whole assemblage looks like bristles or brush-head (*kūrcākāra*); its ovary, however, is inferior. (v) *Cucurbitacea* (*tripuṣpagaṇīya*): these plants bear epigynous flowers which are sometimes unisexual and which have five sepals, five petals, three stamens, and a style with three-pointed stigma; the ovary is tri-locular and produces many seeds. (vi) *Apocynaceae (Mallikāganiya):* these bear *hermaphrodite (samāṅga)* flowers, have a united calyx and corolla, each being five-membered with fine stamens, etc., and their seeds have tufts of fine long hairs. According to modern botanists, the descriptions of the characters of such families are remarkably similar to those of modern botany. The nomenclature of plants had two dimensions: one, based on their botanical characteristics; and the other, on their medicinal or allied properties. For example, *Sesbania grandiflora* was called *vakra puṣpa* because of its curved corolla, and *vṛṇāri,* because it was used for curing inflammation of boils.

Anatomy and Physiology

The *Vṛkṣāyurveda* states that in the plant-tissue there are mechanisms or systems which are meant for the transmission of sap and nutrients. The term used for the vascular system in this text is *sarvasrotāmsi*, meaning that which helps or is conducive to the flow of all fluids. Two categories of this system have been described: one called *syandanī* in the stem, which indicates the transmission of the nutrient fluid (*rasa*) absorbed from the soil to all parts of the body; and the other, known as *sirājālani*, which helps in the redistribution or circulation of nutrient sap from the leaf to other parts of a plant. Modern botanists think that the term *syndanī* corresponds to the xylem. More importantly, the *Vṛkṣāyurveda* has described the internal structure of a leaf as consisting of innumerable compartments that are filled with sap, thus serving as its storehouse. It states further that these compartments are covered by a boundary wall in the form of a cell membrane. The concept of five elements also finds its expression here inasmuch as the sap and its associates are said to be constituted of the five elements along with a colouring principle; they are stated to be invisible to the naked eye, since they are extremely minute. The boundary, which is equally thin, is associated with a fluid which, as some botanists interpret in modern botanical nomenclature, may correspond to protoplasm, while the colouring material could be the green chlorophyll. The *Vrkaṣāyurveda* describes the leaf and the presence in the leaf of colouring matter (*rañjaka*). It also mentions digestion with the help of colouring matter. According to some Indian botanists, these indicate that our ancient thinkers knew about food production in the leaves. There is no denying that descriptions such as these would not have been possible but for prolonged observations of the behaviour and growth of a plant, aided perhaps by some contrivance that could magnify the observed parts, leading to a close understanding of their internal structure. However, no details in this respect are available in the literary sources.

There is also an interesting explanation of the ascent of sap and its mechanism, the tree sucking water from its roots with a force, and water drawn up along with air in much the same way as the water is forced up when a petiole of a lotus reed is cut, dipped in water and sucked from the mouth—a transpiration pull theory according to some modern botanists. On the analogy of the metabolic process in a human body in which *jaṭhārāgni* (the fire element) and its components, called *dhātvagnis* play a pivotal role, the *Śāntiparva* of *Mahābhārata* also explains plant metabolism by stating that the nutrient fluid (water) absorbed by a plant is transformed into food under the influence of *agni* and air, due to which plants grow. Obviously, *agni* here corresponds to solar rays or energy, and some modern botanists opine that the idea of *agni* may be interpreted in terms of photosynthesis in plants. Interestingly, the imagery of the human body and a plant had been, then as now, an integrated one in Indian thought. Even the diseases of plants and their cures were thought of in the same manner as those of human beings.

Plant Diseases

The nature and character of several diseases that could afflict different kinds of plants were also known in ancient India, and there are several references to them in literary sources. The *Atharvaveda* mentions the insect pests that would destroy plants. Varāhamihira's *Bṛhatsaṃhitā* (sixth century) provides an account of plant diseases and points out that they are caused by deleterious environmental conditions, including variations in temperature and air. Four types of diseases—the yellowing of leaves or chlorosis, falling of buds, drying up of branches, and sap exudation—were recognized. Besides, the non-production of flowers and fruits was also regarded as a disease. The *Bṛhatsaṃhitā's* prescription for curing this barrenness consists of the preparation of a mixture of horse gram, black gram, green gram, sesame, and barley in hot milk, and after it cools, sprinkling it on the barren trees. Among the methods of treatment of the general diseases of plants as suggested in this text, is the application of a fine paste made of *ghee*, *viḍaṅga* (*Embelia ribes*), and mud to the infected parts, and the subsequent sprinkling of diluted milk over such parts. The *Bṛhatsaṃhitā* also provides some details about the manure of plants and trees for their healthy growth. One of the methods recommended by this text describes the preparation of dried and powdered dung of sheep, mixing it with half its measure of *til* or sesame, as well as small quantities of naturally occurring zinc compound and cow's flesh. To this mixture, the requisite quantity of water is added and kept for a week. On the eighth day it should be applied to infected trees or creepers.

This encyclopaedic text also deals with the treatment of seeds with a view to making them germinate properly and grow to their full stature. For this purpose, according to this text, the seeds should be smeared with *ghee* and kept immersed in milk for some time. They are then taken out, dried, and the process repeated again and again for ten days. Thereafter, they are mixed with cow dung, flesh of a pig or deer, and sown into the soil, which should be treated with diluted milk. The *Arthaśāstra* of Kauṭilya speaks of the treatment of seeds in a similar way, but adds that, when sprouted, the seeds are to be fed with fish-meal along with the milk of *snuhī* (*Euphorbia nerrifolia*).

The *Vṛkṣāyurveda* of Surapāla, (c. tenth cent. CE) provides several interesting ideas about the diseases of plants. Classifying them into two groups, namely, the internal ones (those arising from the body of the plant) and the external ones (those caused by outside agencies like worms and insects), this text even thinks of internal diseases of plant tissues in terms of *vāta*, *pitta*, and *kapha* (*tridoṣas*), and prescribes some curative methods accordingly. For example, it states that the diseases of *vāta* type could be cured by the administration of flesh, lymph and fat including *ghee*. Trees are cured of bilious (*pitta*) disorders, according to this text, by watering them with the decoctions of liquorice, honey, and *madhuka* or milk and honey. Fumigation of the tree with fumes of ignited white mustard, black pepper, etc., is also recommended for destroying colonies of insects infecting the trees. Even 'healing' the broken parts of trees, or those whose branches have fallen or been damaged by lightning was thought of, and their remedial measures prescribed.

Arbori-Horticulture

Laying-out and maintenance of gardens and parks not only for health and recreation but also as a vital way of dealing with environmental concerns were given due importance right from the Vedic times. In the Buddhist literature, there are descriptions of the pleasure gardens of Kings Bimbisāra and Aśoka used for recreation. The tree at Bodh Gaya under which, after intense penance, Gautama became the Buddha, the Enlightened, was a peepal tree (*Ficus religiosa*). The *Rāmāyaṇa* mentions the confinement of Sītā by Rāvaṇa in a grove of Aśoka trees (*Saraca indica*), while Kṛṣna's activities were intimately connected with the tree called kadaṃba (*Anthocephalus indica*). The *Amarakośa* refers to a variety of gardens or parks and the types of forests, natural as well as sylvicultured, kitchen gardens, pleasure groves, and royal gardens. Varāhamihira, in the chapter titled *Vṛksāyurvedādhyāya* of his compendium, the *Bṛhatsaṃhitā*, deals with several horticultural practices. It gives a list of plants that are suitable for a garden intended to beautify the environment and also provide shade.

In his *Kāmasūtra,* Vātsyāyana has recorded that palaces and houses of wealthy people had enchanting gardens with an amazing variety of trees and flowering plants. Public gardens were looked after by superintendents appointed by the state. An expertise in grafting, which was one of the qualifications of a gardener, was raised to the level of one of the sixty-four *kalās* or art forms. A gardener was required to be knowledgeable about the methods of plant propagation through seeds, cutting, layering and, more importantly, in grafting (*kandāropana*) as well as transplantation (*pādapa-saṃropana*). As regards transplantation, Varāhamihira has listed the names of various trees, whose branches could be cut off either at the base or the nodal region, a paste of cow dung applied, and the tree planted in a pit filled half its depth with fine soil, and adequately watered so as to nourish the transplant effectively. He has also mentioned that the plants which are very young with no branches, and those which have developed either small or substantial branches, should be transplanted in early winter, late winter, and the rainy season, respectively. Before transplantation, Varāhamihira says, the tree should be anointed with a paste made of equal quantities of *ghee*, sesame, honey, milk, cow dung, some fruit pulp as well as roots of medicinal grass (*usīra*), and also worshipped with reverence.

The *Upavanavinoda* (a branch of *Vṛksāyurveda*) is an informative chapter in the comprehensive compendium of Śārṅgadhara, titled *Śarṅgadharapaddhati* (thirteenth century). It deals, among others, with the construction of garden-houses, methods of nourishing plants and treatment of their diseases. The royal families as well as noblemen had pleasure and kitchen gardens attached to their palatial buildings and it was perhaps to cater to their requirements that the selection of suitable soil, sowing seeds, watering plants, their protection and nourishment as well as treatment of their diseases were presented in considerable detail.

In the medieval period, the Delhi Sultanate rulers encouraged the creation and preservation of gardens. As is well known, the Mughal emperors were extremely fond of huge gardens and canals, which are still a great attraction to Indians and foreigners alike. These renowned gardens still display not only flowering plants but also diverse fruit trees, some of which are exotic.

Introduction of New Plants

Though India, a tropical country, has been the home of a wide variety of flora, some of the present-day plants were not natives of the country. During Mughal rule as well as the Portuguese colonization, several new plants brought from outside began to be nurtured. It was during Akbar's time that the cherry (*Prunus avium L.*) was brought from Persia and introduced in the Kashmir valley. Tobacco (*Nicotina tabacum L.*) entered India from South America through Arabia when Muslim pilgrims from South India returned home after the Haj in Saudi Arabia, bringing tobacco back with them. From South India, the cultivation of tobacco spread to North India in the time of Akbar, when one of his lieutenants looking after South India took with him the cuttings of tobacco plant along with some quantity of tobacco to Delhi and presented these to the monarch. Though Akbar tried tobacco once, he was disinclined to smoke. But his son, Jahangir, even passed a decree prohibiting its use. Despite such unfavourable and antagonistic response from the emperors, tobacco cultivation spread far and wide in India in the succeeding decades.

The Portuguese were instrumental in the introduction of the South American native pineapple (*Ananas cosmus L.*) in the area they had occupied in the sixteenth century. In the course of time, its cultivation expanded to different parts of India. Soon it became so popular that it also found a place in the imperial garden at Agra during Jahangir's time. Among the other important South American plants that were introduced by the Portuguese were papāya (*Carica papaya L.*) and cashew nut (*Anacardium occidentalis L.*). However, their spread initially was not as spectacular as the pineapple, but they were cultivated extensively in the areas of Portuguese occupation. The now widespread use of chilli (*Capsicum frutescens L.*) as an important ingredient of Indian cuisine, particularly in the preparation of South Indian dishes, was noted for its absence till about the seventeenth century. It was pepper (*Piper nigrum L.*), a native of India, that was invariably used for hot preparations. It was also an important article of commerce between India and the Roman world in the early centuries of the Christian era, and was referred to as 'black gold'. Roman historian Pliny even lamented that the trade in pepper was a drain on Roman economy. The commercial value of pepper was so high that at one time it used to be weighed against gold and silver. In fact, the high price of pepper was one of the motivating factors for Westerners to find a sea-route to India to find cheaper stocks of this spice. The Arab traders were specially active in carrying Indian pepper from the west coast of India to Europe through the Persian Gulf, Mesopotamia, Syria or the Gulf of Suez.

Since the *Āin-i-Akbarī* does not mention the word or the plant, coffee, it can be inferred that it was not introduced during Akbar's time. The available evidence indicates that the coffee plant began to be grown in South India around the middle of the seventeenth century, having been introduced there from Arabia.

COMMERCIAL OR CASH CROPS

Another plant product of commercial as well as domestic importance is cotton, a crop first cultivated in the Indus Valley.

Archaeologically, some fragments of cotton and threads are stated to have been found at Mohenjo-daro. These have been identified as lint from *Gossypium arboreum* which is extensively grown in India even now. There are early references to cotton cloth in Vedic literature and also in later works like *Manudharmaśāstra*. According to Greek writers such as Heradotus (fifth century BCE) and Theophrastus (fourth century BCE), cotton is a produce of India. Spinning and weaving were known to the people of the Indus Valley Civilization and in the historical period, even in the early centuries of the Christian era, Indian textiles had attracted the attention of Western countries.

India has been home to sugarcane since even the Vedic times, and it has been extensively cultivated all over the country ever since. Dye-yielding plants like indigo (*Indigofera tinctoria L.*) and henna or *mehendi* (*Lawsonia inermis L.*), as well as what is called al (*Morinda tinctoria Roxb*) and areca-nut (*Areca catechu L.*), were among the other major cash crops of the Mughal period. However, with the appearance and extensive use of the synthetic blue dye, the cultivation of indigo was discontinued. This had an adverse effect on the soil in which it was grown, as its nitrogen-fixation ability was lost. The fibre-yielding plants—hemp (*Crotalaria juncea L.*) and jute (*Corchoms spp.*)—were also grown, the former being more extensively cultivated than latter. Narcotics like opium poppy (*Papaver somniferum, L.*) and *bhāṅg* (*Cannabis sativa L.*), too, were cultivated extensively. It is interesting to note that Aurangzeb had banned the cultivation of these plants and the use of their products because of their harmful effects.

Commercially viable fruits like the mango (*Mangifera indica L.*), grapes (*Vitis vinifera L.*), dates (*Phoenix spp.*), pomegranate (*Punica granatum L.*), plantain (*Musa paradisiaca L.*), lemon (*Citrus spp.*), and coconuts (*Cocos nucifera L.)* were grown on a considerably large scale in the medieval period. Firoz Shah Tughluq endeavoured to establish as many as 1,200 fruit gardens in Delhi area alone and over forty in the Chittor area, and Mughal monarchs encouraged the cultivation of flower plants and fruit-bearing trees in many gardens.

Ecological Concerns

The realm of flora—its sustenance, conservation as well as judicious usage—and thus, the protection of forests as well as bio-diversity have been fostered over the millennia since Vedic times. Their ecological and environmental dimensions were viewed and even internalized, as it were, in the totality of the Indian way of living, a holistic sense of *man in harmony with nature*. The Vedic approach to life and its articulation in the form of a wide spectrum of hymnal prayers laid the foundation for ecological and environmental concerns both at the mundane and divine levels. Such concerns also received the deserved attention of the state-craft that came up later. Kauṭilya's *Arthaśāstra* has described vividly the role of governance in relation to ecology and conservation. In the chapter titled *Janapadaniveśa,* Kauṭilya sheds light even on the types of trees to be grown to mark the boundary between villages, such as Śalmalī (*Salmalia malabaricum*), Śami (*Prosopis spicigera*), and Nyagrodha (*Ficus bengalensis*), besides the use of uncultivated land as grasslands for grazing the cattle. He also recommended the appointment of an authority (*Vivitādhyakṣa*) for the maintenance and management of grasslands as well as the plantation of appropriate trees which, in turn, would yield revenue to the state.

Of special value, then as now, were the forests. Kauṭilya, in the chapter *Bhūmichidravidhāna* of his treatise, has stipulated the raising and maintenance of forests for different purposes: (i) forests for the performance of sacred rituals and meditation; (ii) forests for wild animals and hunting; (iii) state parks to protect all kinds of animals; (iv) mono-culture forests with a wide variety of trees; and (v) forests for protecting wild elephants, enclosed on one side by hilly terrain and with plentiful water sources. A penalty was prescribed for killing an elephant and no plant or animal was to be felled; the death penalty was to be imposed for poaching, and heavy fines had to be paid for stealing ivory even from dead tuskers. The chief of the forests was called *Kupyādhyakṣa*, an important official with the vital task of looking after the precious forests; in fact, Kauṭilya has devoted a separate chapter to forests in recognition of their immense value not only for the ecological harmony to which they greatly contributed, but also as a source of revenue for the state.

The *Amarakośa*, in its chapter titled *Vanauṣadhivarga*, lists several types of natural and sylvi-cultural forests including kitchen gardens, pleasure groves, and royal gardens. It enumerates about 270 types of plants—herbs, shrubs, trees, creepers, etc., with their synonyms. The *Vṛkṣāyurveda* of Parāśara, too, gives names of several types of forests with their descriptions and distribution in the other parts of the country. Forests were held in high veneration, not for exploitation but for their role in stimulating human high thinking and plain living in a serene, salubrious, and invigorating environment. One of the constituents of the Vedic literature goes by the name *Āraṇyaka* (literally, forest treatise) which was a prelude to the pristine Indian metaphysical or philosophical treatises called the Upaniṣads. Interestingly, the *Bṛhadāraṇyaka* is also referred to as *Bṛhadāraṇyākopaniṣad*. In the Indian way of living and the mode of thinking, philosophical or otherwise, ecology and environment have been accorded a distinct position and expounded in several ways. The ecology and environment were not merely natural phenomena consisting of flora, fauna and geological structures, but were also sources of inspiration for a harmonious and meaningful life. The duality of man and nature has never appealed to the Indian mind-set. Rather, it has been the concept of man *in* nature, in complete harmony *with* nature, that has been a goal ever in sight, inspiring a sustained endeavour to reach that goal.

As noted before, Indian ethos, in general, is fortified by

the faith that man is an integral component of nature, and the realm of plants, particularly, has drawn both adoration and veneration from man. Some rituals and festivals too have been intertwined with specific plants and trees. The worship of *tulasī (Occimum sanctum)* and the rearing of this plant in orthodox houses particularly are not only in recognition of its medicinal properties, but also a way of preserving the species. The neem is another tree which is worshipped along with the *Asvattha* (peepal) tree, and both of them have an honoured place in the lives of Indian people from time immemorial. The mango tree, apart from its delicious fruits, is revered for its leaves which, infused with water, are used for sprinkling on occasions such as house-warming as an anti-pollutant, air-freshener, and cleanser. There are other rituals involving the use of plants, like milk-weed (*arka*) which is placed on the shoulders while bathing on the *Rathasaptamī* day. In the worship of deities like Gaurī and Ganeśa, a wide variety of leaves and flowers are an abiding part of the rituals, which are incomplete without them. In this way, these species have been preserved over the ages. The conservation of flora of diverse descriptions, protection of forests and the growing of plants around human dwelling places, even in an urban ambience, have been inseparable aspects of Indian tradition.

'Scientific' ideas and practices in India, as briefly discussed earlier, were integral to cultural traditions over the centuries. These had the innate potential to catalyse and contribute substantially to the emergence of specific areas of astronomy, mathematics, medicine, plant science (briefly dealt with already), chemistry, and chemical techniques. From the early centuries of the Christian Era, therefore, independent texts on these subjects began to appear, but within the traditional cultural background. An historical presentation of these traditional sciences, based upon textual and other sources, has been attempted, as we continue our journey through the history of science in India.

3

ASTRONOMY: ITS COMPUTATIONAL AND OTHER DIMENSIONS

The word 'astronomy' owes its origin to the Latin term *astronomie,* which in turn is derived from the word *astro-nomos,* meaning 'star-arrangement', or the knowledge of the observed regular motions of celestial bodies, in particular, of the sun, the moon and the planets. But the observations of the motions of the moon and the sun date back to antiquity, ever since man attained his erect posture and thus was able to turn his eyes upwards to the sky. Then as now, the occurrence of day and night in the context of the rising and setting sun, the waxing and waning of the moon, the innumerable glimmering stars on the vast blue sky during the night, as well as some luminaries now called planets, have had an enchanting, exciting appeal for the human mind. The early food-gathering and wandering man who was fascinated by the celestial objects and their movements had also developed an awed, worshipful attitude towards them. In the course of time, when man became a food-producer, he was able to observe constantly the motions of certain bright objects, and arrive at some kind of an understanding, if not precise knowledge, of their periodicity—especially that of the sun, the moon and the planets. His prolonged observations would have led him to perceive some form of relationship between the movement of the Sun, the varying weather, and the rainfall, thus marking definitive seasons like hot, rainy and cold, which were necessary for his agricultural operations and the planning of his seasonal activities. The broad division of cyclic seasons and the recurring divisions of day and night enabled him to think of the practical divisions of time in relation to the motions of the moon and the sun. Endeavours such as these continued in advanced human settlements or civilizations in the river valleys—those of the Nile (Egyptian) and the Euphrates and Tigris (Mesopotamian) in the fifth-fourth millennium BCE. It was only centuries later that some recordings began to be attempted and the duration of a year determined with a type of calendrical computation. The occurrences of solstices (both summer and winter) as well as equinoxes were also recognized through keen observations of the motions of the sun and the moon. Planetary astronomy, however, appeared much later, although some planets like Jupiter and Venus as bright objects of the night sky were known to people of these civilizations.

The Indus Valley Civilization or the Harappa Culture, which was the youngest but by far the largest of the three ancient Indian civilizations, also would have developed certain astronomical ideas centering on the sun and the moon, as well as some stars. No records, however, of such ideas, nor their calendrical computations have been available so far archaeologically. The so-called 'Indus Script' remains still undeciphered, which is why we have little or no knowledge about their methods of time-reckoning and divisions into months, seasons, and years. On the other hand, a clear picture of early astronomical and calendrical ideas in the Vedic period has come down to us through the ages.

THE VEDIC PERIOD

Several types of sacrifices performed at prescribed times in relation to the movements of the sun and the moon were an integral part of Vedic life. The word *jyotiṣa* in the Vedic texts connotes astronomy (and later, astrology) and was reverentially recognized as the foremost of the six auxiliaries of the Vedas. The tradition of observing the motion of the moon, the sun and some star-groups is as old as the *Ṛgveda* itself. However, the *Vedāṅga-jyotiṣa* is the earliest text on astronomy (see p. 132). It asserts: 'Just like the comb of the peacocks and the crest jewel of serpents, so does *jyotiṣa* stand at the head of the auxiliaries of the Vedas' (*Ṛgveda Vedāṅga-Jyotiṣa,* 35). Such a recognition continued over the millennia as recorded by Bhāskarācārya II (twelfth century CE), who, in his *Siddhāntaśiromaṇi,* explains figuratively thus: 'Of the six *Vedāṅgas*, *Vyākaraṇa* is the face of the person of the Veda; *Jyotiṣa*, his eyes; *Nirukta*, his ears; *Kalpa*, his hands; *Śikṣā*, his nose; and *Chandas*, his feet.' (*Siddhāntaśiromaṇi*, 1.1.1.10)

The goal of astronomy in India right from the Vedic times has been the determination of time associated with the performances of sacrifices (*yajña*). The *Vedāṅga-jyotiṣa* states unequivocally: 'The Vedas have been indeed revealed for the sake of the performances of sacrifices; but these sacrifices have been set out (to be performed) according to the determined sequences of time. Therefore, one who knows astronomy, the science of time, really understands the proper performance of sacrifices.' (*ṚV-VJ,* 36)

Integrated with the relationship between astronomy and the performance of sacrifices lay the lofty concept of the consonance of microcosm (man) and macrocosm (universe), or terrestrial and celestial concordance, in which movements of the sun, the moon and the twenty-seven to twenty-eight *nakṣatras* or asterisms (twenty-eight if Abhijit* is included) loomed large. The Ṛgvedic idea of Aryaman's (divinity) path seems to suggest

*This is the *nakṣatra* reckoned between *Uttarāṣāḍha* and *Śravaṇa*, in some Vedic texts including the *Atharvaveda* and also in the *Sūryasiddhānta*. But *Vedāṅga Jyotiṣa* and the other *Siddhāntic* texts do not mention it.

that the Ṛgvedic priests knew the sun's path (the ecliptic), although it is still debatable whether it refers to the Milky Way or the ecliptic. However, they were well aware of the sun's annual revolution, that it moved northward (*uttarāyaṇa*) during half of the year and southward (*dakṣiṇāyana*) in the remaining half. The apparent beginning of such motion marked the solstices of winter and summer. As for the equinoxes, both vernal and autumnal, there is no explicit expression. However, the coincidence of the vernal equinox with the star-group Kṛttikā (now known as the Pleiades) at the head of the twenty-seven asterisms, leads to the conclusion that the phenomenon of the equinox was known in the Vedic period. The *Śatapatha Brāhmaṇa*, *Taittirīya Saṃhitā* and the *Atharvaveda* list the twenty-seven/twenty-eight *nakṣatras* beginning with the Kṛttikās, while the later list of the *Vedāṅga-jyotiṣa* starts with Bharaṇī, and the much later *Sūryasiddhānta* has Aśvinī as the head, thus revealing knowledge of the precession of equinoxes over a period of time.

The natural units of time were understandably 'day' and 'night'. A day was further divided into five parts: (i) the period of the rising sun; (ii) the gathering of cows; (iii) midday; (iv) afternoon; and (v) sunset. A day was also thought of in terms of thirty *muhūrtas* (one *muhūrta* being equivalent to forty-eight minutes), as evidenced by a statement in the *Śatapatha Brāhmaṇa* that in a year (360 days) there are 10,800 *muhūrtas*. A lunar day was called *tithi*, then as now, which is approximately the thirtieth part of one lunation. More precisely, it is the period during which the moon's longitude leads that of the sun by 12°. A half-lunation (*pakṣa*) consists of fifteen *tithis*, which are named by ordinals in Sanskrit like *prathamā* (first), *dvitīyā* (second), and so on till the last of the fifteenth, which is called *pūrṇimā* or *amāvāsyā*, depending upon whether the *pakṣa* is the bright or dark half of the lunation, a tradition that continues even to this day. In addition to the solar and lunar days, it would appear that the Vedic people also conceived of a *nakṣatra* day as the thirtieth part of the time required for the moon to complete one revolution through twenty-seven *nakṣatras* or asterisms.

The *Ṛgveda* speaks of the twelve-spoked wheel, in a figurative manner, of a year of 360 days comprising twelve months of thirty days each. Then as now, two systems of month-reckoning, that is, one ending with the full moon, and the other ending with new moon were in vogue. In fact the moon was called the 'maker of the month' (*māsakṛt*) just as the English word, month, is derived from '*mooneth*'. The nomenclature of the lunar months was related to the *nakṣatras* in which or near which the full moon occurred. Thus, for example, the names of the months of Caitra, Vaiśākha, Jyeṣṭha, and so on were derived from the *nakṣatras*, Citrā, Viśākhā, Jyeṣṭhā, etc., in which or near which the full moon occurred. Since there will be a difference in the number of days between the lunar and solar months, which were indicative of seasons, and in order to adjust the lunar months with the seasonal months, it became necessary to introduce either an intercalary month at regular intervals or by adding five or six days to one or more months. Both these methods, more so the former, were in practice during the Vedic period. The *Ṛgveda* refers to a year of thirteen months, as does the *Atharvaveda*. This type of additional month was called *saṃsarpa* (a creeping month) in the *Kṛṣṇa Yajurveda*, and by other names in the *Atharvaveda* and *Kāṭhaka Saṃhitā*. A close examination of the Vedic reckoning of a year as well as month reveals that the Vedic priests recognized a lunar year of 354 days (12 X $29\frac{1}{2}$), which was synchronized with the sidereal solar year of 366 days either by adding twelve days each year or a thirteenth month of thirty days approximately every two and a half years.

The Vedic calendar was not a monolithic one but a diverse one inasmuch as five types of years appeared to have been followed in the Vedic period: (i) Sidereal lunar year of 324 days (twelve months of twenty-seven days each); (ii) Sidereal year of 351 days (thirteen months of twenty-seven days each); (iii) Synodic Lunar year of 354 days (six months of thirty days each and six months of twenty-nine days each); (iv) Civil or *sāvana* year (twelve months of thirty days each); and (v) a type of solsticial year (consisting of 378 days obtained by adding eighteen days to the third year after two civil years of 360 days each, which would lead to a correspondence between a sidereal and a solar year of 366 days).

Nakṣatras

At the core of Vedic astronomy was the integrated trinity of the sun, the moon and *nakṣatra*. Being astute observers of the luminaries of the sky and believers in the inseparable relationship between terrestrial and celestial occurrences, the Vedic priests were able to identify twenty-seven or twenty-eight *nakṣatras* in the context of the path of the moon as well as that of the sun. The Mesopotamians and the Chinese, too, had made similar observations in ancient times, and much later, the Arabs also had evolved their star-paths. But, as will be seen below, the Indian concept of *nakṣatras* had its own originality both in its content and context.

The term *nakṣatra* as used in astronomical and other texts connotes (i) a star in general, (ii) twenty-seven equal parts of the zodiac (corresponding to 13°20' each), and (iii) an asterism or group of stars in the zodiacal path. The *Ṛgveda* mentions some *nakṣatras* and also the so-called seven sages (*saptarṣi*) or Ursa Major. But a complete list of twenty-seven (twenty-eight, if Abhijit is included) appears in several Vedic texts like the *Taittirīya Brāhmaṇa*, the *Atharvaveda, Kāṭhaka Saṃhitā, Maitrāyaṇī Saṃhitā* and the *Vedāṅga-jyotiṣa*. However, the context is not merely astronomical, but also describes them in relation to the deities supposed to be governing them as well as their auspicious or inauspicious characteristics vis-à-vis sacrificial and other rites. The *Śatapatha Brāhmaṇa* speaks of some *nakṣatras* as either fit or unfit for the consecration of the sacred fire. By and large, the *nakṣatras* had religious significance, a tradition which is still alive in the Hindu society, apart from their presumed role in the making of horoscopes, astrologically.

An interesting aspect of the *nakṣatras* is that their nomenclature owed not a little to their imagined appearances or shapes. The linkage between the names and the shapes might have been formulated by the Vedic priestly astronomers, although there is no direct evidence for it. However, a much later astronomical text, the *Vaṭeśvara Siddhānta*, gives a clear exposition of this linkage as the following table shows:

Nakṣatras and their shapes (*Vaṭeśvara Siddhānta*, VII, 2.14-15)

Nakṣatra	Shape	*Nakṣatra*	Shape
1. Aśvinī	Horse's head	15. Svātī	Coral
2. Bharaṇī	Uterus	16. Viśākhā	Arched doorway
3. Kṛttikā	Hoe	17. Anurādhā	Heaps of offering
4. Rohiṇī	Cart	18. Jyeṣṭhā	Ear ring
5. Mṛgaśirā	Deer's head	19. Mūla	Lion's tail
6. Ārdrā	Jewel	20. Pūrvāṣāḍhā	Bed
7. Punarvasu	House	21. Uttarāṣāḍhā	Elephant's tusk
8. Puṣya	Arrow	22. Abhijit	Elephant's ear
9. Āśleṣā	Wheel	23. Śravaṇa	Three feet
10. Maghā	Wall	24. Dhaniṣṭhā	Drum
11. Pūrvā-Phālgunī	Bed	25. Śatabhiṣak	Circle
12. Uttarā-Phālgunī	Dais	26. Pūrva-Bhādrapadā	Dais
13. Hasta	Hand	27. Uttara-Bhādrapadā	Bed
14. Citrā	Pearl	28. Revatī	Drum

Nakṣatra-s and their Shapes (Diagrammatic)

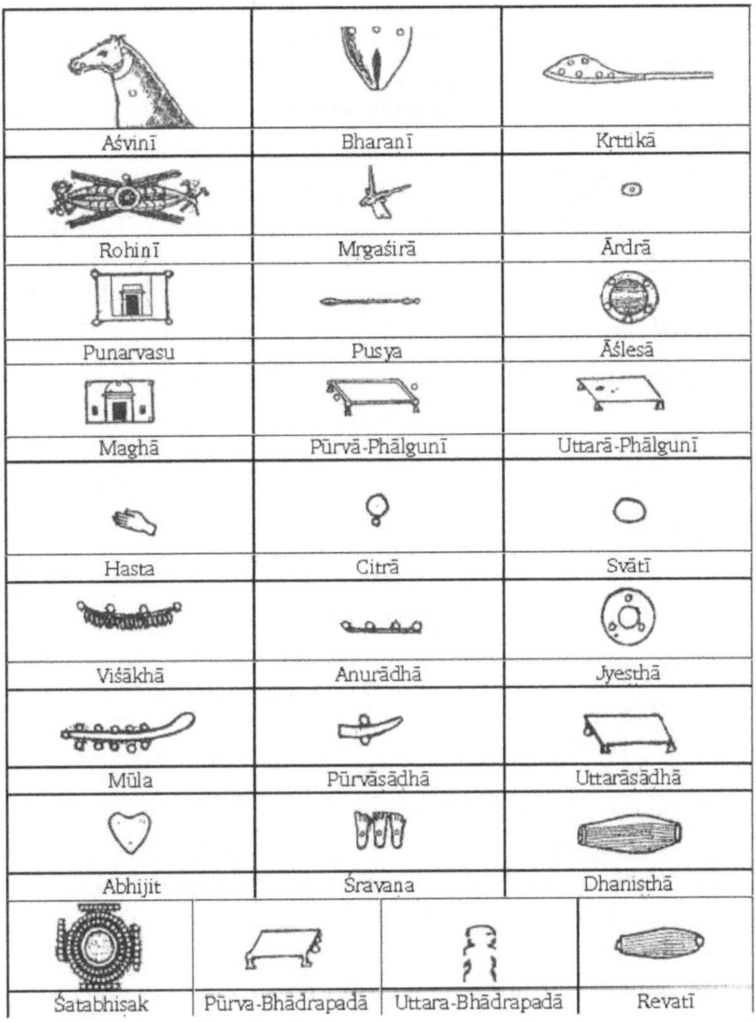

Aśvinī	Bharaṇī	Kṛttikā
Rohiṇī	Mṛgaśirā	Ārdrā
Punarvasu	Puṣya	Āśleṣā
Maghā	Pūrvā-Phālgunī	Uttarā-Phālgunī
Hasta	Citrā	Svātī
Viśākhā	Anurādhā	Jyeṣṭhā
Mūla	Pūrvāṣāḍhā	Uttarāṣāḍhā
Abhijit	Śravaṇa	Dhaniṣṭhā

Śatabhiṣak	Pūrva-Bhādrapadā	Uttara-Bhādrapadā	Revatī

Nakṣatra-s and the objects they resemble (diagrammatic)
Source: *Vaṭeśvara siddhānta*, K.S. Shukla (ed.), VIII, 2.14-15
(Courtesy: Project of History of Indian Science Philosophy and Culture, Centre for Studies in Civilizations)

Meticulous and prolonged observations of the *nakṣatras* by Vedic astronomers even resulted in the enumeration of the number of stars in each *nakṣatra*. The *Atharvaveda* has a *pariśiṣṭa* (supplement) called *Nakṣatrakalpa*, which has recorded the numbers as follows (the number of stars in each *nakṣatra* is given in brackets):

Kṛttikā (6); Rohiṇī (1); Mṛgaśiras (3); Ārdrā (1); Punarvasu (2); Puṣya (1); Āśleṣā (6); Svātī (1); Viśākhā (2); Anurādha (4); Jyeṣṭhā (1); Mūla (7); Āṣāḍha (8): Pūrva (7), Uttara (1); Abhijit (1); Śravaṇa (3); Śraviṣhṭā (5); Śatabhiṣak (100; or 1 according to Brahmagupta); Proṣṭhapada (4): Pūrva (2), Uttara(2); Revatī (1); Aśvinī (2); and Bharaṇī (3). There were, however, some differences in the recognition of the number of stars constituting a *nakṣatra* even among later Indian astronomers. In the Vedic astronomical literature including the *Vedāṅga-jyotiṣa*, there is no mention of what are called *yoga-tāras* or bright junction stars by means of which the conjunction with planets could be considered. The reason is that planetary astronomy was not the forte of Vedic observers. The concept and elucidation of *yoga-tāras* appear later in the Siddhāntic astronomy (see below) around the fifth century CE.

Indigenous Character: Some historians of astronomy, particularly of the Mesopotamian and Chinese civilizations, have inclined to the view that the Indian *nakṣatra* system owed its inspiration to the Babylonian lunar zodiac or to the Chinese *hsius*. An objective examination of the nature and characteristics of the *nakṣatra* system as developed in India reveals its indigenous development in a context that is exclusively Indian. Around 1100 BCE or perhaps a little earlier, Babylonian observers had divided the sky into three zones of twelve sectors each in order to understand the motions of the moon against the backdrop of the constellations in relation to its path. They had given names for constellations, by attempting to correlate the constellations with months. By about 700 BCE, a zodiacal scheme also came to the fore, as evidenced by several clay tablets which speak of as many as eighteen constellations more or less along the ecliptic. In the succeeding century or two, the number increased to thirty-three and thirty-six, sixteen of which were in agreement with Arabian *manāzils*, which were a much later development in the history of early astronomy.

There is a correlation of ten of the *nakṣatras* and the Babylonian ecliptic stars, but none where Mṛgaśiras, Ārdrā, Āśleṣa, Hasta, Mūla, Abhijit and Śraviṣthā are concerned. The Babylonian list comprises thirty-three or thirty-six stars, while the Indian list of twenty-seven or twenty-eight *nakṣatras* appeared in some Vedic texts much earlier, at least around 1200 BCE. More importantly, as mentioned before, the *nakṣatras* were interwoven with the performance of sacrifices, a context that is totally absent in the Babylonian concept of stars. It is true that the motions of the moon as well as of the sun and the background star-groups had an irresistible appeal to all ancient cultures—but what differentiated one culture from the other was the approach towards such celestial luminaries, and the purpose that motivated such observations.

Vedic people conceived of these motions and the star-groups as the pathway of divine-human communion attainable through the mediation of sacrificial rites performed during auspicious times determined by cyclic movements of the moon and the sun in the background of the ever-shining star-groups. The *nakṣatras* were inseparable from the sacrificial performances which were in the nature of spiritual pursuits. The nomenclature of the some of them, such as Ārdrā, Punarvasu, Āśleṣā and Maghā, even had agricultural significance. Thus, the *nakṣatras* were related in several ways to the material and spiritual life of the people—a tradition that continues even to this day. Such a relationship did not appear to have been integrated with the Babylonians.

The Chinese astronomers, according to Sinologist, Joseph Needham, had evolved a system of what are called *hsius* to mark the celestial equator, and to facilitate the observation of certain stars. A series of circumpolar stars were observed in the context of equatorial stars, which was different from the Indian context of observation. Moreover, the full list of twenty-three or twenty-eight *hsius* appeared for the first time in a text of the second century BCE, that is, long after the Indian *nakṣatra* system emerged. Significantly, several of the *hsius* differ from the *nakṣatras*, only nine *hsius* being in agreement with the junction-stars or the determining stars. It is indeed hard to find any mutual influence among the Indians, Mesopotamians and the Chinese in respect of the zodiacal star-system. Each civilization developed this system in its own way and, as Needham states, 'Indians have little to yield in this respect.'

Vedāṅga-jyotiṣa

The tradition of *Vedāṅga-jyotiṣa* continued from Vedic times to the period when Siddhāntic astronomy began. The text consists of two recensions—the *Ṛk* containing thirty-six verses and the *Yajus*, comprising forty-three verses—which deal with astronomical ideas in a precise manner. The text lists the now familiar twenty-seven *nakṣatras* (asterisms), but beginning with Bharaṇī, thus indicating that the vernal equinoctial point was associated with this *nakṣatra* when the text was formulated. As noted earlier, the *Yajurveda* and the *Atharvaveda* list the *nakṣatras* beginning with Kṛttikā, since the vernal equinoctial point was associated with this *nakṣatra* then. The *Vedāṅga-jyotiṣa* expounds a five-year luni-solar cycle called *yuga*, at the beginning of which the sun and the moon are stated to lie at the starting point of the *nakṣatra*, Dhaniṣṭhā. The astronomical elements in the quinquennial cycle, according to the *Vedāṅga-jyotiṣa* are: five solar revolutions; sixty-seven sidereal revolutions or months of the moon; sixty-two moon synodic revolutions or months; 1,830 civil (*sāvana*) days; 1,835 sidereal days; 1,800 solar days; and 1,860 lunar days or *tithis*. The text divides a civil day into thirty *muhūrtas*, one *muhūrta* into two *nāḍikās*; one *nāḍikā* into $10\frac{1}{20}$ *kalās*; one *kalā* into 124 *kāṣṭhās*, and one *kāṣṭhā* into five *akṣaras*. The duration of a *nakṣatra* works out to 610 *kalās*. It is possible that the *nakṣatra* space (13°20′) was subdivided into 124 parts, and the positions of the new and the full moon were arrived at in terms of the subdivisions of *nakṣatra* space.

The *Vedāṅga-jyotiṣa* also deals with the length of day. The shortest day is reckoned as the duration of twelve *muhūrtas* at the winter solstice, and the longest day as eighteen *muhūrtas* at the summer solstice. The text also indicates that in one *ayana* of 183 days, the day length would increase or decrease by six *muhūrtas*. Further, it records the location of the summer and winter solstices in the middle of Āśleṣā and the beginning of Dhaniṣṭhā. From the present positions of the solstices and considering the precessional rate of about 1^0 for every seventy-two years, it can be calculated that such a location of the two solstices would have occurred around 1400 BCE, which could be the date of the *Vedāṅga-jyotiṣa*. However, the two extant recensions (the *Ṛk* and *Yajus*) of the text are of a much later date, about 400 BCE, and the work is attributed to one Lagadha who might have reformulated them from an earlier text.

Both the lunar days (*tithi*) and asterisms (*nakṣatras*) played an important role in the Vedic calendar. It was much later, around the time of Āryabhaṭa (fifth century CE) and perhaps a century or two earlier, during the period of the five *Siddhānta*s (which were redacted by Varāhamihira in the sixth century CE), that the three other constituents, namely, weekdays (*vāra*), *karaṇa*, and *yoga* were reckoned along with the *tithi* and *nakṣatra*, and these five constituted a *pañcāṅga*, then as now.

SIDDHĀNTAS: MATHEMATICAL ASTRONOMY

There was apparently a long interregnum between the tradition of the *Vedāṅga-jyotiṣa* and what came to be characterized as Siddhāntic astronomy, which emerged with well-formulated texts dealing with mathematical astronomy around the fourth or fifth century CE. However, the Siddhāntic tradition appeared to be much older, perhaps going back to a few centuries before the Christian Era. Traditional lore speaks of as many as eighteen recognized *Siddhāntas*, their authors or redactors being Sūrya, Pitāmaha, Vyāsa, Vaśiṣṭha, Atri, Parāśara, Kāśyapa, Nārada, Garga, Marīcī, Manu, Aṅgīrasa, Lomaśa or Romaka, Pauliśa, Cyavana, Yavana, Bhṛgu and Śaunaka. The works of several of them have been lost, except some general stray references to them. Nevertheless, five *Siddhāntas—Saura, Paitāmaha, Vaśiṣṭha, Pauliśa*, and *Romaka Siddhānta*—have come down to us largely due to the scholastic endeavours of Varāhamihira. He meticulously redacted them, and the text is known as the *Pañcasiddhāntikā* (a name given later by Bhaṭṭotpala) which belongs to early sixth century CE. Even so, it is not certain that the original forms of these five *Siddhāntas* were the same as the redacted ones. For, the *Pauliśa* found in the *Pañcasiddhāntikā* is different from the one used by Bhaṭṭotpala. More importantly, the *Paitāmaha* of Varāhamihira is also different from the one found in the *Viṣṇudharmottara Purāṇa*. It needs, however, to be recognized that these five *Siddhāntas,* as well as the thirteen which are practically lost to us, ushered in a tradition of mathematical astronomy in the early centuries of the Christian Era. From the fifth century onwards, right up to the middle of the nineteenth century, this tradition continued unabated through the scholarly attempts of several luminaries who have bequeathed to us their treatises on Indian astronomy, mostly in Sanskrit.

At this stage, four aspects merit our attention. First, Indian astronomy is both observational and computational, sometimes more computational. Second, by the fourth century or so, the decimal place-value system using nine digits and zero had been developed in India, and it provided a new dimension to computational astronomy. Third, since numerical forms or numerals could not be used as such in the emerging texts (in Sanskrit verses) on astronomy, certain innovations had to be adopted for expressing numbers, but in tune with metrical compulsions of the textual verses. The use of word-numerals (*bhūta-saṅkhyā*) and alphabets (*kaṭapayādī*) in a methodical way was helpful to Indian astronomers to carry out computational as well as trigonometrical exercises, for which they had to possess certain qualifications. Fourth, observational astronomy also necessitated the use of various types of instruments (see below).

Jyotiṣa or the astronomical pursuit in India was not peripheral knowledge of the mediocre. In the Vedic socio-cultural life, accurate determination of time in relation to movements of the sun as well as the moon, and its adoption for ritual purposes, were the important functions of priestly-astronomers. But from the Siddhāntic period onwards, an astronomer was expected to acquire expertise in computational methods of diverse nature with associated mathematical aspects, the two together constituting the core of the *Siddhāntas*. Varāhamihira, in his *Bṛhatsaṃhitā* (2.1-12) states:

> 'He (an astronomer) must be thoroughly acquainted with the Earth's rotation (on its own axis) and its revolution along the circle of constellations, its shape and such other details, the latitude of a place and its complement, the difference in the lengths of the day and night (*lit.* diameter of the day-circle), the *carakhaṇḍas* of a place, rising periods of the different signs of the zodiac at a given place, the methods of converting the length of shadow into time (in *ghaṭikās*), time into the length of shadow and such other things as well as those to find out the exact time (in *ghaṭikās*) that has *elapsed* since sunrise or sunset at any required time from the position of the Sun or from the ascendant (*lagna*) as the case may be, (besides other prescriptions).'

Vaṭeśvara's prescriptions for the expertise of an astronomer are indeed elaborate. They include: (i) calculating the lunar days from the civil days; (ii) then, the sidereal days; (iii) then, the omitted days from the intercalary days, and vice-versa; (iv) finding the longitudes of the sun and the moon without making use of the *ahargaṇa* (number of elapsed civil days) and then obtaining the longitude of a desired planet; (v) deriving the sun's longitude from that of the moon and vice versa; (vi) determining the longitudes of the faster and slower planets from each other in ways more than one; (vii) deriving the sun's longitude from that of a planet; (viii) calculating the retrograde (or the westward) longitudes of planets and knowing their mean motion accurately; (ix) rising and setting of asterisms, planets, the sun, and the moon; and (x) the ability to compute so as to correct the mean motions of planets, *manda* (mean anomaly) and *śīghra-kendra* (anomaly of conjunction), and to arrive at motions of planets with precise trigonometrical knowledge or even without making use of *R* sines. Vaṭeśvara has rightly stated that a good astronomer is one who correctly knows the eight varieties of planetary motion, namely, very fast (*atiśīghra*); fast (*śīghra*); natural or mean (*samā*); slow (*manda*); very slow (*mandatara*); retrograde (*vakra*), very retrograde (*anuvakra*), and re-retrograde or transverse (*kuṭila*).

The Five *Siddhāntas*

Sūryasiddhānta: Of the five *Siddhāntas* redacted by Varāhamihira, the *Saura* or the *Sūryasiddhānta* was rated by him as the most accurate for the calculation of *tithi*. Unlike the other four, the *Sūryasiddhānta* does not have human authorship associated with it. Varāhamihira's *Saura* does not, in addition, correspond exactly to what is extant now under the name *Sūryasiddhānta*. Possibly, the current one with 500 verses under fourteen chapters might have taken its present form during the eighth to twelfth centuries CE, since it has certain elements of calculation adopted by Brahmagupta (seventh century CE) and later astronomers. Varāhamihira's *Saura* deals with solar and lunar eclipses, mean and true positions of planets, retrograde motions, heliacal rising of planets, and a few other topics. On the other hand, the current *Sūryasiddhānta* presents, among others, the different ways of reckoning time, the *Caturyuga* (later called *mahāyuga*), the mean motions of planets; their true positions; the three problems (*tripraśna*) of direction, space, and time; solar and lunar eclipses; asterisms; heliacal rising and setting of planets, the moon's cusps, cosmological ideas and astronomical instruments. Its cosmogony is in the nature of a recapitulation of the *Sāṅkhya* (*Seśvara* or with godhead), which, however, does not add much to the knowledge of astronomy.

The *Sūryasiddhānta* subscribes to the traditional view that the planetary motions are due to a wind called *pravaha* and that an invisible or divine force pushes or pulls them at their perigees so as to make their motions become variable. Otherwise *pravaha* was supposed to make the planets move uniformly, and they were believed to have been attached to invisible cords of air at the perigees and nodes. As for the motion of equinoxes, the text has adopted the theory of the libration of equinoxes and explained that, in a *Caturyuga* or 43,20,000 years, the number of oscillations of equinoxes about a fixed point, (like the swing of a pendulum eastward and westward) would be 600, and that the maximum eastward or westward deviation would be 27°. This works out to an annual precession rate of 54" (modern value is 50.25").

There is one important difference between the *Saura* of Varāhamihira and the current *Sūryasiddhānta*. The former deals with a minor *yuga* of 1,80,000 solar years during which there would be 66,389 intercalary months and 10,45,095 omitted lunar days, which would mean 6,57,46,575 civil days in that *yuga* period. Thus, according to this text, there would be 21,60,000 (12 × 1,80,000) solar months, 22,26,389 synodic months (21,60,00+66,389 intercalary months), and 24,06,389 lunar revolutions in a *yuga*. The current *Sūryasiddhānta*, on the other hand, thinks of *Caturyuga* (*Kṛta*, *Treta*, *Dvāpara* and *Kali*) in the descending order of 4:3:2:1, and arrives at a huge period of 12,000 divine years or 43,20,000 solar years (360 × 12,000).

This is extrapolated into a thousand times, called the *Kalpa* (43,20,000,000) which is regarded as the day-part of the Creator's (Brahmā's) life, his night part being of the same duration. The span of Brahmā's life (100 divine years) too is computed by multiplying it with 100, that is, 43,20,000,000 × 2 × 100 or 864,000,000,000 divine years or 311,040,000,000,000 solar years (× 360). Such computations, though it may appear fanciful to us, was in the nature of forging the relationship between the Creator and the created, the terrestrial and the celestial, which was the pristine idea of Indian astronomy. Such ideas as these found their application too in one way or the other in the *pañcāṅga.* Some of these computations also owed their inspiration to the Purāṇic texts as well as *Manudharamaśāstra*, and the concept of *Manvantaras*.

For the first time in the history of Indian astronomy, Varāhamihira's *Saura* employed the method of epicycles for determining the true positions of planets from their mean positions, and thus marked the beginning of mathematical astronomy in India. Whether such methods were developed or improved upon independently by Indian astronomers or they were borrowed from the Hellenistic culture is a moot point.

Over the centuries, the *Sūryasiddhānta* has been an important textual source for traditional Indian astronomers or almanac-makers, though its computations or calculating methods are being constantly refined by its users. This text is even considered a sacred one and believed to have been revealed by the sun to a demon named Maya who had desired to know the celestial movements. Attributing a divine origin to one or the other kind of human knowledge is not uncommon in the Indian ethos. For example, Ayurveda's origin is traced to Brahmā and the *Rasaśāstra* to Śiva and Pārvatī. Such an approach was also in the nature of forging a human-divine interrelationship.

Of the other four *Siddhāntas* of the *Pañcasiddhāntikā*, the *Paitāmaha* (also called the *Brahmasiddhānta*) is inaccurate in its expositions. The basic principles of this text have been given by Varāhamihira in the twelfth chapter of his compilation in the form of only five couplets. The *Paitāmaha* accepts the luni-solar *yuga* of the *Vedāṅga-jyotiṣa*, and its procedure of intercalation by adding one more month after every thirty synodic months, and an omitted *tithi* for every sixty-two lunar days. It also contains a rule for the calculation of *tithi* as well as the duration of a day and *vyatīpāta* (one of the *yogas*). Varāhamihira regarded the *Vāsiṣṭha Siddhānta* as far from accurate. The text, however, contains certain rules for the determination of day-lengths, *lagna*, gnomonic shadow, mean longitudes from the length of the mid-day shadow, and the determination of the true position of the sun by taking the apogee of the sun as fixed. The length of the solar year, according to this *Siddhānta*, is 365.36 days (when computed). The *Vāsiṣṭha sidhānta* now available is in no way similar to that of the *Pañcasiddhāntikā*. Both Brahmagupta and al-Bīrūnī opine that this *Siddhānta* was recast by one Viṣṇucandra.

The *Pauliśasiddhānta* occupies a considerable portion of the *Pañcasiddhāntikā*. It deals with the true motions of the sun and the moon, equation of the centre, computations concerning the *nakṣatras*, the moon's cusps, eclipses, parallax, method of finding accessional differences from the shadow on the equinoctial day, and calculations of the length of the day. Bhaṭṭotpala, in his commentary on the *Bṛhatsaṃhitā* has quoted several couplets from *Pauliśasiddhānta*, and noted that there are differences between this text and Varāha's. There is also a *Pauliśasiddhānta* in the *Viṣṇudharmottara Purāṇa* (fifth century CE); al-Bīrūnī too refers to this *Siddhānta* in his *Indica*, but it had undergone some modifications by the time al-Bīrūnī had access to it. The *Romaka* of the *Pañcasiddhāntikā* was regarded as the least reliable for its astronomical ideas and, as the *Pauliśa* itself points out, the study of *Romaka* needs expatiation! It should, nevertheless, be recognized, that the five *Siddhāntas* have a position of their own in the history of Indian astronomy, particularly before the advent of India's foremost astronomer and mathematician, Āryabhaṭa, towards the close of the fifth century.

TIME AND OTHER MEASURES

In the Indian context, astronomy and time-reckoning are inseparable. Apart from the concept of the timeless Brahman of the Upaniṣads and absolute time, the practical divisions of time necessary for regulated living and periodic performances of rituals as well as understanding of the past, present, and the future, have been formulated and used in an unambiguous but interrelated manner from the Vedic times. The general ideas about the periods of time, both long and short, as found in the *Vedāṅga-jyotiṣa* have been indicated earlier in this chapter. The later texts on astronomy have shed light not only on the various divisions of time but also computations of very long periods in a methodical manner, a brief account of which is essential for a proper appreciation of Indian astronomy. The units of the nine varieties of time-reckoning are as follows:

Units of the nine varieties of time-reckoning.

	Reckoning	*Unit used*		
i.	Sidereal	Sidereal day	=	One star-rise to the next
ii.	Lunar	Lunar month	=	One new moon to the next
iii.	Solar	Solar year	=	Period of one solar revolution
iv.	Civil	Civil day	=	One sunrise to the next
v.	Brāhma	Day of Brahmā	=	Period of two *kalpas*
vi.	Jovian	Jovian year	=	Period of Jupiter's motion through a sign (30°)
vii.	Paternal (manes)	Day of manes	=	One lunar month
viii.	Divine	Day of gods	=	One solar year
ix.	Demoniacal	Day of demons	=	One solar year

The sun's northerly or southward course, seasons, and the larger period of *yugas* were determined on the basis of the solar year; months, fortnights, and *tithis* were based on the lunar year; austerities and associated rites were performed on the basis of the civil year, and the smaller divisions of time like *ghaṭī* were based on the sidereal year. The six seasons and the solar and lunar months were integrated in the following way:

Integration of lunar and solar months and seasons

Lunar months	**Solar months**	**Seasons**	
Caitra	***Madhu***	***Vasanta***	**Spring**
Vaiśākha	***Mādhava***		
Jyaiṣtha	***Śukra***	***Grīṣma***	**Summer**
Āṣāḍha	***Śuci***		
Śrāvaṇa	***Nabha***	***Varṣā***	**Rainy**
Bhādrapada	***Nabhasya***		
Āśvina	***Iṣa***	***Śarad***	**Autumn**
Kārttika	***Ūrja***		
Mārgaśīra (Agrahāyaṇa)	***Saha***	***Hemanta***	**Dewy**
Pauṣa	***Sahasya***		
Māgha	***Tapa***	***Śiśira***	**Winter**
Phālguna	***Tapasya***		

The *Aitareya Brāhmaṇa* (1.1), however, recognizes five and not six seasons by integrating the winter and cool seasons (*Hemanta* and *Śiśira*). This text also enumerates the fifteen day-times of the bright fortnight; likewise, those of the dark fortnight. More importantly, it has a long list of names of *muhūrtas* of both the day- and night-times of each day of the fortnight, in all sixty names, each *muhūrta* being equivalent to (in modern reckoning) forty-eight minutes. There is also rather a curious division in terms of what is called *prati-muhūrta* for each *mūhurta*. The

Śatapatha Brāhmaṇa (X.4.2.18; xii.3.2.5) clearly states that there are 10,800 (360 X 30) *muhūrtas* in a year, that is, thirty *muhūrtas* in a day. The duration of a *muhūrta* (two *nāḍikās*) has been given for the first time in the *Vedāṅga-jyotiṣa*. The whole idea of a *muhūrta* and its importance in several ways have had a perceptible influence on the life of people from early times, and even the *Purāṇas* and the *Dharmaśāstra* texts lent full support to them. Each *muhūrta* was supposed to have its presiding deity and its auspicious or inauspicious character. The so-called *Brāhma-muhūrta* (early morning or the ultimate duration of two *nāḍikas* of the night) has been regarded as the most auspicious and salubrious, since Brahmā the Creator is believed to be its presiding deity. Varāhamihira in his *Bṛhatsaṃhitā* and *Bṛhad-yogayātrā* has provided detailed accounts of the *muhūrtas* in the context of their being auspicious or inauspicious, beneficial or harmful in an astrological perspective. Some *Dharmaśāstra* texts have also dealt with the *muhūrtas* substantially in relation to the performance of auspicious functions and observances in the frame of the twelve astrological *bhāvas* or houses in a horoscope.

Time Measurements

Vaṭeśvara in his treatise (*Vaṭeśvara Siddhānta*) has given the following general time measures and their inter-relatedness:

Time measures in the *Vaṭeśvara Siddhānta*

Lotus-pricking time	= 1 *truṭi*
100 *truṭis*	= 1 *lava*
100 *lavas*	= 1 *nimeṣa* (twinkling of eye)
4 $^1/_2$ *nimeṣas*	= 1 long syllable
4 long syllables	= 1 *kāṣhṭā*
2 $^1/_2$ *kāṣhṭās*	= 1 *asu* (respiration); 4 seconds
6 *asus* (*prāṇas*)	= 1 sidereal *pala* (*caṣaka, vināḍī* or *vighaṭikā*)

60 *palas*	= 1 *ghaṭikā* = 24 minutes
60 *ghaṭikās*	= 1 day
30 days	= 1 month
12 months	= 1 year
43,20,000 years	= 1 *yuga*
72 *yugas*	= 1 *manu*
14 *manus*	= 1 *kalpa*
2 *kalpas*	= 1 day of Brahmā (including night)
30 days of Brahmā	= 1 month of Brahmā
12 months of Brahmā	= 1 year of Brahmā
1 year of Brahmā	= 72 × 14 × 2 × 30 × 12 *yugas* = 7,25,760 *yugas*
100 years of Brahmā	= Life of Brahmā or *Mahākalpa*

It may be noted that in this enumeration a *kalpa* is taken to be 1008 (72 × 14) *yugas* as against the *Sūryasiddhānta's* 1000 *yugas*.

Linear measures: The *Bṛhatsaṃhitā* of Varāhamihira and the *Vaṭeśvara Siddhānta* have given a progressive relationship of linear measures starting with *aṇu*, which corresponds to the *trasareṇu* of the *Nyāya-Vaiśeṣika* exposition, namely, the size of a mote in a sunbeam. This was supposed to be the minutest visible measure from which, step by step, even the largest measure of the circumference of the circle of the sky is calculated. These measures are presented by *Vaṭeśvara* as follows:

Linear measures in the *Vaṭeśvara Siddhānta*

8 *aṇus*	=	1 *kacāgra*
8 *kacāgras*	=	1 *likṣā*
8 *likṣās*	=	1 *yūkā*
8 *yūkās*	=	1 *yava*

8 *yavas*	=	1 *aṅgula* (digit)
12 *aṅgulas*	=	1 *vitasti*
2 *vitastis*	=	1 *kara* (cubit)
4 *karas*	=	1 *nṛ*
1000 *nṛ*	=	1 *krośa*
8 *krośas*	=	1 *yojana*
1,24,74,72,05,76,000 *yojanas*	=	1 circle of sky

Measures of angles: *The Āryabhaṭīya* has enumerated the *R* sines and *R* sine-differences at intervals of 225′ or 3°45′ (see ch. 4).

Computation of Mahāyuga

It has been mentioned before that the *Sūrya (Saura) Siddhānta* of the *Pañcasiddhāntikā* recognizes a period of 1,80,000 solar years and states that this period contains 66,389 intercalary months and 1,04,50,95 omitted lunar days. Thus a *'yuga'* of Varāha's *Sūryasiddhānta* is one-fourth of a *Mahāyuga*, that is, 43,20,000 solar years. The word, *Mahāyuga*, however, is not at all used even in the modern *Sūryasiddhānta*, in which *Caturyuga* (quadruple age) is generally used, as being equal to 12,000 divine years (12,000 × 360 solar years). The modern *Sūryasiddhānta* also employs the word *yuga* to denote a *Mahāyuga*. According to this text, the computation of a *Caturyuga* is as follows:

Kṛtayuga	4,000 divine (*divyābdas*) years
Tretāyuga	3,000 divine years
Dvāparayuga	2,000 divine years
Kaliyuga	1,000 divine years
Total	10,000 divine years

The text implies further that the tenth part the *Kṛtayuga* and others (it does not explicitly state the other three *yugas*) belongs to its dawn, and another tenth part, to its twilight. Thus 400 +

400; 300+300; 200+200; and 100+100 or 2000 divine years are to be added; hence a *Caturyuga* is 10,000+2,000=12,000 divine years. One divine year is regarded as being equal to 360 (solar) years of human reckoning, and thus one *Caturyuga* (*Mahāyuga*) consists of 12,000 × 360 = 4,320,000 human years as the following table reveals:

Number of solar years in a *Caturyuga*

Age	Number of divine years	Dawn	Twilight	Total (Divine years)	Number of Solar years
Kṛtayuga	4000	400	400	4800	1,728,000
Tretāyuga	3000	300	300	3600	1,296,000
Dvāparayuga	2000	200	200	2400	864,000
Kaliyuga	1000	100	100	1200	432,000
Total	10000	1000	1000	12000	4,320,000

There is a further computation, namely, that 1,000 *Caturyugas* would constitute one *Kalpa*, the end of which, according to the modern *Sūryasiddhānta,* brings about the destruction of all that exists, but is regarded as a day of Brahmā, the Creator, His night being of the same length and His age being 100, i.e.,100 × (1000+1000) × 360 *Caturyugas*.

Planetary and other revolutions in a *Mahāyuga*

Planet	*Sūrya-siddhānta summari-zed by* **Varāha**	*Āryabhaṭīyam; reproduced in* **Brahmagupta's** *Khaṇḍakhādy-aka and* **Bhāskara I's** *Mahābhāsk-arīya*	*Śiṣyadhīvṛ-ddhidata-ntra of* **Lalla**	*Vaṭeśvara-siddhānta*	*Modern Sūrya-siddhānta*
Sun	4,320,000	4,320,000	4,320,000	4,320,000	4,320,000
Moon	57,753,336	57,753,336	57,753,336	57,753,336	57,753,336
Mars	2,296,824	2,296,824	2,296,824	2,296,828	2,296,832

Jupiter	364,220	364,224	364,224	364,220	364,220
Saturn	146,564	146,564	146,564	146,568	146,568
Mercury	17,937,000	17,937,020	17,937,020	17,937,056	17,937,060
Venus	7,022,388	7,022,388	7,022,388	7,022,388	7,022,376
Moon's apogee	488,219	488,219	4,88,219	4,88,211	488,203
Moon's node	2,32,226	2,32,226	2,32,226	2,32,234	2,32,238
Number of civil days in a *Mahāyuga*	1,577,917, 800	1,577,917,800	–	1,579,917,560	1,577,917,828
Intercalary months	–	–	15,93,336	15,93,336	–
Omitted *tithis*	–	–	25,082,580	25,082,250	–

NOTED ASTRONOMERS AND THEIR WORKS

Āryabhaṭa I

In the history of Indian astronomy, Āryabhaṭa has been regarded as the first and the most renowned astronomer and an accomplished mathematician. In general, Indian astronomers were able mathematicians, too (see next chapter). Āryabhaṭa is referred to as Āryabhaṭa I since his namesake of the tenth century CE was also a reputed astronomer who is known as Āryabhaṭa II. In the concluding verse of the last section of his succinct treatise, Āryabhaṭa I called his work the *Āryabhaṭīya*, and stated that 'this work, *Āryabhaṭīya* by name is the same as the ancient *Svāyambhuva*' (that is, revealed by Svayambhu or Brahman) and it is true for all time. He has also cautioned that 'one who imitates or plagiarizes it shall lose his good deeds and progeny'. It is interesting to note that the name Āryabhaṭa was given to India's first satellite when it was launched into orbit by the Indian Space Research Organization on 19 April 1975.

The *Āryabhaṭīya*, an extraordinary text known for its brevity and precision, contains 121 stanzas in the *Āryā* meter, except for ten of them which are in the *Gītikā* meter. This pioneering text on mathematical astronomy has four chapters: (i) *Gītikā;* (ii) *Gaṇita-pāda;* (iii) *Kālakriyā-pāda;* and (iv) *Gola-pāda*. The *Daśagītikā* (10+3) and the *Āryāṣṭaśata* (108) are two dimensions of this text, as pointed out by later astronomers like Bhāskara I and Brahmagupta. The *Daśagītikā* is intended to provide necessary instructions to beginners and also enable them to understand the different astronomical parameters, while the *Āryāṣtaśata* deals with both mathematics and astronomy which could be understood by those who have acquired a proficiency in mathematics. The *Daśagītikā*, as the name connotes, should have consisted of only ten stanzas; but it has thirteen including the one of invocation, the other on Āryabhaṭa's novel alphabetical system of numeration and the third on the nature of *prayojana* or benefit.

The first chapter is also known as *Daśagītikā-sūtra*, meaning ten aphorisms, in the *Gītikā* stanzas. Eleven of them, from two to twelve, are actually in that style. 'Knowing this *Daśagītikā-sūtra*, which gives details of the movements of the Earth and the planets on the celestial sphere, one attains supreme Brahman, after piercing through the orders of the planets and stars'— avers one of the stanzas. The general opinion among scholars is that stanzas 2-11 are the *Daśagītikās*. Be that as it may, the four-chaptered *Āryabhaṭīya* heralded the dawn of mathematical astronomy in India.

The *Gītikā-pāda*, the first chapter, sets the tone by offering definitions and important astronomical constants: large units of time like *Kalpa*, *Manu* and *Yuga*, circular units of degrees and minutes as well as linear measures like *yojana*, *hasta*, and *aṅgula*. This chapter also gives the computed number of revolutions of the sun, the moon, the planets, and more importantly, the number of rotations of the earth, in a huge period of 43,20,000 years. Among the other ideas of astronomical significance presented in this section, are the elapsed time from the beginning of the current *Kalpa* to the commencement of the *Kaliyuga* (3102 BCE), positions of apogees, ascending nodes of planets, orbits of the sun, the moon, and the planets, their diameters, the obliquity of the ecliptic, and others. The *Gaṇita-pāda,* the second section, as the name signifies, deals with mathematics (see next chapter). The third, *Kālakriyā-pāda,* is important from the point of view of astronomy. It presents several types of time measures; the procedure for obtaining the true positions of the sun, the moon, and the planets; divisions of a year and a circle; hours and days; methods for the computation of true longitudes of the sun, the moon, and the planets, besides eccentric and epicycles for explaining their motion. The last chapter, *Gola-pāda*, presents rules relating to spherical astronomy, graphical representation as well as calculation of eclipses, and the visibility of planets. It also describes motion relating to the celestial sphere as observed by those on the equator as well as by those on the north and south poles. In addition, various circles of the celestial sphere have been dealt with.

Āryabhaṭa I subscribed to geocentric astronomy, but he did not accept the geostatic notion. He courageously advanced a theory of the earth's rotation about its own axis, and computed the number of the earth's rotation in a *mahāyuga* (43,20,000 years) as being 1,58,22,37,500. This would give the period of one sidereal rotation of the earth, in the modern reckoning, as being equal to $23^h\ 56^m\ 4^s.1$, which is very close to the corresponding modern value of $23^h\ 56^m\ 4^s.091$. The rotation of the earth about its axis is explained by Āryabhaṭa I as an important factor for the apparent westward motion of the stationary asterisms. He has illustrated it by the well-known simile of a man moving in a boat, to whom a fixed object appears to move in the opposite direction. The diurnal rotation of the earth was also known to the Greeks. Heracleides of Pontus (388-315 BCE), a pupil of Pluto and Aristarchus of Samos (310-230 BCE), who was later known as the 'Copernicus of Antiquity' for his heliocentric concept, had held this view. But, Āryabhaṭa I was the first among the Indian astronomers to have indicated that it is due to the rotation of the earth that there is an apparent daily motion of the stars. His concept of the earth's rotation was not viewed favourably by Varāhamihira, Brahmagupta, Lalla, and Bhāskarācārya II. Commentators such as Pṛthūdakasvāmin and Makki Bhaṭṭa, however, accepted this idea.

Āryabhaṭa I wrote another treatise called the *Āryabhaṭa-siddhānta,* which is not extant, but to which references have been made by several astronomers like Bhāskara I, Govindasvāmin, Mallikārjuna Sūri, Rāmakṛṣṇa Ārādhya, and Tammaṇṇa Yajvan. Varāhamihira in the *Pañcasiddhāntikā* has pointed out that Āryabhaṭa I has adopted the reckoning of a day as beginning at midnight at Laṅkā, and also said that the day commences from sunrise at Laṅkā. The former is referred to as of the *ārdharātrā,* and the latter, the *audayika school.* Brahmagupta who in his youth was a bitter critic of Āryabhaṭa I, was so influenced by the popularity of the *Āryabhaṭa-siddhānta* that he incorporated the *ārdharātrika* system also in his *Khaṇḍakhādyaka* (literally, food prepared with sugar candy) long after he wrote the *Brāhmasphuṭasiddhānta.* He thus preserved for posterity his rival's *ārdharātrika* system. The *Āryabhaṭa-siddhānta* reckons the day from one midnight to the next (*ārdharātrika,* like the *Sūryasiddhānta).* Āryabhaṭa I's able commentator, Bhāskara I, has examined in his *Mahābhāskarīya* the important differences between the two works of Āryabhaṭa I. The methods as well as some astronomical parameters found in the *Āryabhaṭa-siddhānta* are nearly the same as those adopted in the *Saura* of the *Pañcasiddhāntikā.* There are, however, noticeable differences between the two in respect of the computed values of the revolutions of Mercury and Jupiter, omitted lunar days, civil days, etc. An important dimension of the *Āryabhaṭa-siddhānta* is that it has dealt with a number of astronomical instruments (shadow-instruments, *yaṣṭi,* water instruments, *ghaṭikā yantra,* gnomons, and so on), and these were brought to light by later astronomers like Rāmakṛṣṇa Āradhya.

The biographical details of Āryabhaṭa I are scanty. In his invocation and introduction to the *Gaṇita-pāda* (second chapter), Āryabhaṭa states that he 'sets forth here the knowledge honoured in Kusumapura... Having bowed with reverence to Brahmā, the Earth, the Moon, Mercury, Venus, Sun, Mars, Jupiter, Saturn and the asterisms...' Bhāskara I, who was the earliest commentator of the *Āryabhaṭīya* has identified Kusumapura with Pāṭaliputra (modern Patna in Bihar) in ancient Magadha. He also points out that in Magadha, the year commenced on the first *tithi* of the dark half (*Kṛṣṇa-pakṣa*) of the month Śrāvaṇa and ended on the fifteenth *tithi* of the bright half (*Śukla-pakṣa*) of the month of Āṣāḍha. Significantly, this has been the tradition in Magadha to the present day. The *Āryabhaṭīya*, as stated by Āryabhaṭa himself, is the same as the one said to have been revealed by Brahmā; Āryabhaṭa and his followers adhered to this kind of thinking which persisted in Magadha for a long time. It therefore stands to reason that Āryabhaṭa I lived in Kusumapura or Pātaliputra or Patna, which is now the capital of Bihar state.

Bhāskara I refers to the *Āryabhaṭīya* as the *Aśmaka-tantra*, and its followers as the Āśmakīyaḥ, which is construed to mean that the original homeland of Āryabhaṭa I was Aśmaka. The *janapada* called Aśmaka which finds mention in Hindu, Bauddha, and Jaina literature, was either a region of northwest India or a territory lying between the Godāvarī and Narmadā rivers, as indicated by the fact that a branch of the Aśmakas had moved and settled on the banks of the Godāvarī. Aśmaka is also mentioned in several inscriptions, which confirms that it became a part of Maharāṣhtra (Nāsik region). There are, however, divergent opinions about the identity of Aśmaka, including the one that the word, *Aśmaka*, means a land of stones. It is a possibility that Āryabhaṭa I belonged to Kerala, since several Kerala savants have written commentaries on the *Āryabhaṭīya* and were devoted followers of Āryabhaṭa I. Amidst all of these controversies and rather wild interpretations, the incontrovertible fact is that Āryabhaṭa I lived and wrote his classic, the *Āryabhaṭīya,* in Kusumapura or Pātaliputra in Magadha. It may be noted that in ancient times Magadha was well known for its centres of learning, including the famous University of Nalanda, which lies not far away from Pātaliputra. Astronomy (*jyotiṣa*) was one of the important subjects taught in this university which attracted students and other knowledge-seekers from far and wide. It would appear that it had an observatory too, and Āryabhaṭa I was a *kulapa* (*kulapati*) or chancellor of this university.

Āryabhaṭa I was born in 476 CE, as evidenced by his own statement made in the *Kālakriyā* section of his work which, according to him, was written when 'sixty times sixty years and three quarters *yuga* had elapsed, twenty-three years had then passed since my birth.' On calculation, this would correspond to mean noon at Ujjainī on 21 March 499 CE and since he was twenty-three years of age then, he was born in 476 CE. Several commentators of the *Āryabhaṭīya* are of the view that, the year 60 × 60 or 3600 of the current *Kali yuga* was mentioned by the author specifically because at that time the mean positions of planets, as computed by the number of their revolutions given in the *Daśagītikā*, did not require any correction. According to Bhāskara I, Āryabhaṭa himself had told his disciples that he wrote his treatise at that time and his disciples included Pāṇḍuraṅgasvāmī, Lāṭadeva, and Niśaṅku. Among them Lāṭadeva became a well-known astronomer and, as Varāhamihira says, he wrote two treatises (which are not available now) in which he reckoned the day from midnight at Laṅkā. He is stated to have been the author of two commentaries on the *Romaka* and *Pauliśa Siddhāntas*, and al-Bīrūnī has recorded that Lāṭadeva was the redactor of a *Sūryasiddhānta* (not the current one).

Āryabhaṭa's is the most revered name in Indian astronomy. His concise but profound astronomical-mathematical masterpiece, the *Āryabhaṭīya*, appealed to astronomers down the ages to such an extent that various commentaries were written on it, the earliest one being that of Bhāskara I (seventh century CE). The other important commentaries now available are by: Someśvara (eleventh-twelfth centuries); Sūryadeva Yajvan (twelfth century) and elucidated by Yallayya (fifteenth century); Parameśvara (fifteenth century); Nīlakaṇṭha and Raghunātha Rāja (sixteenth century); and Kodaṇḍarāma (nineteenth century). More importantly, the Āryabhaṭan school had a large number of followers. Even today, the *Āryabhaṭīya* is one of the major texts, the basic rules of which are employed in Indian calendar computation.

Varāhamihira

The well-known erudite scholar and astronomer-astrologer, Varāhamihira, was a junior contemporary of Āryabhaṭa I. His birthplace was Kapitthaka in the modern Farrukhabad district of Uttar Pradesh, but all of his attainments in astronomy, astrology, and a wide variety of other subjects were in Avanti (modern Ujjain), where he worked under the benign patronage of rulers belonging to the Aulikara clan. The name, Varāhamihira, is stated to be an altered version of the Iranian name, Mihirvarāz. For, his ancestors belonged to the Iranian Maga priestly class, and his grandfather, who was believed to have migrated to India, had embraced Hinduism. The word *Varāhamihira* combines the name of the divine boar (Varāha), an incarnation of Viṣṇu, and the Sun god (Mihira). In fact, Varāhamihira was a devoted worshipper of the sun, and he was even regarded by common people as an incarnation of the sun. Further indicating his family's sun-worshipping tradition, was his father's name, which was Āditya Dāsa (servant of the sun).

Varāhamihira's important works are: the *Pañcasiddhāntikā* (redaction of the then prevalent five *Siddhāntas* as noted before); *Bṛhajjātaka* (a major work on astrology); *Bṛhatsaṃhitā* (an encyclopaedic work); *Laghu-jātaka* (an abridged version of the *Bṛhajjātaka*); *Laghu-vivāhaphala*; *Mahāyātrā, Tikaṇikā-*

yātrā, and *Yoga-yātrā* (astrology relating to travel). His leaning towards astrology notwithstanding, Varāhamihara showed a scientific approach as an astronomer, and he had a rational understanding of the occurrence of eclipses.

The *Pañcasiddhāntikā* (redacted in 505 CE) has eighteen chapters. The aim of this work, as stated by Varāhamihira, was to 'state in full the best of the secret lore of astronomy extracted from the different schools of ancient teachers rendering it easy and clear' (*PS.* I.1-2). To realize such an objective required a versatile intellect, which Varāhamihira certainly possessed. It is largely through him that the astronomical concepts and computational methods that were in popular use in India in the early centuries of the Christian Era, have been known and handed down to us.

As a competent astronomer, Varāhamihira was vehemently against the concept of the traditional or mythological *Rāhu* (demon) 'devouring' the sun and the moon, and emphatically stated that *Rāhu* is not at all the cause of eclipses. He ridiculed the then persistent belief even among the learned men that an eclipse could take place, except when there is a combination of five planets in the same zodiacal sign and that a week before the eclipse, its characteristics could be inferred from the behaviour or appearance of a drop of oil poured on the surface of water. In his codification of the five *Siddhāntas,* Varāhamihira had assimilated their mathematical or computational dimensions. But he was noticeably inclined towards astrology along with its Hellenistic or Ptolemaic ideas. He even praised the Mlecchas *(Yavanas)* or the Hellenistic practitioners of astrology. His *Bṛhajjātaka* is the first important treatise in India on astrology, with its emphasis on the twelve *rāśi*s or zodiac signs, their effects and several nuances of predictive astrology which are more or less similar to those found in Ptolemy's (second century CE) *Tetrabiblios,* a major book on astrology at that time. There are over thirty Greek words in the *Bṛhajjātaka* and other astrological works.

Varāhamihira's encyclopaedic *Bṛhatsaṃhitā* deals with a wide range of topics: iconography, architecture, cosmetics and perfumery, water-divining, omens, geology, gems, horticulture, characteristics of animals and the like, all in a precise and lucid manner. His writings are marked by linguistic elegance and poetic flavour. Bhaṭṭotpala (eleventh century CE), the able commentator on the *Bṛhatsaṃhitā,* has praised Varāhamihira as an incarnation of the sun, who expounded *jyotiṣa* in its triple aspects (*tantra* or mathematical astronomy, *jātaka* or horoscopy and *saṃhitā* or mundane astronomy). Al-Bīrūnī, the well-known Islamic savant who was in India in the first half of the eleventh century CE, refers frequently to Varāhamihira in his magnum opus, *Kitāb-al-Hind,* but is also critical of him as well as Brahmagupta.

Brahmagupta

An accomplished astronomer and mathematician, Brahmagupta (b. 598 CE) lived in the reign of the King Vyāghramukha of the Cāpa dynasty in Gujarat. Even before al-Bīrūnī came to India, he was already conversant with Brahmagupta's work in its Arabic version, and stated that Brahmagupta belonged to a place called Bhillamāla (which has now been identified with the present Bhinmāl or Śrīmāla in northern Gujarat). Varuṇa, one of the commentators of Brahmagupta's treatise on astronomy, refers to the astronomer as Bhillamālakācārya. There are no further biographical details about Brahmagupta.

Brahmagupta wrote two texts: (i) the *Brāhmasphuṭasiddhānta* in Śaka 550 (or 628 CE), as stated by him; and (ii) *Khaṇḍakhādyaka*, a *Karaṇa* work in Śaka 587 (or 665 CE), that is, when he was sixty-seven years old. The *Brāhmasphuṭasiddhānta* is a voluminous treatise comprising 1,008 verses (1,022 if those in the last chapter on generalities are included). Divided into twenty-four chapters, it presents an extensive range of topics: mean planetary motions; their true positions; the three problems of direction, space, and time;

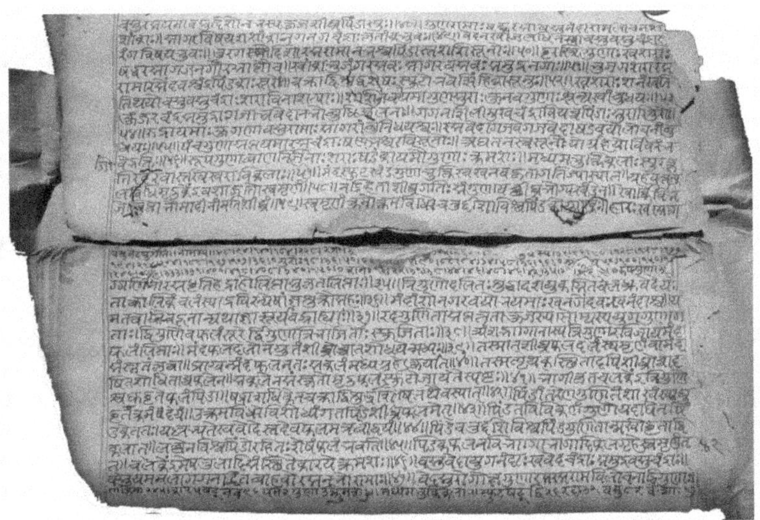

Brahmasphuṭasīddhānta

Facsimile of three pages from paper manuscript in Bombay University Library (Courtesy: Project of History of Indian Science Philosophy and Culture, Centre for Studies in Civilizations)

lunar and solar eclipses; heliacal risings and settings of planets; the moon's cusps and shadow, conjunction of planets and stars, astronomical instruments and related topics. Chapters twelve and eighteen deal with mathematics (see chapter on Mathematics in this volume). Interestingly, the author has mentioned the number of verses under each chapter. The astronomical elements discussed by Brahmagupta reveal his originality and critical approach, and they were fortified with his mathematical acumen.

Brahmagupta rejected the quinquennial *yuga* cycle of the *Vedāṅga-jyotiṣa*. He also viewed with disdain the strange Jaina concept of two suns and two moons. He was even critical of the Greek methods of determining planetary positions that could have entered Indian astronomy. He was also obstinately opposed to the theory of the rotation of the earth put forward by Āryabhaṭa I, and posed some odd questions. He observed: 'If the earth were to be spinning at the rate of one *kalā* (one minute) in one *prāṇa* (approximately four seconds, as stated by Āryabhaṭa I) then, what, where, and why would birds live? In a rotating earth, how would the arrows shot up fall in the same place?' Despite his mathematical expertise and astronomical computational techniques, Brahmagupta appeared to have shown his inclination to the mythological *Rāhu* postulate in the explanation of the occurrence of an eclipse. He was not favourably disposed towards Āryabhaṭa I and even chose to make rather uncharitable remarks about him.

The nineteenth chapter of the *Brāhmasphuṭasiddhānta*, titled *Tantraparīkṣā*, is exclusively devoted to the examination of the correctness or otherwise of different schools of astronomy prevailing during his time. A majority of the verses in this chapter point towards the Āryabhaṭan school, the most popular one at that time, and had the potential to attract to its fold several followers in the future. Brahmagupta has, however, dealt with what he called the inaccuracies in some of the parameters and the conceptual framework of the Āryabhaṭan system. A follower of the *sāyana* (precession of equinoxes) method, Brahmagupta developed (i) methods for finding out instantaneous motions of planets vis-à-vis their *manda* (slow motion) and *śīghra* (fast motion) inequalities; (ii) precise equations for determining parallax in longitude as well as latitude; and (iii) appropriate formula for the movement of equinoctial point. He had assumed that the solar ingress into Aries (*Meṣa*) occurred on an equinoctial day; hence the length of the solar year, according to his calculations, works out to 365 days, eleven *ghaṭis*, thirty *palās*, and twenty-two *vipalas*, which is slightly less than that found in the other Siddhāntic texts. Āryabhaṭa I had divided a *mahāyuga* (43,20,000 years) into four equal parts, each of 10,80,000 years duration. Brahmagupta, on the other hand, did not accept this, though he conceded that a *mahāyuga* comprises 43,20,000 years, following the Purāṇīc ideas. Āryabhaṭa I reckoned the first day of the *Kalpa* (1000 *mahāyugas*) as Thursday, but Brahmagupta regarded it as Sunday by his own calculation. Brahmagupta's undue bitterness towards Āryabhaṭa was unjustifiable, though his methods and results were reliable and accurate. However, as he grew old, his rivalry with the Āryabhaṭan astronomy began to diminish. As noted before, he showed a positive attitude when he, at the age of sixty-seven, wrote a *karaṇa* work called the *Khaṇḍakhādyaka*, adopting the Āryabhaṭan *ārdharātrika* system.

The *Khaṇḍakhādyaka,* in its eight chapters, deals with *tithis*; mean and true places of planets; the three problems of direction, space, and time; lunar and solar eclipses, etc. But as a *karaṇa* text, it has adopted Śaka 587 (665 CE), the year he wrote this text, as the epochal year for working out his own rules with associated mathematics, leading to more accurate results. This text received wide acclaim and was commented upon by several succeeding astronomers like Pṛthūdakasvāmin, who also produced a valuable and insightful commentary on the *Brāhmasphuṭasiddhānta*. Both the treatises of Brahmagupta found favour with the West Asian Islamic astronomers.

They were rendered into Arabic at the Abbasid Caliphate in Baghdad towards the close of the eighth century and later. Like Āryabhaṭa I, Brahmagupta, too, laid a solid foundation for astronomical pursuits in India, particularly in computational, mathematical astronomy.

Bhāskara I

The first and the foremost commentator of *Āryabhaṭīya* was Bhāskara I (to distinguish him from his namesake, Bhāskarācārya II, who lived about five centuries later), and was also a contemporary of Brahmagupta. Practically no information is available about his parentage, nor about his place of birth. It is believed, on the basis of some meagre evidence in his works, that he might have lived in the Aśmaka region and written his commentary on the *Āryabhaṭīya* in Vallabhī (in Saurashtra or Kathiawar). Interestingly, Bhāskara I refers to the beginning of a new year with its month and day from *Kārtika-śukla-pratipad*, still a living tradition in Saurashtra. Bhāskara I's well-known treatises are: (i) the *Mahābhāskarīya* (also called the *Mahātantra* by later astronomers); (ii) the *Laghu-bhāskarīya*; and (iii) the *Āryabhaṭīyabhāṣya.*

In the beginning of the *Mahābhāskarīya*, the author offers salutations to the god Śambhu or Śiva; likewise, in his commentary on the *Āryabhaṭīya*, he hails Śiva as the creator and destroyer of the universe and as effulgent as thousands of suns. Nevertheless, to express his allegiance to Āryabhaṭa I, Bhāskara I pays obeisance to Brahmā also in the beginning of his commentary on the first and the fourth chapters of the *Āryabhaṭīya*. Bhāskara I, however, was not a direct pupil of Āryabhaṭa, since he lived more than a century later. But he was a great admirer of the astronomical acumen of Āryabhaṭa, as evidenced by the second stanza of his *Laghu-bhāskarīya,* in which he states: 'Victorious is Āryabhaṭa whose excellent fame has crossed the bounds of the oceans and whose science leads to accurate results even in a far off place after the lapse of so much time.' He even said in the commentary on the *Āryabhaṭīya* that, of all the *Siddhāntas* known, it was Āryabhaṭa's text that was honoured.

The *Mahābhāskarīya* is a large compendium comprising 403 stanzas on mathematical astronomy and allied aspects. It has, as one of its foundational ideas, the sunrise (*audayika*) system formulated by Āryabhaṭa I. It deals with the usual subjects presented by the other major *Siddhāntas*: mean motions of planets; their true positions; the three problems; eclipses; heliacal rising and conjunction of planets with asterisms. But the order in which these are presented is somewhat different from that of the other Siddhāntic texts ,except the later *Siddhānta-śekhara* (eleventh century CE) of Śrīpati. Bhāskara I gives equal importance to both the epicyclic and eccentric theories, and has provided lucid explanations of the planetary motions in an authoritative manner. He also developed a method of his own for finding the mean positions of the planets, besides using certain technical terms in an unusual manner. For example, he has used the word, *Jyau* for Jupiter, which was probably derived from the Greek word, Zeus (the name of the Greek god). Bhāskara's ingenuity consisted in the application of indeterminate equations of the first degree to astronomical problems, and the calculation of sine or cosine of a given arc by using a formula developed by him, instead of using the tabulated ones. Among his other innovations were: (i) the method of ten *R* sines, which helped in calculations concerning an eclipsed body (though this was considered as being not so accurate; (ii) discussions on planetary conjunctions; and (iii) the application of parallax in the calculation of solar eclipse.

The *Mahābhāskarīya* appealed to several scholiasts in the succeeding centuries. Among them were Govindasvāmin (Bhāskara I's own pupil), Śaṅkaranārayaṇa (ninth century), Sūryadeva (twelfth century), Makki Bhaṭṭa (fourteenth century), Parameśvara (fifteenth century), and Sundararāja (sixteenth century), who produced valuable commentaries on this large work. The abridged version, the *Laghu-bhāskarīya*, was also popular because of the clarity and straightforward presentation of the basic ideas of mathematical astronomy. It became a textbook for the beginners in Indian astronomy in South India, particularly in Kerala, where an important school of astronomy came up from the fourteenth century onwards. Bhāskara I's *Āryabhaṭīya-bhaṣya* is a masterpiece because it has taken into consideration different shades of opinion on various astronomical paradigms, and established the superiority of Āryabhaṭa's position in each case, besides introducing some new ideas to supplement the *Āryabhaṭīya*, all in an incisive style of questions and answers. Bhāskara I's innovative methods and authoritative expositions had a catalytic effect, as it were, on the dissemination and popularity of the Āryabhaṭan school.

Devācārya

Contemporaneous with Bhāskara I and Brahmagupta, who both lived in the Gujarat region, was a Keralite (southern) astronomer, Devācārya. He lived in a region that was far away from that of his contemporaries. Yet it is amazing that Devācārya knew the works of both and, in his own way, was an avid follower of Āryabhaṭa I. He produced an authentic manual or handbook of 254 verses titled *Karaṇaratna*, which was intended to facilitate computations. He records in this book: 'Having taken a deep plunge into the entire ocean of the *Āryabhaṭa-śāstra* with the help of the boat of intellect, I have brought out this jewel, the *Karaṇaratna* with the rays of all planets.' By the metaphoric phrase of the entire ocean of the *Āryabhaṭa-śāstra,* Devācārya meant not only the works of Bhāskara I but also the *Khaṇḍakhādyaka* of Brahmagupta, which had adopted the Āryabhaṭan *ārdharātrika* system; indeed, he was influenced by it to a considerable extent. For, the divisors used in Devācārya's *Karaṇaratna* for calculating intercalary months and the omitted lunar days in the computation of *ahargaṇa* (elapsed civil days during any interval of time) are exactly those adopted in the

Khaṇḍakhādyaka. Moreover, Devācārya used the same rule as given in the *Khaṇḍakhādyaka* for the type of correction needed in the case of the moon's motion. What is astonishing and almost inexplicable is how Indian astronomers, separated by vast distances and a long span of time, were able not only to know about the astronomical endeavours of each other, but also to use the same terminology. It may be noted that Āryabhaṭa I lived in the fifth century CE in Pātaliputra (eastern region), Brahmagupta and Bhāskara I in the seventh century in western India, and Devācārya in the southern tip of India.

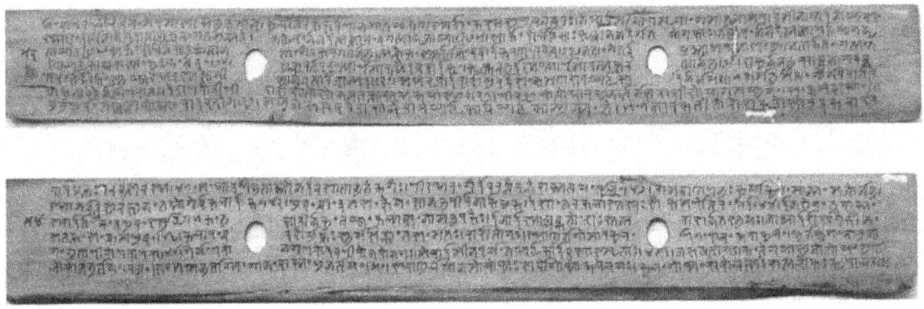

Karaṇaratna
Facsmile of two pages from palm leaf manuscript in Oriental Research Institute, Mysore
(Courtesy: Project of History of Indian Science Philosophy and Culture, Centre for Studies in Civilizations)

The *Karaṇaratna* has, as its zero point or the epoch, the beginning of *Caitra* in the year Śaka 611 or 689 CE. It deals with practically the entire range of Indian planetary astronomy but does not shed any light on the instruments. It also provides an account of the elements of *pañcāṅga*, especially the movable (*cara*) and immovable (*sthira*) *karaṇas* as well as the *vyatīpatayoga* (one of the 27 *yogas* of *pañcāṅga*). However, its rule for determining the value of the precession of equinoxes, on the assumption that the motion of equinoxes is oscillatory, is far from being accurate, since it leads to a precessional rate of 47" per year, while the actual modern value is about 50" per year. Devācārya knew that the corrections for planetary positions would be necessary from time to time. He enunciated three types of corrections, namely *Śakābda*, *Manuyuga*, and *Kalpa*. Besides, he evolved a new method for the computation of the mean longitudes of planets, particularly of the sun, the moon, the moon's apogee and its ascending node from the omitted lunar days elapsed, and the residual lunar days, a procedure that was not noticeable even in the other major works on Indian astronomy, except in the case of the moon. A feature of the *Karaṇaratna* is its brief account of topics dealt with in a chapter, at the end of it also.

Haridatta

Probably a contemporary of Devācārya, Haridatta was well known for introducing what came to be known as the *Parahita* system, which led to easy methods of calculations. His *Grahacāranibandhana* and *Mahāmārganibandhana* were the foundational texts of this system. He also adopted the *Kaṭapayādi* method of expressing numbers in his texts, instead of the rather cumbersome alphabetical (letter) numerals as conceived and adopted by Āryabhaṭa I, even though he was a great admirer and follower of Āryabhaṭa I's work. Haridatta was also well known for his corrections using a method of his own creation (*Bhaṭa-saṃskāra* or *Śakābda-saṃskāra*), a new approach that was appreciated and adopted by subsequent astronomers, especially of Kerala.

Lallācārya

Āryabhaṭa I, Bhāskara I, and Brahmagupta had no doubt laid a solid foundation of Indian astronomy, but Lallācārya (in the beginning of the eighth century CE) was such a versatile exponent that his text rekindled the astronomical talent of those who wished to pursue their studies. He aptly titled his treatise *Śiṣyadhīvṛddhidatāntra*, meaning 'that which increases or enhances the knowledge of the students of astronomy'. The author says in the beginning (invocation) of the first chapter of his treatise, 'Having made obeisance to Brahmā, Hari, Śiva, the Sun, the Moon, Mars, Mercury, Jupiter, Venus, Saturn and Ganeśa, Lalla now depicts the motion, etc., of the heavenly bodies, a subject which has been dealt with by Ācārya Āryabhaṭa. But the subject is rather difficult. Therefore, Lalla is composing the work *Dhīvṛddhida*, that is, to enlighten the intellect and to enhance the knowledge of students of astronomy.'

His main objective was to structure the procedures of principles enunciated by Āryabhaṭa. It may be noted that in his invocation, Lallācārya has recorded the order of the sun, the moon, and the planets, not according to their distances or motions, but corresponding to the order of the week days. For, by his time, the concept of seven days a week and the nomenclature of each day after a planet had entrenched itself in the daily life of common man as well as religious people of all walks of life. No details of his life are known, except that his father's name was Trivikrama and that he belonged to Lāṭadeśa (present southern Gujarat). In addition to his famous treatise, Lallācārya is believed to have written another text, the *Siddhānta tilaka*, which is not available, but which is mentioned by later astronomers like Yallayya and Rāmakṛṣṇa Ārādhya. According to some historians of astronomy, he had also produced a text on mathematics which was probably a synthesis of Āryabhaṭa I's *Āryasiddhānta* and Brahmagupta's *Brāhmasphuṭasiddhānta*. Bhāskara II makes a reference to Lallācārya's work on mathematics, but no such text is available. However, his compendium called the *Ratnakośa*, which deals with astrological matters, is well known.

Lallācārya's masterpiece *Śiṣyadhīvṛddhidatantra* has twenty-two chapters under two sections: (i) *Gaṇita* (mathematical astronomy: thirteen chapters); and (ii) *Golādhyāya* (spherics: nine chapters). The first section makes a lucid presentation of the usual topics on astronomy like the mean and true positions of planets, eclipses, rising and setting of planets and their conjunction, cusps of the moon and the three problems of direction, space, and time. The second section has a graphical representation of the motion of planets, description of the earth, false motions, astronomical instruments and problems. The armillary sphere and its construction have also been delightfully depicted. An ardent follower of Āryabhaṭa, Lallācārya has divided the quadrant into arcs of 225′ each. He also constructed his own table which he called *laghu-jyā,* dividing the quadrant into arcs of 10°. Some of his formulae, especially those relating to the computation of the mean longitudes of the sun, the moon, and planets are also found in Brahmagupta's works. In general, it is not wide off the mark to think that the *Śiṣyadhīvṛddhidatantra* was in the nature of a synthesis of the methods of Āryabhaṭa I and Brahmagupta in a clear and easily understandable way.

Vaṭeśvara

An astute astronomer, Vateśvara lived in a town called Nāgarapura in northern Gujarat. He was born in about 880 CE as he says in his enchanting treatise, the *Vaṭeśvara Siddhānta,* and that 802 years had elapsed since the beginning of the Śaka era when his birth took place, and twenty-four years of his life had gone by when he wrote his treatise. The date of his work, therefore, was 904 CE (802+24+78). His father was Mahīdatta, who belonged to Ānandapura, which is identified with Vaḍnagar in northern Gujarat. Vaṭeśvara first produced a *karaṇa* work under the title *Karaṇasāra* to help the *pañcāṅga* makers of his time in their computations. But his fame rests upon the extensive text called the *Vaṭeśvara Siddhānta,* which is available in its full form. He is stated to have prepared another text known as *Gola,* which is not extant in its complete form, but the first five chapters have been ably brought to light by K.S. Shukla from the fragments available. A voluminous astronomical treatise, next only to the *Sīddhāntatattva-viveka* of Kamalākara and the *Siddhāntadarpaṇa* of Sāmanta Chandrasekhara, which came up several centuries later, the

Vaṭeśvara Siddhānta (excluding *Gola*) comprises as many as 1,326 verses. It is interesting to note that the *Āryabhaṭīya* has 121 verses; *Brāhmasphutasiddhānta*, 1,008 verses; Lallācārya's work, 322; the *Mahābhāskariya* of Bhāskara I, 403 verses; the *Sūryasiddhānta*, 500 verses; *Siddhāntaśiromaṇi* of Bhāskara II, 460; and Sripati's *Siddhāntaśekhara*, 890 verses.

The *Vaṭeśvara Siddhānta* adheres to the Āryabhaṭan or the Brahmā school of Indian astronomy and reckons the day from sunrise at Laṅkā (*audayika*). This text also presents and discusses the usual topics that are normally found in the Siddhāntic works: mean positions of planets; their true positions; eclipses; heliacal rising and setting of planets; cusps of the moon; conjunction of planets and the like. It has eight chapters, each divided into several sections in which appropriate details of mathematical astronomy, different methods for the calculation of *ahargaṇa*, parallax, etc., have been given. The format of the chapters and the succinct sections along with examples that would shed light on the methodology associated with the determination of astronomical parameters has made this text not only unique but also a definitive landmark in the history of astronomy in India. Vaṭeśvara has provided simple methods for the computation of derived Rsine and Rversed-sine tables for intervals of 56′15″, besides well-related time measures. He has also adopted some new word-numerals and technical terms which, however, do not alter the general understanding associated with them. In the history of Indian astronomy, the *Vaṭeśvara Siddhānta* stands as a comprehensive treatise, heralding innovative mathematical methods and refinements in astronomical parameters. These features also marked the new mood of later astronomers.

Muñjāla

Among the early pioneers of the new mood was Mañjula or Muñjāla (literally, lovely or beautiful), as he was referred to by the astronomers who came after him. Both Bhāskara II (twelfth century) and the thirteenth-century astronomer, Sūryadeva

Yajvan called him Muñjāla, while Yallayya and Tammaṇṇa Yajvan (fifteenth century) referred to him as Mañjula. He lived in Prakāśa Paanṭ which, according to Mallikārjuna Sūri, was near Patna. Significantly, in Prakāśa Paanṭ was a famous temple of the Sun god and, in the regional dialect, Mañjula was used as a synonym of the sun. The *Laghumānasa* is an important text written by Muñjāla (tenth century) and has a special position in the history of Indian astronomy. Possibly, he might have composed another bigger work under the title *Bṛhanmānasa* or *Mahāmānasa,* since the extant text goes by the name of *Laghumānasa*.

Muñjāla was by and large a follower of Āryabhaṭa I, although he had not identified himself with any school or tradition. The *Laghumānasa* comprises just sixty verses in the *Aṇuṣṭubh* meter and is in the nature of a *karaṇa*; as a result, it is of value in almanac-calculations. It deals with the positions of the sun and the moon, the apogees of planets for the beginning of the year (*Caitra* month), mean positions of planets and the like, besides the lunar corrections and the methods of determining the elements of a *pañcāṅga.* It also throws light on the construction of a diagram of an eclipse. More importantly, Muñjāla includes the 'evection' term for the moon for the first time in the history of astronomy in India. His concise and lucid text enjoyed such a reputation that several commentaries came up by scholars such as Praśastidhara (his own contemporary), Sūryadeva Yajvan (thirteenth century), Parameśvara (fifteenth century), Yallayya (late fifteenth century), and Bhūdhara (sixteenth century). Not a few succeeding astronomers referred to him or his methods in one way or the other.

Śrīpati and Śatānanda

In the beginning of the eleventh century, an astronomer by the name Śrīpati was noted for his systematic studies. Practically nothing is known about his life, except that his father's name was Nāgadeva. He produced two elegant works: (i) *Dhīkoṭī*, a *karaṇa* work adopting Śaka 961 or 1039 CE as the epochal year for calculations; and (ii) *Siddhāntaśekhara*, which has twenty chapters encompassing traditional topics like the mean and true places of planets, eclipses, heliacal rising and settings of planets and their conjunctions in a systematic manner. The nineteenth chapter deals with astronomical instruments. Śrīpati also wrote a text on mathematics under the title *Gaṇitatilaka*. His most important contribution to astronomy consists in his method of calculation leading to an understanding of the moon's second inequality.

Śatānanda was a junior contemporary of Śrīpati and produced an elegant and popular *karaṇa* work titled *Bhāsvatī* with Śaka 1021 (1099 CE) as the epochal date. This text is largely based on the *Sūryasiddhānta* and also Varāhamihirā's *Saura* in the *Pañcasiddhāntikā*. The planetary positions calculated by Śatānanda are those obtained by applying Varāhamihira's corrections to positions that are derived from the *Saura* of the *Pañcasiddhāntikā*. Such calculations, however, are based on the number of years elapsed, but not on *ahargaṇa* or the number of elapsed days. Apart from this novelty, Śatānanda adopted a centesimal system for the epochal positions as well as using multipliers and divisors in an ingenious way for the calculation of planetary positions. He has also assumed that Śaka 450 (528 CE) was the zero precession year. Among the other important aspects of astronomical value in the *Bhāsvatī* are the enunciation of rules for the accurate determination of the occurrence of eclipses and the computations of epochal positions in relation to Pūrī (Odisha) where the author lived, but not of Ujjayanī, the usual paradigm for Indian astronomers. Śatānanda's *Bhasvatī* won the recognition of subsequent astronomers, some of whom produced commentaries on it.

Bhāskarācārya II

In the twelfth century lived Bhāskaracārya, an astronomer-mathematician of extraordinary calibre, and reputed for his texts on astronomy and mathematics. He was called Bhāskara II to distinguish him from his namesake who had lived earlier in the seventh century. There is a stone inscription (Epigraphia Indica I. 340) in a temple of a village called Pātan, which is not far away from Chalisgoan in Khandesh area (Maharashtra), which enumerates the following lineage: Kavicakravartin Trivikrama, his son Bhāskarabhaṭṭa (who received the title of *Dyupati* from King Bhoja), his son Govinda or Govindasarvajña, his son Prabhākara, his son Kavīśvara Maheśvarācārya and his son Bhāskara (astronomer), his son Lakṣmīdeva, his son Caṅgadeva. The inscription states further that in the court of King Siṅghana of the Yādava dynasty, the grandson of Bhāskara (Caṅgadeva), built a *maṭh* (monastery) at Pāṭan for teaching the works of Bhāskara and those of his descendents. King Saideva of the Nikumbha dynasty made an endowment for the maintenance of this *maṭh* in Śaka 1128 (1206 CE), and his brother, Hemādideva, made a grant as well; but only traces of the *maṭh* exist now. Bhāskara II's acumen in both astronomy and mathematics (see next chapter) was of such a high order that not only his lineage but also the royal patronage extended to the dissemination of his knowledge of astronomy has been testified by this inscription.

A native of Vijjaḍavīda or Bijjaḍabīḍa (perhaps modern Bijapur), Bhāskara II produced his renowned treatise, the *Siddhāntaśīromaṇi* at the age of thirty-six. Indeed a lucid exposition of the various dimensions of astronomy and mathematics of his time with his own inventions, this book has four parts: (i) *Līlāvatī* (arithmetic); (ii) *Bījagaṇita* (algebra); (iii) *Grahagaṇitam;* and (iv) *Golādhyāya* (both dealing with astronomy). Later, he wrote a smaller text on astronomy under the title *Karaṇakutūhalam*. In the *Grahagaṇitam* and *Golādhyāya,* Bhāskara has dealt with mean and true motions of planets; the three problems of direction, space, and time; eclipses; risings, setting and the conjunction of planets, and others, in a way similar to topics discussed in the other major

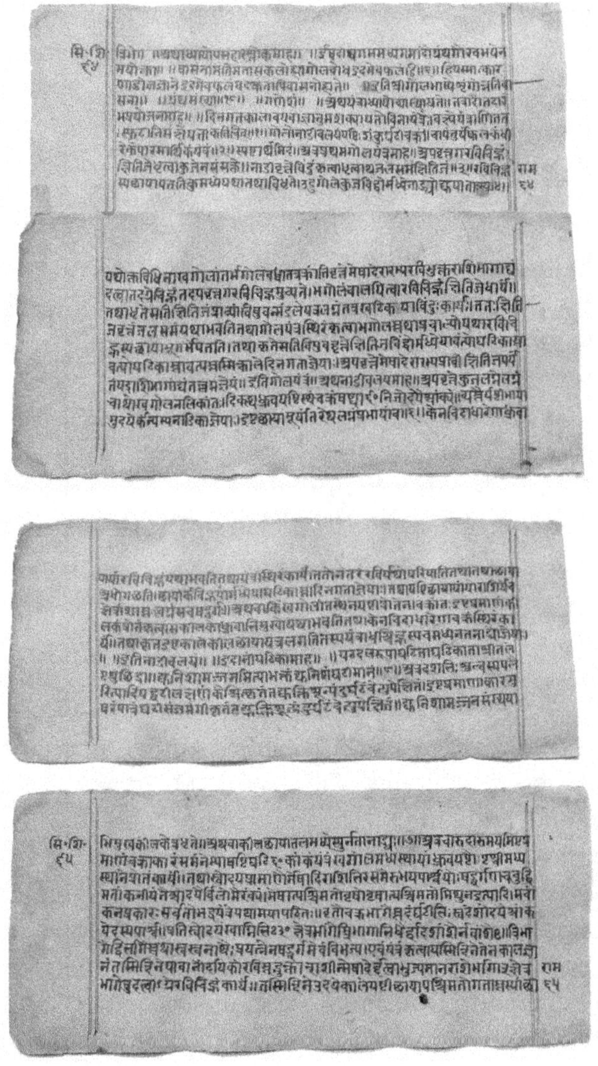

Śiddhāntaśiromaṇi
Facsimile of four pages of a manuscript in Oriental Research Institute, Mysore.
(Courtesy: Project of History of Indian Science Philosophy and Culture, Centre for Studies in Civilizations)

Siddhāntas, with appropriate mathematical astronomy. In the *Golādhyāya* particularly, the epicyclic-eccentric theories of planetary motion have been presented in detail. The astronomical instruments described in this text are more in number with their clear descriptions as well. The year Śaka 1104 is the epochal year for its *Karaṇakutūhala,* also called *Grahāgama kutūhala* or *Brahmasiddhāntatulya.* It has 139 verses under ten chapters. The popularity and the authority of Bhāskara II was such that, in the succeeding centuries, several commentaries on his works appeared, and these held aloft the tradition of Indian astronomy with new intellectual vigour.

Bhāskara II wrote his own commentary under the title *Vāsanābhāṣya*. All of his works are noted for clarity of thought, felicity of expression, and rationales of several formulae. The *Siddhāntaśīromaṇi* is an important text for students of Indian astronomy or *jyotiṣa,* which is being taught in almost all Sanskrit Vidyāpīṭhas (some of them are now deemed universities).

The Kerala School

There is a general view that Indian astronomy and mathematics had reached their zenith with Bhāskarācārya II in the twelfth century and, thereafter, it showed signs of decline, with the result that there were no notable achievements in these fields. Contrary to this view, both astronomy and mathematics witnessed new intellectual vigour, especially mathematics, in which there were attainments that antedated similar developments in Europe. Active schools of astronomy and mathematics flourished in both the Kerala and Mahārashtra regions. In Kerala, which had already witnessed noted astronomers like Haridatta and Devācārya (seventh century CE), there was an efflorescence of astronomers who were also able mathematicians: Mādhava and Parameśvara (fourteenth-fifteenth centuries); Damodara (fifteenth century); Nīlakaṇṭha Somayājī (fifteenth-sixteenth centuries); Jyeṣṭhadeva (sixteenth century); Śaṅkara Vāriyar (sixteenth century); Acyuta Piṣāraṭi (seventeenth century); Putumana Somāyājī (seventeenth-eighteen centuries); Kṛṣṇadāsa (eighteenth century); and Śaṅkaravarman (nineteenth century)—an unbroken preceptor-disciple lineage of astronomers-mathematicians for about five centuries. The Kerala School nurtured a tradition which, by and large, owed allegiance to the Āryabhaṭan system. It is mainly for this reason that the best and the most erudite commentaries on the *Āryabhaṭīya*, after that of Bhāskara I, have come out in this region. A significant aspect of the Kerala School was the ceaseless endeavours of its astronomers to arrive at the consonance of astronomical computations with observations by applying the needed corrections.

Mādhava

Mādhava of Saṅgamagrāma was the doyen among Kerala astronomers-mathematicians. He devised a versatile method for the determination of the position of the moon precisely to the second at intervals of two hours forty minutes each. He called this method *Veṇvāroha* metaphorically, in view of a parallelism between this method and the knots in a bamboo. This method was appreciated by the later Kerala astronomers like Putumana Somayājī, who wrote a text titled *Veṇvārohāṣṭaka*.

By this time, Vararuci had originated a formulation couched in the form of a number of mnemonic words, phrases or short sentences called *vākyas*. These, when deciphered in terms of the *Kaṭapayādi* system of numeral notations of decimal place-value, would yield the desired astronomical tables. Known as *Vākyakaraṇa*, this system is popular in South India, and is being used extensively in Tamil Nadu for the computation of *pañcāṅga*. Mādhava's *Veṇvāroha* and *sphuṭacandrāpti* improved the accuracy of Vararuci's *vākyas*.

Parameśvara

An erudite astronomer of Kerala, Parameśvara formulated what is called the *Dṛk* system, or calculations based on the observed data and involving multipliers and divisors with the associated trigonometrical factors.

Parameśvara was a native of Aśvatthagrāma (Ālattur being the local name) near the confluence of the Nīlā river with the Arabian Sea. He was a keen observer of celestial bodies and their motions. He has left behind a record of his observations of eclipses, both solar and lunar, over a period of fifty-five years in his well-known text, the *Sīddhāntadīpikā,* which is a scholarly commentary on the *Mahābhāskarīya* of Bhāskara I. Parameśvara's son, Dāmodara, was a learned teacher in Hindu astronomy. His pupils included Jyeṣṭhadeva, the author of the *Yuktibhāṣā,* and Nīlakaṇṭha Somayājī, the author of *Tantrasaṅgraha.* Both *Yuktibhāṣā* and *Tantrasaṅgraha* are texts that give a clear understanding of the planetary motions (both mean and true), parallax, eclipses, the moon's horns, and related ideas. They also vividly describe the rationale for different procedures adopted in Indian astronomical texts for accurate computations.

Nīlakaṇṭha Somayājī

A devoted astronomer of extraordinary capabilities, Nīlakaṇṭha had several reverential appellations, such as Somayājī, Somasut, Somasutvan, and their Malayalam derivative *comātri*, indicating that he was a performer of the *soma* sacrifice. He was a follower of Āśvalāyana-sūtra of the Ṛgvedic tradition and well versed in Sanskrit. He belonged to village Tṛk-kaṇṭi-yūr of the Ponnani taluk in South Malabar (now part of Kerala), and was a Śaivite. Nīlakaṇṭha's acumen in astronomy was recognized and encouraged by the religious head of the Nampūtiri Brāhmins of Kerala. But the teacher who inspired him was Dāmodara, son of the celebrated astronomer, Parameśvara, whom Nīlakaṇtha used to call his *Paramaguru* (the most exalted preceptor). He devoted his entire span of life (he is stated to have lived over a hundred years!) to astronomical pursuits and produced path-breaking treatises: *Golasāra*; *Siddhāntadarpaṇa* and *Vyākhyā*; *Āryabhaṭīya-bhāṣya*; *Candracchāyā gaṇita* and *Vyākhyā*; *Tantrasaṅgraha*; *Grahṇanirṇaya*; *Sundararājapraśnottara*; *Grahaparīkṣākrama* and *Jyotirmīmāṃsā*.

The *Golasāra*, a short text of fifty-six verses, has three sections and deals lucidly with several topics like epicycles of the planetary motions, diameters of the orbits of the sun and the moon, celestial globe and planetary motions, various circles, divisions of the ecliptic and others, all by and large being the computational ones. On the other hand, Nīlakaṇṭha's *Siddhāntadarpaṇa* and *Vyākhyā* present several astronomical constants verified through his own observations. The *Vyākhyā* is meticulously instructive with logical reasoning and demonstrative examples. Nīlakaṇṭha's exquisite exposition of the way in which an armillary sphere is to be constructed and used is indeed exemplary. The *Candracchāyāgaṇita*, as the name indicates, is mainly concerned with the ascertainment of the exact time on the basis of the shadow of the moon, or the computation of shadows with reference to time. In Kerala, this text was popular, and the Kerala astronomers were adepts in the computations based on shadow. In view of its importance, Nīlakaṇṭha himself wrote a commentary on his own work.

The most important text penned by Nīlakaṇṭha is the *Tantrasaṅgraha,* which, like the *Āryabhaṭīya*, has taken the beginning of *Kaliyuga* (17-18 February 3102 BCE) as the zero point for computation purposes. In its 432 verses under eight sections, this enchanting text also follows the tradition of presenting the usual topics like the mean and true longitudes of planets, lunar and solar eclipses, gnomonic shadow, elevation of the moon's horns, *vyatīpāta* or the situation when the sum of the longitudes of the sun and the moon is 180° and others. This text is a landmark in the history of Indian astronomy inasmuch as it is here that Nīlakaṇṭha, based on his computations, puts forth his planetary model, including the correct application of the equation of centre for the inner planets Mercury and Venus.

In his *Āryabhaṭīya-bhāṣya*, he has explained how, purely on the observation of latitudinal deflection, it is possible to conclude that Mercury and Venus also revolve round the sun, and the sun along with the interior and exterior planets, revolves round the earth.

Indian astronomers, till the time of Nīlakaṇṭha (1500 CE) adopted two types of formulae for determining the latitudes of planets. Bhāskara II who also followed this duality, however, noted that the computation of the latitudes for the interior planets by the usual procedure was inappropriate. It was felt that the two different methods—one for the interior, and the other for the exterior planets—needed a unified formulation in the context of the motions of planets as a whole. Nīlakaṇṭha addressed himself to this need, and endeavoured to arrive at a unified formulation for the computation of latitudes of all planets. In his commentary on the *Āryabhaṭīya,* he demonstrated how the latitude of a planet should be a measure of the deflection of the planet from the ecliptic, and not from any other entity. He proposed that what was considered as *śīghrocca* (higher apsis, or the uppermost point of the epicycle employed in what is called *śīghra* correction, or the apex of the planet moving faster) in respect of an interior planet in the traditional way, should be identified with the planet itself. He also argued that the mean sun needed to be considered as the *śīghrocca* of all planets, and those of Mercury and Venus are really the mean planets themselves. Though Nīlakaṇṭha's planetary model was arrived at by computational methods, it represented a marked departure from the traditionally accepted total geocentric view over a long time. In the history of astronomy Nīlakaṇṭha has a distinct position; for, it was he who for the first time presented a model of this type in the geocentric framework. It was about ninety years later that such a picture was put forward by Danish astronomer Tycho Brahe (1546-1601). Tycho's model was a result of his protracted observations over a considerable time, and an alternative to the entirely heliocentric model expounded by the Polish astronomer, Nicolas Copernicus (1473-1527), who had argued that the heliocentric model would offer better explanations of the revolutions of the celestial bodies and make the associated calculations simpler. Even so, like the model of the Greek philosopher Aristotle, Copernicus believed that the universe was spherical and finite, and assumed that the movements of celestial bodies were always circular with uniform velocities. He employed several circles including the epicyclic-eccentric methods, to explain mathematically the movements of celestial bodies. Tycho Brahe's model, though similar to that of Nīlakaṇṭha, expounded the theory that the sun would revolve round the earth once in twenty-four hours, carrying with it all the planets, of which Mercury and Venus have smaller orbits while the other three planets have larger orbits.

Tycho Brahe's scheme of planetary motion was the last effort in the history of astronomy in the West to represent geocentric as well as circular motions of planets (solar system). The Copernican heliocentric hypothesis began to gain ground with some modification largely through the efforts of Galileo (1564-1642) and Johannes Kepler (1571-1630), the German Protestant who demolished the circular motion as well as uniform velocities of planets. He demonstrated through his planetary laws, that the planets move round the sun not in circles, but in ellipses, with the sun being one of the foci, and that a planet does not move uniformly but in such a way that a line drawn from it to the sun sweeps equal areas of ellipses in equal times. He also mathematically showed that the squares of the period of revolution of planets round the sun are proportional to the cubes of their distance. Such a revolutionary approach did not surface in India even after Nīlakaṇṭha's unified formulation of planetary movements. The result was that Indian astronomy continued to be geocentric and confined to traditional methods of computations.

An ardent follower of Āryabhaṭa I, Nīlakaṇṭha produced an erudite commentary on the *Āryabhaṭīya,* and even called it *Mahābhāṣya* (a great commentary). In his commentary, Nīlakaṇṭha left out the *Gītikāpāda* of the *Āryabhaṭīya* as such, but its ideas were to some extent covered in the commentary on the other three *pādas*. Another work of Nīlakaṇṭha was the *Grahaṇanirṇaya*, which deals with both lunar and solar eclipses. The manuscript of this text is not available, but some other succeeding astronomers have made references to it. Likewise, the manuscript of *Sundararājapraśnottara*, which is stated to be in the nature of detailed answers given to the questions of Sundararāja, an astronomer from Tamil Nadu, about the almanac-computations, is not traceable. Nīlakaṇṭha, like his *Paramaguru,* Parameśvara, laid emphasis on observation and verification of computations that would yield accurate results. His *Jyotirmīmāṃsā*, probably the only one of its type in Indian astronomical literature, expounds admirably the importance of observation and verification. He has explained in this text that 'divine instruction' signifies only the sharpening of intellect through divine grace towards observation and inference, which are essential for obtaining accurate results. He also stressed the need for applying corrections from time to time in respect of planetary computations. Besides energizing the Kerala School of Astronomy, Nīlakaṇṭha strengthened the tradition of Indian astronomy in such a way that his name stands out as an exemplary one.

Among the other notable Kerala astronomers were Jyeṣṭhadeva, also a disciple of Damodara and the author of *Yuktibhāṣā* and *Dṛkkaraṇa*; Śaṅkara Vāriyar who produced two commentaries, the *Laghu vivṛiti* and *Yuktidīpikā* on the *Taṅtrasaṅgraha* of Nīlakaṇṭha; Acyuta Piṣāraṭi, who produced an important *karaṇa* work titled *Karaṇasāra* and *sphuṭanirṇayatantra*; Putumana Somayājī who was well known for his work *Karaṇapaddhati* and *Veṇvārohāṣṭaka;* Azvānecri Tamprakkal, who produced two works, *Jyotiḥśāstrasaṅgraha* and *Saṅgrahasādhanakrīya*; and Śaṅkaravarman, who authored an elegant treatise, *Sadratnamālā*.

The Maharashtra School

Apart from the emergence of the Kerala School of Astronomy, there was similar development in the region of what is now known as Maharashtra. A sixteenth-century astronomer, Jñānarāja, about whose life not much is known, wrote a comprehensive treatise titled *Siddhāntasundara* consisting of eighteen chapters. This work deals with the usual astronomical topics, but is known for its methodical presentation. The most popular astronomer in the region was Gaṇeśa Daivajña, whose father, Keśava Daivajña, was also an accomplished astronomer. The latter's text *Grahakautuka* laid emphasis on observational astronomy and the associated methods of calculation. Gaṇeśa (b. 1507) was a prolific writer, and his major work is a *karaṇagrantha,* with the title *Grahalāghavam.* His other texts comprise the *Bṛhat-* and *Laghu-tithicintāmaṇi; Buddhivilāsinī* (which is a commentary on the *Līlāvatī* of Bhāskarācārya II); *Grahalāghava-vivṛti*; *Pātasāriṇi* (astronomical tables); *Pratodayayantra* (description of an astronomical instrument); *Vivāhavṛndāvanaṭīkā* and *Muhūrtatattvaṭīkā* (auspicious time-determinations for marriage, etc.); and *Śrāddhanirṇaya* (time for performing death anniversaries). As a *karaṇa,* the *Grahalāghava* has taken the Śaka epoch, which corresponds to 19 March 1520. A significant aspect of the *Grahalāghava* is its attempt to make calculations easier. Besides, Ganeśa has dispensed with trigonometric functions and thus helped the almanac-makers who are not well-versed in the trigonometrical methods found in the major Siddhāntic texts. This approach of Gaṇeśa Daivajña is found especially in his *Laghucintāmaṇi,* which is also known as *Pañcāṅgasādhana* and is very popular among *pañcāṅga*-makers in several parts of India, even beyond Maharashtra. In this direction, a commentary by Kamalākara on Gaṇeśa Daivajña's *Grahalāghavam* also enhanced the latter's popularity. Several succeeding astronomers like Mallāri, Gangādhara, Nārāyaṇa Daivajña, Viśvanātha and, lately,

Sudhākara Dvivedi, shed fresh light on the *Grahalāghavam* and even elaborated upon it, further enhancing its popularity.

The Maharashtra School was dominated by a family of astronomers led by Divākara, a native of Golagrāma, which was on the northern bank of the river Godāvarī. He was a reputed disciple of Gaṇeśa Daivajña. Divākara's sons, grandsons and great-grandsons carried on the tradition of astronomy for four generations as an exemplary lineage. This devoted tradition breathed fresh life into astronomical endeavours more or less on par with the Kerala School. Of the five sons of Divākara, Viṣṇu Daivajña was the author of a *karaṇa* titled *Saura-pakṣa-gaṇita*, also called *Sūryapakṣasaraṇakaraṇa* (early seventeenth century), based obviously on the *Sūryasiddhānta*; Mallāri, another son, produced two works on eclipses, *Grahaṇa* and *Grahasāriṇī*, besides his commentary on the *Grahalāghava*. Divākara's grandson, Nṛsiṃha, received extensive training in astronomy from his uncles, Viṣṇu and Mallāri, and moved to Benaras where he wrote two commentaries on the *Sūryasiddhānta* and the *Siddhāntaśiromaṇi* of Bhāskara II. Kamalākara, the great-grandson of Divākara, was a well-known scholar in astronomy who produced several works, the most important of which was the *Siddhāntatattvaviveka.* This is a voluminous work (thirteen chapters and about 1,500 verses) written by him in 1658. It deals exhaustively with the usual topics of a Siddhāntic work, but has in it a chapter called *Bimbādhikāra*, which is not found in the other texts. It also qualitatively describes such topics as optics, obscuration of the sun by Mercury and Venus (which in a way implies an understanding of the heliocentric motion of these planets), and variations of the visible area of a celestial object with distance.

The *Siddhāntatattvaviveka* has chosen a place called Khāladātta as a reference point on the earth from which the longitudes are specified, in addition to the standard reference point of Laṅkā. Its *Bimbādhikāra* has also stressed the need for studying the shape and size of the celestial objects, the sources of their light, their characteristics, etc. However, in this text, Kamalākara has rejected the theory of Āryabhaṭa I regarding the rotation of the earth about its own axis. But refreshingly, he has dealt with the causes of rainbows, earthquakes, circular motions of planets, apparent motion of the celestial sphere and the like, explaining some of them with the aid of diagrams. His explanations, though sometimes not satisfactory, were a marked departure from the traditional ones to some extent. Kamalākara had an open mind and he did not hesitate to criticize some of the ideas of even the great Bhāskarācārya II on the one hand and, on the other, to praise Islamic astronomy that had taken roots in India during his time.

Kamalakara's youngest brother, Ranganatha, wrote a commentary on the *Sūryasiddhānta* with the title *Siddhāntacūḍāmaṇi,* besides another work called *Laghubhaṅgīvibhaṅgi* in which he refuted the methods adopted by another contemporary astronomer, Munīśvara, for finding out true planetary positions. Munīśvara was a capable astronomer and produced an elegant work called the *Siddhāntasārvabhauma* in eight chapters, supplementing them with his own commentary. His extensive commentary on the *Siddhāntaśiromaṇi* of Bhāskara II under the title *Marīcī,* elucidated the importance of the former. But his attempts to include in it here and there his own methods for the computations of true positions of planets were inaccurate and criticized by other astronomers. Both the Kerala and the Maharashtra Schools of astronomy were the torchbearers of the Indian astronomical tradition for nearly six centuries after Bhāskara II, the former owing allegiance by and large to the Āryabhaṭan system, while the latter promoted the *Sūryasiddhānta* principles.

Sawai Jai Singh II (1686-1734)

Towards the latter half of the seventeenth century appeared a prince who soon became an astronomer-Mahārājā. He was Sawai Jai Singh II of Jaipur. He was not a follower of any particular school of Hindu astronomy. There is no authentic information either about which Hindu astronomical texts the Mahārāja had procured and studied. But he was ably instructed in Hindu astronomy by his court astronomer-preceptor, Jagannātha Paṇḍita. There is, however, enough evidence to indicate that Jai Singh II was acquainted with *Almagest* (in the Arabic version), an astronomical treatise of Ptolemy (second century CE) of Alexandria. It was this text that was a landmark in the history of Graeco-Roman geocentric astronomy, which was dominant in Europe till the advent of the Copernican heliocentric theory in the sixteenth century. Jai Singh had also studied several other works on astronomy: the astronomical tables of Ulugh Beg of Samarqand; the works of Jamashid al-Kāshī, (an associate of Ulugh Beg) and of al-Sūfi, another Islamic astronomer; the tables of astronomy (*Tabulae Astronomicae*) of La Hire; *Historia Coelestis Britannica* of the British Astronomer-Royal John Flamstead, a contemporary of the famous Isaac Newton; Euclid's *Elements* (geometrical) and also the then emerging logarithmic tables. More importantly, he had an intimate knowledge of the work of Nasīr-ud-Din-al-Ṭūsi, one of the renowned Islamic astronomers who lived in the thirteenth century and played a pivotal role in the emergence of what later came to be known as the Marāgha School of Islamic astronomy.

Jai Singh's spectrum of astronomical knowledge was a broad one. Deeply interested as he was in astronomy, he was constantly pondering over the manner in which astronomical observations could be pursued with great accuracy. He knew the primal importance of the day-to-day observations of the motion of the sun, the moon and the planets for precise calculations. He was well aware of the prolonged observations of the planets made by Islamic men of astronomy at Samarqand and other places, and the tables or Zījes that were scrupulously prepared by them.

Jai Singh was also interested in the methods of European

astronomers and the expertise of Jesuit missionaries who were in India then. He invited some of them, such as Father Figuerado, a Portuguese Jesuit, and even sent him to Europe to bring the records and books on European astronomy to India. In addition, he invited Father Boudier and his associates who were in the French territory, Chandernagore near Calcutta, to come to Jaipur. Don Pedrode Sylva, who was both a physician and astronomer, also worked in the court of Jai Singh for some time. Also in the Mahārāja's court was the ablest Hindu astronomer of that time and his preceptor, Jagannātha Paṇḍita, who was well-versed not only in Sanskrit and Sanskritic texts on Hindu astronomy, but also in Arabic and Persian (the court language of the Mughals). On the suggestion of the astronomer-Mahārāja, he translated Ptolemy's *Almagest* (Arabic) and Euclid's *Elements* (Arabic) into Sanskrit under the titles *Samrāt-siddhānta* and *Rekhāgaṇita*, respectively. Above all, the *Paṇḍita* had realized the paramount importance of observational astronomy and the recording of observations in a methodical way. In fact, the Mahārāja was particularly interested in observational astronomy.

Jai Singh's name has been immortalized in the huge, impressive observatories erected by him in Benaras, Mathura, Ujjain, Delhi, and Jaipur. The source of inspiration for such a gigantic endeavour lay, to some extent, in the Marāgha School of observatories in Central Asia in general and Samarqand in particular where Ulugh Beg and Jamashid al-Kāshi were active. He was well aware of the inaccuracies of the observations recorded by Islamic astronomers like Said Gurgāni and Khāgani who lived in Persia (modern Iran), because they had used small instruments mostly made of brass with the possible inaccuracy of the divisions of degrees into minutes. However, it was his own disdain for inaccurate observations, made tardy and unmethodical by small instruments, that prompted him to plan systematic observations with the help of several huge masonry instruments. For, he was convinced that such large instruments would lead to an accurate compilation of the observed data of celestial motions.

Maharaja Sawai Jai Singh's Observatory, Jaipur: A view

W. Hunter, a British administrator who was conversant with the endeavours of Jai Singh, has recorded Jai Singh's work as follows:

> 'Therefore he (Jai Singh) constructed in Dar al-Khalafak Shah Jahānābad (Delhi) which is the seat of empire and prosperity, certain instruments of his own invention such as *Jai Prakash*, *Rāṃyantra* and *Samrāṭ-yantra*, the semi diameter of which is of eighteen cubits, and one minute of it is equivalent to barley corn and half. They were of stone and lime in perfect stability with attention to rules of geometry and adjustment to the meridian as well as to the latitude of the place, and with care in measuring and fixing them, so that the inaccuracies from the shaking of the circles and the wearing of their

> axes and displacement of their centers, as well as the inequality of the minutes, might be corrected. Thus an accurate method of constructing an observatory was established, and the difference which had existed between the computed and the observed positions of the fixed stars as well as planets by means of observing their mean motions and observations, was removed.' (*Asiatick Researches*, vol. V, pp. 177ff)

The instruments, which are preserved in Jai Singh's observatories, especially in Jaipur, Delhi and Ujjaini, are enchanting. They are a constant reminder of the dedication and zeal of the astronomer-king for the accuracy of observations. The astronomical work carried out under the direction of Jai Singh and ably supervised and recorded by Jagannātha Paṇḍita have been compiled in the form of tables, named after the then Mughal Emperor, Muḥammad Shah, of whom Jai Singh was a vassal. Titled *Zīj Muḥammad Shāhī*, the tables are available both in Persian (complete) and Sanskrit (incomplete). The Persian manuscript is now preserved in the British Museum, while the Devanāgarī one, which begins with a homage to Ganeśa, is in Jaipur. The *Zīj* has also a catalogue of forty-eight constellations, besides other computations. It is rather intriguing that, despite his knowledge of European astronomy, which had by then accepted the heliocentric theory of Copernicus, Jai Singh continued to subscribe to the traditionally fostered geocentric astronomy. He knew the use of telescope in observational astronomy and he has recorded his observations through a telescope, of the moons of Jupiter, the surface of Mars, etc. Yet he did not appear to have shown any inclination towards giving due importance to telescopic observations. On the other hand, he relied on his huge masonry and other instruments in his observatory. Jai Singh breathed his last in 1734, after which the adventurous momentum he gave to observational astronomy in India in the eighteenth century began to recede.

Sāmanta Chandrasekhara

In India, the nineteenth century witnessed the advent and growth of Western science, including modern astronomy as well as liberal education with English as medium of instruction. Some European savants who were employees of the colonial government evinced a keen interest in the study of Hindu astronomical texts. Amid this influx and influence lived a versatile Hindu astronomer named Sāmanta Chandrasekhara (b. 1835), of the royal family of Odisha. Even as a boy of four, he was a keen observer of the sky, and able to recognize the bright planet of Venus. By the time he was fifteen years old, he had studied the *Sūryasiddhānta* and the *Siddhāntaśiromaṇi* of Bhāskara II, both of which provided him with a critical understanding of Siddhāntic astronomy. He devised his own instruments—one of them was known as *Mānayantra*—for an accurate observation and measure of time in relation to the motion of planets. Sāmanta noticed that the positions of planets as computed in accordance with the Siddhāntic texts and the actually observed ones did not tally. He was convinced that traditional ideas about planetary motions needed a different approach and should be based on observations that he pursued himself by devising his own instruments for nearly two decades. In 1869, when he was thirty-four, Sāmanta wrote a comprehensive, treatise (2,500 verses) on palm leaves in Oriya script, titled *Siddhāntadarpaṇa*. But this important work of a young astronomer had to wait for thirty years before it was published (1899) in the Devanāgari script in Calcutta with the financial support of the kings of Mayurbhanj and Athmalik. Jogesh Chandra Ray of Cuttack College in Odisha (presently known as Ravenshaw College) and Mahesh Chandra Nyāya-ratna, principal of Sanskrit College, Calcutta, recognized the merit of Sāmanta's work, and the former's efforts led to the publication of the *Siddhāntadarpaṇa*.

Sāmanta belonged to a royal family, yet his life was not a

bed of roses, as he was plagued by inadequate finances and ill health, with chronic dyspepsia and recurring stomach aches. Even so, he had an insatiable thirst for astronomy, ascribing his passion to the blessings of Lord Jagannātha of Puri—and, according to folklore, Sāmanta breathed his last while devotedly bowing before Lord Jagannātha. He did not have any formal education in modern astronomy, but it is not improbable that he may have acquired some knowledge of the heliocentric motions of the planets either by reading the available books on this subject, or through interaction with those well versed in modern astronomy.

Conjectures of this type apart, Sāmanta's expertise in astronomy is marked by a certain originality of its own. In his *Siddhāntadarpaṇa*, he states clearly at the end of some chapters that his treatise is an educative textbook, and a tally of calculations and observations (*gaṇita-akṣi-siddhi*). The text comprises twenty-four chapters under five sections: (i) mean motions of planets; (ii) their true places; (iii) the three problems of direction, space and time; (iv) spherics; and (v) time-calculations. Of the 2,500 verses, 334 deal with the description of the earth, 111 with instruments, and the last but one chapter, a non-astronomical one, is in the form of prayer to God. Sāmanta has raised several important questions in chapter sixteen of his treatise regarding conventional ideas of the shape of the earth, the velocity of planets, dimensions of epicycles and the like. He sought to introduce some changes in the very process of calculations according to the traditional methods. Such an approach in Sāmanta's treatise, which involved questions, is somewhat unique. Of the sidereal periods relating to the planets, the value he arrived at even by naked-eye observations of Saturn is almost accurate.

Sāmanta was the first Indian astronomer to have recognized all the three irregularities of the moon's motion: evection, variation, and annual equation. With the application of certain corrections as determined by him, he was able to calculate the occurrence of the lunar and solar eclipses accurately. He also attempted to improve upon the methods of calculating with respect to eclipses by increasing the ratio of the distance of the sun to that of the moon, since he noticed that the earlier astronomers had thought of it as being far less. Sāmanta's most important contribution was the depiction of a planetary model similar to that of Nīlakaṇṭha, perhaps developed by him independently. At the time when Sāmanta's *Siddhāntadarpaṇa* was published (1899), and even earlier when he was engaged in its preparation, the Copernican heliocentric system was known in India. The Madras Observatory, in which European astronomers who were conversant with his system worked, had been established in 1794. Sāmanta too, like Jai Singh II, did not evince interest in heliocentric astronomy; he still preferred the geocentric one. Charles Whish, an East India Company employee, had already made known, among others, the works of Nīlakaṇṭha who, as noted before, had developed practically the same model—that is, the heliocentric five planets and the sun, along with them, going round the Earth. There is a view that Sāmanta had no knowledge about it, but this view needs re-examination. In any case, he was hailed as greater than Tycho Brahe by the prestigious journal *Nature,* when a report on Samanta's treatise was published by Jogesh Chandra Ray, which was a tribute to a devoted Indian astronomer at an international level.

Occurrence of Eclipses

The authors of the Siddhāntic texts have scientifically dealt with the occurrence of eclipses, their duration, parallax involved in their observation, and other aspects including the possibility of an eclipse, by computational method. They also appeared to have applied corrections needed for the prediction of the exact time of an occurrence of an eclipse. Twelfth-century inscriptional evidence reveals that in the past, traditional astronomers were capable of predicting the exact day and time of the occurrence of an eclipse. For example, the inscription dated 880 of the Kālachuri era (1128 CE) records as follows:

> 'Padmanābha, son of Kulacandra (of Śrīvatsa gotra and the *pravara*) belonging to Madhydeśa, was proficient in astronomy and knew two *siddhāntas* (probably the *Sūryasiddhānta* and the *Brāhma-sphuṭa Siddhānta*). In the presence of all astronomers in the assembly of King Ratnadeva, he (Padmanābha) asserted that there would be a total lunar eclipse when three quarters of the night had passed, and the Moon would be in the *nakṣatra Rohiṇi*, on Thursday, the *Pūrṇimā* of *Kārtika* in the year 880 (*Kalachuri*). When the eclipse actually occurred at the predicted time, the King became pleased and donated the village *Cincātalai* to Padmanābha.'

But the present *pañcāṅga*-makers do not bother to arrive at the exact time of the occurrence of an eclipse and its duration by adopting the Siddhāntic method of calculations with the appropriate corrections. Instead, they reproduce the eclipse-related data of the Indian ephemeris in the *pañcāṅga*. They prescribe, however, traditional methods of propitiation and subscribe to the Rāhu-Ketu postulate concerning an eclipse, which the common people continue to believe in, viewing such occurrences, especially of a total solar eclipse, with horror—such has been the inhibiting impact of an eclipse on the public from time immemorial. But the Siddhānta texts do not subscribe to the superstitious beliefs associated with an eclipse.

ISLAMIC ASTRONOMY AND INDIA

Like Unāni medicine (Graeco-Arabic), which entered India through the Muslim monarchs during the Delhi Sultanate (twelfth-fifteenth centuries) and flourished in the Mughal period (sixteenth-eighteenth centuries), Islamic astronomy, too,

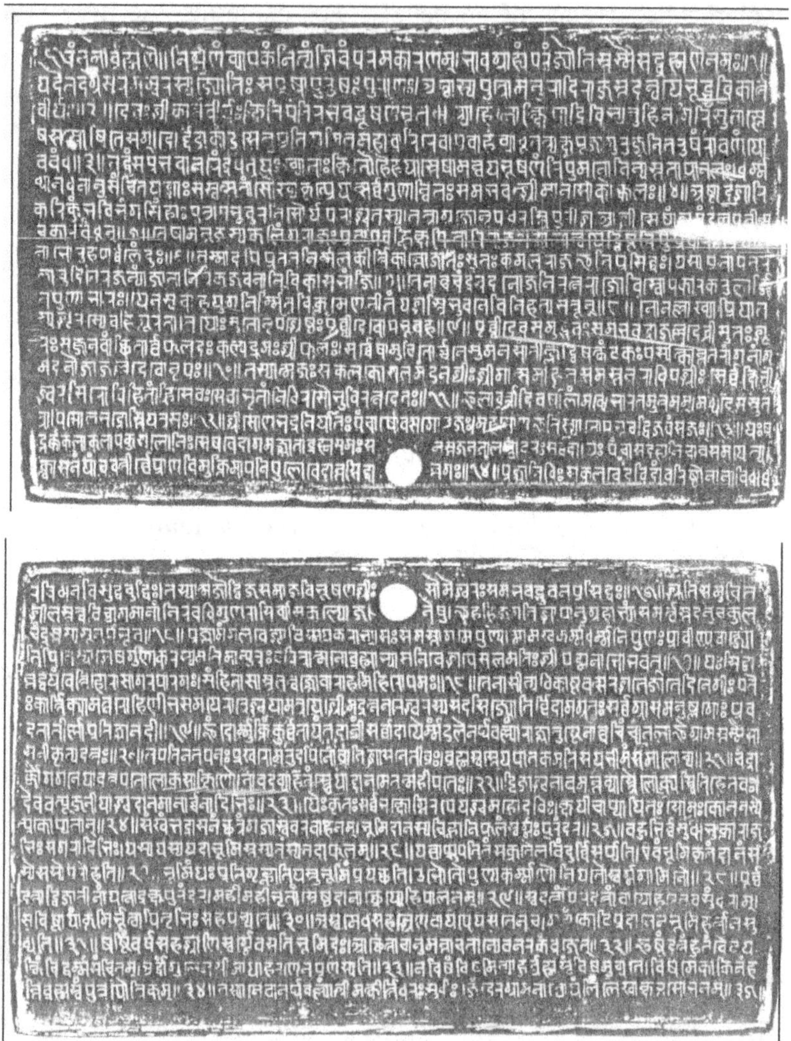

Inscription dated 880 of the Kalachuri era depicting the donation made by the king to the astronomer Padmanābha, for his predicting exactly the occurrence of a total lunar eclipse on the *Pūrṇimā* of Kārtika, 880 Kalachuri era (November 8, 1128 CE). (Courtesy: Project of History of Indian Science Philosophy and Culture, Centre for Studies in Civilizations)

was pursued by Muslim religious savants who were engaged in the furtherance of Islamic religious observances. Islam had emphasized the acquisition of knowledge, and Islamic education, including astronomy, was promoted in an integrated manner, especially in the *madrasas* or institutions of higher learning, right from the advent of Islam in West Asia. In the latter half of the eighth century, during the reign of the second Abbasid Caliph al-Manṣūr (753-774), astronomy received substantial attention. It was during this period that Ptolemy's *Megale Syntaxis* (a seminal work on Greek astronomy), was rendered into Arabic under the title *Almagest*. Alongside, the *Brāhmasphuṭa Siddhānta* and the *Khaṇḍakhādyaka* of Brahmagupta were also translated into Arabic. Later, under the reign of al-Māmūn (the descendant of al-Maṇṣūr), two observatories were built at Baghdad and Damascus. The next four centuries (ninth to the twelfth) witnessed impressive endeavours in the study of astronomy by several Islamic astronomers.

Islamic astronomy was somewhat a modification of Greek astronomy, expressed in the Arabic language. It regarded the universe as consisting of several concentric spheres. The outermost sphere was assumed to rotate on its axis once in twenty-four hours, causing the cyclic succession of day and night. Islamic astronomy absorbed to a great extent the Aristotelian model of the universe along with the later Alexandrian or Hellenistic astronomy developed by Hipparachus and Ptolemy.

Despite the devastation caused by the Tartars in West and Central Asia in the middle of the thirteenth century, Islamic astronomy did not take a back seat. On the other hand, it had found fresh impetus in the well-known Marāgha observatory built by an outstanding astronomer, Naṣīruddīn-al-Ṭūsī. This versatile astronomer wrote *Tadhkirah* and *Zīj-i-Ilkhānī,* or astronomical tables, which paved the way for the later emergence of several *Zījes*. These were in the nature of an index of the continued efforts of Islamic astronomers, especially in Central Asia. Ṭūsī also produced an encyclopaedia titled *Durrat-ut-tāj*. Another Islamic astronomer, Chāgminī, wrote an important text titled *Al-Mulakhkhaṣ fi'l-Hai'ah*. The monarch Ulugh Beg (fifteenth century) was also a famous astronomer-mathematician, who established an academy in Samarqand which was devoted to the pursuit of science of the times. There were four renowned members of this academy who made important contributions to astronomy: Qāzī Zādeh Rūmī (the commentator of Chagminī's *Mulakhkhas fi'l-Hai'at*); Ghaiyāth ud-dīn Jamshed Kāshī (the author of *Zīj-i Khāqāni*); Mu'īnuddin Kāshī and Alāuddīn Qushjī (the *de-facto* author of *Zīj-i Ulugh Beg*). Ulugh Beg was personally involved in the erection of an observatory at Samarqand (1420 CE). The extensive observations made at this observatory were compiled by Ulugh Beg with his associates, and these came to be known as *Zīj-i Ulugh Beg*, a landmark in Islamic astronomy.

After the death of Ulugh Beg and his associates, the astronomical activities in Samarqand began to wane. India was slowly emerging as an important region for the practice of Islamic astronomy, and for the preparation of a number of *Zījes*. According to Mullā Farīd, the court astronomer of Emperor Shāhjahān, the Arabic word *raṣad* means an observatory in which observations of different celestial bodies are made with the help of specifically fabricated instruments. Such observations comprised determination of the positions of the luminaries in the sky, measurement of their movements, their distances from one another, distance from the earth, their sizes and the like. The observed data are carefully entered in a register, and that register is called a *Zīj*. Mullā Farīd categorized the *Zījes* into (i) *Zīj-i Riṣadī,* or tables recording the data observed through instruments, and (ii) *Zīj-i Ḥisābī,* or computational tables, which were supposed to be improvements over the one compiled already, but through calculations. A majority of tables in Arabic and Persian come under *Zīj-i-Ḥisābī*. It may be noted that the first ever *Zīj* among Islamic astronomers was the one prepared by Al-Fazāri, the court astronomer of the Caliph al-Manṣūr

(753-774). It was this astronomer who, with the help of Hindu *paṇḍits* in Baghdad, rendered into Arabic the *Brāhmasphuṭa Siddhānta* of Brahmagupta. His *Zīj* was based upon the Indian *Siddhānta,* though the years employed for its computation were the Islamic Hijiri ones. Al-Fazārī's colleague, Y'aqūb ibn Ṭārīq, who also produced an Arabic version of Brahmagupta's *Khaṇḍakhādyaka,* prepared another *Zīj* titled *As-zīj-al-Maḥlul min-as Sindhind,* based on the siddhāntic texts. The scholastic environment at Baghdad was so catholic that the binding element among the astronomers was the spirit of astronomy, not the religion which they professed.

The Islamic educational *(madrasa)* system was patronized with financial support by Sultans and Mughal Emperors as well as by local chieftains and aristocratic nobles. There were about a thousand *madrasas* or similar institutions for specialized learning during the reign of Muḥammad Shah Tughluq (1325-51). His successor, Firoz Shah Tughluq, established about thirty new *madrasas.* During the Mughal rule, a large number of them were still in existence with the patronage also of private residencies (*dargāh*). It was in these *madrasas* that the Islamic rational sciences—astronomy, mathematics, as well as medicine and philosophy, including ideas on the physical world and logic—were being taught. Several Islamic scholars in astronomy even came to India from Central Asia, particularly from Persia. The classical texts of astronomy were those by Chaghminī and Ṭūsī as well as the commentaries on them, besides Ṭūsī's recension in Arabic of *Almagest* (Ptolemaic astronomy).

Islamic astronomy and Sanskritic Hindu astronomy began to coexist without any confrontation. Refreshingly, there were transmissions too through translations or adaptations of some Arabic and Persian astronomy texts into Sanskrit carried out by both Muslim and Hindu scholars. In the court of Firoz Shah Tughluq was a Hindu astronomer, Mahendra Sūri, who produced a work in Sanskrit on the astrolabe (1370), a versatile astronomical instrument which the Islamic astronomers had brought and used in India. Published copies of this text with the commentary of another astronomer, Malayendu Sūri, are still available. The astronomical tables *Zīj-i Ulugh Beg* were translated into Sanskrit by a team of scholars in the court of Akbar (sixteenth century). An eighteenth-century manuscript of this work has been preserved in the city Palace Museum Library at Jaipur. *Zīj-i Shāhajāhāni*, which was compiled by Farīuddīn Munajjim, an astronomer in the court of Shāhjahān, was translated into Sanskrit by another court-astronomer, Nityānanda (1639). A manuscript of this translation, titled *Siddhāntasindhu* (scribe: Gaṅgārāma) is available in the City Palace Museum Library, Jaipur. It would appear that Nityānanda's smaller text, *Siddhāntarāja*, is based on the *Romakasiddhānta,* but its planetary model and cosmology seem to be Islamic. An astronomer, Nayanasukhopādhyāya, translated another work on astrolabe from Persian into Sanskrit. As noted before, Jagannātha Paṇḍita, the preceptor of Jai Singh II, translated the Arabic version (Ptolemy's) of *Almagest* into Sanskrit, titled *Samrāt-siddhānta*. There were a few other Arabic and Persian sources on astronomy which were either translated into Sanskrit or their contents were used by Hindu astronomers. In all of these endeavours, the astronomers transcended their religious labels.

The first observatory attempted in India by Islamic rulers was the one at Bālāghāt in 1407 by the order of Sultān Firūz Shāh Bahmanī, even before Ulugh Beg had erected his observatory in Samarqand in 1420. But it could not be completed, as its architect, Ḥasan Jīlānī, soon passed away; it is in ruins now. The Mughal Emperors, Akbar and Humāyūn, encouraged the construction of observatories, but with little success. But it was during the Mughal rule of Muḥammad Shah (1719-48), that Sawai Jai Singh II took energetic steps, as noted earlier, for the construction of observatories in Benaras, Mathura, Ujjain, Delhi and Jaipur. These are impressive observatories with colossal masonry as well as metallic instruments like astrolabes. There was a family of astrolabe-makers in Lahore (among them was Lāhorī Mullā Allāh-dād and Zia-ud-din) who also produced globes. There are a large number of astrolabes preserved in museums all over the world, which were produced by this Lāhorī family. The astrolabe was perhaps the most popular astronomical instrument in medieval India and elsewhere. It was called *yantrarāja* (king of instruments) by Indian astronomers. The doyen of the Marāgha School, Naṣīr al-Dīn Ṭūsī, had given such an importance to this instrument that he wrote a tract in Arabic: *Twenty Chapters on Astrolabe.* Nearly sixty manuscripts of this work are still available in the libraries of the Indian subcontinent. Interestingly, Sultan Firoz Shah Tughluq is stated to have been highly skilled in astrolabe-making.

Zījes

The first *Zīj* prepared in India, called *Zīj-i Nāṣirī*, was dedicated to Nāṣīr al-Din Iltutmish, the Sultan of Delhi, (dated around 1260). However, it was the Mughal period that witnessed the emergence of important Zījes. During Akbar's reign, Mullā Chānd produced a simplified version of Ulugh beg's tables under the title: *Tashīl Zīj-i Ulugh Beg*. It deals with different calendars as well as eras and the determination of the positions of stars and planets. In the reign of Shāhajahān was a noted Islamic astronomer, Mullā Farīd who wrote several works, of which the *Śirājul-istikhrāj* dealing with astronomical ideas, and *Zīj-i Shāhajahāni*, (astronomical tables) are well known. Mullā Farīd collaborated with his brother Mullā Ṭayyib and some Hindu *paṇḍits* in astronomy to compile the *Zīj-i Shāhajahāni*. Though based on the *Zīj-i Ulugh Beg*, the *Zīj-i-Shahajahānī* is a refined one with accurate computations of planetary positions and it was preferred even to the tables of Ulugh Beg. Shāhajān, impressed as he was with the *Zīj* of Mullā Farīd, ordered that it be translated into Sanskrit. Divided into four *maqālas* or informative sections, *Zīj-i Shāhajahānī* is an important source for understanding the nature and extent of astronomical ideas prevalent during the time of Shāhajahān. Different calendars were also discussed: the *Ilāhi* calendar (introduced by Akbar); *Hijri* calendar; Greek, Persian, and Malikī calendars; transformation or equivalence of time-elements from one calendar to another; Indian or Hindu Śaka *pañcāṅga*, and its transformation into Hijirī and vice versa; Chinese and Tibetan (Uighur) almanacs; and important days of festivals and observances as determined by all these calendars.

The *Zīj-i Muḥammad Shāhī* of Jai Singh II, is divided into three *maqālas*. The first one, in its four chapters, deals with calendars, relating to the Hijirī, the Hindu *Saṃvat,* Christian ones and, in honour of the Emperor Muḥammad Shah, the era bearing his name. The second *maqāla,* with nineteen chapters, expounds in considerable detail the trigonometrical aspects, astronomical ideas of the ecliptic and celestial equator, determination of ascendants, azimuth, longitudes and latitudes of important places, except the direction of *Qibla* (Mecca). The third *maqāla* is devoted to planetary motions and positions (longitudes and latitudes), eclipses, rising and setting of the 'lunar mansions' or star-groups of the lunar zodiac. *Zīj-i Muḥammad Shāhi* contains 147 tables (according to the manuscript available in the Aligarh Muslim University Library): first *maqāla*: ten; second: sixty-four (including trigonometrical tables); third: sixty-seven; and the concluding section (*khātima*): six.

Two other important works (nineteenth century) on Islamic astronomy in India are those of Maulāna Ghulām Husain of Jaunpur, titled *Jāmī Bahādur Khānī* (1835) and *Zīj-i Bahādur Khānī* (1844). He dedicated to the former Rāja Bahādur Khān of Tikari. A voluminous mathematical and astronomical compendium, *Jāmi Bahādur Khānī* deals extensively with geometry, arithmetic, trigonometry, optics, astronomy, and the preparation of *Zījes*. It is regarded as the next best to *Zīj-i Muḥammad Sḥāhī* of Jai Singh II. Ghulām Husain meticulously observed the motion of the star-planets, determined their longitudes for fifteen years, and recorded them systematically.

The *Zīj-i Bahādur Khāṇī* comprises an introduction and seven chapters or *maqālas*. They deal with (i) computations using Indian numerals as well as the decimal place-value, along with the sexagesimal system wherever necessary; (ii) different eras and the conversion of one into another; (iii) determination of the time of ascendant and associated calculations; (iv) motion of celestial bodies, longitudes and latitudes, solar and lunar eclipses, visibility of the new moon, etc.; (v) elements of Hindu Pañcāṇga; (vi) relative positions of planets; and (vii) astrology. The *Zīj* literature produced in India by Indian Muslim astronomers has a distinct place in the history of Indian astronomy. Not all of them have been studied critically with the attention which they deserve; more studies may throw light on the mutual interaction and adaptability among Hindu and Muslim astronomers alike.

Al-Bīrūnī

In the eleventh century, a rare Islamic scholar of proven acumen and intensive knowledge in mathematics, astronomy, physics, medicine, minerals and other fields, was in India for about a decade and became the most able transmitter of Indian astronomy to West Asia. He was Abū Raihān Muḥammed ibn Ahmad al-Bīrūnī (973-1048). Born in 973 in a suburb called Bīrūn of al-Khwārizm or modern Khiva in Central Asia, al-Bīrūni even in his young age had read extensively the books on various subjects in Persian and Arabic, and developed an encyclopaedic mind. In 1016, when Sultan Muḥaṃmūd of Ghazni invaded and conquered Khwārizm, he brought the accomplished scholar to his court in Ghaznavi. Al-Bīrūnī lived in a period which was intellectually vibrant with celebrated Islamic savants like Ibn Sīnā (980-1037) who had produced a classic medical treatise, *Al-Qānūn*, and Ibn al-Haitham (965-1038) who had blazed new trails, especially in optics. Al-Bīrūni was well aware of Hindu astronomy through the

Arabic versions of Brahmagupta's *Brāhmasphuṭa Siddhānta* and *Khaṇḍakhādyaka* at the Caliphate in Baghdad. He also knew the Hindu numerals and the decimal place-value system as well as medicine since the *Caraka* and the *Suśruta Saṃhitās* were also available in Arabic translations. When Sultan Muḥammud invaded the northwestern parts of India, al-Bīrūnī accompanied him and, after the invader left for home, al-Bīrūnī travelled in western India for nearly ten years. He had an insatiable thirst for knowledge, and was fascinated by Indian astronomy, mathematics, medicine, religion and customs. He had an open but critical mind, and he learnt Sanskrit, a language totally foreign to him, and soon acquired proficiency in it.

Al-Bīrūnī was deeply interested in the Sanskrit text on astronomy by Vijaya Nandi (early tenth century). The manuscript that was available to him was titled *Karaṇatilaka,* and he rendered it into Arabic. The original text in Sanskrit is now not available in any of the repositories either in India or abroad. Al-Bīrūnī's Arabic version (though in a summary form and not entirely literal) of this work is the only source of a *karaṇa* work that was written between the time of *Khaṇḍakhādyaka* of Brahmagupta or the *Karaṇaratna* of Devācārya (seventh century) and the *Bhāsvatī Karaṇa* of Śatānanda (eleventh century). A re-copy of the Arabic version (sixty-eight pages) has been preserved in one of the libraries in Ahmedabad (library of the dargāh of Pir Muḥammad Shah).

According to al-Bīrūnī, the author of *Karaṇatilaka* is Vijaya Nandi, son of Jaya Nandi, who resided in the city of Benares, and Hindus held this text in esteem from the religious point of view. Al-Bīrūnī is stated to have procured a copy of the manuscript of this work in Multan from a local astronomer called Durlabha, whose work he describes in his enchanting treatise, *Kitāb-al Hind* (see below). The Arabic version of *Karaṇatilaka* prepared by al-Bīrūnī is titled *Ghurrāt-uz-zījāt*. With a view to making his translation intelligible, he added some useful information in its beginning about various calendars and their interrelationship, since *Karaṇatilaka* was a source for calendar-computations by Hindu astronomers. The text now available has been rearranged by a scholar in Aligarh under fourteen chapters which are in the nature of traditional themes presented in the major Siddhāntic works, with corrections wherever they were necessary. The periods in civil days for the sun, the moon, and the star-planets in the Arabic version of *Karaṇatilaka* are fairly accurate when compared with modern values. Students of Hindu astronomy are indeed indebted to al-Bīrūnī for the resurrection and preservation of this rare astronomical text.

The most seminal work that has added a new approach to the understanding of Indian (Hindu) astronomy is undoubtedly *Kitāb Tā'rīkh al-Hind* in Arabic, which al-Bīrūni has bequeathed to succeeding generations. It was translated into English with notes and indices by E.C. Sachau, and published in London in two parts in 1910. This voluminous treatise speaks eloquently of the Islamic scholar's catholicity of outlook, his learning in Sanskrit, and passion for understanding Hindu astronomy. However, al-Birūnī is critical of the attitude of Hindu paṇḍits. Although he unhesitatingly denounced the meaningless destruction caused by Ghazni's invasion, he has some misgivings about the Hindu paṇḍits he encountered and was critically sarcastic about their 'foolishly vain', 'self-conceited' and 'stolid' character. But the knowledge-seeker in him asserted itself and sustained him during his stay in the northwestern region of India. He had thus no inkling of the scholastic potential that existed in the eastern and particularly in the southern parts of India.

Al-Bīrūnī cites Varāhamihira, but misinterprets him. In his *Bṛhatsaṃhitā*, Varāha says: 'The Yavanas (Greeks) are *mlecchas* (untouchables?). This science (astrology) is well established among them. (Therefore) even they are honoured as sages. Then what to say of an astrologer (knower of Fate) if he is a twice-born (a *brāhmaṇa*)'. While making this statement, Varāha had in his mind Greek astrology and not astronomy. Al-Bīrūnī's statement is in the nature of a distortion of Varāha's, when he writes: 'The Greeks, though impure, must be honoured since they were trained in sciences, and therein excelled others. What, then, are we to say of a Brāhman, if he combines with his purity the height of science?' He adds: 'In former times the Hindus used to acknowledge that the progress of science due to the Greeks is much more important than that which is due to themselves. But from this passage of Varāhamihira alone you see what a self-laudatory man he is, whilst he gives himself airs as doing justice to others.'

Possibly al-Bīrūnī's was a pre-conditioned mind and he did not seem to have appreciated the spirit in which Varāha acknowledged Greek astrology. If al-Bīrūnī thought that Varāha indulged in self-laudation, he himself was not above it. For, he says with pride:

> 'I began to show them (Hindus) the elements on which this science (astronomy) rests, to point out to them some rules of logical deduction and the scientific method of all mathematics, and they flocked together round me from all parts, wondering and most eager to learn from me, asking me at the same time from what Hindu master I had learnt these things, whilst in reality I showed them what they are worth, and thought myself a *great deal superior* to them, disdaining to be put on level with them...' (Sachau, *Alberuni's Indica*, (reprint: Oriental Books, Reprint Corporation, New Delhi, 1983) vol. I, p. 23).

Al-Bīrūnī is noticeably vituperative in his utterances about Hindu mathematical and astronomical literature; he compared it to 'a mixture of pearl shells and sour dates, or of pearls and dung, or of costly crystals and common pebbles'. Stating that both are equal in their eyes since, according to al-Bīrūnī, 'Hindu astronomers cannot raise themselves to the methods of a strictly scientific deduction.' (Sachau, vol I, p. 25) These half-truths apart, al-Bīrūnī was the earliest Islamic erudite to have studied Hindu astronomy of the time—first the Arabic translations of Brahmagupta's works before he came to India, and later the Sanskrit texts, while he was in India.

Al-Bīrūnī's knowledge of Hindu astronomy is really amazing. In his *Kitāb Tā'rīkh al-Hind,* he throws light on *adhimāsa, ūnarātra, ahargaṇa* (Ch. LI), Yamakoṭi; Rūm (Rome) and Laṅkā; Hindu methods of determining longitude, calculation of *deśāntara* (XXIX–XXX); divisions of time (Ch. XXXIV); *kalpa, caturyuga* etc. (L); stations of the moon (Ch. LVI); order of the planets (Ch. LV); heliacal risings of the stars (Ch. LVII); and lunar as well as solar eclipses (Ch. LIX) (the chapter numbers in brackets correspond to those in the translation by Sachau). On the eclipses, he quotes from the *Brāhmasphuṭa Siddhānta* of Brahmagupta: 'Some people think that the eclipse is not caused by the Head (*Rāhu*). This, however, is a foolish idea; for, it is *he* who in fact eclipses, and the generality of the inhabitants of the world say that it is the Head who eclipses. The *Veda*, which is the word of God from the mouth of Brahman, says that the Head eclipses; likewise, the book *Smṛti*, composed by Manu and the *Saṃhitā* composed by Garga, the son of Brahma' (Vol. 2, p. 110). In al-Bīrūnī's view Brahmagupta's explanation of the cause of an eclipse was not scientific. However, Brahmagupṭa, after giving a scientific explanation of the occurrence of an eclipse, tries to reconcile it with the mythological view. In any case al-Bīrūnī was aware of the scientific explanation offered by Āryabhaṭa I, Varāhamihira, and Viṣṇucandra, when he quotes Brahmagupta.

Al-Bīrūnī meticulously examined the various astronomical computations of the Siddhāntic astronomy, particularly of the *Pauliśasiddhānticā* and the *Brāhmasphuṭa Siddhānta*. He does not seem to have studied the work of Āryabhaṭa, but refers to him as Arjabhar. Even in Baghdad and other parts of the Islamic region then, Āryabhaṭa's works were imperfectly known. There is no denying that al-Bīrūnī had made in-depth studies of Hindu astronomy, and his *Kitāb Tā'rīkh al-Hind* is indeed a valuable source for a student of the history of astronomy.

INDIAN ASTRONOMY AND EUROPEAN SCHOLARS

Europe witnessed a resurgence of learning during the twelfth and thirteenth centuries. This was due to the translation activities which resulted in the availability of several Arabic and Greek scientific texts in Latin. During this period Latin gradually emerged as the language of science. Indian astronomical ideas could well have been known to interested European scholars through the Arabic works. However, direct studies of original sources on astronomy in Sanskrit by Western scholars became possible from the nineteenth century during the colonial rule. Even a century before, there were some attempts in this direction, like that of J.D. Cassini, director of the Paris Observatory, who had examined a Sanskrit manuscript dealing with the computation of *ahargaṇa* sent to him by the French Ambassador in Siam; it was published in the *Memoirs* of the French Academy in 1691. It is now known that the Siamese manuscript was based on the *Pauliśasīddhānta* included in the *Pañcasiddhāntikā* of Varāhamihira. In the 1760s, Le Gentil of the Paris Observatory came to India to observe the transit of Venus in 1761 as well as in 1769. His attention was drawn to Hindu astronomy, especially the *pañcāṅga*-makers in Tamil Nadu and Pondicherry, which was a French territory. He was also interested in the twenty-seven *nakṣatras* of the lunar zodiac. Another noted French astronomer, Sylvain Bailly, studied a few Sanskrit texts on astronomy with the help of paṇḍits, and published a book in French titled *Traite' de l' Astronomie Indienne et Orientale*, in which he appreciated the Hindu methods of determining planetary positions and the associated rules of computation. He went to the extent of recording that astronomy originated in India and was subsequently transmitted to Mesopotamia (Chaldeans) and also to the Greeks. The Edinburgh astronomer,

John Playfair, who had examined Bailly's book and who had carried out his own studies, added his appreciation, particularly about the impact of the Hindu astronomical tables on European astronomical pursuits. Some of the European scholars who came to India either as employees of the East India Company or as missionaries evinced interest in what they thought was Hindu astronomy.

The Asiatic Society, which was founded in 1784 in Calcutta, encouraged Indological studies. Its first founder-president, William Jones, gave a lead in this direction by studying and publishing articles on the antiquity of the Indian zodiac, Indian chronology and the lunar year of the Hindus in the Society's journal, *Asiatick Researches*. Samuel Davis studied the *Sūryasiddhānta* and its computations. He too published some articles on Hindu astronomical computations in *Asiatick Researches*. The high antiquity as well as originality of Hindu astronomy was advocated by Bailly, Playfair, Jones, and Davis. There were, however, opponents as well as moderate critics of such an advocacy. John Bentley adduced his own arguments against the Hindu originality in astronomy, and even pointed out what he considered errors in the Hindu methods of computation, while orientalists H.T. Colebrooke and E. Burgess were somewhat cautious in their evaluation of Hindu astronomy.

Colebrooke made in-depth studies of the major texts of Hindu astronomy, especially the *Sūryasiddhānta*, *Brāhmasphuṭa Siddhānta*, *Siddhāntaśiromaṇī,* and *Grahalāghavam*. He examined especially the Hindu methods relating to the position of the junction stars, the precession of equinoxes, twelve signs (*rāśis*) of the zodiac and some astronomical instruments, particularly the armillary sphere vis-à-vis those of the Greeks. He opined that Hindus had their originality in the observation and determination of the junction-stars among each of the *nakṣatras*. He argued that the twelve signs in relation to the *nakṣatras* were not just borrowed from the Greeks, but reconciled and adapted to the division of the ecliptic into twenty-seven *nakṣatra* parts in an ingenious manner. He noticed that the armillary sphere as described in the *Sūryasiddhānta* and the Greek model as described in the Arabic translations, were different in detail. He thus concluded that the Hindu model was not a replica of the Greek one. Colebrooke was also cautious about his opinion on the transmissions of the twelve signs of the zodiac from the Greeks to India. He noticed that several Indian astronomers had more or less accepted the libration theory of the equinoxes which the *Sūryasiddhānta* had advocated, in contradistinction to the Greek approach of the precession of equinoxes as postulated by Hipparachus (second century BCE) and the degree or rate of precession determined by him.

Ebenezer Burgess, an American missionary who lived in the Bombay Presidency, published an English translation of the *Sūryasiddhānta* (1860) with critical notes and comprehensive tables. He was scholastically assisted by W.D. Whitney, and both of them thought that the astronomical computations of the *Sūryasiddhānta* were based on the Greek sources with some modifications by using a huge cyclic period of 43,20,000 years. In their view, the transmission of the Greek methods probably occurred during the maritime trade between the west coast of India and Alexandria (in Egypt) where Greek mathematical astronomy was vigorously pursued by Hipparachus, Ptolemy, and others.

However, after carefully analyzing the contents of the *Sūrysiddhānta,* Burgess compared the planetary system of the Hindus with that of Greeks. He observed that the two were more or less the same basically, but was of the opinion that the Hindu system had certain dimensions which were not encountered in Greek astronomy. In his view, neither the Hindus nor the Greeks borrowed from each other, but both obtained their ideas from a common source. Burgess also thought that Hindus were original in regard to most of the principles of astronomy, and in their cultivation of astronomy, were not indebted to the Greeks. But his associate Whitney held the contrary opinion, and did not give the credit due to the Hindus.

The question of foreign influences on Hindu astronomy attracted the attention of such scholars as G.R. Kaye, Otto Neugebauer and his disciple, David Pingree. Kaye in his writings exhibited a strong bias in favour of the Greek influence on Hindu astronomy and even mathematics, without adducing tangible evidence. Neugebauer, too, was inclined to this view, and even suggested the possible Babylonian influences on Hindu astronomy. David Pingree went a step further and pointed out that several ideas like the huge cyclic period of 43,20,000 years (*mahāyuga*), epicyclic models, precession and trepidation of equinoxes were of Greek origin. He also did not give Āryabhaṭa I the credit due to him. French Indologist Roger Billard, on the other hand, showed the observational basis of Āryabhaṭa I as well as of Brahmagupta. He developed a novel statistical method for the evaluation or assessment of the mean longitudes of the sun, the moon, and the five star-planets. Swiss Indologist and historian of mathematical astronomy, Van der Waerden, also adduced evidence to show in unmistakable terms the observational basis of the *Brāhmasphuṭa Siddhānta* of Brahmagupta.

The inclination of Western scholars, in general, was to compare the astronomical parameters found in Babylonian or Greek texts with those in Sanskrit texts. If any such contents were found in the former chronologically earlier, they jumped to the conclusion that there was transmission to or borrowing from the former by the latter. Such a conclusion, to be valid, needs positive and incontrovertible evidence of well-identified pathways of transmission, besides the exact sources which influenced such a transmission. It is a recognized principle in the history of science that there were somewhat similar developments in different cultures independently of one another from time to time. For example, the atomism of the Greeks and that of the Indian Vaiśeṣika came up more or less at the same time around the fifth century BCE. Kaṇāda in India expounded atomism in the context of the reality of the knowable world confining the atomic nature to the four elements: *pṛthivī, ap, tejas* and *vāyu* being eternal, while they are not eternal in their produced states. Greek atomism was postulated in a different context. An examination of the transmission of ideas from one culture to another needs a sound and objective historiography that examines both the contents and the contexts in which such ideas germinated.

Insofar as it concerns Hindu astronomy, there is no definitive evidence in respect of the textual channels which were involved in the transmission; nor is there even remote evidence of the translation into Sanskrit of any Greek source on astronomy like that of Ptolemy's *Megale Syntaxis* (which found its Arabic version in the Islamic culture-area). On the other hand, there is evidence and even admission by an open-minded scholar like Varāhamīhira, that Greeks (*Yavanas*) were adept in astrology, and there are several Greek words in Hindu astrological works, unlike in the Sanskrit astronomical texts. One can also see the astrological ideas presented by Ptolemy in his *Tetrabiblios*, reflected *mutatis mutandis* to some extent in Varāhamihira's *Bṛhajjātaka*. But the Greek astronomical ideas like the precession of equinoxes are not found in Hindu astronomical texts like those of Āryabhaṭa I, Varāhamihira, Bhāskara I, and Brahmagupta. P.C. Sengupta, who studied scientifically the Greek and the Hindu methods in spherical astronomy, has conclusively shown how the Hindu methods were independently developed in India. The translator of the *Sūryasiddhānta,* E. Burgess, states that there is no evidence to indicate that the works of Hipparachus and Ptolemy influenced Hindu astronomy. But he opines in a general way, without adducing any concrete example, that some Hellenistic astronomical elements could have influenced the Hindu calendar-making, and this could have been around the first or second century CE during the maritime trade between India and Rome. It is true that there was commercial intercourse of this type, as confirmed by the *Periplus of the Erythrean Sea* by an unknown author, and the concept of a week and the naming of a week-day after a planet were transmitted to India, and these found an entry into the Indian calendar. But to aver that Indian astronomy did not develop mathematical techniques on its own is an oversimplification. The five *Siddhāntas* redacted by Varāhamihira early in the sixth century CE need to be looked at from a different perspective. There are divergent views about the *Romaka* and the *Pauliśa Siddhāntas* about their foreign origin. Even so, *Romaka* was considered inaccurate and, in the words of Varāhamihira, 'its study itself is to be expiated'. As for the *Pauliśa*, there are several versions. The one which, according to David Pingree, was the work of a Greek scholar, Paulus Alexandricus, deals with astrology but not astronomy. Pingree, however, admits that several astronomical parameters or elements of Indian mathematical astronomy are modifications or improvements and these permit us to characterize this astronomy as *Indian* rather than Babylonian, Greek, or Islamic. It has been shown how the Vedic *nakṣatra* system had its originality associated with its religio-cultural dimensions and was uninfluenced by Babylonian ideas. There was practically no influence of Islamic astronomy in the medieval period, since Hindu astronomical endeavours were continued with renewed vigour, especially by the Kerala School of astronomers. Both Islamic and Hindu astronomy existed side by side, then as now.

The intellectual climate in India in the five or four centuries preceding the Christian Era was one of high attainments in such fields as philosophy, problems of matter and motion, space and time, and epistemological excursions. These also influenced the foundational ideas of Ayurveda, the science of life. *Yoga* with its eight constituents was methodically structured. Sanskrit linguistics had attained new heights. The altar-geometry had its mathematical propositions like the conception of irrational numbers, the equivalence of the square of the hypotenuse of a right angled triangle to the sum of the squares of the other two sides. Indian atomism of the *Nyāya-Vaiśeṣika* the Jaina and Bauddha schools was noted for its originality and it in no way owed its inspiration to that of Leucippus and Democritus. The Nyāya epistemology came up in its own characteristic manner, and had no resemblance to Aristotelian logic. It would not be, therefore, wide off the mark to suppose that Indian astronomy too was uninfluenced by Hellenistic astronomy. There were as many as eighteen *Siddhāntas* during this period, and only five of which were available and redacted several centuries later by Varāhamihira. There is little or no information about the thirteen others, but it was not improbable that at least one of them would have contained the seed-idea of epicyclic or eccentric postulate. The influence of Greek astronomy on Hindu astronomy continues to be a debatable issue from the standpoint of the history of the dynamics of ideas.

INDIAN ASTRONOMY IN IRAN AND WEST ASIA

The Sassanian emperors (third century CE), Ardshīr I and Shāpūr I, were interested in Hindu and Greek astronomy and encouraged the translations into Pahlavī of some important texts on astronomy in Greek as well as in Sanskrit. Indian astronomical computations in the form of tables known as *Zik-i-Hinduk* were being used in Iran for a considerable time. Around the middle of the sixth century a set of astronomical tables appeared in Pahlavī titled *Zik-i-Shariyāran*, several elements of which were derived from the *Paitāmahasiddhānta* as found in the *Viṣṇudharmottara Pūrāṇa*. The old *Sūryasiddhānta's ārdharātrika* system (midnight to midnight reckoning of a day) also found its way into Pahlavī under the title *Zīzāt-Arkaṇd*, the word *arkaṇḍ* being perhaps an outlandish form of *ahargaṇa* (the number of elapsed days). It is on record that Kusro Anūshirwan, the Sassanian emperor in the middle of the sixth century CE, had asked his court astronomer-astrologers to compare the tables in the Pahlavī's version of Ptolemy's *Megale Syntaxis* with *Zījat-arkand*. They expressed their opinion that the latter (Hindu one) was preferable to the former. Hindu astronomical tables were used in Iran for astrological purposes, too.

Islamic Arabs who had established trading centres in the Malabar coast in the seventh century and conquered the Sind province in the eighth century, were well aware of Indian drugs and medical practices as well as Hindu calendars. Hindu astronomical texts had their appeal to the Caliphs at Baghdad and the Hindu methods of determining time as accurately as possible were found useful, especially for saying Islamic prayers at appropriate times, five times a day. As noted earlier, Caliph al-Manṣūr was fascinated by Hindu planetary astronomy, and ordered that the *Brāhmasphuṭa Siddhānta* and the *Khaṇḍakhādyaka* of Brahmagupta be translated into Arabic. With the assistance of Hindu *pandits* who were associated with the Caliphate at Baghdad, Muḥammad al-Fazāri translated the former, and Ya'qūb ibn Ṭāriq rendered the latter into Arabic, under the titles *Sindhind* and *Arkaṇḍ* respectively. There was another work in Arabic, a translated one based on the astronomical elements of the *Pauliśasiddhāntā* contained in the *Viṣṇudharamottara purāṇa*, and the *Brāhmasphuṭa Siddhānta*, referred to as the *Mahāsiddhānta* (which al-Bīrūnī called *Sindhind al-Kabir*).

The most outstanding Islamic astronomer-mathematician was Abu Ja'fer Muḥammad ibn Mūsa al-Khwārizmī, popularly known as al-Khwārizmī (early ninth century). He came from the Central Asian region of Khwārizm, and worked in Baghdad. He had developed new methods for computing the mean motion of planets based on *Sindhind* (Indian); their equations, on *Zīj-i-Sharyāran* (Persian); and ecliptic of the sun, based on Ptolemy's *Almagest*. He also produced authentic tables called *Zīj-al Sindhind* on the advice of the Caliph Al-Mamūn, and a seminal arithmetical text under the title *Hind-hisāb* using Indian (Brāhmī) numerals and the associated decimal place-value system. Indian astronomical methods received the high appreciation of Islamic astronomers in Baghdad and elsewhere because of their viable canvas of decimal place-value numeration, and the ease with which the *Zījes* could be prepared.

INSTRUMENTS

The principal objective of Indian astronomers was to determine accurate time by careful observations of the movements of the sun as well as the moon, and the planetary positions. They were aware that the computed positions and the calculated timings of these luminaries needed to be verified by observations, since the astronomers accorded the deserved importance to the observed data. Naturally, the observational endeavours necessitated the use of certain devices or instruments. The Vedic priestly-astronomers have given expression to their continuous observations of the movements of the sun and the moon. The question is: Did they use any type of instrument for their observations? The *Ṛgveda* speaks of *turīya,* which has been interpreted by Vedic scholars as being a device, a quadrant. The *Aitreya Brāhmaṇa* appears to indicate the earliest use of gnomon in India in the context of the summer solstice and the performance of the *Viṣuvat*. However, there is no tangible evidence in the Vedic literature to prove beyond doubt the observations of the gnomonic shadow or the use of a quadrant. Naked-eye observations were the forte of the Vedic priestly-astronomers. The earliest use of gnomon for the fixation of cardinal points is found in the description given in the *Kātyāyana* Śulba sūtras around the fifth century BCE. A century or two later, the *Arthaśāstra* of Kauṭilya gave vivid details of the variation of the gnomonic shadow by stating that in the month of Āṣāḍha, the midday loses the shadow, and in the next six months beginning with Śrāvaṇa, the shadow gradually increases by two *aṅgulas* (roughly an *aṅgula* is three-fourths of an inch) in each month, while in the six months starting from Māgha, it decreases by two *aṅgulas* every month. The height of the gnomon is stated to be twelve *aṅgulas*, a measurement of the gnomonic tradition that persisted even in the Siddhāntic period.

Armillary Sphere: Indian astronomers devised and used basically twelve types of instruments: the *gola; bhagaṇa; cakra; dhanus; ghaṭī*; *śaṅku*; *śakaṭa; kartarī; pīṭha; kapāla; śalākā;* and *yaṣṭi*. But it was the *gola* or the armillary sphere that earned a prime position among all the instruments. Lallācārya, the astronomer of the ninth century, stressed the importance of the knowledge of the armillary sphere thus:

> 'Just as a disputant without the knowledge of grammar, a sacrifice (*yajña*) without the knowledge of the Vedas, Purāṇas and the like, so is an astronomer totally unsuccessful without the knowledge of armillary sphere. [...] An astronomer who has a comprehensive knowledge of the armillary sphere sees in front of his eyes, as it were, the whole universe with a variety of celestial luminaries, and would also become the recipient of spiritual fruition, wealth, fame and salvation. But if one had to be proficient in the construction of an armillary sphere and use it efficaciously, one should be an expert in its geometrical representation and the several spheres associated with it.' (*Śiṣyadhīvṛddhidatantra*; 14, 1-5)

The armillary sphere fabricated by Indian astronomers comprised an inner starry sphere and an outer celestial sphere; such a juxtaposition would help in comparing the positions of these inner and outer spheres. The technical terms coined by them for these spheres convey the astronomical imagery associated with them. The armillary sphere is described in terms of *Nakṣatragola* (or *Bhagola*) and *Khagola* along with other details, such as the solstitial colure (*dakṣiṇottora*); the celestial equator (*ghaṭikā-maṇḍala* or *ghaṭikā-vṛtta*); equinoctial colure and the crosses (*viṣuvad*); ecliptic (*apama-vṛtta* or *apamaṇḍala*); and several diurnal circles (*ahorātra-vṛttas*) proved to be valuable to the Indian astronomical tradition. The orbits of the moon and related circles constitute the *Nakṣatragola*. The *Khagola* or the celestial sphere, which is constructed outside the *Nakṣatragola* has a horizontal circle, the prime vertical (*samamaṇḍala*); and

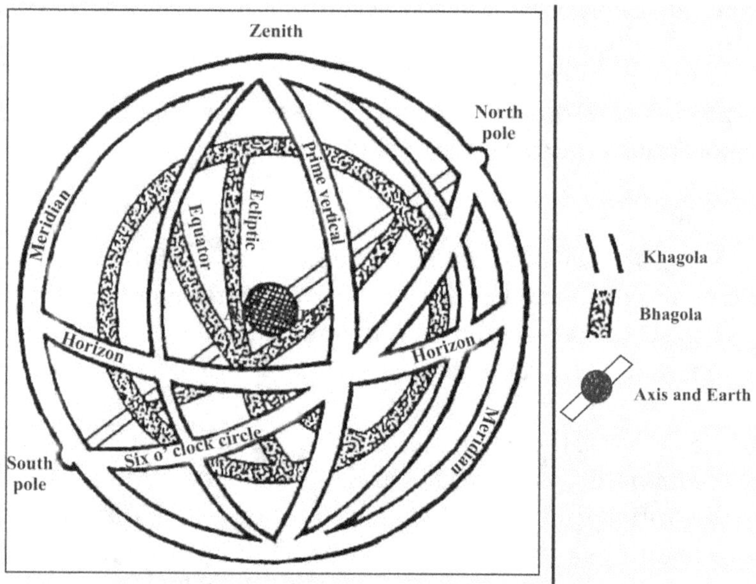

Armillary Sphere (some circles: diagrammatic)
(after Yukio Ohashi, *Indian Journal of History of Science*, 29(2), 1994, p.270)

Hindu astronomers recognize (i) the prime vertical (*Samamaṇḍala*);
(ii) meridian (*Yāmyottaramaṇḍala*); (iii) horizon (*Kṣitija*);
(iv) celestial equator (*Viṣuvadvṛtta* or *nāḍīvṛtta*); and
(v) the six o' clock circle (*Unmaṇḍala*) as fixed for the stars and planets.
(Courtesy: Project of History of Indian Science Philosophy and Culture, Centre for Studies in Civilizations)

a circle passing through the north-south poles and the east-west points, called the *unmaṇḍala* along with the north-south circle or meridian. A model of the earth was prepared from clay or some other plastic material in the shape of a sphere and placed at the centre of the axis of the armillary sphere. As regards the related graduations, the six circles of the celestial sphere were each divided into 360° while the celestial equator (*viṣuvadvṛtta*) was divided into sixty *nāḍīs* or *ghaṭkās* (hence called *nāḍīvṛtta*). The ecliptic was to be to the north of the equator to the extent of 24°, and the first point of Cancer (*Karka*) was to be situated there. Similarly, to the south of the equator to the extent of 24° would lie the first point of Capricorn (*Makara*). The armillary sphere was so designed that the sun, the moon (nodes), and the planets were to move along the ecliptic. The shadow of the earth also would move on the ecliptic, but at a distance of six signs (*rāśis*) from the sun. The oriental ecliptic point or *lagna* was the place on the ecliptic where the eastern horizon would meet it, while the setting point (*astalagna*) was the one on the ecliptic where the western horizon would meet it. Indian astronomers recognize five circles which are fixed for the stars and planets at any given latitude, and several other cycles that are not fixed. The major texts have a separate chapter on instruments and describe the ways of observation.

Of the other instruments, special mention needs to be made of an automatically rotating sphere to keep pace with time; a shadow instrument for the determination of both time and the elapsed degrees since the sunrise at any time of the day; a water-instrument to indicate the flow of time; a needle instrument to find out the rising of a planet; and a gnomon to fix the cardinal directions.

Jai Singh II's instruments: In addition to his well-known compendium, *Zīj-i Muḥammad Shāhī*, Jai Singh II also produced a work in Sanskrit called the *Yantraprakāra* based on some Islamic sources, including the Arabic version of Ptolemy's *Almagest*. This elegant and comprehensive text (tr. S.R. Sarma), deals not only with the methods of constructing several types of instruments but also the manner of computations of astronomical tables. The following nineteen kinds of instruments have been described, the material used for each of them being indicated in brackets: (i) *Jayaprakāśa* (masonry or metal or wood); (ii) *Yāmyottara-yantra* (metal); (iii) *Yamyottarabhitti-yantra* (masonry/stone/wood); (iv) *Dhāt al-Ḥalaq* with five rings (metal); (v) *Dhāt al-Ḥalaq* with six rings (metal); (vi) *Dhāt al-Shu'ybatayn* (wood); (vii) *Dhāt al-Thuqbatyan* (wood); (viii) *Yantrarājā* (metal); (ix) *Krānti-yantra* (metal); (x) *Digaṃśa-yantra* (masonry); (xi) *Agrā-yantra* (masonry and metal); (xii) *Śara-yantra* (metal); (xiii) *Krāntivṛtta-yantra* (masonry and metal); (xiv) *Krāntivṛtta-yantra* for each sign of the zodiac (masonry and metal); (xv) *Suds Fakhri* (masonry); (xvi) *Nāḍīvalaya-yantra* (masonry and metal); (xvii) *Palabhā-yantra* (masonry); (xviii) *Sarvadeśīya-kapāla-yantra* (masonry); and (xix) *Cūḍā-yantra* (metal). The *Yantraprakāra* provides sequentially the details of construction of each instrument, except the *yantrarājā* or the astrolabe and the *śarayantra*, an instrument for determining celestial latitude.

Among the impressive masonry instruments erected by Jai Singh, three instruments: (i) the *Samrāt-yantra;* (ii) the *Ṣaṣṭhāṃsā;* and (iii) the *Dakṣinottara-bhitti,* can be regarded as higher precision instruments. As many as six *Samrāt-yantras*—two in Ujjain, one each in Varanāsi and Delhi, and two in Jaipur—were built as equinoctial sundials for the measurement of local time, the hour-angle and declination of celestial objects. They can still be used for these purposes. The *Ṣaṣṭhāmśa* was built in Delhi (one) and Jaipur (four). It is in the form of a sixty-degree meridian chamber, which can be used for measuring declination, zenith distance and even the diameter of the sun's disc through a pin-hole. Jai Singh constructed the *Dakṣiṇottara-bhitti*, one each at the five places—Delhi, Jaipur, Ujjain, Varanāsi and Mathura. It is in the form of a quadrant or a semicircle inscribed on a north-south wall (*bhitti*), and a rod is fixed at the centre of the arc. This device is useful for determining the meridian altitude, especially of the sun, and also the zenith distance of a celestial luminary, though the results may not be very accurate.

A unique masonry instrument which is found only in the Delhi Jantar Mantar, but not in the other four places (including the extensive Jaipur observatory), goes by the name of *Miśrayantra*. It was erected not during Jai Singh's time, but by his second son, Madho Singh. As its name implies, *Miśrayantra* is a combination, composed as it is of several instruments in one structure: the *Dakṣiṇottara-bhitti*, *Samrāt-yantra* in two halves, *Karka-rāśi-valaya*, four *Niyata cakras*, and a quadrant arc. Though the *Dakṣinottara-bhitti* was capable of being used for determining the altitude of the midday sun and the zenith distance, its efficacy and accuracy are doubtful since it is too close to the ground to permit a clearly discernible shadow. The *Samrāt-yantra,* which is a part of the *Miśrayantra,* has two halves, one to be used before the noon hour, and the other at the noon hour to determine the local time. The scale-markings on the two gnomons have been blurred over time and by weathering. Likewise, the circular *Karka-rāśi-valaya* which is on the back wall of the *Miśrayantra* ensemble faces north and is parallel to the plane of *Karkāṭaka,* the Tropic of Cancer. Such a contrivance could be used once in twenty-four hours to measure directly the longitude of a celestial object or the summer solstitial point of the ecliptic. However, the *Karka-rāśi-valaya* is now in a dilapidated condition. The *Niyata cakra* which comprises four semicircular scales on marble is positioned on either side of the gnomon. Thcsc could be used for measuring the declination of a celestial object at the desired intervals when it traverses from east to west in the course of a day. The quadrant arc was probably intended to measure directly the latitude as well as the longitude of a celestial luminary once in twenty-four hours when the summer solstitial point on the ecliptic or the first sign of Cancer (*Karkāṭaka*) would be at the meridian.

Mānamandira: Among the other observatories, the Mānamandira at Kāśī merits attention. Though first constructed by Mānasiṃha of Amer City (near Jaipur) in 1566, it was Jai Singh II, his descendent, who installed in it (in 1720) several instruments: the *Yāmyottara*; *Samrāt-yantra*; another *Bhitti-yantra, Nāḍīvalaya*, a smaller *Sāmrāt-yantra*; *digaṃśa*; and an aperture gnomon which corresponds to *Suda-al Fukhrī*. Details of each of these instruments have been given by Bāpudeva Śāstrī in his scholarly text, called *Mānamandira*. The tradition of using astronomical instruments had reached its pinnacle in the eighteenth century.

PAÑCĀṄGA

The practical face of Indian astronomical tradition is *pañcāṅga*. The word *pañcāṅga* connotes five constituents, namely (i) lunar day *(tithi);* (ii) asterism (*nakṣatra*); (iii) name of the week-day (*vāra*); (iv) an element related to the lunar day (*karaṇa*); and (v) the one related to the longitudes of the sun and the moon on that day (*yoga*). *Pañcāṅga* is a calendar of the Indian astronomical tradition including Hindus, Jainas and Buddhists. Even now it is an important annual compendium being prepared by traditional almanac-makers; and on each new year's day, it is reverentially worshipped and read in orthodox Hindu households. In addition to its five components, a *pañcāṅga* now provides a wide variety of information: days of religious ceremonies, festivals, car-ceremonies in temples, auspicious and inauspicious times, good days for travel, propitious times *(muhūrtas)* for the performance of marriages, thread ceremony, inaugurations, etc. A *pañcāṅga* also provides the planetary positions and the time of occurrence of eclipses. A large number of *pañcāṅgas* are being prepared and published annually in different parts of India, each one with its own characteristics, but maintaining the five core constituents. They can be broadly categorized into Luni-solar and Solar.

In a luni-solar *pañcāṅga,* the lunar month (twelve in a year) begins with the first day (*pratipat*) of the bright half (śūkla pakṣa) and ends with the last day of the dark half (*kṛṣṇa pakṣa*) or *amāvāsyā*. Known as the āmānta system, this is followed mostly in the Karnataka, Andhra Pradesh, Maharashtra, and Gujarat regions. Several of the North Indian regions follow the other system, the *pūrṇimānta* (ending with a full-moon day) system in which the *Caitra* month commences a fortnight earlier than in the āmānta one. As noted before, the twelve lunar months in a *pañcāṅga* are: Caitra, Vaiśākha, Jyeṣṭha, Āṣāḍha, Śrāvaṇa, Bhādrapada, Āśvina or Āsvayuja, Kārtika, Mārgaśira, Puṣya, Māgha, and Phālguṇa. The names of the lunar months are derived from the name of the *nakṣatra* in which the full moon generally or nearly occurs.

In a solar *pañcāṅga,* the solar year commences when the sun enters the first zodiacal sign, *Meṣa rāśi,* and it is known as *Meṣa saṅkramaṇa* (ingress). The solar year has also twelve months, but named after the zodiacal signs into which the sun enters: *Meṣa, Vṛsabha, Mithuna, Kaṭaka, Siṃha, Kanyā, Tula, Vṛścika, Dhanus, Makara, Kumbha and Mīna,* corresponding to Aries, Taurus, Gemini, Cancer, Leo, Virgo, Libra, Scorpio, Sagittarius, Capricorn, Aquarius, and Pisces which are in vogue in the West in relation to the Gregorian calendar, but with different dates of their commencement and ending.

One of Jahangir's coins on which a zodiacal sign has been portrayed

Tithi (lunar day): It is common experience that in the bright half of a lunar month, the visible part of the moon goes on increasing on successive days until it is nearly half on the seventh or eighth day after the new moon, and becomes fully visible and bright on the *pūrṇimā* day. In the dark half of the lunar month, the changes or phases of the moon diminish in the reverse order, and the moon becomes invisible on *amāvāsyā*. A lunar month is divided into thirty *tithis*, the duration of one *tithi* being the time taken by the moon to move nearly 12^0 relative to the sun. Because of the vagaries in the moon's motion, the duration of each *tithi* is not the same. It is for this reason that one notices in a *pañcāṅga* the mention of a *tithi* as well as the next one with reference to the different timings on the same day. The nomenclature of fifteen *tithi*s for each fortnight is in terms of Sanskrit words for numbers in a serial sequence like *prathamā, dvitīyā, tṛtīyā,* etc., the fifteenth one being *pūrṇimā* (bright-half) or *amāvāsya* (dark-half). Some of the *tithi*s are regarded as being inauspicious for the performance of marriages and other types of celebrations. Generally, the second, third, fifth, seventh, and thirteenth *tithi*s are considered to be auspicious, provided the other factors of an astrological nature are found to be favourable. The eighth, ninth, eleventh and the fifteenth (*pūrṇimā* or *amāvāsyā*) are normally avoided, although some festivals or worships are specially performed on these days according to paurāṇic or śāstraic tradition.

Nakṣatra: In the context of the lunar zodiac, there are twenty-seven *nakṣatra*s (Aśvinī...Revati; the extent of a *nakṣatra* being 13°20′ (360° divided by 27) or 800′). In a *pañcāṅga,* the ruling *nakṣatra* of the day occupies pride of place and is intimately associated with the *tithi* and its extent. Like the *tithis, nakṣatras* are also thought of in terms of auspicious and inauspicious ones in predictive astrology concerning performance of rites, celebration of marriages, birthdays and the like. From the Vedic times till about the fourth or fifth century CE, the *nakṣatra-vidy*ā held the fort. Even after the advent and influence of horary or planetary astrology of the twelve *rāśīs* (around the sixth century CE), the importance of *nakṣatras* in the daily as well as socio-cultural life of the people has not diminished; on the other hand, it has gained an additional dimension.

Week-day: The word *vāra* generally means the duration of seven days or a week. But it is also associated with the name of a weekday, each day of the week carrying the name of the sun, the moon, Mars, Mercury, Jupiter, Venus and Saturn respectively. A week of seven days has no significance astronomically; nor are the sequential names of the sun, the moon, and the planets representative of their distances from the earth. The *Āryabhaṭīya* has thrown light on the order of the weekdays and their determination. In the *Kālakrīyā-pāda* of this text, Āryabhaṭa I states that the *nakṣatras* are the outermost and beneath them lie the planets: Saturn, Jupiter, Mars, the Sun, Venus, Mercury and the moon. The rationale adduced by him is that they are arranged in increasing order of their velocity: Saturn the slowest, and the Moon, the fastest. He enunciates further that in this order, they are also the lords of the successive hours, and the planets occurring fourth in the order of increasing velocity are also the lords of successive days which are reckoned from the sunrise at Laṅkā.

The order of the planetary names for weekdays is governed by the division of a day into twenty-four hours so that the lords of each hour (seven planets consecutively) are determined. If the reckoning is continued with a planet (lord) of the hour, the first, eighth, fifteenth, and twenty-second hours will have the same planet as their lord. Hence, the twenty-third and twenty-fourth would have the lords following the stated planetary order. Then the twenty-fifth, or the first hour of the next day, will have the fourth planet as the lord, and its name would be associated with the next day. For example, if the lord of the first hour is the sun, that is, if it is Sunday, then the lord of the first, eighth, fifteenth, and twenty-second hours would also be the sun; and lord of the twenty-third hour (according to the order: Saturn-Jupiter-Mars-Sun-Venus-Mercury and Moon) would be Venus; of the twenty-fourth hour, Mercury; and of the twenty-fifth, or the first hour of the next day, would be the moon. Hence, after Sunday would be Monday. It would appear that the order of weekdays was suggested by the Egyptian reckoning of a day as comprising twenty-four parts, each part being ruled by one planet or the other.

However, if the day is divided into sixty *ghaṭikās* as conceived by Indian astronomers, and each *ghaṭikā* is assigned to one planet, placed in the reverse order (in relation to their decreasing velocities of planets (that is, the Moon-Mercury-Venus-Sun-Mars-Jupiter-Saturn) and we begin with Saturday, the lord of the first *ghaṭik*ā will be Saturn; after 7 × 8 times (fifty-six), the fifty-seventh will also be of Saturn; fifty-eighth the moon; fifty-ninth Mercury; sixtieth, Venus; and the sixty-first, that is, the first *ghaṭikā* of the next day will have the sun as its lord; hence after Saturday comes Sunday.

The origin of the concept of a week as a period of a certain number of days goes back to perhaps three or four centuries before the Christian Era, among the Mexicans, Jews, Babylonians, Romans, and the Incas of South America, with varying number of days constituting a 'week': Mexicans (five days); Romans (eight days); Babylonians and Jews (seven days). It is well known that according to the Old Testament, God is believed to have been engaged in the act of creation for six days, and rested on the seventh day, which was blessed by God. God is also said to have commanded the Jews to work hard for six days and to rest on the seventh day. The Old Testament does not, however, mention the names of days, although *Sabbath* was the last day (of rest) and was observed at the end of six days. The *Sabbath* was not observed uniformly in different cultures; for example, the Jews observed it on Saturday and the Egyptians, on Thursday. The idea of *Sabbath*, the Biblical creation theory, the Egyptian time-reckoning, and Chaldean astrology appeared to be behind the idea of a week of seven days. Since the Greeks did not follow the planetary week, possibly the origin of the association of the sun, the moon, and the five planets with the names of weekdays might be traced to either Egyptians or Chaldeans around the fourth century CE. The question of the origin of the order of planetary weekdays is still a debatable one. In the succeeding two or three centuries, the concept of the planetary week spread far and wide, including in Greece and India.

In India, the Eran stone pillar inscription of King Buddhagupta, which mentions *Bṛhaspativāra* (Thursday) along with *dvādaśī* (twelfth day) of the *Śukla pakṣa* of Āṣāḍha in the Gupta year of 165, that is, 484 CE, seems to be the earliest record available about weekdays. A little later in 499 CE, the *Daśagītikā* section of the *Āryabhaṭīya* mentions Thursday in connection with the commencement of the revolution of all planets at the beginning of the sign of Aries (*Meṣa*) at sunrise at Laṅkā.

Yoga: The fourth component of a *pañcāṅga* is *yoga*, which represents a division of the sum of the *nirayaṇa* (not taking into consideration the precession of equinoxes) longitudes of the sun and the moon by twenty-seven. Hence there are twenty-seven *yogas* enumerated in all the *pañcāṅg*as: (i) Viṣkambha; (ii) Prīti; (iii) Āyuṣmān; (iv) Saubhāgya; (v) Śobhana; (vi) Atigaṇḍa; (vii) Sukarma; (viii) Dhṛti; (ix) Śūla; (x) Gaṇḍa; (xi) Vṛddhi; (xii) Dhruva; (xiii) Vyāghata; (xiv) Harṣaṇa; (xv) Vajra; (xvi) Siddhi; (xvii) Vyatīpāta; (xviii) Variyān; (xix) Parigha; (xx) Śiva; (xxi) Siddha; (xxii) Sādhya; (xxiii) Śubha; (xxiv) Śukla; (xxv) Brahman; (xxvi) Aindra; and (xxvii) Vaidhṛti. The extent of a *yoga* is regarded as the time during which the sun and the moon together accomplish 13°20′ or 800′; otherwise *yoga* does not have any significance astronomically. The first, Viṣkambha *Yoga*, is said to end at 13°20′; the second, Prīti, at 26°40′, and so on. To determine or identify the *yoga* that is running at a particular time, the sum of the longitudes of the sun and the moon are converted into minutes, and this sum is divided by 800. The quotient is the number of elapsed *yogas*, and by adding one to the completed number of *yogas*, the current *yoga* is identified. If, however, the sum of the longitudes of the sun and the moon exceeds 360°, that is, when one cycle is over, 360 needs to be subtracted from the sum, the resultant converted into minutes and divided, as before, by 800.

As for the *yoga*, Vyatīpāta, it is referred to as Lāṭa-vyatīpāta, if the sum of the longitudes of the sun and the moon is 180^0 (six *rāśis*), and if it is 360°, it is known as *Vaidhṛta*. In the case of the former, the *ayanas* of the sun and the moon are different and declinations are equal, while in respect of the latter, both the *ayanas* and the declinations are equal. Both Vyatīpāta and Vaidhṛta are also reckoned in terms of *rāśis* as follows: when the sum of the true longitudes of the sun and moon is twelve rāśis, and they are in the same hemisphere of the ecliptic, but are on the opposite sides of the solstitial point, there is Vyatīpāta *yoga*; on the other hand, Vaidhṛta *yoga* occurs when the sum is twelve *rāśis* and both the sun and the moon are in different spheres of the ecliptic, but on the same side of the solstitial point. Both the Vyatīpāta and the Vaidhṛta have been branded as extremely inauspicious; but no reason has been given by any author or text for such a consideration. It is believed that during this period, offering gifts, saying prayers and performance of *yajña* would ward off evil effects. The *Sūryāsiddhānta, Brāhmasphuṭa Siddhānta, Vaṭeśvara Siddhānta* and Lalla's *Śiṣyadhīvṛddhidatantra* have dealt with these two *yogas* under the heading *pāta*, pointing out their astrological significance in a *pañcāṅga*.

Karaṇa: The fifth constituent of a *pañcāṅga* is *karaṇa,* which is related to a *tithi* inasmuch as the first half of a *tithi* is regarded as one *karaṇa*, and the second half, another *karaṇa*. Thus there are sixty *karaṇas* corresponding to thirty *tithis*, but in the form of repetition of only eleven types of *karaṇas*. They are (i) Bava; (ii) Bālava; (iii) Kaulava; (iv) Taitale; (v) Gara (je); (vi) Vaṇika; (vii) Bhadre or Viṣi; (viii) Śakuni; (ix) Catuṣpāt; (x) Nāgavān; and (xi) Kiṃstughna. The first seven are believed to be movable (*cara*), and the other four, immovable (*sthira*). The origin of the concept of *karaṇa* is not clear; but a *karaṇa* is calculated to be situated at 6^0 of the angular distance between the sun and the moon; this is understandable because a *tithi* is reckoned in terms of the moon being about 12^0 in relation to the sun and a *tithi* has two *karaṇas*. Like the *yoga, karaṇa* too has some astrological import. The *Bṛhatsaṃhitā* of Varāhamihira has provided a list of performances that could or could not be carried out during the *karaṇas*. Bava is considered to be auspicious for all rites; religious acts during Bālava are said to be beneficial for Brāhmins. Kaulava is believed to be good for choosing a bride and establishing a relationship or friendship; Taitale is recognized to be favourable for house-building and the like, while Gara and Vaṇika are stated to be beneficial for agricultural operations and trade, respectively. Bhadre or Viṣi is considered to auger terrible occurrences; but Kiṃstughna is supposed to be auspicious for marriages and other religious performances. The whole concept of *karaṇa* and its presumed effects are irrational and superstitious alike. Yet, it continues to be given importance in a *pañcāṅga*.

Siddhāntic texts provide rules for the calculation of *tithis*, *nakṣatra, karaṇa* (not to be confused with the type of *karaṇa* texts) and *yoga* to determine their duration in a day. Among the other important calculations involved in a *pancāṅga* are those pertaining to the intercalary months (*adhikamāsa*), omitted days (*kṣaya*), and the elapsed days since an epoch (*ahargaṇa*). An intercalary month at regular intervals brings about adjustments between the lunar year and the solar year. In the solar calendar, there is what is called *saṅkramaṇa* or *saṅkrānti*, that is, the solar ingress into *rāśi* (zodiacal sign). If, however, a lunar month is within the period of the corresponding solar month, there will be no *saṅkramaṇa* during that solar month. Such a lunar month is *adhi* or *adhikamāsa*. Generally, an *adhikamāsa* occurs between the months of Phālguna and Āśvina. In addition, one month (*kṣayamāsa*) is deleted once in 141 years, or sometimes once in nineteen years for adjustments. The intercalary months need to be taken into consideration while calculating the number of elapsed days since a chosen epoch. A *pañcāṅga* is concerned with all of these.

Standardization: In tune with the differing local traditions and diverse practices of religious nature, there is a plethora of *pañcāṅgas* all over the country. Though it is professed that they are by and large based on the Siddhāntic texts, there is a lack of uniformity in the computations of *tithi,* etc., as well as planetary positions. Moreover, the planetary parameters change over a period of time and corrections are to be applied to planetary positions from time to time. The need for the preparation of a standardised *pañcāṅga* with modifications or methods that would be followed by all *pañcāṅga*-makers, leading to the formulation of a single type of *pañcāṅga* annually, was keenly felt even in the early decades of the last century. This issue was discussed at several Oriental conferences. But no consensus could be arrived at, and the traditional *pañcāṅga*-compilers continued to move along the already determined grooves. The common people by and large continue to accept their local *pañcāṅga*, and are lukewarm to any reformation or refinement of the calculations, particularly of the planetary positions. The precession of equinoxes does not seem to be the prime concern, as it should be, of the traditional *pañcāṅga*-calculators, although in one of the conferences held in Sangli in 1919, it was recommended that the rate of precession of 50.2 *vikalās* (the modern value) be adopted by all the *pañcāṅga*-makers, and that the junction-star of the asterism Revatī (Zeta Piscium) be deemed as the beginning of the zodiacal belt. Some attempts were made in this direction; for example the *Tilak pañcāṅga* (1924) which, however, did not produce the expected results; nor did it find general acceptance.

Soon after Independence, the Government of India set up, in 1952, the National Calendar Reforms Committee under the distinguished chairmanship of Meghnad Saha. The Committee deliberated on this issue for over three years, interacted with many traditional astronomers as well as *pañcāṅga*-makers and submitted its recommendations in November 1955. Among its major recommendations were: (i) The Śaka year to be used in the unified national calendar—the Śaka year 1876 corresponds to 1954-55 CE; (ii) The commencement of the year was to be from the day following the vernal equinox day; (iii) The normal year was to consist of 365 days, while a leap year would have 366 days (the procedure is that, after adding 78 to Śaka year, if the sum of the digits of the year is divisible by 4, then that year becomes a leap year, with 366 days; if, however, the sum becomes a multiple of 100, it would be a leap year only when it is divisible by 400; otherwise it would be a normal year with 365 days); (iv) Caitra would to be the first month of the year and the lengths (in days) of twelve months would be as follows:

Caitra : 30 days (31 days in a leap year)	Āśvina : 30 days
Vaiśākha: 31 days	Kārtika : 30 days
Jyeṣṭha : 31 days	Margaśira : 30 days
Āṣāḍha : 31 days	Pauṣa : 30 days
Śrāvaṇa : 31 days	Māgha : 30 days
Bhādrapada : 31 days	Phālguṇa : 30 days

More importantly, the Committee made several recommendations for the religious calendar, keeping in view the seminal role being played by *pañcāṅga* in various religious activities and festivals, determining the timings relating to the day, month, and *nakṣatras,* particularly for marriages and religious observances. The Committee was of the view that its recommendation was to be regarded only as a measure of compromise, so that a violent break with the established practice would be avoided. However, this and other recommendations of the committee have not received any support from the traditional almanac-makers.

The Eras: *Pañcāṅgas* refer to some eras like the Śālivāhana and Vikrama-śaka. Since the eras relate to royal years or the reigning period of a king, they are also recorded in several inscriptions. Probably no era-reckoning was in vogue during the early Indian dynasties including the Aśokan rule. According to Aśoka's Girnar inscription, he was in diplomatic correspondence with the Greek kings as well as the Ptolemys (royal families) of Egypt where the Selucidian era was in use then. The Vikram era is believed to have begun in 57 BCE, the year in which King Vikramāditya of Ujjain was supposed to have defeated the Śaka or the Scythian hordes. However, there is no proven historical evidence for this or that it even existed. Alongside were the Gupta era (320 CE) and Harṣa era (early seventh century CE). It may be noted that the Gupta emperors had an 'Āditya' title and some of them were even referred to as 'Vikramāditya'.

The most popular era in India has been the Śālivāhana-śaka which has been reckoned from 78 CE. It goes by various names: Śaka-kāla, Śakendra-kāla, and Śaka-saṃvat. In the Śālivāhana Śaka era, its year begins with Caitra and ends with Phālguṇa for luni-solar reckoning, and Meṣādi for solar-reckoning. Generally in the north, the months are *pūrṇimānta* (ending with full-moon day) and in the south, they are *amānta* (ending with new moon day). The use of the Śālivāhana-Śaka year in inscriptions dates from the time of Pulakeṣin I of the Chālukya dynasty, as evidenced by an inscription of 405 of the *Śaka* era. Among the other major eras are: the Kollam era (in Kerala), also called the era of Paraśurāma and the Jovian era, also called the Bṛhaspati era, of sixty years, beginning with the year called Prabhava and ending with the year named Akṣaya, in a cyclic manner. Two cycles of Jovian years have been recognized, one consisting of twelve Jovian years, and the other of sixty years. The years of the former bear the same names as those of the twelve lunar months. The twelve-year cycle, however, begins with the Āśvina or Āśvayuja month. The commencement of the sixty-year cycle could also be Vijaya in the place of Prabhava. If a Śaka year is to be converted into the Jovian era, eleven is to be added to the former, and the sum to be divided by sixty. The quotient, if any, is to be ignored. The remainder gives the year in Bṛhaspati era as counted from Prabhava. For the conversion of the Kollam year into Kaliyuga year, 3926 is to be added to the latter. The Siddhāntic texts reckon the Kaliyuga era for their computation, and it is stated to have commenced on 17-18 February 3102 BCE. An extrapolated era, it began to be adopted probably for the first time in the *Āryabhaṭīya* (fifth century CE). The earliest inscription in which this era has been mentioned is of King Pulakeṣin II (634-35 CE).

A *pañcāṅga,* which is a repository of calculations of its five components based on certain rules enunciated in the Siddhāntic and *karaṇa* texts, has been the torchbearer of the long Indian astronomical tradition which, however, does not easily accept any reformation in the name of uniformity or modernity. *Pañcāṅga* also plays a pivotal role in planetary astrology, notwithstanding anyone's opinion or judgment whether astrology is scientific of unscientific. Few will deny that astrology is very much a part of the Indian way of thinking and living, in one form or the other.

4

MATHEMATICS: APPROACH AND ACHIEVEMENTS

The dictionary definition of mathematics is that it is an abstract science of space, number and quantity. Communicable language along with a number-reckoning system with its numerical forms have been the warp and weft of any civilizational fabric. These two along with agricultural operations, arts and crafts, and the purposive use of minerals and metals, have meaningfully shaped diverse human settlements in different parts of the world right from the Neolithic times. Between the two, the concept and formulation of numbers are indeed a triumph of the human mind. Man is the creator of numerals and numbers, since they do not exist concretely in nature, to be discovered. More importantly, over the millennia, man has acquired such a mastery of number-reckoning and its usage that man and numbers have become inseparable. There is rather a subtle distinction between numerals (*forms*) and numbers (*quantity*), though the latter is also used in the sense of the former. Numbers are both art and science: art, in their notational forms; and science, as they have rationally engendered both theoretical and quantitative intellectual exercises.

The origin of number-reckoning can be traced to remote antiquity, to the activities of early man as food-gatherer who was perhaps intuitively able to identify 'one', 'two' and the several, as 'many' by pointing to the objects before him. Some experiments with birds like the crow have revealed that they do possess a number-sense of 'one', 'two' to a limited extent. But man, born with pointers in the form of ten fingers, and later, toes too, was able to denote numbers through them. His early efforts, however, were still circumscribed by the act of correspondence between the objects before him and his number-pointers. In course of time, some type of recording, generally a vertical line to start with to denote one, two lines to indicate two, etc., or an assemblage of pebbles, sticks, or notches in wood or bone, or even tying the desired number of knots in a string of rope, were employed for tallying purposes. Gradually, appropriate vocal sounds began to manifest themselves as words-tally.

The most significant development in the history of early number-reckoning was the 'grouping' of numbers. It is surmised that initially the first attempt at 'grouping' was in sets of 'four', probably because of the set or row of four fingers of the human hand. Later, the thumb was included to identify a 'group' of 'five', which eventually led to the group of 'ten' and 'twenty'. The first was based on the ten fingers of both the hands, and the second, including the ten toes. The ten fingers gave rise to the ten or the decimal system, and with the addition of toes, what was known as the vigesimal system emerged. But it is the decimal system that has prevailed and become an integral part of human life. The Egyptians adopted the decimal scale; likewise the Greeks and the Romans. The Babylonians, however, adopted the decimal scale up to ten and, thereafter, what is known as the sexagesimal system with number sixty as the base. The Mayas (Mexico region) had adopted the vigesimal system with twenty as thc base.

The ingenuity of number-reckoning attained new dimensions with the emergence of specific numerical forms or notations in advanced human settlements or early civilizations. The urbanization which was steadily coming to the fore had its social, economic, and bureaucratic compulsions—the maintenance of records of agricultural production and

Egyptian numerals

Number	Hieroglyphic	Hieratic
ONE	|	|
TWO	||	||
THREE	|||	[hieratic sign]
FOUR	||||	[hieratic sign]
FIVE	|||||	7
SIX	||| |||	2
SEVEN	|||| |||	[hieratic sign]
EIGHT	|||| ||||	[hieratic sign]
NINE	||||| ||||	[hieratic sign]
TEN	∩	∧
TWENTY	∩∩	[hieratic sign]
THIRTY	∩∩∩	×
FORTY	∩∩∩∩	[hieratic sign]
FIFTY	∩∩∩∩∩	7
SIXTY	∩∩∩∩∩∩	[hieratic sign]
SEVENTY	∩∩∩∩∩∩∩	[hieratic sign]
EIGHTY	∩∩∩∩∩∩∩∩	[hieratic sign]
NINETY	∩∩∩∩∩∩∩∩∩	[hieratic sign]
HUNDRED	𓍢	[hieratic sign]
TWO HUNDRED	𓍢𓍢	[hieratic sign]
THREE HUNDRED	𓍢𓍢𓍢	[hieratic sign]
FOUR HUNDRED	𓍢𓍢𓍢𓍢	[hieratic sign]
EIGHT HUNDRED	𓍢𓍢𓍢𓍢 𓍢𓍢𓍢𓍢	[hieratic sign]
NINE HUNDRED	𓍢𓍢𓍢 𓍢𓍢𓍢 𓍢𓍢𓍢	[hieratic sign]
THOUSAND	𓆼	[hieratic sign]
TWO THOUSAND	𓆼𓆼	[hieratic sign]
NINE THOUSAND	𓆼𓆼𓆼 𓆼𓆼𓆼 𓆼𓆼𓆼	[hieratic sign]

Source: Cajori, F.: *A History of Mathematical Notations*, 2 Vols, Chicago, 1928, p. 12

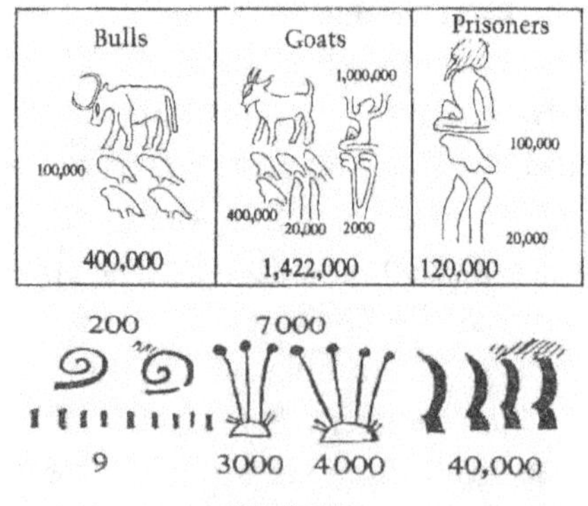

Source: Ifrah, Georges, *From One to Zero: A Universal History of Numbers*, New York, 1985, pp. 204-05

management, adoption of standards for weights and measures, payment of wages, maintenance of trade accounts, counting of the days, months, and years—all of which needed a viable system of numbers and numerical forms for their precise communication. Each civilization tried to develop its own numerical forms and the methods of recording as well as the media through which such records were stored: papyrus of the Egyptian civilization, clay tablets of Mesopotamia, palm leaves or birch bark of the later Indian culture or steatite tablets of the Indus Valley civilization. The numerical forms of the Egyptian, Mesopotamian, Chinese, Greeks, Early Romans, and the Mayan have been presented here.

Mesopotamian numerals

	Classical Sumerian	Assyrian-Babylonian or Akkadien
One		
Two		
Three		
Four		
Five		
Six		
Seven		
Eight		
Nine		
Ten		
Twenty		
Thirty		
Forty		
Fifty		
Sixty		

Source: Labat, Rene: *Manuel D'Epigraphie Akkadienne: Signes, Syllabaire, Ideogrammes*, Paul Geuthner, Paris, 1986; pp. 242-247

An example of the Mesopotamian numeration from an economic tablet, belonging to the nineteenth century BCE.

60 40 ----> 100 | 2 ME ----------> 200 | 1 ME 3 ----------> 103 | 1 ME 50 4 ----------> 154

A northern Babylonian account-tablet about the number of sheep, (seventeenth century BCE):

1 SHU-SHI 3 ----------> 63 | 60 10 3 --------> 73 | 60 20 5 --------> 85 | 1 ME 1 SHU-SHI 8 ----------> 168

Source: Ifrah, Georges, *From One to Zero: A Universal History of Numbers,* Viking Penguin Inc., New York, 1985, p. 327

Chinese numerals (oracle-bones and bronzes).

	Shang Oracle- (14th-11th Cent. B.C)	Bronze and Coin forms (10th-3rd Cent. B.C.)	Other forms on Coins (6th-3rd Cent. B.C.)
1	—	—	—|
2	=	=	=‖
3	≡	≡	≡|||
4	≣	≣	≣ XX Ⅲ XX
5	X	X	≣ XXX
6			⊥ T
7	+	+	T ⊥
8	)(	)(	X
9			
10	|		+
11			
12			
13			
14	Analogous, but not found in inscriptions		

	Shang Oracle- (14th-11th Cent. B.C.)	Bronze and Coin forms (10 th-3rd Cent. B.C.)	Other forms on Coins (6th-3rd Cent. B.C.)
15	X		X
20	U		
30			
40			
50			
60			
100			
200			
300			
500			
1,000			
3,000			
4,000			
5,000			

Source: Needham. J., *Science and Civilization in China*, Cambridge University Press, III, pp. 22-23

Chinese rod numerals

FIRST SERIES

SECOND SERIES

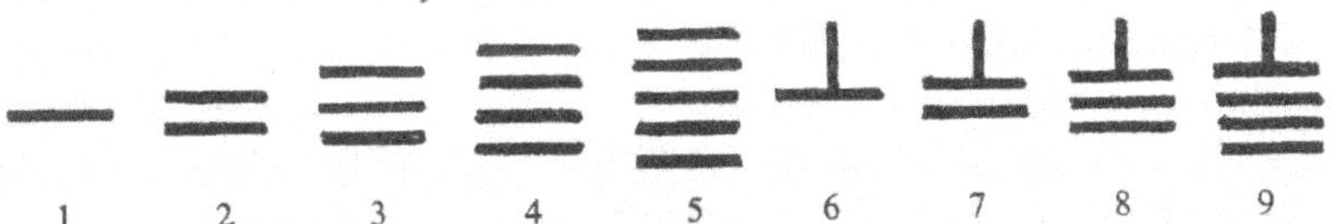

Greek numerals

Numbers	1	5	10	50	100	500	1,000	5,000	10,000
Old Celtic (1500 B.C.)	) or I	))))) or IIIII	• or —		\ or /		◇		
Attic Greeks (600 B.C.) and Herodionic	I	Γ	Δ	𐅄	H	𐅅	X	𐅆	M

Source: Cajori, F: *A History of Mathematical Notations,* 2 Vols, Dover Publications, Chicago, 1928, p 21; Crozer, V.S.: *A Survey of Mathematics,* New York, 1968, p.40

Roman numerals

Modern	1	2	3	4	5	6	7	8	9	10	50	100	500	1,000	5,000
Roman	I	II	III	IIII	V	VI	VII	VIII	IX	X	L	C	D	M	ↁ

Source: Crozer, V.S, *A Survey of Mathematics,* New York, 1968, p.41

Mayan numerals

Modern	1	2	3	4	5	6	7	10	11	15	16	19	20
Maya	•	••	•••	••••	—	• —	•• —	= 	• =	≡	• ≡	•••• ≡	• ⊖

Source: Cajori, F, *A History of Mathematical Notations*, 2 Vols, Dover Publications, Chicago, 1928, p. 41; Crozer, V.S, *A Survey of Mathematics*, New York, 1968, p. 51

INDIAN NUMERATION

In the Indian subcontinent, the earliest numerical signs (vertical strokes) for one to twelve have been found on what are called the Indus seals and other inscribed objects. Whether all the Indus script forms are numerals is a moot point. The Indus script has not been deciphered so far. There was, however, a long interregnum of over eighteen centuries between the usage of vertical strokes on the Indus inscribed objects and some definitive numerical forms which appeared first in the Aśokan Brāhmī and Kharoṣṭi inscriptions.

Kharoṣṭi inscriptions

ONE	/
TWO	//
THREE	///
FOUR	X or ////
FIVE	/X or /////
SIX	//X
SEVEN	///X
EIGHT	XX

NINE	\XX
TEN	ↄ
TWENTY	3
FORTY	33
SIXTY	333
EIGHTY	3333
HUNDRED	ʎ or ʎl
TWO HUNDRED	ʎll

Brāhmī numerals

Number	Aśokan	Nānāghāt	Nāsik	Āndhra. Kuśān Mathura	Gupta	Pallava	Place-value Notations copper plates and Gwalior
ONE	[illegible]	[illegible]	[illegible]	[illegible]	[illegible]	[illegible]	[illegible]
TWO	[illegible]	[illegible]	[illegible]	[illegible]	[illegible]	[illegible]	[illegible]
THREE		[illegible]	[illegible]	[illegible]	[illegible]	[illegible]	[illegible]
FOUR	[illegible]	[illegible]	[illegible]	[illegible]	[illegible]	[illegible]	[illegible]
FIVE			[illegible]	[illegible]	[illegible]	[illegible]	[illegible]
SIX	[illegible]	[illegible]	[illegible]	[illegible]	[illegible]	[illegible]	[illegible]
SEVEN		[illegible]	[illegible]	[illegible]	[illegible]	[illegible]	[illegible]
EIGHT			[illegible]	[illegible]	[illegible]	[illegible]	[illegible]
NINE		[illegible]	[illegible]	[illegible]	[illegible]	[illegible]	[illegible]
TEN		[illegible]	[illegible]	[illegible]	[illegible]		[illegible]
TWENTY		[illegible]	[illegible]	[illegible]	[illegible]		[illegible]
THIRTY				[illegible]	[illegible]		[illegible]
FORTY			[illegible]	[illegible]			[illegible]
FIFTY	[illegible]			[illegible]			[illegible]
SIXTY		[illegible]		[illegible]	[illegible]		[illegible]
SEVENTY			[illegible]	[illegible]			[illegible]
EIGHTY		[illegible]		[illegible]	[illegible]		[illegible]
NINETY				[illegible]	[illegible]		[illegible]
HUNDRED		[illegible]	[illegible]		[illegible]		[illegible]
TWO HUNDRED	[illegible]	[illegible]	[illegible]		[illegible]		[illegible]
FOUR HUNDRED		[illegible]	[illegible]		[illegible]		[illegible]
THOUSAND		[illegible]	[illegible]				[illegible]
FOUR THOUSAND		[illegible]	[illegible]				[illegible]
TWENTY THOUSAND		[illegible]	[illegible]				[illegible]

Source: B.V. Subbarayappa, *Indus Script: Its Nature and Structure*, New Era Publications, Madras, 1996, p. 33

Nevertheless, during this long period, the number system found expression in the form of words in the Vedic literature. Vedic Indians adopted ten as the base and evolved an appropriate terminology for expressing even large numbers on the ascending decimal scale (*daśaguṇottara*). The *Taittirīya* and the

Vājasaneyī Saṃhitā (*Yajurveda*) have enumerations as follows: *eka* (one); *daśa* (ten or 10^1); *śata* (10^2); *sahasra* (10^3); *ayuta* (10^4); *niyuta* (hundred thousand $=10^5$); *prayuta* (ten hundred thousand = 10^6); *arbuda* (hundred hundred thousand = 10^7); *nyarbuda* (ten hundred hundred thousand = 10^8); *samudra* (hundred hundred hundred thousand = 10^9); *madhya* (ten hundred hundred hundred thousand = 10^{10}); *anta* (hundred hundred hundred hundred thousand = 10^{11}); and *parārdha* (ten hundred hundred hundred hundred thousand = 10^{12}). The *Pañcaviṃśa Brāhmaṇa* enumerates large numbers in the context of sacrifice and the merits that would accrue to a performer of sacrifice as follows (S.N. Sen, *A Concise History of Science in India*, pp. 178-179):

> 'By sacrificing with the *agnihotra,* he reaches in one day, ten "House-lords" (i.e. he becomes equal to, gains the merits of ten H.); by sacrificing with the *agnihotra* during ten years, he becomes equal to one who (regularly) performs the sacrifices at full and new moon, he becomes equal to one who performs the sacrifices of *soma*. By offering the *agniṣtoma* sacrifices, he becomes equal to one who performs a sacrifice of a thousand cows as sacrificial fee. By offering ten of these, he becomes equal to one who performs a sacrifice with ten thousand *dakṣiṇās*. By offering ten of these, he becomes equal to one who sacrifices with a sacrifice of a hundred thousand *dakṣiṇās*. By offering ten of these, he becomes equal to one who sacrifices with a sacrifice of a million *dakṣiṇās*. By offering ten of these, he becomes equal to one who sacrifices with a sacrifice of 10 million *dakṣiṇās*. By offering ten of these, he becomes equal to one who sacrifies with a sacrifice of 100 million *dakṣiṇās*. By offering ten of these, he becomes equal to one who sacrifices with a sacrifice of 1,000 million *dakṣiṇās*. By offering ten of these, he becomes equal

> to one who sacrifices with a sacrifice of 10,000 million *dakṣiṇās*. By offering ten of these, he becomes equal to one who sacrifices with a sacrifice of 100,000 million *dakṣiṇās*. By offering ten of these he becomes the cow.' (*Pañcaviṃśa Brāhmaṇa*, XVII, 14-1-2)

Another Vedic text, the *Kāṭhaka Saṃhitā* lists the same, but with the exchange of places for *niyuta* and *ayuta*. Some texts follow this enumeration up to *nyarbuda*, then introduce the words *nikharavaka, bhadra, akṣita* and *go*.

A progressive arithmetical series involving odd (*ayugma*) and even (*yugma*) numbers is also found in the Vedic literature, again in the context of performing sacrifices. For example, the *Vājasaneyī Saṃhitā* of the *Yajurveda* (18.24-25) has a chant as follows: 'May my *one* and *three*, my *three* and *five*, my *five* and *seven*...my *thirty-one* and *thirty-three* prosper by sacrifice; May my *four* and my *eight*; my *eight* and *twelve*...my *forty-four* and *forty-eight* prosper by sacrifice.' It would appear that the number-reckoning as well as the manipulation of desired numbers was a favourite pastime of Vedic priests.

Operations of addition, subtraction, multiplication and division too needed the usage of precise number-names or appropriate words. The Vedic scholars were proficient in developing a vocabulary that facilitated these operations. The first nine digits were named *eka, dvi, tri, catur, pañca, ṣat, sapta, aṣṭa* and *nava*. The second group of nine numbers were obtained by multiplying each of the nine digits by ten, that is, *daśa, viṃśati, triṃśat, catvāriṃśat, pañcāśat, ṣaṣṭi, saptati, aśīti* and *navati*. This method was also adopted beginning with 100. The additive aspect was accomplished by naming a number in which the first and the second groups were the components: *eka-daśa*; (ii) *sapta-viṃśati* (27); *aṣṭā-triṃśat* (38). Higher numbers were denoted by juxtaposing the concerned words: *sapta śatāni viṃśatiḥ* (720), *ṣaṣṭiṃ-sahasra navatiṃ nava* (60,099). A type of subtractive principle was also in vogue in respect of the number nomenclature: for 19, one less than 20 (*ekānna viṃśat*); for 99, one less than 100 (*ekānna-śata*). The word *ekānna* or one less, later became *ekona* and *ūna*.

Vedic people knew simple fractions; for example, *pāda* $\left(\frac{1}{4}\right)$; *ardha* $\left(\frac{1}{2}\right)$; *tripāda* $\left(\frac{3}{4}\right)$; *śapha* $\left(\frac{1}{8}\right)$; *kuṣṭḥa* $\left(\frac{1}{12}\right)$; and *kalā* $\left(\frac{1}{16}\right)$. Later, words such as *aṃśa* and *bhāga* were in use for denoting fractions—*tryaṃśa* or *tribhāga* $\left(\frac{1}{3}\right)$; and *pañcamabhāga* $\left(\frac{1}{5}\right)$, *pañcādaśa-bhāga* $\left(\frac{1}{15}\right)$. There were other expressions like *triaṣṭama* or *tryaṣṭa* $\left(\frac{3}{8}\right)$ and *pañcamasya caturviṃśa* $\left(\frac{1}{24} \text{ of } \frac{1}{5}\right)$. Though the learning process was by and large oral, there would have been some ways of writing the numbers; but we have little or no information on this.

The Śulba sūtras deal with addition, subtraction, multiplication, division, and even the squaring of fractions. For instance, the *Baudhāyana* Śulba sūtra (iii, 219-24) states as follows:

> 'Then he measures the area of this square-shaped *citi* whose side is three *puruṣas* less one-third. At the western side of this square is the handle whose east-west length is half a *puruṣa* plus ten *aṅgulas,* and the north-south breadth is one *puruṣa* less one third. This makes the area of the *agni* sevenfold plus two *aratnis* and one *prādeśa* (this is equal to 60 *aṅgulas* or half *puruṣa*) (in all $7\frac{1}{2}$).'*

---

*If this is expressed in numbers, the formula is: the area of the *agni* =

$$\left(3-\frac{1}{3}\right)^2+\left(\frac{1}{2}+\frac{1}{12}\right)\left(1-\frac{1}{3}\right)=7\frac{1}{2}.$$

(The foregoing details are based on: Sen, S.N., *A Concise History of Science in India*, pp. 181-182)

Such methodical exercises involving numbers are found in the Śulba sūtras, which are also repositories of geometrical knowledge of the Vedic priests who were well versed in the performance of sacrifices by using different forms of altars (as will be seen later).

However, Vedic texts do not provide any significant clue to the numerical forms that were in use then, the main reason being the assiduously fostered oral tradition. But the numerals needed a recording tradition. The earliest records are the Aśokan inscriptions (third century BCE) in which some of the numerals could be found, apart from the notations of numerical values for one, two...up to twelve (vertical strokes) which are noticed on the Indus seals and other inscribed objects. The numerals of Aśokan and later inscriptions like those of Nāṇeghāt and Nāsik are called Brāhmī numerals since the inscriptions are in the Brāhmī script. Practically contemporaneous with the Brāhmī numerals were those in the Kharoṣti script (hence called Kharoṣṭi numerals) found also in the inscriptions of the Śaka, Pārthian, and Kuśāna royal dynasties from the third century BCE to the second century CE. These numerals were used on the basis of what is known as the additive-multiplicative principle.

Ciphered System

In the history of numbers or numerals, what is called the Ciphered System was governed by the additive-multiplicative process and had a long innings in the three early civilizations as well as in the Chinese civilization. In this system, generally, the base b employed is ten, and hence its numeration is decimal in its conceptual matrix. There are different symbols for the numerals one to nine ($b - 1$); a new symbol for *ten*; different symbols for multiples of ten ($2b$, $3b$...$9b$) but derived from the symbol for ten; a new symbol for *hundred* which is the square of the base b: (b^2); multiples of hundred ($2b^2$, $3b^2$...$9b^2$) the symbols being derived from that of hundred; and a fresh symbol for *thousand*, the cube of base b, (b^3); which is followed by its multiples. The desired numbers were reckoned by adopting the additive or multiplicative or both the processes. For example, for expressing the number 26, the symbol for 20 or two symbols for 10 were used along with the symbol for six placed side by side. If the number, say, 300 was to be expressed, the symbol for 100 was fused with that of three to indicate multiplication. The ciphered system, therefore, had a plethora of symbols and, for expressing a large number, different symbols could be used employing the additive-multiplicative process. For example, to express the number 420, one of the following ways was adopted:

(i) the symbol for 400 or,
(ii) the symbol for four fused with that of 100 or,
(iii) Symbol of 100; four of them placed side by side (indicating addition);

Similarly for twenty:

(i) Two symbols of ten placed side by side or,
(ii) The symbol of twenty or,
(iii) The symbol of five fused with that of four.

The total of the numerical symbols would be additive.

Though cumbersome, the ciphered system would reflect the exact nature of the numbers (quantitative measures) involved. But, to express large numbers a multitude of symbols had to be used.

The *Vedāṅgas* and Early Mathematics

Of the six *Vedāṅgas,* the Kalpa sūtras are in the nature of sacerdotal manuals concerning the performance of sacrifices. The Śulba sūtras of the Kalpa manuals contain the precise rules relating to the geometrical aspects of the sacrificial altars. Although the Śulba texts have not explicitly stated certain mathematical ideas, the following can be inferred: the rules assume that a circle can be divided by drawing its diameters; diagonals bisect each other; a rectangle can be divided by its diagonals into four parts equal in all respects; an isosceles triangle can be divided into two equal halves by a straight line which joins the middle point of the opposite side with the vertex. The Śulba rules also deal with parallelograms and the area of the square inscribed within a circle and touching the circumference. Besides, they provide several rules which enable one to construct a straight line perpendicular to another straight line; drawing a square with a given side, or a rectangle with given sides; a parallelogram and rectangle on the same base and within the same parallels having the same area; the diagonals of a rhombus bisecting each other at right angles; a quadrilateral formed by the lines joining the middle points of the sides of a rectangle being a rhombus whose area is half of that of the rectangle; and the construction of a square equal to the sum of two different squares. The geometrical instructions given in the Śulba texts are lucid and accurate.

A special feature of the Śulba sūtra*s* is the set of rules given by them for the combination and transformation of rectilinear figures, particularly the rectangles and squares.

The Śulba texts have also dealt with methods of constructing a square equal to the sum of two different squares. The so-called theorem of Pythagoras, namely, the square on the hypotenuse is equal to the sum of the squares on the two sides of a right-angled triangle, is stated precisely in terms of the sides and diagonal of a rectangle in the Śulba sūtra*s*. The *Baudhāyana* Śulba (i, 48) says: 'The diagonal of a rectangle produces both areas which its length and breath produce separately. In a rectangle, the diagonal results in the formation of two right-angled triangles.' The *Kātyāyana Śulba* specifies it by stating that this is applicable to plane figures. In general, the Śulba sūtra*s* give this theorem in an accomplished form for plane figures. But, it needs to be recognized that the Śulbavid*s* would have arrived at it through valid proof and concrete examples. This has been evidenced by a statement in the form of a rule in the *Baudhāyana Śulba* (i.49), that this is valid for rectangles having their sides, 3 and 4; 12 and 5; 15 and 8; 7 and 24; 12 and 35; 15 and 36.

Hence, $3^2+4^2=5^2$; $12^2+5^2=13^2$; $15^2+8^2=17^2$; $7^2+24^2=25^2$; $12^2+35^2=37^2$; and $15^2+36^2=39^2$. It may be noted that the relationship between the diagonal (or the hypotenuse) and the two sides of rational rectangles, or triangles, was utilized in practice for finding out the perpendicular directions (north-south) at the east or the west point of the east-west line in the construction or setting up of the altars.

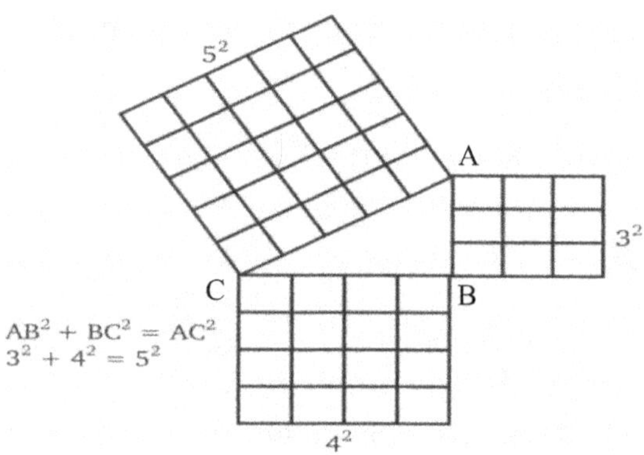

The 'Baudhāyana' theorem

In the history of mathematics, the Pythagorean theorem has no doubt an honoured position. But whether the Pre-Socratic Greek philosopher–mathematician, Pythagoras (*c.* sixth century BCE) was the actual propounder of this theorem or not is still an unsolved question. He was no doubt an outstanding Pre-Socratic thinker who even thought that numbers were the primordial attributes of all things, and 10 was the perfect one which comprised the whole nature of numbers. His name as associated with this theorem appears in the much later works of Cicero (first century BCE), Diogenes Laertius (second century CE), Heron (third century CE), and others. A knowledge of the relationship of the type expounded in the Pythagorean theorem could also be found among the Egyptians and the Chinese, although it is not so well delineated as in the Indian Śulba sūtras.

In the history of science, it is essential to understand the idea or thought-structure in its context. The so-called Pythagorean theorem arose out of the general ideas of Pythagoreans. In number they saw many things numerically—even reason and justice. They regarded the whole heaven as a musical and numerical scale, and conceived of the 'harmony of the spheres'. On the other hand, the Śulba sūtras presented the theorem in terms of the square on a diagonal (in a rectangle) being equal to the sum of the squares on the two sides of the right-angled triangle thus formed. Each of these two approaches needs to be understood in its context to appreciate its originality.

The *Kātyāyana Śulba* provides a general rule for the construction of a square n times a given square. This can also be done in a sequence of steps. The diagonal of a square produces the double square, the diagonal of a rectangle formed of the previous diagonal and the side of a given square give a square which is three times large, and so on.

In the history of mathematics, one of the most difficult problems for a long time was how to construct a square (using only ruler and compass) whose area is that of a given circle. Known as 'squaring the circle', this intriguing problem was taken up seriously by the śulbavids and they enunciated practical methods for such constructions, and even vice versa. In these exercises, the value of π, a constant number, (the ratio of the circumference of any circle to its diameter) needed to be approximated to 3.0885, though it was less than the modern approximated value (3.1416 approx). It is now known that an exact construction of 'squaring a circle' or 'circling a square' in the sense of equivalent areas is impossible.

To construct a circle whose area is equal to that of a given square (approximately), Baudhāyana's instructions are as follows:

'If you wish to circle a square, draw half of its diagonal about the centre towards the east-west line; then describe a circle together with a third part of that which lies outside the square.' (i, 58)* The *Āpastamba sūtra* rightly states that the circle thus constructed is not exact (*anitya*).

From Vedic geometry of the construction of different types of sacrificial altars, certain algebraic ideas emerge like those of quadratic equations, indeterminate equations of the first degree, surds, and their approximate values. The standard *mahāvedi* has the shape of an isosceles trapezium, with its base, face, and altitude of 30, 24, and 36 units respectively having a total area of 972 sq. units. The geometrical format of the

---

*

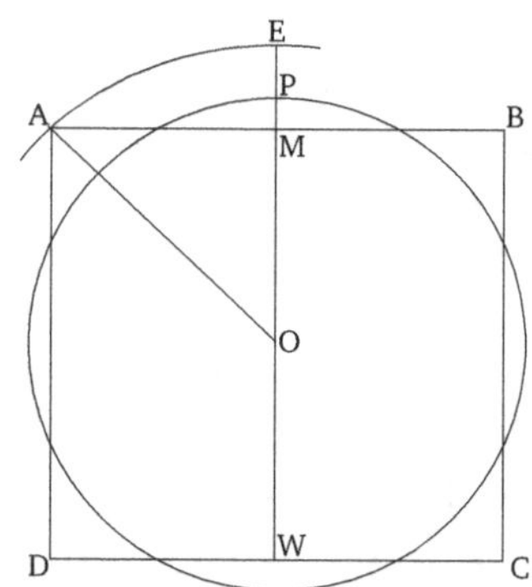

Circling the square

Let ABCD be a square which is to be transformed into a circle so that the areas of both are the same. Let O be the central point of the square. Join OA. Then draw a circle intersecting the east-west line at E, with O as centre and OA as radius. Divide EM at P such that EP=2 PM, where M is the midpoint of the side AB as shown in the figure. Now, with O as centre and OP as radius draw a circle. This circle is nearly equal in area to that of the square ABCD. (*Source:* S. Balachandra Rao, *Indian Mathematics and Astronomy,* Bharatiya Vidya Bhavan, Bangalore, (rev. 3rd ed.) 2004, p. 20).

standard *Śyena cit* (falcon-shaped altar) is as follows: a central portion or trunk of 4 sq. units, two rectangular side wings, each measuring 1 by $1\frac{1}{5}$ or $1\frac{1}{5}$ sq. units, while the tail part measures 1 by $\frac{1}{10}$ or $1\frac{1}{10}$ sq. units, the total area of a *Śyena-cit* being $7\frac{1}{2}$ sq. units. However, some parts of the altar could be proportionately increased, retaining the same measurements of the other parts. The units for the central portion could also be changed proportionately while the measurements of the other parts could be left unchanged. Such exercises were meticulously carried out with great precision.

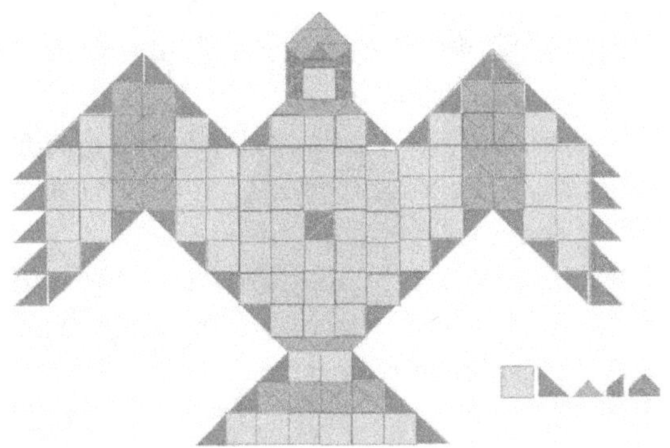

Śyena cit.
(Courtesy: K. Ramasubramanian)

The sizes and the number of bricks were important for the construction of different layers of the altars, and these would involve the solution of simultaneous indeterminate equations of the first degree. For instance, according to *Baudhāyana Śulba*, a *gārhapatya vedi* should be constructed with five layers of bricks, each layer containing twenty-one bricks, which are to be arranged in such a way that their rifts in two consecutive layers do not coincide. For an altar of the shape of a square in which square bricks are to be used, according to this *Śulba*, three kinds of bricks with their sides $\frac{1}{6}, \frac{1}{4}$ and $\frac{1}{3}$ of the side of the altar would provide the solution. Since the number twenty-one is not a square number and since the rifts of the consecutive layers should not coincide, square bricks of at least three different sizes will have to be used. Thus the constructions of several types or forms of altars for the performance of sacrifices of different objectives throw ample light on the geometrical knowledge of Vedic times. The *Śulba* texts are noted for their clarity and brevity of mathematical expressions.

Operations with $\sqrt{2}$, $\sqrt{3}$, $\sqrt{18}$, etc., have been dealt with in the Śulba texts. They are called *dvi-karaṇī, tri-karaṇī and aṣṭādaśa-karaṇi* respectively. The word *karaṇi* is used in the sense of surd or root. The *śulbavids* (experts in Śulba geometry) were also aware of the irrationality of quantities like $\sqrt{2}$, and $\sqrt{3}$, and attempted to understand them, besides giving the value for such irrational numbers to a high degree of approximation. For the value of $\sqrt{2}$ the *Kātyāyana śulba* (ii. 13) states as follows (translation of the passage in English).

'Increase the measure by its third, and this third by its own fourth less the thirty-fourth part of that fourth. This is the value with a special quantity in excess.'

If 1 is taken as a measure of side of a square, the above formula leads to the value of the diagonal as follows:

i.e. $\sqrt{2} = 1+\frac{1}{3}+\frac{1}{3.4}-\frac{1}{3.4.34}=1.4142156$

(Likewise, the value of $\sqrt{3}$ is $1+\frac{2}{3}+\frac{1}{3.5}-\frac{1}{3.5.52}=1.7320513$ could be arrived at).

However, the method or the manner by which such a formulation was arrived at has not been presented in the Śulba texts. Some scholars have reconstructed the possible methods adopted by the *śūlbavids*. A critical study of the Śulba texts would reveal that in the mathematical exercises of the Vedic period, geometrical diagrams and methods were adopted for solving algebraic problems or equations.

Permutations and Combinations

If the Śulba sūtras of the *Kalpa* (one of the *Vedāṅgas*) are a repository of Vedic geometry and the associated algebraic notions, another *Vedāṅga* known as *Chandas* (prosody) represents an important mathematical pursuit in permutations and combinations. This is evident in certain chantings of the Vedic priests in the form of permutations and combinations. Such a pursuit was also noticeable among the Jaina thinkers. The Vedic metres like *Gāyatrī* (6 syllables), *Anuṣṭubh* (8), *Bṛhatī* (9), *Triṣtubh* (11) and *Jagatī* (12), and their variations were in a way related to precise permutations and combinations. Vedic priests being specialists in proper metres, they were capable of producing different types of metres by changing the long and short sounds with each syllable group. Thus, the Vedic metres assumed new nuances as well as forms as a result of definitive permutations and combinations. Later, such an approach found its application in the possible combinations of six tastes (in Ayurveda) taking one, two, three, etc., at a time. Varāhamihira, in his *Bṛhatsaṃhitā*, has adopted this method for enumerating the possible total number of perfumes that could be made from sixteen different aromatic substances taking one, two, three, or four basic ingredients at a time. The Jaina thinkers, too, adopted this approach to determine the possible number of philosophical categories from *n* fundamental categories taking one, two, three, etc., at a time. But attempts such as these were more theoretical than practical.

Pīṅgala's *Chandaḥ-sūtra* (c. 200 BCE) is a landmark in the linguistic permutations and combinations. It presents a method called *meru-prastāra*, a triangular array, or pyramidal expansion for finding the number of combinations of *n* syllables taking 1, 2, 3...*n* at a time. Halāyudha, a South Indian commentator in the tenth century CE has succinctly explained this method as follows:

> 'After drawing a square on the top, two squares are drawn below (side by side) so that half of each is extended on

either side. Below it three squares, below this again four squares are drawn and the process is repeated till the desired pyramid is obtained. In the topmost first square, the symbol for one is to be marked. Then in each of the two squares of the second line, the symbol one is to be placed. Then in the third line, symbol for one is to be placed in each of the two extreme squares. In the middle squares of this third line, the sum of the figures in the two extreme squares immediately above are to be placed; this is the meaning of the word *pūrṇa*. In the fourth line, the figure one is to be placed in each of the two extreme squares. In each of the middle squares, the sum of the figures in the two squares immediately above, that is three, is to be placed. Subsequent squares are to be filled in the same manner. Thus the second line gives the expansion of the combinations of the short and long sounds forming one syllable; the third line provides the same for two syllables, the fourth line for three syllables and so on.' (Sen, S.N., *A Concise History of Science in India*, pp. 196-197.)

Meru-prastāra can be expressed graphically, as follows:

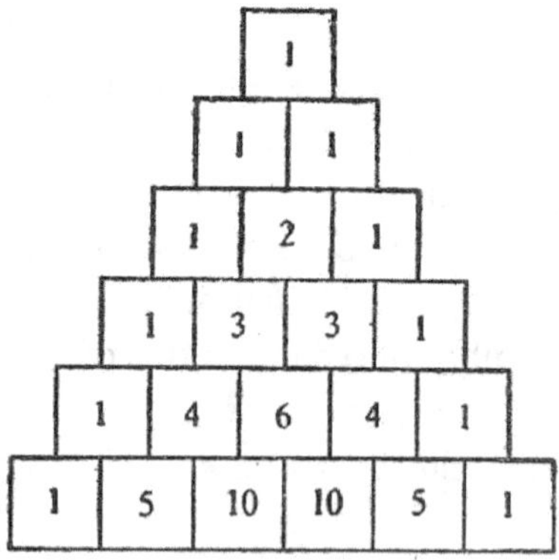

Meru-prastāra (diagram)

Vedic linguist-mathematician Piṅgala's *meru-prastāra* is the same as the triangular array which is referred to in Europe as Pascal's triangle and which first appeared in Europe on the title page of a book on arithmetic by Apianus in 1527 CE and employed subsequently by some European mathematicians in the sixteenth century. Blaise Pascal's posthumous publication titled: *Traite' du triangle arithmetique* which appeared in 1665 dealt with this type of triangular array involving numbers. Piṅgala's method in the form of *meru-prastāra* was in the nature of a ready-reckoner for finding the number of combinations of metrical syllables. More importantly, it envisaged the computation of the binomial terms.

It is important to note that three of the six *Vedāṅgas*—the *Kalpa, Chandas* and *Jyotiṣa*—nucleated early Indian scientific endeavours, while the other *Vedāṅga*, called *Vyākaraṇa* (grammar) had its own influence on the linguistic aspects in a rational manner.

Jaina Contributions

In the Jaina technical terminology, the word *sthāna*, is used for connoting the decimal place. After the words (given in brackets) like ten (*daśa*), hundred (*śata*), and thousand (*sahasra*) in the Jaina enumeration, the order preferred is in terms of tens of thousands, hundreds of thousands, tens of thousands of thousands (*koṭi*); and in terms of *koṭi*, hundreds of *koṭi* and the like, unlike the Hindu enumeration wherein special words are used for higher numbers on the ascending decimal-scale. However, as in the Hindu system, Jainas also classified numbers into odd (*oja*) and even (*yugma*). Numbers are also thought of as being numerable (*saṅkhyeya*); innumerable (*asaṅkhayeya*); and infinite (*ananta*), with further divisions in terms of the lowest, highest, intermediate, etc. The *Anuyogadvāra-sūtra* has dealt with number-groups of a higher order, but such an exercise appeared to be of an imaginative nature. Two notable Jaina texts, the *Samavāyāṅga-sūtra* (fourth century BCE) and the later text, *Prajñāpanā-sūtra* (second century BCE), refer to *aṅkalipi* and *gaṇitalipi*, implying that the Jainas of that time were using certain types of numerical forms, for which there is no precise epigraphic evidence so far. In fundamental operations like division and multiplication involving numbers, Jaina mathematicians had their own simplified way. They would neglect whenever the fractional part was less than half; but if it was greater than half, it would be rounded off to unity.

The Jaina name for geometry was *rajju* (cord) which was similar to the word, *śulba*. The other names for the geometrical forms like circle, diameter, squares, etc. were more or less the same as those in Sanskrit. Generally, the Jaina astronomers or mathematicians took the value of π as being $\sqrt{10}$ or even the number 3 as the *Sūryaprajñapti*, a Jaina astronomical work, has used it. They were adept in permutations and combinations, since they were also the linguistic (meter) specialists naturally involved in this exercise.

Jainas were greatly interested in mathematics, as evidenced by the list of several mathematical topics that appear in the *Sthānāṅga* sūtra of about the first century BCE. The list encompasses the science of numbers (*saṃkhyāyana*); *rajju* (geometry); fundamental operations (*parikarma*); subjects of practical treatment (*vyavahāra*); solid mensuration (*rāśi*); fractions (*kalāsavarṇa*); equations or algebra (*yāvat-tāvat*); square and quadratic equations (*varga*); biquadratic equations (*varga-varga*); cube and cubic equations (*ghana*); and permutations and combinations (*prastāra* and *vikalpa*). It would appear that the early Jaina mathematicians, about whom no authoritative information is available, had their influence on the Brāhmanical schools. It is not unlikely that the origin of some of the developments in geometry and arithmetic of these schools could have been influenced to some extent by early Jaina sources. But more systematic studies are necessary in this direction.

The Bakhshālī Manuscript

A rare manuscript dealing with mathematics was discovered in 1881 in a small village called Bakhshāli near Peshawar which is now in Pakistan. This manuscript of about seventy leaves is written on birch bark using the script known as Śarada, and its language is known as Gāthā, which is a refinement over old Prākṛt, but closer to Sanskrit. European scholars like Augustus Hoernle, Johann Georg Buhler, and Florian Cajori as well as Indian mathematician B.B. Dutta , have opined that this manuscript could be dated to about the third or fourth century CE. But G.R. Kaye, whose approach to Indian science is generally biased, has argued that its date could not be earlier than the twelfth century CE. He even doubted its Indian origin; nevertheless, he has left behind its photographic facsimiles and translation into English, which are valuable for further in-depth studies. The general view of Hoernle is that the arithmetical operations presented in the *Bakhshālī manuscript* are much earlier than the date of the manuscript. Recently, a Japanese scholar, Takao Hayashi, has edited and translated the *Bakhshālī Manuscript* into english.

The *Bakhshālī manuscript* presents several rules involving arithmetical, algebraic, and geometrical operations. The major part of the work, however, is concerned with the arithmetic of square-root extraction, fractions, rule of three, progressions, summation of series, computations of income and expenditure, profit and loss as well as interests. The manuscript also deals with solving simple and simultaneous linear equations, quadratic equations, value of surds, negative signs, mensuration and others. However, it does not throw any light on indeterminate equations either of the first or second order like the *Āryabhaṭīya* or the *Brāhmasphuṭa Siddhānta*; nor does it deal with the geometry even of the shadow cast by a gnomon. These are the reasons which indicate that the *Bakhshāli Manuscript* could be earlier than the *Āryabhaṭīya*.

Another notable aspect of this work is that it enunciates rules, provides examples, and demonstrates the application of the rules. This type of mathematical presentation is in sharp contrast with that found in the other works, including the *Āryabhaṭīya*. The manuscript has other original features too, like compound series and certain arithmetical formulae which are stated to be more efficient and easily adoptable. The use of decimal place-value notation is very much in evidence in this manuscript.

DECIMAL PLACE-VALUE NUMERATION

Numeration: As urbanization progressed with increasing commercial activities, the need arose for inventing a simple method of accounting by using the minimum number of numerals or digits. Such a method was invented in India by using the numerical forms of one to nine and that of zero, perhaps around the first or second century CE. Known as the decimal place-value system, it was capable of expressing very large numbers since the same numeral has a place-value depending upon its placement in the sequence. More importantly, the ingenuity of the system lies in its according a place-value to zero also. For example, if the symbol for zero is placed in the first position, its value is zero, in the second position its value would be ten; and in the third position, hundred and so on. It was this Indian ingenuity that made the decimal place-value system unique in computations, leading to its adoption by other cultures and becoming universal by about the fifteenth century.

It is no exaggeration to say that the invention and the use of zero in the context of the decimal place-value principle marked the most fruitful stage in human adventure in number-reckoning and commercial activities. The use of zero freed a counter from the shackles of the tiresome counting board or abacus. The new algorithm generated by it catalyzed the progress of modern mathematics and mensuration as well as commercial transactions.

It is now known that a place-value system on the sexagesimal scale (with the number sixty as the base) was in vogue among Babylonian mathematicians or accountants. The view that the Indian decimal place-value system owed its inspiration to the Babylonian one is only a conjecture. For, although right from the Harappan times there was contact between India and West Asia, including Babylonia, and barring just a few instances of Babylonian sexagesimal parameters appearing in Indian astronomical texts (for example, time-measures and certain computations), the sexagesimal system was never adopted in India. The extensive decimal ascending scale articulated in the Vedic literature, the ingenious method of using word-numerals or alphabets in the decimal place-value system in Indian astronomical texts over the centuries, unmistakably point out the independent Indian invention using nine digits and zero. The mathematical cuneiform texts no doubt use a system of numeration based on place-value notation on the sexagesimal scale; but they do not use zero. Similarly, the Chinese method of number-writing as evidenced in the Shang oracle bones (1400-1000 BCE), and later by what are known as rod-numerals, is also without the use of zero. There is a view that the concept of 'emptiness' in Chinese Taoist mysticism could have contained in it the idea of zero in the same way as the Indian concept of 'void' (*sūnya*) and contributed to the invention of a symbol for zero. In any case, it cannot be denied that the word *zero* in use now is derived from the Indian word, *śūnya*.

Indian astronomical-mathematical texts were rendered into Arabic in the Abbasid Caliphate at Baghdad in the eighth to tenth centuries. The word *śūnya* used for zero in the Sanskrit texts was translated as *al-ṣifr* which literally means *void* in Arabic. In the twelfth century, when the Arabic works were translated into Latin by European scholars like the English monk, Adelard of Bath, the Arabic *al-ṣifr* gave rise to such words as the French *chiffre*, the German *ziffer,* and the Italian *cifre*, all of which meant a numeral. There were words like *cifra, cyfra* and *cyfre* which were used with the meaning of zero. The Italian mathematician, Leonardo Pisano (*c.* thirteenth century) called it *zephirum* in his well-known work titled *Liber Abaci*. In the course of time, this word underwent some alterations and became the Italian *zefiro*, and finally *zero*, which appeared for the first time (perhaps in the way it is used now) in the text *De Arithmetica Opusculum*, by Philippi Calandri, and printed in 1491 in Florence (Italy). Since then, the use of the symbol for zero has become universal along with the decimal place-value numeration, and the word, zero, became part of the medieval Latin vocabulary only since the fifteenth century.

Word-numerals: Practically all Indian texts on astronomy and mathematics are in verses which have their own metrical compulsions. But numerical forms cannot be used as such in versified texts. The decimal place-value system no doubt provided a new tool for computational astronomy. However, it needed to be tuned to the metrical requirements of the texts, and this was meticulously accomplished by the use of word-numerals and alphabets. The word-numerals are called the *bhūta-saṃkhyā,* which were generally symbolic of the numbers associated with them and also drawn from the *Purāṇas* and other texts. For example, all synonyms of 'sky' were used for denoting 'zero'; those of the earth and the moon for 1; eyes and hands for 2; fire and worlds for 3; ocean for 4; arrow and the *bhūtas* for 5; seasons, tastes for 6; horse and mountain for 7, and so on. The following are among the important *bhūta-saṃkhyās* used especially in Indian astronomical literature.

Bhūta-saṃkhyās

0 ananta, antarikṣa, abhra, ambara, ākāśa, kha, gagana, jaladhrapatha, nabha, pūrṇa, bindu, randhra, viyat, viṣṇupada, vyoma, śūnya: all synonyms of 'sky';

1 abja, ādi, indu, ilā, urvarā, kalādhara, ku, kṣapākara, kṣiti, kṣmā, go, candra, jagati, tanu, dharaṇī, dharā, nāyaka, pitāmaha, pṛthivī, prāleyāṃśu, bhū, mahī, mṛgāṅka, rajanīkara, rūpa, vasudhā, vasundharā, vidhu, śaśadhara, śaśāṅka, śaśin, śītakara, śītaraśmi, śītāṃśu, śveta, sudhāṃśu, soma, himakara, himagu, himāṃśu: all synonyms of 'earth' and 'moon';

2 akṣi, ambaka, ayana, aśvin, īkṣaṇa, oṣṭha, kara, karṇa, kuca, kuṭumba, gulpha, cakṣu, jaṅghā, jānu, dasra, dṛṣṭi, dvandva, dvaya, naya, nayana, nāsatya, netra, pakṣa, bāhu, bhuja, yama, yamala, yugala, yugma, ravicandrau, raviputra, locana: mostly synonyms of 'eye' and 'hand';

3 agni, anala, kāla, kṛśānu, guṇa, gṛha, jvalana, tapana, trikāla, trigata, triguṇa, trijagat, trinetra, dahana, pāvaka, pura, bhuvana, ratna, loka, vaiśvānara, vahni, śikhin, haranetra, hutabhuk, hutabhuj, hutāśa, hutāśana: mostly synonyms of 'fire' and 'worlds';

4 abdhi, ambudhi, ambhodha, ambhodhi, ambhonidhi, arṇava, āya, āśrama, udadhi, kaṣāya, kṛta, kendra, koṣṭha, gati, ghana, caraṇa, jala, jaladhi, jalanidhi, turya, diś, payodhi, payonidhi, bandhu, yuga, lavaṇoda, varṇa, vāridhi, viṣanidhi, veda, śruti, samudra, salilākara, sāgara, sukha: mostly synonyms of 'ocean';

5 akṣa, artha, indriya, iṣu, karaṇīya, tattva, parva, pavana, pāṇḍava, prāṇa, bāṇa, bhāva, bhūta, mahābhūta, rāga, ratna, viṣaya, vrata, śara, śastra, sāyaka: all synonyms of 'arrow', of basic elements and senses;

6 aṅga, ari, ṛtu, kāya, kāraka, kumāravadana, khara, tarka, darśana, dravya, māsārdha, rasa, rāga, lekhya, ṣaṇmukha, śāstra;

7 aga, acala, atri, adri, aśva, ṛṣi, kalatra, giri, graha, chandaḥ, tattva, turaga, dvīpa, dhātu, dhī, naga, pannaga, parvata, bhaya, bhūbhṛt, mātṛkā, muni, yati, vājin, vāra, vyasana, śaila, svara, haya, all synonyms of 'horse' and 'mountain';

8 anīka, anuṣṭubha, ahi, ibha, karman, kuñjara, gaja, takṣa, tanu, danti, dik, diggaja, durita, dvīpa, dvirada, nāga, puṣkarin, bhūti, maṅgala, mada, mātaṅga, mati, vasu, sarpa, siddhi, sindhura, hastin: all synonyms of 'elephant' and 'serpent';

9 aṅka, anilāhva, upendra, keśava, gīr, go, graha, chidra, tārkṣyadhvaja, durgā, dvāra, nanda, nidhi, padārtha, randhra, labdha, labdhi;

10 avatāra, aṅgulī, āśā, kakubh, karman, dik, diś, diśā, paṅkti, rāvaṇaśira;

11 akṣauhiṇī, īśa, īśvara, bharga, bhava, mahādeva, mṛda, rudra, śaṅkara, śiva, śūlin, śvargeśa, hara: all synonyms of god 'Śiva';

12 arka, āditya, ina, tīkṣṇāṃśu, dinanātha, dinapa, divākara, dyumaṇi, bhānu, bhāskara, maṇḍala, mārtaṇḍa, māsa, ravi, rāśi, vyaya, sūrya: all synonyms of 'sun';

13 aghoṣa, atijagatī, karaṇa, viśva, viśvedevāḥ;

14 indra, manu, loka, śakra, śarva;

15 ahan, tithi, dina, pakṣa;

16 aṣṭī, kalā, nṛpa, bhūpati;

17 atyaṣṭi;

18 dhṛti, purāṇa, vidyā;

19 atidhṛti;

20 kṛti, nakha;

21 svarga, prakṛti, mūrchanā;

22 jāti;

23 vikṛti;

24 arhat, gāyatrī, siddha, jina;

25 tattva;

26 utkṛti;

27 nakṣatra, bha, uḍu;

32 danta, rada;
33 amara, tridaśa, deva, sura, surādhipa;
40 naraka;
48 jagatī.

The above list is only indicative, but not exhaustive since there are *bhūta- saṅkhyās* for other numbers too.

Āryabhaṭa's method: There was, however, another method that was devised and adopted by Āryabhaṭa I, the noted astronomer of fifth century CE, which may be called the alphabetical system of enumeration. This eminent author has given a novel system of expressing numbers with the help of consonants and vowels (Sanskrit alphabets) based on the decimal place-value principle. The *Paribhāṣā* stanza in the *Gītikā* section of the *Āryabhaṭīya* enunciates this system; its English translation is as follows:

> 'The *varga* letters (*k* to *m*) (are to be written) in the *varga* (square) places and the *avarga* letters (*v* to *h*) in the *avarga* (non-square) places; the *varga* letters take the numerical values 1, 2, 3, etc., from *k* onwards; (the numerical values of the initial *avarga* letter) *y* is equal to *ṅ* plus *m* (i.e. 5+25). The nine vowels should be written (one vowel in each pair of the *varga* and the *avarga* places.'

Āryabhaṭa I's method of numeration

***Varga* and *Avarga* letters with their numerical values**										
Varga	k	kh	g	gh	ṅ	c	ch	j	jh	ñ
	1	2	3	4	5	6	7	8	9	10
	ṭ	ṭh	ḍ	ḍh	ṇ	t	th	d	dh	n
	11	12	13	14	15	16	17	18	19	20

	p	ph	b	bh	m					
	21	22	23	24	25					
Avarga	y	r	l	v	ś	ṣ	s	h		
	30	40	50	60	70	80	90	100		

Vowels

Vowel	a	i	u	ṛ	lṛ	e	ai	o	au
Varga	10^0	10^2	10^4	10^6	10^8	10^{10}	10^{12}	10^{14}	10^{16}
Avarga	10^1	10^3	10^5	10^7	10^9	10^{11}	10^{13}	10^{15}	10^{17}

Ka-ṭa-pa-yā-di: Another alphabetical system which is believed to have been in use perhaps even earlier than Āryabhaṭa, was what is called the *ka-ṭa-pa-yā-di* system, a schematic presentation of which is given below:

Ka-ṭa-pa-yā-di

Value	1	2	3	4	5	6	7	8	9	0
Consonants	k	kh	g	gh	ṅ	c	ch	j	jh	ñ
	ṭ	ṭh	ḍ	ḍh	ṇ	t	th	d	dh	n
	p	ph	b	bh	m					
	y	r	l	v	ś	ṣ	s	h	ḷ	

In this system, *ñ*, *n* and vowels when standing alone denote zero. The consonants beginning with *k*, *ṭ*, *p* and *y* denote the digits according to the numerical values given to them. In a conjoint consonant, the value of the last consonant is to be taken. The vowel suffixed to a consonant is to be ignored. The digits (values) are to be written from right to left to arrive at the number in the decimal place-value reckoning. Sometimes the letters of the chronogram are even read from left to right to get the decimal place-value numeration. The *visarga* at the end of a chronogram is also left out of consideration, if the latter is followed either by a word or another chronogram.

For example, while dealing with the mean planets, Āryabhaṭa II states (*Mahāsiddhānta*, 2, 1-18):

> *'Ghumatisa' guṇitam 'Kenananenai' vibhajedyadatra tasya syāt* |
> *'saranacagaghadhamabharakedhā' Kalyādau dyugaṇa eṣaḥ* |

Multiply (the number of revolutions of any planet) by ('ghumatisa'→ gh m t s →) 4 5 6 7 and divide by (*'Kenananenai'*→ k n n n n →) 10,000. The quotient is its position at the beginning of Kali. From the (*dyugaṇa*) in Kali, all (the mean positions in a given day) can be obtained.

The *dyugaṇa* at the beginning of the present Kali is (*'saranacagagadhamabharakedhā'* → s r n c g g dh m bh r k dh). This means 720, 634, 954, 219. Even much larger numbers can be expressed in the *ka-ṭa-pa-yā-di* system.

SPREAD OF THE DECIMAL PLACE-VALUE SYSTEM AND HINDU NUMERALS

As noted earlier, al-Khwārizmī produced the famous *Zīj al-Sindhind* in the court of the Caliph al-Mamūn in Baghdad. He also wrote another text, *Hind-Hisāb.* Though its Arabic original is lost, a Latin translation of it has been preserved. It was al-Khwārizmī who played a pivotal role in the dissemination of Hindu numerals in the Arabic world. The decimal place-value system as developed in India using nine digits and zero was known to mathematicians outside of India as early as the seventh century. The Syrian Bishop and scholar, Severus Sebhokt, not only mentioned the Indian numerals, but also praised it as a 'computing that surpasses all description'. In their conquests from the seventh century onwards, the Arabs took over the forms of writing numbers which were used in the conquered territories: in the Eastern Mediterranean, the Greek alphabetic system; in Sind, Hindu numerals and so on. Hindu (Brāhmī) numerical forms were also being used in trade transactions along the Indo-Arabic trade routes. It was, however, al-Khwārizmī's treatise on 'Arithmetic' and 'Algebra' that proved seminal for the use and transmission of Hindu numerals with some modified forms in view of the style of writing in mathematical texts in Arabic. The arithmetical treatise of al-Khwārizmī is the oldest known Arabic work in which the Indian decimal place-value system and operations with it have been described. Unfortunately, as noted before, his Arabic manuscript is not available. However, its later Latin version based on an earlier translation into Latin can be accessed.

Recently in 1997, two German scholars, Menso Folkerts and Mitarbeit von Paul Kunitzsch, brought out an edition of al-Khwārizmī's work on Indian arithmetic based on the available manuscript in Latin. The following account is based on the summary in English provided by the editors.

Chapter One of the book explicitly says that the Indian procedure with numerals is to be demonstrated so that the writing of numbers and operation with them become easier. It also deals with the place-value principle which the Indians had worked out. First the central ideas of the decimal place and of the zero are introduced. The differences in the forms of the fifth, sixth, seventh, and eighth numerals have been noted (perhaps refers to East Arabic numerals). In the entire text (Latin version), zero is represented by the word *circulus* (Arabic *dāira*). Al-Khwārizmī explained in his 'Arithmetic' how to write large numbers using zero and the place-value. 1,180,703,051,492,863 is the example given in the text for writing large numbers. This huge number is divided into groups of three digits (from right to left) to facilitate reading. It is stated that numbers are to be pronounced from right to left while being written, in a tradition that was alive for a long time in the Islamic culture-area.

In Europe, the twelfth century witnessed a new beginning of translations of Arabic texts into Latin. Apart from al-Khwārizmī's 'Arithmetic' which was rendered into Latin, another treatise on Indian arithmetic was written in the twelfth century, of which two manuscripts are available and one of them gives the author's name as H. Ocreatus and the title as *Helceph Sarracenium*, the word 'Helceph' coming from the Arabic *al-ḥisāb* (calculation). It is well-known that al-Khwārizmī dealt with what he called *'al-hind hisāb'* (Hindu calculation).

In the first half of the thirteenth century, two well-known authors, Alexander de Villedieu and John Sacrobosco, wrote treatises titled *Carmen de algorismo* and *Algorismus Vulgaris* respectively. The word, algorism, is derived from 'al-Khwārizmī'. Leonardo Pisano produced an arithmetical classic under the title *Liber Abaci*. These works were soon adopted by the medieval universities in Europe and became standard textbooks prescribed by the universities. Through them, the Hindu-Arabic numerals as well as the decimal place-value system entered the curricula of European schools too. The numerals, however, underwent some changes in their forms, especially in Spain (Western Islam) around the eleventh century and, in the succeeding three or four centuries, with the advent of printing techniques, they took the forms more or less in the way they are written now.

LUMINARIES OF INDIAN MATHEMATICS

In the early centuries of the Christian Era, certain new developments came about, particularly in astronomy and mathematics—the latter being in many ways a catalytic component of the former. Indian astronomers, as noted earlier, were also able mathematicians. For this reason the texts of the torchbearers of Indian astronomy contain sections on mathematics; in fact, mathematical astronomy was their forte. Problems relating to the determination of the correct or true positions of planets from the mean ones, their motions, and the like necessitated the refinements in such areas as algebraic solutions of indeterminate equations. They also needed a new mathematical technique of analysis, including the plane and spherical trigonometry, and Indian astronomers-mathematicians rose to the occasion. Their attainments were indeed of a high order and some of them ante-dated similar mathematical accomplishments in Europe by several centuries. Apart from the needs of astronomy, the expanding urban settlements and commercial activities, too, stimulated developments in mathematics, particularly in arithmetic. The following is a bird's eye view of the accomplishments of Indian mathematicians and the readers can find their details in books on the history of Indian mathematics.

Āryabhaṭa I

Among the illustrious Indian mathematicians, Āryabhaṭa I (fifth century CE) has a distinct position. His exemplary mathematical investigations and achievements were a beacon to the generations of mathematicians who followed him. His concise and lucid text, the *Āryabhaṭīya*, though cryptic at times, is a text on astronomy and mathematics, the latter being dealt with in the *Gītikāpāda* and the *Gaṇītapāda* sections. The author begins the work with obeisance to Brahman in the first stanza of the *Gītikā* section; in the very second stanza itself he expounds the method of writing numbers in the decimal-place value mode. He has enunciated R sine and R sine differences in the twelfth stanza. But it is in the second section of his masterpiece, rightly called the *Gaṇitapāda*, which has thirty-three stanzas, that Āryabhaṭa's acumen and brevity of expression in mathematics stand out. He defines a square in terms of an equilateral quadrilateral with equal diagonals. Extraction of square and cube roots also engaged his attention. More importantly, Āryabhaṭa I has stated that the value of π could be arrived at as follows:

'Add four to one hundred, multiply by eight and then add sixty two thousand; the result is approximately the circumference of a circle, the diameter of which is twenty thousand.' (translation of the Sanskrit verse, *Gaṇitapāda*, 10)

$$\text{i.e. } \pi = \frac{\text{circumference}}{\text{diameter}} = \frac{8(100+4)+62{,}000}{20{,}000} = \frac{62832}{20{,}000} = 3.1416$$

(approx.)

Āryabhaṭa was well aware that the ratio of the circumference of a circle to its diameter is always an approximate one. It is hard to understand why there is no equivalent word in Sanskrit for π in any of the Indian astronomical-cum-mathematical texts. In any case, Āryabhaṭa's value for π was adopted by Lalla and Bhaṭṭotpala, the commentator of Varāhamihira's *Bṛhatsaṃhitā*. Bhāskara II, used this value in its abbreviated form $\frac{3927}{1250}$. Al-Khwārizmī, the renowned Islamic mathematician who, as noted before, played an important role in the dissemination of Hindu numerals and the decimal place-value system, practically reproduced Āryabhaṭa's method in his work on algebra, as he records as follows: 'You multiply the diameter by sixty-two thousand eight hundred and thirty-two and then divide the product by twenty-thousand, the quotient is the periphery (or the circumference).'

Gaṇitapāda deals with several mathematical aspects: the area of a triangle (the product of the perpendicular dropped from the vertex to the base and half the base); of the volume of a right pyramid (half the product of the area of the triangular base and the height which, however, is not so accurate); the area of a circle, the volume of a sphere (though the expression given by him is not accurate); the area of a trapezium (in terms of the lengths of the perpendiculars on the base and face, and multiplying half the sum of the base and the face by the height); a general rule for the calculation of the area of plane figures (determination of the adjacent sides of the rectangle into which that figure can be transformed, and finding out the area by taking their product); in determinate equations of the first degree, etc.

In the *Gaṇitapāda*, Āryabhaṭa has dealt with the computations of *R* sine tables and *R* sine differences. The trigonometrical ratio are with reference to an arc of a circle rather than an angle. In the *Gītikāpāda*, he has in four lines also provided values for *R* sine differences cryptically as follows:

makhi bhaki phaki dhakhi ṇakhi ñakhi
ṅkhi hasjha sphaki kiṣg śdhaki kidhva
ghlaki kigra hakya dhaki kica
sga sjha ṅva kla pta pha cha kalārdhajyāh (*Gītikā*, 12)

When converted into numbers on the alphabetical system developed by Āryabhaṭa himself, the following table emerges for *R* sine differences at intervals of 225 minutes of arc. For purposes of comparison, the corresponding modern values correct to three decimal places have been given, and these indicate the extremely near-accuracy of Āryabhaṭa's values.

The twenty-four *R* sines given in the *Sūryasiddhānta* (ii. 17-22) are exactly those in column 2 of the table below. It would appear that the author of the *Sūryasiddhānta* (modern) might have based his *R* sines on the values enumerated by Āryabhaṭa I.

R sines and *R* sine differences at intervals of 225' or 3° 45'

Āryabhaṭa I's values			**Modern values**	
Arc	***R* sines**	***R* sines difference**	***R* sines**	***R* sines difference**
225′	225′	225′	224′.856	224′.856
450′	449′	224′	448′.749	223′.893
675′	671′	222′	670′.720	221′.971
900′	890′	219′	889′.820	219′.100
1125′	1105′	215′	1105′.109	215′.289

Āryabhaṭa I's values			Modern values	
Arc	*R* sines	*R* sines difference	*R* sines	*R* sines difference
1350′	1315′	210′	1315′.666	210′.557
1575′	1520′	205′	1520′.589	204′.923
1800′	1719′	199′	1719′.000	198′.411
2025′	1910′	191′	1910′.050	191′.050
2250′	2093′	183′	2092′.922	182′.872
2475′	2267′	174′	2266′.831	173′.909
2700′	2431′	164′	2431′.033	164′.202
2925′	2585′	154′	2584′.825	153′.792
3150′	2728′	143′	2727′.549	142′.724
3375′	2859′	131′	2858′.592	131′.043
3600′	2978′	119′	2977′.395	118′.803
3825′	3084′	106′	3083′.448	106′.053
4050′	3177′	93′	3176′.298	92′.850
4275′	3256′	79′	3255′.546	79′.248
4500′	3321′	65′	3320′.853	65′.307
4725′	3372′	51′	3371′.940	51′.087
4950′	3409′	37′	3408′.588	36′.648
5175′	3431′	22′	3430′.639	22′.051
5400′	3438′	7′	3438′.000	7′.361

Source: Shukla and Sarma, *Āryabhaṭīya: Gītikā*, 12: pp. 29-30

Indian mathematicians did not deal directly with the trigonometrical ratios. On the other hand, they preferred to deal with the length pertaining to the several elements of an arc of a circle of given radius.

Indian mathematicians were also adepts at calculating the sine of any arc as well as their multiples and sub-multiples, which led to their construction of sine tables for different arcs generally between 0° and 90°. Such tabulations were required for astronomical computations, especially of the planetary positions. In the *Āryabhaṭīya* and the *Sūryasiddhānta*, the tables of *jyā* or half-chords have been provided in respect of a quadrant divided into twenty-four equal parts so that the smallest arc becomes equal to 3° 45' or 225'. However, the sines as conceived and recorded in the Sanskrit texts dealing with mathematical astronomy are not the ratios of the corresponding half-chord and the radius. Instead they represent the half-chords themselves, and their values are governed by the length of the radius chosen. Āryabhaṭa I (*Āryabhaṭīya*), Brahmagupta (*Brāhmasphuṭasiddhānta*) and Vaṭeśvara (*Vaṭeśvarasiddhānta*) adopted different values for computation.

Bhāskara I, Brahmagupta, Bhāskara II and others have given a formula for the calculation of a half-chord directly from the given arc, that is, the angle.

Although details of the exact method of arriving at Bhāskara I's formula are not delineated, as in the case of others too, there is no denying that the erudite Indian astronomer-mathematicians knew the mathematical processes involved in arriving at the formula concerned.

Bhāskara I: The earliest exponent of the *Āryabhaṭīya* and its astronomy as well as mathematics was Bhāskara I. Āryabhaṭa I had dealt with indeterminate equation (now called the Diophantine equation), which is referred to as *Kuṭṭaka* (literally, pounding or pulverizing). Bhāskara I threw fresh light on the subject. He also elucidated Āryabhaṭa's methods in trigonometry as well as geometry. Like Āryabhaṭa I, Bhāskara I adopted a geometrical approach to certain problems of algebra and arithmetic.

Brahmagupta: In the history of Indian mathematics and astronomy, Brahmagupta has an honoured position. Though an uncompromising and even virulent critic of Āryabhaṭa I,

Brahmagupta followed Āryabhaṭa and introduced two chapters on mathematics in his (twenty-four-chaptered) magnum opus, the *Brāhmasphuṭasiddhānta*, the twelfth one being called *Gaṇitādhyāyaḥ* and the eighteenth one, *Kuṭṭādhyāyaḥ*. The former deals with several types of arithmetical operations including square and cube roots, fractions; rules of three, five, seven, etc.; barter series; plane figures; stocks and what are called the eight practical ones (*vyavahāras*). Brahmagupta's conception of a mathematician was a practical one, that is, a scholar who has a mastery over arithmetical operations and the eight practical activities. Brahmagupta is well known for his theorem that the rectangle contained by two sides of a triangle is equal to the rectangle contained by the circum-diameter and the altitude to the base. He was probably the first mathematician to have provided an equation for finding out the area of a (cyclic) quadrilateral. He states that the exact area of a cyclic quadrilateral is the square root of the four sets of half the sum of the sides respectively diminished by the sides. This means:

If a, b, c and d are the sides of a cyclic quadrilateral $ABCD$ and $2s$ is their sum, it follows that the area of $ABCD$ = $[(s-a)(s-b)(s-c)(s-d)]^{1/2}$, s being equal to $\frac{(a+b+c+d)}{2}$.

It may be noted that this formula appeared nearly a thousand years later in 1619 in the work of W. Snell. Brahmagupta, however, did not seem to have specifically mentioned that his formula was valid only for a cyclic quadrilateral.

Brahmagupta also enunciated a theorem for the diagonals of a quadrilateral inscribed in a circle as follows:

> 'Divide mutually the sums of the products of the sides attached to both the diagonals and thereafter multiply the quotient by the sum of the products of opposite sides. The square-roots of the results would be the diagonals of the quadrilateral.' (*Brāhmasphuṭasiddhānta*, XII, 28)

The cyclic quadrilateral with rational sides is recognized as Brahmagupta's quadrilateral and this became known to the West through Arab mathematicians who were conversant with Brahmagupta's mathematics.

Brahmagupta's fame as an extraordinary Indian mathematician rests on his innovative contribution to algebra in respect of the solution of the indeterminate equation of the second order: $Nx^2 + 1 = y^2$ i.e. to find out the values of x and y in integers, for a given constant (integer) N. This equation is called *Vargaprakṛti*. Brahmagupta gave a method for solving such equations and from the lemma discovered by him, a number of integral solutions could be obtained. The Sanskrit word used for the lemma is *bhāvanā* with its further categorizations.

In the seventeenth century, the equation with the interpolator was called the Pellian equation (after John Pell, the younger contemporary of Fermat). But, historians of mathematics rightly point out that the full credit for discovering the lemma to solve the indeterminate equations of the second order should undoubtedly go to Brahmagupta. It may be noted that Brahmagupta's lemmas were not only rediscovered in Europe a thousand years later, but also their methodological importance was recognized by the well-known mathematicians, Leonhard Euler in 1764 and Joseph Louis Lagrange in 1768, nearly eleven centuries after they emanated from the fertile mind of Brahmagupta (seventh century). Some historians of mathematics, like D.E. Smith, were of the view that it was European mathematician Pierre de Fermat who in 1657 showed for the first time that the equation, $Nx^2 + 1 = y^2$, where N is a non-square integer, has infinite solutions. However, it is now recognized that it was Brahmagupta who, nearly a thousand years earlier in India, had shown that an indeterminate equation of this type had many solutions. His method saw perfection in the *Cakravāla* method of Bhāskarācārya II in the 12th century and even earlier in the enunciation by Jayadeva.

Brahmagupta has provided as many as eighteen theorems in

the geometrical part of the *Gaṇitādhyāya* (twelfth chapter of the *Brāhmasphuṭa Siddhānta*). His able scholiast, Pṛthūdakasvāmin (ninth century CE), and succeeding astronomer-mathematicians have shed light on them in their commentaries. Much later, the devoted European savant, H.T. Colebrooke, rendered them into English early in the nineteenth century, and this translation enabled the German mathematician, Ernst Kummer to analyze Brahmagupta's work in the middle of the nineteenth century. He was full of admiration for the mathematical acumen of Brahmagupta. Chasles, a French mathematician, even praised Brahmagupta's expression for the rational cyclic quadrilateral and its diagonals as well as the area of a cyclic quadrilateral, since they marked a significant development of basic geometry independent of, and different from, that of the classical Euclidean geometry.

Brahmagupta used the word, *Kuṭṭaka* or *Kuṭṭaka-gaṇita* for algebra. His scholiast, Pṛthūdakasvāmin, was perhaps the first to have used the compound term *bījagaṇita* in the sense of calculation or operation with the unknown, and Bhāskara II adopted it too. In Hindu mathematical texts, to denote unknown quantity several terms are used such as, *avyakta, gulikā, yāvat-tāvat, yadṛcchā, varṇa, vāñcchā* and *kāmika*. For the equations, *sama-karaṇa, samī-karaṇa, sadṛśī-karaṇa* and the like have been used. Brahmagupta has classified the equations into three major groups: i) equations with one unknown (*eka-varṇa-samī-karaṇa*); (ii) equations with many unknowns (*aneka-varṇa-samī-karaṇa*); and (iii) equations with products of unknowns (*bhāvita*). And each class can also encompass linear, quadratic, or even equations of higher power.

The use of some kind of symbols for unknown quantities as well as other operations, is characteristic of algebra, just as diagrams with letters are in geometry. The use of the word *varṇa*, indicates that letters of alphabets had a place in algebra. In Indian mathematical texts, generally, to denote addition, *yu* (for *yuta*); for subtraction, the symbol + (perhaps derived from Brāhmī symbol for *k* in *kṣaya* or minus)*; gu* (for *guṇa*) for multiplication; and *bhā* (from *bhāga*) for division were in use. *va* was used to denote squaring (*varga*); *gha* for cubing (*ghana*); *va-va* (squaring the square); and *mu* for finding the square root (*mūla*). The *Bakhshāli Manuscript* also sheds light on the use of such symbols.

Mahāvīra: A Jaina by birth and hailing from Karnataka, Mahāvīra (ninth century CE) rose to be an illustrious mathematician. While mathematics was an integral part of many major works on astronomy, Mahāvīra's *Gaṇitasārasaṅgraha* is perhaps the only large treatise devoted exclusively to mathematics, the reason being that the Jaina tradition encouraged the study of mathematics for its own sake, while the Hindu tradition developed it in association with astronomy. Mahāvīra, nevertheless, was well aware of the Hindu mathematical attainments particularly of Brahmagupta. In the invocation to his *Gaṇitasārasaṅgraham,* Mahāvīra 'bows to the glorious Lord of the Jinas, who as the shining lamp of knowledge of numbers made the whole universe to shine'. A mathematician to the core, he has expressed the seminal importance of mathematics as follows: '...whatever is there in all the three worlds which are possessed of moving and non-moving beings, all that indeed cannot exist without mathematics.' (GSS. 1.16). He was also full of praise for the enlightened *Rāṣṭrakūṭa* king, Amoghavarṣa Nṛpatuṅga (815-878 CE) who was the benevolent ruler of a fairly large kingdom and who was reputed for his learning and patronage of scholarship alike. The *Gaṇitasārasaṅgraha* is a rich blend of mathematics and poetic expressions of enthralling problems and examples. It deals with arithmetic, algebra, geometry and other aspects in its nine chapters of nearly 1,100 verses, and enjoyed great popularity, especially in South India.

Mahāvīra has given some special products of numbers which remain the same when read either from left to right or right to left. These are stated to be in the form of a garland (*mālā*).* He has also dealt with the cube of natural number and the finite series for n^3. Besides, he has discussed the method of expressing a fraction as the sum of several fractions, which have unity in their numerators. Mahāvīra was the first in the history of mathematics to have given a formula for the determination of the number of ways of selecting r objects at a time out of n objects. He has also shown that (i) the square of any number, whether positive or negative, would always be positive; (ii) the square of zero is necessarily zero; and (iii) a negative number cannot have a square root.

Mahāvīra's enunciation of problems, with living examples, is indeed fascinating. For example: 'One third of a herd of elephants and three times the square root of the remaining part of the herd of elephants, were seen on the mountain slope; and in a lake was seen a tusker along with three female elephants. How many elephants were there?' (GSS. IV. 41)**

Another example of Mahāvīra's exposition of a problem in an enchanting manner relates to a solution of simultaneous equation with two or more unknowns. The statement runs as follows:

---

*For example: 139 × 109 = 15151; 152207 × 73 = 11111111; 12345679 × 9 = 111111111; 333333666667 × 33 = 11000011000011.

**Let x be the number of elephants in the herd. The problem now corresponds to the equation:

$$\frac{x}{3}+3\left(\frac{2}{3}x\right)^{\frac{1}{2}}+4=x$$

Or, $x-\frac{x}{3}-4=3\left(\frac{2}{3}x\right)^{\frac{1}{2}}$

By squaring both sides and simplifying, we obtain

$(2x-3)(x-24)=0$

Accepting the positive integral solution for x as valid, we get $x=24$ elephants.

(Source: S. Balachandra Rao, *op. cit*, pp. 138-142.)

> 'A wizard having magical powers sees a cockfight going on and speaks privately to both the owners of the cocks. To one he says, "if your bird wins, you give me your stake money; but if it does not win, I shall pay you two-thirds of it." Thereafter to the other owner, he promises the same way, but three-fourths, if he loses it. "Tell me, O ornament of the first rate mathematicians, the stake money of each of the cock-owners if from either of them, the wizard would earn a profit of twelve gold coins."'(GSS, VI. 270-72).

Mahāvīra appears to be the first Indian mathematician to express a fraction as the sum of several fractions which have unity or (1) in the numerators. Calling them *rūpakāṃśa rāsi*, Māhāvīra described the method of obtaining such expressions, one that was not attempted by any Indian mathematician either before or after him, according to some historians of Indian mathematics likewise, is his significant contribution to permutations and combinations .

The *Gaṇitasārasaṅgraha* had extensively dealt with what is known as the *Kṣetragaṇita-vyavahāra* (areas of several shapes). It has also thrown light on the five types of quadrilaterals, namely, rhombus and square (*sama*); opposite sides equal (*dvi-dvisama*); two sides equal (*dvisama*); three sides equal (*trisama*); and all sides different (*viṣama*).

His formula for the area of a trapezium/quadrilateral is: $\sqrt{(s-a)(s-b)(s-c)(s-d)}$, where *a, b, c, d* are the sides, and *s* is the semi-perimeter. But Mahāvīra categorically states that this formula does not hold good for the area of a non-cyclic quadrilateral. Although he appeared to be the only Indian mathematician who dealt with the problem of an ellipse and provided a formula for the determination of the circumference of an ellipse, his formula is found to be only a close approximation.

Āryabhaṭa II: The author of the well-known treatise on astronomy, the *Mahāsiddhānta*, has also discussed the indeterminate equation of the first degree and suggested improved methods for solving such an equation. Like in Brahmagupta's *Brāhmasphuṭasiddhānta*, the 18th chapter of the *Mahāsiddhānta*, has a section titled *Kuṭṭakādhyāya*. Āryabhaṭa II has also provided rules for the solution of quadratic equations more or less on lines similar to those followed by Āryabhaṭa I and Brahmagupta. Besides, several arithmetical operations like the extraction of square and cube-roots, rule of three, fractions and operations with zero, have also been presented in this work.

In the late tenth and early eleventh century, there were two able mathematicians, Śrīdhara and Śrīpati. The former produced two works on mathematics, *Pāṭīgaṇita* and *Triśatīkā*, in which several arithmetical topics have been discussed. For multiplication, Śrīdhara presented a method called *Kapāṭa-sandhi* (the modern Gelosia method) and it would appear that this Hindu method was transmitted to the West through Arabic savants. His algebra is known from the references made to it a century or so later by Bhāskarācārya II, according to whom Śrīdhara was the discoverer of a method of solving quadratic equations. Śrīpati was the author of an arithmetical text called *Gaṇita-tilaka*, which was popular because of its lucid exposition of the usual arithmetical topics.

Bhāskarācārya II: A versatile luminary of Indian astronomy and mathematics, Bhāskara II's great treatise, the *Siddhāntaśiromaṇi*, has been mentioned already. He named the first part of his monumental work, *Līlāvatī*, after his only daughter, who became a widow at the early age of ten years. It is said that this calamity occurred despite the erudite father's efforts to avoid it based on astrological calculations in which he was supposed to be proficient. Be that as it may, Bhāskara II decided to teach the terse subject of mathematics, marshalling his poetical talents and composing enchanting verses embellished with figures of speech. He made mathematics as natural as possible by introducing in his exposition, birds and other animals, such as peacocks, swans, bees, monkeys, and snakes. They were also enlivened by interspersing anecdotes from the epics. Earlier Mahāvīra had also made some attempts in this direction; but Bhāskara II excelled in this mode of presentation, thus driving away the dreariness of mathematics.

The *Līlāvatī,* a popular text, is essentially an arithmetical and geometrical text, along with some aspects of mensuration and elementary algebra. Bhāskara II's arithmetic does not contain further novelties than those that are found in the earlier works. But the *Līlāvatī* is marked by its own lucid and beautiful presentation as evidenced by a few examples given below:

(i) 'A beautiful pearl necklace of a young lady was torn apart in a love-quarrel, and the pearls were all scattered on the floor; one third of the number of pearls was on the ground, one fifth was on the bed, one sixth was found by the pretty lady, one tenth was collected by the lover, and six were seen hanging in the thread. Tell me the number of pearls in the necklace' (*Līlāvatī*, 54).*

(ii) 'The square-root of half of a swarm of bees went to a rose-bush to suck honey, followed by eight-ninth of the entire swarm; one she-bee was caught in a lotus flower which closed at night. She was humming in response to another

---

***Solution (i) :**

Let us assume that the number of pearls in the necklace is x.

Then, $\frac{x}{3}+\frac{x}{5}+\frac{x}{6}+\frac{x}{10}+6=x$

Taking the LCM and solving this equation, we obtain the value of x as 30.

i.e., total number of pearls in the necklace is 30.

male bee. My daughter, tell me the number of bees in the swarm.'* (*Līlāvatī*, 72)

Bhāskara II enunciated a verse for the value of π in two ways: (i) If the diameter of circle is multiplied by 3927 and divided by 1250, the circumference of a circle would be obtained accurately, i.e. 3927/1250= 3.1416 (which, like Āryabhaṭa I, he regarded as the value of π correct to four decimal places); and (ii) if the diameter is multiplied by 22 and divided by 7, i.e., 22/7 = 3.1428, which is approximate.

Bhāskara II also dealt with arithmetic progression and permutations as well as surds. But, he is noted for his method of solving an indeterminate equation of the second order, which is an improvement over that of Brahmagupta. Known as *Cakravāla* or cyclic method, it was hailed by the noted German

---

***Solution (ii):**

Let x be the number of bees in the swarm

The number of bees that went to the rose-bush is = $\sqrt{\frac{x}{2}}+\frac{8}{9}x$

Number of bees in the lotus and outside = 2

$$\therefore \sqrt{\frac{x}{2}}+\frac{8}{9}x+2=x$$

$$\therefore \sqrt{\frac{x}{2}}=x-\frac{8}{9}x-2$$

or $\sqrt{\frac{x}{2}}=\frac{x}{9}-2=\frac{x-18}{9}$

Squaring both sides $\frac{x}{2}=\frac{(x-18)^2}{81}$

i.e., $2(x-18)^2-81x=0$

or $2x^2-153x+648=0$

or $(x-72)(2x-9)=0$

This means x = 72. Hence the swarm consisted of 72 bees.

(Source: Balachandra Rao, *op.cit.*, pp. 164-165.)

mathematician, Hermann Hankel, as 'the finest thing achieved in the theory of numbers before Lagrange'. This method was earlier developed by Jayadeva nearly a century before; but Bhāskara II couched it more elegantly. Mathematicians of the calibre of Brahmagupta and Bhāskarācārya II ingeniously attempted to solve the indeterminate equation of the second order in a conceptually refreshing manner, and they have a distinct position in the history of world mathematics.

Kerala School

From the fourteenth century onwards, mathematical activities, like those in astronomy, were pronounced in Kerala, where mathematics began to attain new heights. The Kerala mathematical tradition owed its roots to, and in fact continued the older tradition, specifically that of Āryabhaṭa I. It is significant to note that practically all the important commentaries on the *Āryabhaṭīya* emanated from Kerala savants. The compact geographical region of the Kerala School of mathematics and astronomy is interesting. All the stalwarts of this school were from the basin of the Nīlā river in central Kerala, between Tirunāvāya and the Arabian Sea, except Mādhava, the leader of this school, who belonged to a place very near the region of the Nīlā river. It is indeed remarkable that this region witnessed a succession of extraordinary scholars not only in astronomy and mathematics, but also in *Vyākaraṇa*, *Mīmāṃsā* and *Śilpa-śāstra*, for well over four centuries, the most fertile period being 1400-1550 CE.

In mathematics or astronomy, the Kerala intellectuals, though they were devoted followers of Āryabhaṭa I, did not adopt his alphabetical system of number-reckoning, probably because they found it rather cumbersome. Instead, they preferred the more flexible *Ka-ṭa-pa-yā-di* scheme that was certainly in usage even during the time of Haridatta there (eighth century CE). The major texts of this school are in Sanskrit verses as elsewhere in India. But some of them are in slightly archaic Malayalam, and the fact that texts on astronomy-mathematics could be written in the local language reveals that even those who were not proficient in Sanskrit could enter these fields and make remarkable contributions.

As noted in the previous chapter, the originator and acknowledged leader of the Kerala School, Mādhava of Saṅgamagrāma (*c.* 1340-1425), authored several astronomical-mathematical works: *Venvāroha, Sphuṭacandrāpti, Laghuprakaraṇa, Agaṇita grahacāra Madhyamā-nayanaprakaraṇa, Mahājyānayanaprakaraṇa,* and *Golavāda*. Mādhava did not publish the general or individual methods used by him to arrive at his results, but seems to have only communicated them orally to his disciples. As noted before, Mādhava was followed by Parameśvara, and the latter's son, Dāmodara, whose pupil was the illustrious Nīlakaṇṭha Somayājī. Jyeṣṭhadeva was another noted pupil of Dāmodara, among whose students was Acyuta Piṣāraṭi. Putumana Somayājī from Trichur, was a noted mathematician.

Value of π: Āryabhaṭa I had made a succinct statement for obtaining the value of π (3.1416), stating at the same time that this value was approximate. Nīlakaṇṭha in his erudite commentary on the *Āryabhaṭīya* clearly explained: 'If measured by whatsoever unit, the diameter yields an integral number; measured by the self-same unit, the circumference is never integral; both cannot be integral for any unit whatsoever far one proceeds, there is always some left over; π is thus incommensurable.' Mādhava had given the value of π up to eleven places as 3.14159265359. In furtherance of this value, an idea occurred to him that one could write down possibly an infinite series whose sum would give an exact value of π, and that a reasonable rational approximation could be obtained by taking a partial sum of sufficiently high order and applying a correction termed as *antyasaṃskāra*. It was this idea that enabled Kerala experts to think of not only an infinite series of representation for π but also others like $\tan^{-1}x$* infinite series including the expansions for sin x and cos x. They seemed to have some notions about the convergence of such expansions, although this aspect does not appear to have been discussed explicitly by them.

Putumana Somayājī, in his *Karaṇapaddhati*, calculated its value as 3.1415926536 correct to ten decimal places, while the *Sadratnamālā* of Saṅkara Varma gave a value 3.14159265358979324 correct to 17 decimal places.

The credit of enunciating the power series expansion for sin x and cos x as well as π also belongs to Mādhava. It may be noted that these series were rediscovered in Europe only in the seventeenth century. Nīlakaṇṭha Somayājī, an outstanding astronomer and mathematician, provided detailed explanation, somewhat on modern lines, establishing the convergence of infinite series.

Two outstanding works of the Kerala School merit special attention: the *Kriyākramakarī*, an extraordinary commentary on the *Līlāvatī* of Bhāskara II, which was produced by the joint

---

* $\tan^{-1}x = X - \frac{X^3}{3} + \frac{X^5}{5} - + \ldots\ldots$ to infinity. (such a series expansion was given later (seventeenth century) by Gregory and Leibniz).

If x is equal to 1, we get

$\frac{\pi}{4} = 1 - \frac{1}{3} + \frac{1}{5} - + \ldots\ldots$ to infinity (Leibniz series).

Or $C = 4D - \frac{4D}{3} + \frac{4D}{5} - + \ldots\ldots\ldots$ to infinity.

Where C is the circumference and D is the diameter. Mādhava enunciated it in the same manner.

efforts of Śaṅkara Vāriyar (1500-60), a pupil of Nīlakaṇṭha, and Narāyaṇa (1494-1570) of the Mahiṣa Maṅgalam family. This text throws enormous light on the mathematical analysis as well as the rationales for the several formulae enunciated in the *Līlāvatī*. Another seminal text was undoubtedly the *Yuktibhāṣā* by Jyeṣṭhadeva, a contemporary of both Śaṅkara Vāriyar and Nārāyaṇa. It is also known by the name, *Gaṇitanyāya-saṅgraha*, the first part of which expounds the rationales for several mathematical propositions, while the second part deals with problems of astronomy of that time. The *Yuktibhāṣā* is written in Malayalam, and it would appear that there was also its version in Sanskrit under the title, *Gaṇita-yuktibhāṣā*. In addition to the two aforementioned texts, the *Sadratnamālā* (1819), by Prince Śaṅkara Varma, is also an important compendium on mathematics and astronomy.

Other Mathematicians: Apart from Kerala mathematicians, Bhāskara II's mathematical expositions were a source of inspiration for later developments. In Mahārāshtra, Ganeśa Daivajña produced the rationale for the conditions governing the four sides of a trapezium. He also added a geometrical proof for the Baudhāyana (the so-called Pythagorean) theorem in his *Buddhivilāsinī*, a commentary on the *Līlāvatī* of Bhāskara II. He even elucidated the type of triangles out of which the quadrilaterals could be constructed.

Nārayaṇa Paṇḍita (fourteenth century CE), son of Nṛsiṃha Daivajña wrote two treatises: the *Gaṇitakaumudī*, and the other on algebra titled the *Bījagaṇitāvataṃsa*. The latter, though influenced by Bhaskara II's work, has its own originality and merit. *Gaṇitakaumudī* (575 verses; 395 verses of examples), is modeled after the *Līlāvatī* (272 verses) of Bhāskara II. It deals with—in a much more detailed manner—geometry or planar figures (Ch. IV), and factorization of numbers and various ways of partitioning of unity into a sum of unit-fractions (Chs XI and XII), besides providing a general mathematical formulation of most of the combinatarial problems (Ch. XIII), and magic squares (Ch. XIV).

Magic Square

The Chinese are regarded as being the earliest to have possessed some knowledge of the magic square. While its tradition in China goes back to hoary antiquity, the first available recorded example belongs to about the first century CE. Later Arab astrologers used magic squares in their predictive calculations. Through Arabs, perhaps, magic squares spread to the West by about the twelfth century.

In India, the first mathematical use of magic square was by Ṭhakkar Pherū in his *Gaṇitasāra Kaumudī* (early fourteenth century). Nārāyaṇa Paṇḍita devoted a chapter to magic squares in his *Gaṇitakaumudī*, a topic that had not received adequate attention by the earlier mathematicians, including Mahāvīra and Bhāskara II. One Mahendrasūri constructed a magic square called *Sarvotabhadra*, with rows, vertical columns and diagonals totalling 170 as follows:

25	80	15	50
20	45	30	75
70	35	60	5
55	10	65	40

Source: S.N. Narahari Pandit, 'Mathematical Recreations in India'; *Scientific Heritage of India* (eds) B.V. Subbarayappa and S.R.N. Murthy, Mythic Society, Bangalore, 1988 pp. 101-07

Nārāyaṇa presents the magic squares under the topic of *Bhadragaṇita* (auspicious mathematics), which is part of *Średhīgaṇita* (or theory of series and sequences). He has provided certain rules to be followed for the construction of magic squares and illustrated them with examples. The construction of 4×4 pan diagonal magic squares with sum 40 is as follows:

8	2	13	17
14	16	9	1
7	3	12	18
11	19	6	4

17	13	2	8
1	9	16	14
18	12	3	7
4	6	19	11

In perspective, there was a continuum of mathematical activity since the time of Āryabhaṭa I. But there is a view expressed by some Indian mathematicians that it could well have reached greater heights than were attained between the fifth and seventeenth centuries, if certain basic concepts in other branches of learning had been pursued with the rigour of mathematics. For example, the practice of denoting numerals by a set of cardinality and the application of the *bhūta-saṅkhyā* system could have led to a theory of sets. The *Navya-nyāya* had initiated certain approaches of symbolic logic, and the *Nyāya-Vaiśeṣika* had tried to explore the meaning and content of a category of reals called *abhāva*. Such abstract ideas could well have stimulated mathematical thinking. The Indian attitude to mathematics in general was radically different from that of the Greeks, particularly in geometry. Indian astronomy, known for its extensive computational character, provided some stimulus to the related mathematical operations. But at the same time, the traditionally fostered computations led the latter to move along the already charted grooves. There is also a view, a debatable one, that the practice of presenting mathematics in verse could have had a retarding effect on its growth. Its pursuit as an auxiliary to astronomy, too, in the frame of geocentricity and circular motions, appears to have had its own limitations.

Arabic and Persian Sources in India

In the medieval period, when some Islamic astronomers/ mathematicians from Western and Central Asia came to India and settled here, they had substantial knowledge of Greek mathematics in general and of Euclidean geometry in particular.

That knowledge, however, was confined to themselves and did not appear to have influenced in any manner contemporary Indian mathematics. Even earlier, the noted Islamic scholar, al-Bīrūnī, was believed to have attempted to translate Euclid's *Elements* into Sanskrit. But his attempt did not meet with the desired success. Euclidean geometry was an essential pre-requisite for Islamic astronomers in India, besides the *Spherics* of Theodosius, *Cylinder* and *Sphere* of Archimedes and *Conics* of Apollonius, all in either Arabic or Persian translations. These and Naṣīr al-Din Ṭūsī's *Taḥrīr-i-Uqlidis* (Arabic version of Euclid's *Elements*), were written outside of India. The earliest commentary on al-Ṭūsī's work was on geometry in Arabic and written in India by Mīr Muḥammad Hashim bin Qāsim al-Ḥusainī in 1635, under the title: *Sharh Taḥrīr-usūl-Handasah wa'l-Ḥisāb*. A Persian translation of al-Ṭūsi's work under the title: *Tarjamah i Taḥrīr Uqlidis*, by Khair-Allāh Khān appeared in 1732. In 1684, Ismatu'llah as-Saharānpuri produced a treatise on arithmetic and algebra, *Ḍabit Qawā'idu'l-Ḥisāb*. In the eighteenth century there were active mathematicians like Amīnu'ddin al Lahori and Faridu'd-Din Aḥmad. In 1718, Paṇḍita Jagannātha, the court astronomer of Sawai Jai Singh II, had already translated the Arabic (Ṭūsī's) version of Euclid's work into Sanskrit under the title *Rekhāgaṇita*. No commentary of any sort on the *Rekhāgaṇita* appeared even later, revealing that the Euclidean geometry continued to remain in the background, and confined to Islamic mathematicians.

It was only in 1873 that the Sanskrit College at Calcutta, under the colonial rule, brought out a partial text in Sanskrit, the first five chapters of Hutton's Euclidean geometry under the title: *Kṣetra-tattva-pradīpikā*. A Greek classic, *Spherics* of Theodosius, was also translated into Sanskrit from its Persian version, with the title *Ukarākhyagrantha* (1730). Around this time, one Nayanasukhopādhyāya, with the help of an Islamic scholar also rendered its Arabic version into Sanskrit. The Arabic recensions of the mathematical treatises of Archimedes on sphere and cylinder, and on conics by Appolonius, were attractive as textual sources among Islamic students in India.

Perhaps the earliest original work on arithmetic in Persian was written by Ābd al-Ḥamīd Ghaznavī Dehlawī (1365). He had a refreshing approach to mathematics, and felt that it was only in mathematics that there would be no religious differences. His book was found useful for governmental administration. Mughal Emperor Akbar ordered that Bhāskārācārya II's *Līlāvatī* be rendered into Persian, and one Persian scholar, Faizī, accomplished this task in 1587. During Emperor Shāh Jahān's reign, a Persian version of Bhāskarācārya's *Bījagaṇita* by Āṭaullāh Rashīdī came out, and it was dedicated to the Emperor. It was this Persian translation (*Tarjamaḥ-i-Bīj-Gaṇita*) that was rendered into English with notes by Edward Strachey Samuel Davis (1813) with the title *Bījagaṇita* or *Algebra of the Hindus*. Thus European mathematicians came to know of Hindu algebra first through the Persian sources. In 1817, H.T. Colebrooke published an English translation titled *Algebra with Arithmetic and Mensuration from the Sanskrit works of Brahmagupta and Bhāskara II.* An Indian Muslim scholar, Najm al-Dīn Khān Thāqib of Kakori (Lucknow Dist.) wrote a mathematical book in Persian under the title: *Risālaḥ dar jabr w Muqābala* (1812), which contained a good deal of algebra. The formulae given in this book are presented in the form of verses (illustration in prose) so as to help students in memorising and understanding them.

Apart from the books on mathematical subjects, several compendia were meticulously prepared by well-known Arabic and Persian scholars. Among them were *Laṭaif al fuyūd* (1594), by Hasan Muḥammad bin Rājī Muḥammad al-Firshauri al-Ābbāsi al-Sarkeji, with forty chapters dealing with the geometry of circles, plane and spherical trigonometry, time-reckoning, mathematical geography, astronomical instruments for measurements and the like; the second one was composed in 1814, which elucidated problems in arithmetic, geometry and astronomy; and the third authored by the noted astronomer, Ghulām Ḥusain Jampuri (1834) was in the nature of an encyclopaedia of mathematics in Persian, dealing with arithmetic, Euclidean geometry, algebra and trigonometry, logarithmic, trigonometric and astronomical tables as well as astronomical instruments.

There have been no systematic studies on mathematical works in Arabic and Persian by Indian Muslim scholars and others. Such studies would be important not only for understanding the history of mathematics in India, especially between the fourteenth and eighteenth centuries, but also in evaluating the nature and extent of mutual interactions among Hindu and Muslim mathematicians.

5

CHEMICAL TECHNIQUES AND ALCHEMICAL PRACTICES

In all ancient cultures, a variety of techniques of a chemical nature played a significant role in embellishing and enriching material life. Different techniques were adopted in the production of pottery and glass, metallurgy and metal-working, dyes and pigments, cosmetics and perfumery, fermented drinks, paper-making and ink. Potters, metal-smiths, dyers, glass-makers, brewers, and perfumers were practical chemists who were empirically successful in operations such as heat treatment, distillation, concentration, calcination, reduction, and oxidation, even though they had no theoretical knowledge of the associated chemical transformations. The technique of baking clay and the consequent creation of a wide variety of pottery forms was an important landmark in the history of chemistry. For, the use of pots for distillation, evaporation, treatment of minerals and other similar purposes, enabled ancient practical chemists to experiment with the processes involved in the generation of useful products.

POTTERY

Among the early inventions of man in the Neolithic Period (from about the seventh millennium BCE) in the West (Jericho and Jarmo) and Central Asia (Mehrgarh), was the production of pottery from clay, hand-made and sun-dried. At a later stage in advanced human settlements, baked and thus hardened pottery appeared in shapes determined by practical needs and functional compulsions. In the Indian context, a distinctive Neolithic culture came up in Kashmir Valley (4000 BCE), the neolithic settlement Burzahom being a representative type, characterized by semi-underground pit dwellings. The Neolithic phase was marked by the appearance of hand-made pottery like bowls and vases, grey or black ware shaped into dishes-on-stands, bowls with ring bases, high-necked jars and the like, some of which were also later made with the use of the wheel.

The technique of making simple pots, in its earliest stage, was physical rather than chemical because a lump of clay was merely pressed by the hand into the desired shape. Large storage vessels were, however, made by placing coils of clay one above the other, after the earlier one had dried up and become hard, and was able to withstand the weight of the next coil, a technique which is in use even today in certain rural areas. But it was the invention of the potter's wheel and the development of the technique of baking diverse pottery forms in a closed open kiln and the application of slip that added new dimensions to the art and the utility of pottery.

Harappan Pottery

Though ancient potters were skilled in their profession, they did not know the nature of chemical processes that would bring about changes in colour, hardening, and binding after the pots were subjected to thermal action. Nor were they aware of the processes or the conditions of reduction or oxidation. Even so, from Harappan times, over the millennia, Indian pottery showed remarkable attainments as Black and Red ware, Painted

Grey ware, Northern Black Polished ware, ochre-coloured ware, Malwa ware and others, some of which were also decorative, with enchanting naturalistic motifs and graffiti.

GLASS-MAKING

The term 'glass' applies to a range of products made of substances (silicates) characterized mainly by a controlled rate of cooling or super-cooling from a state of fusion to a solid state so that it does not undergo crystallization. In practice, glass is produced by melting a mixture of silica with sand (about 75 per cent), soda (about 15 per cent), and calcium compounds (lime or calcium carbonate, about 10 per cent). Some amounts of metallic oxides like copper or cobalt, are added to serve as colouring agents. The glassy products are chemically a mixture of silicates; they are also known for their physical properties like transparency, translucency, reflection, and refraction, depending upon their chemical composition and the mode of their production.

The knowledge or technique of glass-making could have been chanced upon, perhaps during early metallurgical operations where sand and other substances were subjected to high temperature through heating in small furnaces, when accidentally emerged certain types of glass objects. In the course of time, glass-making became an accomplished technique, especially in Egypt and Mesopotamia, even in the third millennium BCE. There were glass-producing units in Egypt in the reign of Amhenotep II (1448-1420 BCE) during the Eighteenth dynasty. Several clay tablets, now preserved in the British Museum in London, show details of glass-making as they existed at the time of Assurbaniphal (seventh century BCE) in Mesopotamia. However, it was during the Graeco-Roman times (from the third century BCE to the fourth century CE) that the glass industry attained new heights, particularly in glass-blowing, sheet-making, and the production of glass vessels of different shapes. In the medieval period, Europe emerged as the centre of a sophisticated glass industry, and Venice was regarded as the home of the art of fabricating exquisite glass objects.

In India, there is no archaeological evidence to indicate that the technique of glass-making was known to the Harappan people, although they were familiar with glazing methods and had to their credit several attainments. There is, however, some archaeological evidence in the form of a few glass beads in a Chalolithic site in Karnataka, dated to about the beginning of the first millennium BCE. The Vedic text, *Śatapatha Brāhmaṇa*, which belongs to this period or a little earlier, refers in a general way to *kāca*, the Sanskrit term for denoting glass. But the available archaeological evidence points out that it was only around the fourth or third century BCE, that glass production in India became significant, and in the next five to six centuries, the glass industry gained momentum. Nevertheless, its expertise was limited to the production of beads, bangles, ear-reels, 'eye-beads' and other such small objects, but did not include any type of vessels. However, the unearthed material has confirmed that glass objects of blue, green, red, orange and other shades, were produced.

Glass beads of a brownish or black colour, which are mainly composed of soda-lime-silicate with traces of phosphates and potassium and varying quantities of iron-oxide, have been found at Hastinapur (700 BCE). Of historical importance are the coloured glass beads of several shapes found in the Bhir Mound of Takṣaśīla, an important centre of the then northwestern province of the Mauryan Empire (fourth century BCE). Perhaps the contact with the Graeco-Roman culture-area acted as stimuli for the production of glass objects. In any case, the later town of Sirkap in this area has yielded remnants of foreign glass like mosaic and milleflori, and blue and white cameo. It was well known that Roman glass-makers were adept at making the milleflori with its characteristic floral and cellular structures. The remnants of glass flasks found at Sirkap also seemed to be from the Graeco-Roman region. Sirkap was known for the technique of making stratified eye-beads, but its technique appears to have been borrowed from outside India.

There is but scanty archaeological evidence that could throw light on the technical details of glass production in India. A circular open-hearth oven, about 75 cm in diameter and half its depth (37 cm), made of burnt clay, has been found in the archaeological site of Nevasa (third-fourth century BCE). The various glass objects—mostly beads, bangles and small objects—that have been excavated in different sites of India, reveal that Indian glass-makers had the necessary expertise in such methods as moulding, folding, twisting, and double stripping. Possibly, the wire-winding method was also used to make beads of different kinds. Chemical analyses of glass objects found in several sites indicate that Indian glass-makers were well aware of the importance of metallic oxides and other compounds to impart the desired colours to glass objects and the use of minerals containing iron, such as haematite (red or green), copper and cobalt (blue), manganese, lead or aluminum (opaque).

The production of glass objects was, no doubt, carried out on a small scale in several regions all over India. There was perhaps a large glass factory in Kopia, on the bank of the river Anoma in the Basti district of Uttar Pradesh (third century BCE-third century CE), as evidenced by a large numbers of glass objects found there, including blocks of glass, some weighing over fifty kilos. The glass industry received royal patronage, particularly during the Mughal period, when Persian glass-makers came to India with their craftsmanship and began to produce glass objects like dishes, dish covers, spittoons, flat-bottomed vessels, mirrors and tiles.

However, in India, glass objects did not enjoy the same status or social value as those of metals or even ceramic objects. Ceramics were preferred for religious functions, the use of glass vessels for such functions being abhorred, then as now. From about the eighth century CE, alchemical ideas transformed themselves into iatrochemical practices. These necessitated the adoption of such processes as distillation, evaporation, and sublimation. But the apparatus was by and large earthen, as glass vessels were rarely used. The opaque earthenware did not permit the iatrochemists to observe the chemical operations carefully, which would have provided them with the insights and stimuli to appreciate and even record their observations. It needs to be recognized that transparent glass of high quality, besides the coloured ones, have played a distinct role in science since the seventeenth century. Lenses and mirrors leading to the manufacture and use of the telescope as well as the microscope opened up new vistas in astronomy and in observations of the otherwise invisible world of minute organisms. As a result, both observational astronomy and biology took on far-reaching new dimensions.

It was the glass apparatus, such as retorts and condensers, that in a way impelled chemistry to attain the status of science after the discovery of oxygen (1745) by Antoine Laurent Lavoisier, in which the glass apparatus enabled him to observe and verify experimental processes, and reproduce them in a quantitative way. Fabrication of glass apparatuses—flasks, retorts, dishes, condensers, thermometers, prisms, tubes, convex and concave lenses—went hand in hand with the progress of experimentation in the physical and biological sciences. In India, the importance of glass was hardly realized, and its role in experimentation went unrecognized. Even today, certain rituals are performed only through the use of pottery, despite the enormous appeal of alternative materials, like attractive glass containers of different shapes for other purposes.

METALLURGY AND METAL-WORKING

In the civilizational development of early man, an important pathway was the use of minerals, the extraction of metals from their ores and the utilization of metals in diverse ways. Of them, extraction or metallurgy was, and has been, a technique of primal importance, which has added a veneer of excellence to the quality of material life. In the history of human progress, metals have played a pivotal role in diverse ways, and ancient civilizations are befittingly categorized as belonging to either the Bronze Age or Iron Age depending on the metal used for enrichment of their material life.

Early farming communities sprang up in the region now known as Central and West Asia, as well as in the steppes of the Ural-Altaic area, as early as in the seventh or eighth millennium BCE. About 3,000 years later, by accident or otherwise, man, the maker (*Homo faber*) was able to develop a method for the extraction of copper from its ores, the earliest metallurgical practice of far-reaching importance. Archaeologists have designated this as the Chalcolithic (*chalco*: copper; *lithic*: stone) stage which was noticeable in the region now known as Iran or Persia. The Chalcolithic cultures, which came up later elsewhere, also heralded the use of copper and its alloy, bronze. This useful alloy was obtained by heating copper with ores containing tin to produce utensils and other appliances, and supplanting those of stone. Advanced Chacolithic settlements with their own cultural accomplishment appeared in the fertile river valleys of the Nile (Egypt), the Euphrates and Tigris (Mesopotamia), and the Indus (India). The Indus Valley Civilization too had developed, like the other two, its skills in copper-bronze technology.

Indus Valley Civilization: An archaeological inventory of the metallic objects found in several settlements of this civilization indicate the use of gold, silver, lead, copper and bronze. Nevertheless, there is no evidence of any furnace or device for the extraction of metals like copper, nor of the manner in which gold and silver were obtained. On the basis of some ore-correlation studies, it has been inferred that silver might have been obtained from Afghanistan and Persia, and gold from the reefs of Mysore (probably the Hatti gold mines); copper ore could have been obtained from Rajasthan as well as Afghanistan, and lead ore from around the Ajmer area. The nature of metallic objects reveals that the metal-smiths of this civilization were well-versed in the techniques of soldering, sheet-making, rivetting, coiling and, possibly, in the lost-wax (*cire purdue*) process of casting. Gold was used mainly for making ornaments like beads, pendants, brooches, armlets, etc. Silver, besides being shaped into ornaments, was used for making some vessels, but in a limited way.

The copper-bronze objects unearthed at Mohenjo-daro and other places comprise axes, daggers, knives, spears, arrow-heads, chisels, drills, fish-hooks and metal mirrors. About 10 per cent of them are found to be practically unalloyed, containing only about 1 per cent of tin in them, but 14 per cent of them are alloyed with tin to obtain bronze (8-12 per cent tin). Since the Harappan metal-smiths were obviously mixing tin ore with copper and heating them to obtain the alloy, they were not able to maintain the correct proportions of the two metals in all cases. Apart from tin-bronze, they also produced an alloy of copper and arsenic (3-4.5 per cent) which could be used in the place of low-grade bronze. They knew how to cold-work as well as anneal metal. Possibly, copper vessels were shaped, raised, or sunken by hammering, and pure copper sheets would have been used for such purposes. Copper ingots (99.5 per cent pure) found at Lothal, a port town, indicate that pure copper was perhaps obtained from the Mesopotamian or Iranian region.

Though the details of smelting operations are not available archaeologically, it would appear, as one of the excavators (E.J.H. Mackay) has pointed out, that the ore (possibly with alternate layers of charcoal, which would act as a reducing agent) was smelted in an open hearth, using additional charcoal as fuel over a cavity in the ground so that the molten metal would flow into the cavity. The skills of Harappan metal-smiths were remarkable, as evidenced by the small bronze statuette called the 'Dancing Girl' fashioned by casting (discovered at Mohenjo-daro), and 'chariot with driver' (Daimabad in Maharashtra). Besides, there

Copper-bronze casting: Dancing Girl (Mohenjo-daro) (Courtesy: Project of History of Indian Science Philosophy and Culture, Centre for Studies in Civilizations)

was another interesting metallic tool: the true saw, with teeth and the adjoining part of the blade set alternatively from side to side—a tool that was unknown elsewhere until the Roman times. The metal arrow-heads found at Mohenjo-daro, too, had their own distinctive characteristics.

Copper Hoards: An archaeological find which goes by the name of 'copper hoards' is interesting and intriguing alike. Interesting, because nearly a thousand copper objects, found in hoards (hence the name), have been reported from about thirty-four sites in a widespread region comprising Uttar Pradesh, Central India, Bihar, Bengal, Odisha and Andhra Pradesh. Intriguing, because there is little or no information about the authors or producers of the almost unalloyed copper objects. The types of hoards include flat axes, shouldered axes, chisels, bar-cells, spearheads, harpoon-heads, antennae swords, and some anthropomorphic objects. The tools could well have been fashioned by closed casting. These hoards have not been found in association with any type of pottery, although what is known as Ochre-Coloured Pottery (OCP) has been noticed in the later excavations of some sites of copper hoards. Their date range is stated to be 1700-1000 BCE.

COPPER

In the Vedic period, copper was an important metal and copper utensils were preferred for religious purposes, even to those of silver or gold. The *Ṛgveda* refers to all the metals as *ayas*, and the *Yajurveda* speaks of *kṛṣṇāyas*, which means 'black metal' and possibly connoted iron. The metallurgy of copper might have been an accomplished technique during the Vedic and post-Vedic periods. However, there is no archaeological evidence that could shed light on the smelting furnaces, and one has to wait till the nineteenth century (see below) for authentic details of the metallurgy of copper. Nevertheless, copper objects have been unearthed in various sites of the ancient period. A solid copper bolt (about 60 cm in length, with a circumference of 35 cm at the centre and 30 cm at the ends) found in the Rampurva Aśoka Pillar, near the Nepal border, reveals the expertise of the smiths of those days.

The macro-technological expertise of coppersmiths was manifested in a remarkable sculpture of the Buddha (about 2.3 m in height and one tonne in weight) probably belonging to the fifth century CE. An engineer of the East India Company discovered it in 1864 in the ruins of a Buddhist monastery at Sultanganj (in Bihar), and it is now kept in the Museum and Art Gallery of Birmingham. According to some experts, this huge statue might have been cast in two layers: the outer layer with all its features by the *cire purdue* process and the inner layer appears to have been cast in segments on a mould using sand, clay, and paddy husk with iron bands for holding the segments together.

Copper statue of the Buddha

Extraction: The extraction of copper from its ores is a simple one, involving a reduction process, with the reducing agent being carbon or charcoal, which would also act as the source of heat, if used additionally. Rajasthan is known for its copper ores, as evidenced by a native copper-smelting furnace at Singhana near Kehtri in Rajasthan, which was still in operation in the early nineteenth century. It is possible that such furnaces were in vogue in ancient times, too. In this metallurgical operation, the copper ore, which has a sulphur content, needs to be roasted and thus converted into its oxide. For this purpose, the raw ore is mixed with cow dung, made into balls and roasted. The roasted ore, charcoal and some iron slag which would act as a flux are placed in a blast furnace and smelted for nine to ten hours. The slag that comes out is drawn off, and the smelted copper that would accumulate at the bottom of the furnace is removed the following day, by which time it would have cooled off. This material is again melted, refined by reheating in an open furnace and finally cast into bar shapes. The yield could vary from 2 $^{1}/2$ – 7 $^{1}/2$ per cent depending upon the type of ore used and its extractable copper content.

Even though copper utensils in desired shapes were used during religious observances and other ceremonial functions, they did not enjoy any special place in the household repertory because of copper's reactions to any liquids, acidic or alkaline in nature, which would be harmful. Earthen vessels were, therefore, preferred to copper vessels for storage purposes. Nevertheless, copperplates, in contradistinction to the earthen or terracotta plates, became the most permanent medium of publicity or legal records of the royal court. Inscriptions on copperplates, meticulously documented, speak of the personality traits of a king through panegyrical lines of poetry engraved on them as well as the generous grants or gifts made by the royalty on ceremonial or religious occasions or other auspicious days either to individual donees or to temples. Several hundreds of copperplate-grants have come down to us. The technique of inscription on thick copperplates was standardized. Generally, the inscriber was a carpenter (*takṣaka*) who, with his precise and sharp tools of iron or any other sharp-edged material, was able to inscribe the text provided to him as accurately as possible on the copperplate. Much more than the script-forms engraved on stone, the characters inscribed on copperplates have been of greater value to palaeographic as well as linguistic studies of regional importance.

The elegant copperplates with royal seals and emblems on a thick ring of copper, which also holds the plates together, testify to the associated technical skills that were fostered over many centuries. Although normally the engraving was done on one side of the plate, there are later examples of engraving on both sides. Such plates were so dexterously inscribed that the impression of one side did not impinge on the other side; possibly, a mild alloy of copper could have been used in such cases. Copperplate inscriptions, now preserved all over the country, have been of historical value for understanding the contributions of several dynasties like the Śātavāhanas, Guptas, Vākaṭakas, Pallavas, Pāṇḍyas, Cholas, Chālukyas and the rulers of Vijayanagar empire. They have high literary merit, too, and are an index of the contemporary literary attainments of the time.

Icon-casting, Iconography and Iconometry

In the Vedic and the post-Vedic period, gods representing natural forces, rather than one Supreme Being, were worshipped, and the hymnal and other forms of prayer were the preferred mode of divine worship. Even though various god-forms were conceived during this period, they were an imaginative personification of natural phenomena or forces, and not concretized into physical, material forms. However, the practice of worshipping forms of the Buddha and the Tīrthankaras (like Parśvanātha and Mahāvīra), by the Bauddhas and Jainas respectively might have nucleated the Hindu *arca* (worshipping physical icons) tradition from the beginning of the Christian Era. The various modes of worship received their sustenance from the *Purāṇas*, which had glorified the extraordinary prowess and benevolence of a plethora of gods and goddesses in their diverse forms. The *arca* practices became an established tradition due to their standardization and codification of the prescriptions by various Āgamic manuals. These manuals, besides giving prescriptions, legitimized the mode of worship of several iconic forms by according them Śāstraic sanction. This development gave a fillip to iconography and iconometry.

Śiva-Tāṇḍava (Chola: ninth-tenth century CE)
(Courtesy: Project of History of Indian Science Philosophy and Culture, Centre for Studies in Civilizations)

Soon, the *arca* tradition ushered in what may be called iconic religion in its twin facets: first, concrete forms of gods and goddesses; and second, transcending these physical forms to an incommunicable experience of the divinity beyond its label. The latter was attained by artistic iconography with accurate iconometry, an idealization of human anatomical features of the divinity. The basic material for the production of exquisite icons were stone and copper-bronze: stone was used for sculpted forms of the main deity or *mūla bhera* in the *sanctum sanctorum* of the temple, while copper-bronze was the material for images of the *kautuka bhera* or *utsava behra* to be taken out in procession either round the temple or along the streets of the temple-township. In both the forms, the sculpted image or icon was so elegant, balanced, and beautiful that its religious worship was pursued as the pathway of spiritual experience, from the vision of *form* to the experience of the *formlessness* of divinity. Behind a wide variety of forms and rituals, both sublime and bizarre, centring round the iconic religion, lay the inner urge of the devout towards the non-duality of the self and the cosmic being or cosmic consciousness.

In addition to the principal or main icon of the deity, copper-bronze icons had a special appeal to the *arca* tradition. Incredible indeed is the number of fabricated bronze images bearing innumerable names of gods and goddesses, some single, and others in twos or threes, usually on a pedestal. Their styles also became characteristic of the Chola, Pāṇḍya or Vijayanagar periods. The canonized saints of both Śaivism and Vaiṣṇavism, the followers of the *Ācārya* tradition and their exposition of divine forms, the Śākta worship in accordance with tāntras and, of course, the Purāṇic divinities—all found expressions in images of both stone and copper-bronze. The devotional (*bhakti*) movement in different parts of India and in different ways became integral to the iconic religion leading to what was known as god-realization, and the associated temple-culture. Even temple architecture was subjected to the measurements stipulated in the manuals. For the Hindu tradition, the oldest manuals were the *Bṛhatsaṃhitā* of Varāhamihira (early sixth century CE) and the *Viṣṇudharmottara Purāṇa* (probably of the same time or a little later). Around the eighth century or so appeared the *Vaikhānasāgama* and its associates, like the *Marīca Saṃhitā*, *Kāśyapa Saṃhitā* and *Vimānārcana kalpa*. The prescriptions or injunctions of these manuals were scrupulously followed in all Hindu temples, a tradition that continues with great vigour even today.

Casting

Icon-casting was carried out generally through the lost-wax or *cire purdue* process. Copper-bronze icons were fabricated according to canonically prescribed mensuration, but featured aesthetic ornamentation too. The traditional unit of mensuration was *aṅgula* (about three-fourths of an English inch) which is the twenty-fourth part of a *hasta*, as measured from the tip of the middle finger to the end of the elbow point. It was also reckoned in terms of the digit of the middle finger of a normal person of ninety-six *aṅgulas* (six feet) height. Traditionally, in iconometry, three kinds of *aṅgulas* have been recognized: (i) *mānāṅgula*; (ii) *mātrāṅgula*; and (iii) *dehalabdhāṅgula*. One *mānāṅgula* is regarded as being equal in length to that of eight barley grains placed side by side. A *mātrāṅgula* is reckoned in terms of the length of the middle digit of the middle finger of the right hand of the *Ācārya*, while the *dehalabdhangula* is calculated by taking into consideration the *total height* of the proposed icon, divided into *ten* parts, and dividing each part into further *twelve* units. The word *tāla* is used to denote a specified height. An icon or sculpted image could be either *daśatāla, navatāla* or *aṣṭatāla*. According to the manual, Kṛṣṇa Vasudeva is to be in *daśatāla*, Brahmā and Śīva to be nine and half *tālas*, while the other divinities need to be *navatāla,* and human figures are to be of *aṣṭatāla*. Likewise, specific parameters have been fixed for *dikpālakas* and other demi-gods.

Indian copper-smiths were adept in the technique of the 'lost-wax' process for casting an icon. A Sanskrit compendium of the twelfth century, the *Mānasollāsa* or the *Abhilaṣitārthacintāmaṇi,* attributed to King Someśvara Bhūlokamalla of the Chālukya dynasty in the Karnataka region, has given a succinct account of the lost-wax process as follows:

> 'According to the *navatāla* measurement, the expert should first prepare the image (that is, the model)

in beeswax complete with all its limbs and others, yellowish in colour, beautiful to look at, with the arms and weapons as prescribed.

After placing the wax tubes of the length of a *dhaturā* flower on the back, on the shoulders and on the neck or head of the image, the artist should besmear the image with refined clay. To the clay should be added charred husk, finely rubbed and purified cotton with a little of finely powdered salt. These are ground well on a smooth stone, and the paste thus obtained should be applied three times over the image.

The first layer of this clay paste should be thin as well as transparent, and dried in shade. After a couple of days, a second layer should be applied on the dried first layer, and allowed to dry in shade. When dry, the third coating should be applied thick.

The whole image or the beeswax model inside these coatings is besmeared with clay, leaving the mouths of the tubes open, and allowed to dry.

The expert should first (before beginning the process) measure the wax of the image which has to be made either in brass, copper, silver or gold. Brass and copper should be taken ten times that of wax, silver twelve times, and gold, sixteen.

Now, one should encase the metal, either gold or any other desired one, with clay in a coconut-shaped crucible which is also dried. Next, one should melt away the wax from the mould by heating it over fire, and afterwards, the crucible containing the metal is heated...

After making a hole with an iron rod on the top of the crucible and holding it tightly with a pair of tongs, one should pour the molten metal into the mouth of the tube in a continuous stream and stop when it is filled to the brim of the tube. The mould with the molten

> metal is allowed to cool naturally. After it gets cooled, the expert should break the clay very carefully. The metal image verily resembles that in wax, endowed with similar limbs and other details.' (Ruth Reeves, *Cire Perdue Casting in India*, New Delhi, 1962; pp. 22-23)

The acceptability of the icon as the end product would depend upon the dexterity with which the wax image was fashioned according to details in the Āgamic manuals. The expert would give finishing touches so that the image would have an aesthetic appeal. The opening of the eyes (*cakṣu-unmīlana*) was a ritual by itself and would be done only in the last or the *pratiṣṭḥa* (installation) stage. The method is called the 'lost-wax' process because, when the molten metal is slowly poured into the mould from the opening at the top, the inner beeswax melts away, and the metal assumes the form of the icon pattern of the wax. The 'lost-wax' process in Sanskrit texts is known as *madhucchiṣṭa vidhāna*, during which several *mantras* need to be chanted at appropriate stages of the process to invoke the divine presence and guidance.

Copper-bronze icons so produced had to be consecrated with elaborate rituals before they could attain the status of *utsava bhera* (processional deity). The same was true in the case of *dhruva* or *mūla bhera*, the main or the fixed diety in the *sanctum sanctorum*. In some temples like the *Ranganāthasvāmi* in Srīrangam or Śrīraṅgapatna or *Govindarājasvāmi* in Tirupati, the *mūla bhera* is in a reclining or sleeping posture, while the god or goddess form in copper-bronze or any other metal must always be in the standing position. Each needed a distinct iconography of its own, but within the Śāstraic framework.

Metallic divine images (copper-bronze, silver or gold) started appearing from the late ninth or early tenth century, particularly in South India. In northern parts of India, the concept of separate processional deities (*utsava bhera*) was not as strong as in the south. In the Hindu tradition, copper-bronze was preferred for casting icons, while the Jainas and the Bauddhas formed the images of their Tirthankaras, the Buddha or the Bodhisattvas, even in brass. Brass, which is an alloy of copper and zinc (while bronze is an alloy of copper and tin), is malleable, and the brass sheets can look like golden sheets in lustre if the percentage of zinc in the alloy is about 30-35 per cent, the rest being copper. But such an alloy is possible only when extracted zinc is mixed with copper in this proportion and melted.

Zinc Metallurgy

The brass objects unearthed at Takṣaśīla (fourth-third century BCE) have revealed, on chemical analysis, that zinc in its metallic form was added to copper in such a proportion that some of them have about 30-35 per cent zinc content, thus imparting a golden lustre to them. It is reasonable to suppose that the extraction of zinc from its ores was an accomplished metallurgical practice at least by about the fourth century BCE. On the other hand, zinc metallurgy began to be practised in Europe only by about the sixteenth century CE. As regards the ancient Indian method, the *Memoirs of the Geological Survey of India* (vol. 92, Ch. 5, pp. 47-54) has this record:

> 'The large heaps of zinc and lead-bearing residues as well as slags, clay retorts and fragments of furnaces in the Zawar area (in Rajasthan) bear mute testimony to an ancient zinc smelting industry of impressive magnitude; but they are only fragmentary and wholly inadequate accounts of this metallurgical industry in written records... Two pyrometallurgical processes were employed in the smelting of Zawar ores, and from chemical analysis and other evidence it is inferred that metallic zinc and zinc oxide were recovered by the process that produced a sinter as waste product, and lead in the glassy slag... Commercial production of

zinc did not take place in Europe until the 16th century, having been introduced there by traders.'

Extraction of Zinc by Downward Distillation Technique: The extraction of zinc from its ores is a complicated one. Generally, its ore is known as sphalerite, a sulphide of zinc, which needs to be converted into zinc oxide to facilitate reduction of the oxide into the metallic state. But zinc, in turn, oxidizes rapidly, and hence an air-tight reducing atmosphere is to be created for the distillation of zinc. Its boiling point is about 900^{0}C, which is lower than the minimum temperature of about 1150^{0}C necessary for the separation of zinc from zinc oxide, but at that temperature zinc vapours would oxidize and form what is known as 'philosopher's wool'.

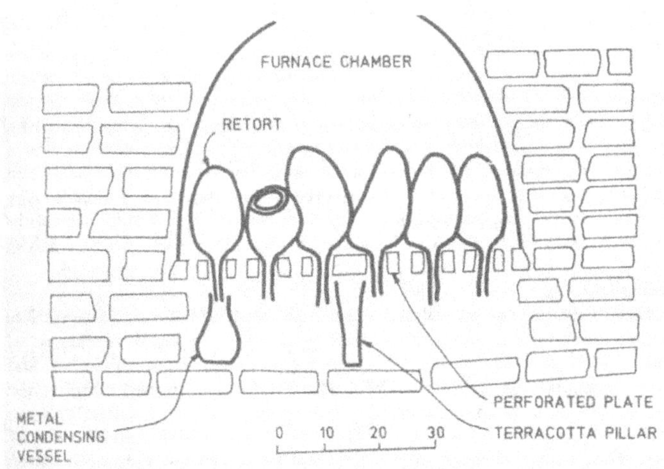

Model of Zinc Smelting Furnace in Zawar in Rajasthan
Source: *Chemistry and Chemical Techniques in India*, p.41
(Courtesy: Project of History of Indian Science Philosophy and Culture, Centre for Studies in Civilizations)

More importantly, the condensation temperature needs to be maintained in the very narrow range of 420°C to 520°C. The chemical reaction: $ZnO + C \rightarrow Zn + CO - 57{,}000$ calories is an endothermic one, and hence certain optimum conditions are to be maintained. It is amazing indeed that, notwithstanding the fact that ancient Indian zinc-smiths lacked any precise knowledge of the various parameters involved in the extraction of zinc, they accomplished it in an empirical but efficient manner and successfully extracted metallic zinc by what is called the *downward distillation process.* This method also required specially designed furnaces, retorts for the roasted ore, the reducing agent, reducing atmosphere, controlled temperature, and a mechanism or device for downward condensation. That the zinc-smiths were able to achieve this is borne out by the remnants now available archaeologically and their skills evoke our admiration, even after 2,500 years.

Extraction Furnace: On the basis of the archaeological evidence, the metallurgy of zinc adopted by ancient zinc-smiths would have been as follows: The ore, sphalerite (zinc sulphide with dolomite) was mined, crushed and concentrated, probably by the sieving method. It was then roasted in order to convert it into its oxide form. This was then shaped into pellets of about one cubic centimetre each, which were mixed with some quantity of common salt (to serve as flux), but a large quantity of carbon or charcoal, and placed in small, thin-walled clay retorts. Funnel-like condensers were luted on to them with clay. Cylindrical reeds were inserted into the retorts with a view to preventing the material inside the retorts from falling down when they were inverted and placed vertically (with their mouths downwards) in the furnace. It has been inferred from archaeological remains that there used to be six rows of such retorts, each row composed of six retorts, making in all thirty-six retorts used in one batch of extraction.

The furnace had two chambers: the upper chamber for heating the retorts with the roasted ore along with the reducing agent (charcoal) and the lower chamber, for condensing the zinc vapours in retorts kept below. The two chambers were separated by a perforated terracotta plate. In all probability thin cow-dung cakes were used as fuel. The furnace was heated and the temperature maintained by a constant supply of fuel and air for some hours, the details of which are not available. The zinc oxide in the upper retorts would get reduced by the charcoal into zinc vapours, which would in turn, condense into retorts placed below in a reducing atmosphere inside the furnace. It is calculated that less than half a kilogram of zinc would be obtained from each retort, that is, in one batch of extraction using 6 x 6=36 retorts, about 18 kg of zinc would be produced—a remarkable metallurgical feat at that time indeed! However, this metallurgical activity in the Zawar mining area in Rajasthan had come to an end by the end of the eighteenth century, probably because of the devastation caused by the Maratha battles that took place then or the poisonous atmosphere caused by cadmium dust (found in the residues of the zinc smelting area), which would have been detrimental to the inhabitants. In any case, Indian metal-smiths were the pioneers in developing the metallurgy of zinc successfully about 2,500 years ago.

IRON AND STEEL

In the history of extractive metallurgy, iron is rather a late entrant inasmuch as the extraction from its ores was developed around the middle of the second millennium BCE, while the origins of copper metallurgy could be traced back to about 4000 BCE. The melting point of iron is high—about 1550°C—while that of copper is much less. Indian iron-smiths were not able to develop the high temperature to reach the molten state of iron, but their furnaces could attain a temperature of 1100-1200°C, with the result that the reduction of iron oxide (ore) to iron by carbonaceous material with the help of a flux, would be accomplished. The resultant metal would, however, be a spongy mass with slag and other impurities. The latter had to be removed by the repeated heating of the spongy iron and hammering it. Indian iron-smiths had the technical expertise of obtaining iron of high purity (even close to cent per cent), now called wrought iron. They were not, however, capable of producing cast iron (with about 4.5 per cent carbon) and developing the casting techniques like their Chinese counterparts. Nevertheless, they showed the rare craftsmanship of producing a high carbon steel, now known as 'Wootz Steel', which was in great demand in West Asian countries from about the sixth century CE. More importantly, Indian iron-smiths were adepts in the forge-welding technique, leading to a macro-technology of massive pillars like that of the Delhi Iron Pillar, which has remained unrusted for over 1,600 years, a record that is unparalleled anywhere in the world. (see below).

Iron was known in the early human settlements, especially the Chalcolithic ones, in two ways: one, the meteoric iron (or 'iron from heaven' as expressed in some ancient texts) that was found in craters formed as a result of meteoric fall, and had a fair amount of nickel in it; the other, during the extraction of copper metal from the chalcopyrites ore when either haematite or iron oxide was used as a flux, a small lump of iron would have been formed in the furnace. Iron, under these situations, was considered rare and precious. Sometimes it was even regarded as having supernatural powers. The questions—when and where did the extractive metallurgy of iron appear first? Was it developed in India indigenously or was the technique borrowed from the West Asian Hittite civilization?—have no definitive answers at present, despite the extensive archaeological data available and the associated radio-carbon datings. The diffusion theory, namely, the discovery or invention of a substance, metal, etc., in one particular place, and its subsequent spread or diffusion into other regions, does not have convincing acceptability in such areas as agriculture, ceramics, dyes and pigments, cosmetics and perfumery, and even in metal-workings, because different human settlements were capable of developing these by themselves to meet their social necessities. The metallurgy of iron is a case in point. A useful metal, with its own characteristics or special properties in the form of its carbon-alloy, known as steel, iron emerged in course of time as an important metal in the material enrichment of human life through its varied uses in agriculture, construction of temples, and even weapons of warfare. Moreover, iron ore occurs in larger quantities than the ores of other metals. Its extraction method—a reduction process of its oxide, in which form the ore occurs—of obtaining iron could have been developed by several later cultures independently, although the transmission of the method of technique from one culture to the other in the earlier stages cannot be wished away.

As for the antiquity of iron, literary references are as valid as the archaeological data. The earliest reference to extracted iron is probably in a record of the Hittites, indicating the metallurgy of iron in West Asia around 1700 BCE. In the Indian context, the records of some early foreign historians and travelers are informative. Heradotus, a Greek historian, mentions the iron heads of cane arrows carried by Indians in the Persian army (fifth century BCE). Another Greek historian Ktesias refers to the gift of two swords made of Indian steel given to him in the Persian Court. Quintus Curtis speaks of a hundred 'talents' (a Greek weight) of steel presented to Alexander the Great (326 BCE) as a tribute by King Porus of Takṣaśīla. As noted earlier, in the *Yajurveda* (1200 BCE) the word, *Kṛṣṇāyas* or *Loha-ayas*, definitely connotes iron. Archaeologically, iron objects or slag have been found in Atranjikheda and Hastinapur, which are datable to around 1100-1000 BCE. Several objects of iron have been unearthed in association with the Painted Grey Ware (800 BCE). But excavations of megalithic sites indicate that iron objects, which had a special appeal to the people of the megalithic culture, were in use in India earlier than 1100 BCE, probably during 1300-1200 BCE.

That the use of iron was widespread throughout India from about the sixth century BCE is evidenced by several types of iron objects found archaeologically. Of the important sites that have delivered iron objects, special mention needs to be made of the following (the types of iron objects found in them are indicated in brackets): Rupar (the Punjab: nails, hooks, bars, spikes, knives, daggers, sickles and spearheads); Ujjain (Madhya Pradesh: arrowheads, crow-bar types, spade, spearheads, knives, etc.); Nagda (Madhya Pradesh: chisels, knives, sickles, arrowheads, hoes, bowls, nails, etc.); Sambhar (Rajasthan: double-edged daggers, crucibles, clasps, fish plates, sickles, braces, rings, bells, ladles, spearheads, arrowheads, etc.); Rairh (Rajasthan: sword-blades, spearheads, knives, sickles, door fittings, etc.); Nasik (Maharashtra: caltrops, choppers, knife-blades, axes, drills, nails, ladles, etc.); Taxila (bowl, spoons, saucepans, different types of dagger, elephant goad, axe, adzes, tongs, anvils, nails, etc.). Similar iron objects have been found in Sisupalgarh and Jaugada (Odisha), Maheswar, Navdatoli, Nagda and Tripuri (Madhya Pradesh), Prakash and Bahal (Maharashtra), Sonepur and Rajgir (Bihar) and Purana Qila (Delhi).

In South India, a large number of megalithic burials have yielded, as grave goods, several kinds of iron objects: flat axes, tanged daggers, spears, arrowheads, wedge-shaped blades, horse-bits, sickles, spades, hoe, pickaxe, tripods, chisels, tridents, knives, rings, etc. It is important to note that the megalithic burials have been found from Adichenallur (in Tirunelveli district of Tamil Nadu) to Nagpur (in Maharashtra), practically in the whole of peninsular India. The Megalithic culture is also noticeable in some parts of North India. Excavations at three habitational Megalithic sites at Naikunda (Vidarbha region in Maharastra), Banahalli (Karnataka: Kolar district) and Kodumanal (Tamil Nadu), have yielded perceptible remains of iron-smelting furnaces, copious iron slags, earthen tuyers and other accessories, probably belonging to the period 700-200 BCE. In addition to other techniques like ceramics, carpentry, basketry, and bead-making, Megalithic craftsmen were known more for their expertise in iron-ore prospecting, smelting, and smithy. In fact, the Iron Age in India assumed new dimensions with the Megalithic people.

Metallurgical Operations

The archaeological excavations in Ujjain as well as certain Megalithic sites have provided some clues to the possible ancient method of the extraction of iron from its ores. The smelting operations consisted of heating alternate layers of the ore (generally haematite or Fe_2O_3) and charcoal in small furnaces or kilns with passage for blowing air from below, escape of gases from above, and holes towards the bottom for the removal of slag, and thereafter the spongy mass of iron. Even now in Madhya Pradesh, Bihar and Maharashtra, the primitive methods of the smelting of iron ore, coupled with repeated forging of the red hot mass to drive out slag and other impurities, are in vogue. They have been adopted by tribals, such as the Agarias, Khuntia Choks, Marias, and Murias (of Madhya Pradesh); Birs and Agaria Asurs (of Bihar); Gonds and Munnewars (of Maharashtra). The metallurgical operation is generally conducted in a vertical or slightly tilted cylindrical kiln made of clay, about 90 cm high and 120 cm wide. The kiln tapers towards the top like a mini blast-furnace, with three openings: a wider one at the top, and two (one slightly large and the other small) at the base. The opening at the top is to insert selected good-quality iron ore and charcoal in alternate layers. A short wall is erected to serve as a protective parapet, and it is sloped towards the opening of the furnace so that the iron ore and charcoal can be slid into it. One of the openings at the base is for blowing air as well as maintaining the blast and receiving the bloom, while the other is for taking out the slag. Bellows are employed for maintaining the flow of air and the temperature. The bloom is taken out and hammered to squeeze out slag and other impurities, and the resulting mass is of fairly high-quality of wrought iron in the form of cakes or lumps.

There is some evidence to indicate that a small quantity of limestone was added to the charge. Perhaps Indian iron-smiths knew empirically that this would act as flux and facilitate the reduction process. While the Agarias generally do not use any flux at all, the tribal smelters in other regions use it. Even the type of the furnace varies from region to region in conformity with the kind of clay used and the plan of the base and its height.

In South India, furnaces were circular or squarish at the base and conical in shape. According to the records *Public Consultations, Indian Iron and Steel Company or Porto Novo Works*, 4 June 1833, Madras Records office (now Chennai), preserved by the East India Company, small furnaces made of red potter's clay mixed with sand were used in considerable numbers by the iron-smiths of the Salem region (Tamil Nadu) for the production of bar iron early in the nineteenth century. The height of the conically shaped furnace was slightly under four feet, the diameter sloping from a little more than 30 cm at the bottom to about 18 cm at the top. The charge consisted mainly of the magnetic iron ore or haematite mixed with the requisite quantity of charcoal, which would serve both as a reducing agent and fuel, to maintain the temperature. The charge was inserted from the top and, as usual, there were two openings towards the bottom—one for letting in the blast of air by the application of bellows (made of goatskin) to attain the needed temperature, and the other, for taking out the slag and, subsequently, the lumps of red hot iron. In about two and a half hours, the operation would be over, yielding lumps of about 50 per cent iron. They were repeatedly forged to obtain bar iron of high quality with some proportion of steel. The record of the East India Company asserts that the worst of Salem bar iron was as good as the best English iron. Moreover, the price of the native product was far less than that of the market price of the cheapest English bar iron in India. It is likely that the ancient methods of iron production were more or less similar to the ones practised by tribal iron-smiths.

The Mehrauli Iron Pillar (Delhi)

Standing tall, majestic and serene, in the Mehrauli area of Delhi, is a historic vestige of the late fourth-early fifth century CE and a shining example of the macro-technology of iron practised and perfected by Indian iron-smiths. It is the famous Iron Pillar, a marvel that has remained practically unrusted for over 1,600 years. It is 7.735 m (with about half a meter below ground) high; its diameter diminishes from about 41.6 cm below to about 34 cm above. The specific gravity of the metal is 7.5, and the Pillar weighs more than 6 tonnes. The overall chemical composition of the Pillar is as follows:

Chemical Composition of the Iron Pillar

	Hadfield	Lal, B.B. (chemist)
	(1912)	(1945)
Iron	99.72	99.67
Manganese	Nil	Nil
Sulphur	0.006	0.007
Carbon	0.08	0.09
Phosphorus	0.114	0.174
Silicon	0.046	0.048
Others	0.246	0.011
Specific gravity	7.81	7.5

The Pillar was forge-welded, but the task was by no means easy to accomplish, considering its height, weight, and uniformity. It is estimated that about 200 furnaces (of the type noted already) would have been operated in tandem, or the same set of furnaces would have been used repeatedly to produce iron of over 99.5 per cent purity with consistency, both qualitatively and quantitatively. The technique of producing individual pieces of pure iron (probably about 20 kg each), repeatedly heating and forge-welding them, has evoked the admiration even of several modern metallurgists. It would appear that the Pillar's fabrication and positioning would have been facilitated by the use of iron clamps at appropriate levels, and the protruding portion of such clamps would have been chiselled away when the entire operation was over. This explanation is supported by the visual proof of the presence of such clamps at two points on the Delhi Iron Pillar.

The Iron Pillar near Mehrauli, Delhi

The Pillar is in two parts: the main body and the decorative bell-capital on top. It also bears an inscription in Sanskrit on the surface of the main body. After forge-welding, the surface of the Pillar must have been smoothed by careful chiselling, and the surface burnished to provide a smooth tapered, cylindrical shape. Thereafter, the Sanskrit inscription would have been die-struck and finally, the decorative bell-capital would have been fitted on to the top portion of the Pillar, and thus got the pillar ready for its placement and erection. There is at present a hollow slot on top of the Pillar.

An inscription has been worked into the Pillar; the technique of inscribing was in the nature of the die-striking operation. In other words, specific die-shapes were used to create the characters that were intended to appear on the chosen surface by striking the die with force, thus transferring the image of the die as a depression on the surface. The inscription in Sanskrit on the Pillar's surface at a height of about 2 m from the stone platform, says that 'Chandra having in faith fixed his mind on Viṣṇu set up the Pillar at Viṣṇupadagiri.' The date of this inscription is stated to be around 400 CE and Chandra II, has been identified with Chandragupta II (Vikramāditya, who reigned between 375 and 415 CE). Possibly, the inscription was die-struck during the time of Chandragupta II himself. It is reasonable to think that Viṣṇupadagiri was the original location of the Pillar, and this place has been identified as modern Udayagiri, which is not far away from Besangar, Vidisha and Sanchi (which are about 50 km east of Bhopal). It is interesting to note that there is an Aśokan Pillar in Sanchi, and at Besanagar the famous Herodotus Pillar, but they are not metallic and belong to a much earlier period. Two Udayagiri cave inscriptions mention the name of Chandragupta II, who is also stated to have built a temple there. The worship of the footprint of Viṣṇu (*Viṣṇu pada*) was also a practice there, besides the tradition of iron-production in that region. The original erection site of the Pillar could have been in front of the passage ahead of cave number 7 at Udayagiri. There is a view that the Pillar had astronomical significance for determining solstices and equinoxes, and that there was an astronomical observatory on the northern edge of the hilltop. But this view needs further investigation.

Of interest is the fact that the Pillar's present location is in the Mehrauli area of Delhi in the courtyard of the Quwwat-ul-Islam mosque. As the name indicates, the mosque was built as a monument to the 'Might of Islam', by Qutubuddin Aibak (as the memento of Turkish slave commander and Viceroy of Muhammad Ghori, the Afghan king) to commemorate the victory over the Rajput king, Prithviraj Chauhan, in 1192. The Muslim monarch Iltutmish who invaded Malwa in 1233, is stated to have carried the Iron Pillar with him to Delhi and installed it in the courtyard of the aforesaid mosque in 1233. It was erroneously thought of by him and others as a 'brass pillar' because of its somewhat brass-like appearance when viewed from certain angles. As for its fixture in a vertical position, it would appear that two iron rods were placed parallel to each other, and another two iron rods were placed above them perpendicular to the former so as to form a grid-like structure. The iron rods were suitably positioned on the lead sheet, and the base of the Iron Pillar was fixed on top of this structure which provided the needed grip.

'The Rustless Wonder': The outstanding characteristic of the huge Pillar has been its resistance to corrosion despite its exposure to rain and the glaring sun for over 1,600 years, earning for itself the epithet, 'The Rustless Wonder'. Metallurgists who have examined the Pillar and conducted chemical investigations, have adduced several reasons for its enduring quality of non-corrosion. One of the reasons is stated to be the not-so-high level of humidity in the atmosphere of Delhi, which is less than 60 per cent for about three-quarters of a year. However, it needs to be understood that during the rainy season for about three months, the humidity is indeed high and the Pillar is also exposed to rain. Other reasons for its chemical resilience are: (i) the near purity (over 99.5 per cent) of the iron of the Pillar; (ii) the presence of rather high phosphorus; (iii) the absence of corroding elements like sulphur and manganese; and (iv) the formation of the slag-enveloped grain structure that has acted as a protective film against the process of corrosion. Yet another reason put forward is concerned with the large mass of the Pillar, which would get heated up relatively rapidly, with the result that this heat would drive off any moisture on its surface, which otherwise would have played a role in causing corrosion.

Possibly all of these reasons might have been responsible for the continued excellent corrosion-resistance of the Pillar. Nevertheless, due credit needs to be given to the ancient iron-smiths for their meticulous endeavours in selecting the ore with its high phosphorus content, its processing before the smelting operation, and forging and welding of the spongy mass of almost pure iron, leading to a corrosion-resistant finished product. The formation of a protective film on the surface of the Pillar, that has warded off corrosion over the centuries, has been well understood scientifically. But how the ancient iron-smiths achieved such a spectacular result by their own methods needs far more systematic studies.

In any case, the formation of a protective film has been scientifically verified through rust-characterization studies on the rust collected from the region just below the bell capital of the Pillar, using X-ray diffraction and spectroscopic methods. The formation of iron-hydrogen phosphate hydrate in the microcrystalline state as well as iron oxide/hydroxide has been identified by these scientific investigations. The high phosphorus content of the Pillar (obviously because phosphoric iron-ores were employed for the production of the metal) has enhanced the Pillar's corrosion resistance.

Some metallurgists argue that alternate wetting and drying conditions play an important role in accelerating the formation of the phosphate phases. They have also pointed out that the large mass of pillar (about six tonnes) has a contributory role in aiding alternate wetting and drying, since this massive object would heat up faster and also cool down faster. Studies on the Delhi Iron Pillar have a relevance to modern iron technology inasmuch as they have revealed the importance of phosphoric iron to ward off corrosion of iron objects under non-congenial atmospheric conditions.

The Delhi Pillar evoked the admiration of an eminent British geologist, V. Ball, who recorded in 1881: 'It is not many years since the production of such a pillar would have been an impossibility in the largest foundries of the world, and even now there are comparatively few places where a similar mass of metal could be turned out.' (*Economic Geology,* III, 338) The empirical method adopted by Indian iron-smiths to achieve corrosion-resistance of macro-iron objects was not short-lived, as the later Dhar Iron Pillar proves.

The Dhar Pillar

There is another iron pillar which is much larger in dimension than the Delhi Pillar. Belonging to the eleventh century, with a length approximately over 13 m, this pillar is now lying in three broken pieces in front of the Lat Masjid at Dhar (near Indore in Madhya Pradesh). Vincent Smith, a British historian-scholar, wrote about it: 'While we marvel at the skill shown by the ancient artificers in forging a great mass of the Delhi Pillar, we must give a still greater measure of admiration to the forgotten craftsmen who dealt so successfully in producing the still more ponderous iron mass of the Dhar Pillar monument with its total length of 42 feet.' ('Dhar Pillar', *Journal of the Royal Asiatic Society*, 1898, pp. 143-46)

Dhāra was the capital of Malwa under the famous king, Bhoja (1010-53 CE). According to local tradition, Raja Bhoja, who was not only a great patron of art and letters but also well-versed in several technical subjects including the production of iron weapons, ship-building, etc., was instrumental in constructing and erecting the pillar. Since Bhoja was a Śaivite, this iron pillar might have been positioned in front of a Śiva temple and it would have had a trident (*triśūl*) on its top. Interestingly, since the region of Dhar was also known as *Lāṭeśvara maṇḍala*, the name of its deity could have been Lāṭeśvara, the word *la* also connoting a pillar. It seems that when Sultan Bahadur Shah of Gujarat overran the Malwa region and wanted to carry the pillar to Gujarat, it fell down in the process and broke into three pieces. It lay in that state till the Archaeological Survey of India placed it horizontally on a platform near the mosque. The style of a late inscription in Persian (1598) on this pillar shows that it was inscribed when the pillar was lying down.

Of the three pieces, the largest is 24 feet 3 inches in length, the second, 11 feet 7 inches and the third, 7 feet 6 inches. The first piece has almost a square section, the second has a rectangular section, while the third part is octagonal, beginning with a cone shape. The total height of the Dhar Pillar is nearly double the height of the Delhi Pillar and its combined weight (of the three pieces) is estimated to be 7.3 tonnes, at least a tonne more than the Delhi Pillar. The first and the second pieces were joined by the usual forge-welding process while the conical top of the second piece and the conical bottom of the third piece were joined by a forged ring. It would appear that a fourth piece of over 3 feet in length is missing, and thus the total height of the Dhar Pillar would have been about 46 feet, if the fourth piece is taken into reckoning.

Though the technique of fabricating the Dhar Pillar was the forge-welding one, its surface is not as smooth as that of the Delhi Pillar. There have been differing views among metallurgists, both Indian and foreign, regarding the production of this pillar, including the level or the intensity of the forge-welding technique employed. More systematic studies need to be carried out on the chemical composition of the material this historic relic is composed of, as well as the precise causes of its corrosion-resistance, similar to the extensive investigations conducted on the Delhi Pillar. In any case, the Dhar Pillar speaks volumes about the most fascinating metallurgical skills of ancient iron-smiths.

The Kodachadri Pillar

Both the Delhi Iron Pillar and the Dhar Iron Pillar were produced in a region (Central India) known for its deposits of rich iron ores and centuries of iron-metallurgical heritage. They are the indices of the peaks of technical excellence achieved by Indian iron-smiths in macro-metallurgy. In southern India, also reputed for its magnificent iron ore deposits, there is a pillar on the Kodachadri hills in Karnataka (pure iron as indicated by X-ray diffraction analysis), which rises about 10 m above the ground, possessing a rectangular cross-section of 8.5 cm x 6 cm. The weight of the Kodachadri Pillar is estimated to be about half a tonne, assuming that about 1.4 m of the pillar lies underneath. According to local tradition, this pillar is the flagstaff (*dhvajasthambha*) of the Ādi Mūkāmbikā temple on the Kodachadri hill, and it is also the top portion of the trident (*triśūl*) of the Mother Goddess by which the demon was 'nailed down to the earth.' The hill is about 40 km from Kollur where the image of goddess Mūkāmbikā now exists. During the rainy season, however, the pillar also acts as a lightning arrester. This function is indicated by the solidification of the flow of molten metal that can be seen on top of the pillar.

The exact date of the pillar's construction is not certain and, if the folk story of Ādi Mūkāmbikā is taken into consideration along with the traditional view that the great Śaṅkarācārya was associated with this goddess, the pillar would have been erected around the eighth century, but its date could also be much later. The amazing fact is that this pillar has also been corrosion-resistant over the centuries, despite the heavy rain that persists in the area for four to six months in a year and its geographical position, which is just about 50 km from the Arabian Sea. The reddish colour of the surface of the pillar suggests that a protective layer similar to that on the Delhi Iron Pillar might have engendered the Kodachadri Pillar's resistance to corrosion. But the phosphorus content of this pillar has not been determined, nor has a detailed analysis of its composition been undertaken so far.

Iron Beams

While the vertical massive iron pillars are marvellous for their fabrication and corrosion-resistance, equally incredible are the horizontal large beams that were used in the construction of temples in the state of Odisha, which lies on the humid east coast. The region of Odisha and its adjoining areas are well known for high-quality iron ore, and the excavated iron objects of antiquity in this region point to the iron-metallurgical activity from remote times. There is evidence to indicate that iron clamps and beams were used for joining the protruding

The Sun Temple at Konark, Odisha (thirteenth century CE)

cornices as well as roof-stones of some temples built at Bhubaneswar as early as the seventh century. The use of a large number of iron beams, however, was more pronounced in the temples at Puri and Konark: the Jagannath temple built at Puri in the thirteenth century, the Sun temple, first built at Konark in the ninth century, and its Naṭamandir, constructed later, in the thirteenth century.

Astonishing it is indeed that in the Gunduchibari garden temple of the Jagannath temple complex, as many as 239 iron beams, each up to about 5 m in length and 15 cm × 10 cm or 12.5 cm × 15 cm in cross-section, have been counted. Some of the iron beams of the Sun Temple (now in ruins) at Konark have even larger dimensions: over 7 m in length with sections of 20 cm × 23 cm; twenty-nine iron beams are still lying on the lawns of the Sun Temple, some nine of them being about 6 m to 7 m long and 30 cm square. The largest of the iron beams is about 11 m long with an 18 cm square cross-section, weighing about 2.5 tonnes. However, an examination of the beams has revealed that their forging is not as perfect as it is in the Delhi Iron Pillar. According to a chemical analysis of one of the beams, it is found to be made of wrought iron (similar to that of the Delhi Pillar) or 99.64 per cent pure iron, with 0.15 per cent phosphorus; traces of carbon and sulphur are present, but without manganese. Its specific gravity is 7.8. As for the corrosion-resistance of beams, they have slightly corroded mainly because of the extreme humidity and salt-laden atmosphere (nearness to sea), although less than modern beams do. It may, therefore, be surmised that the corrosion-resistance of the beam is due to its protective layer of iron oxide and phosphorous compound. But more systematic studies of many more beams are needed to arrive at a clear picture of their production and corrosion-resistance.

Steel

In addition to their extraordinary excellence in the macro-technology of iron, Indian iron-smiths had displayed rare expertise in the micro-technology of high-quality crucible steel, known as Wootz steel (the word, *Wootz*, being an outlandish or corrupt form of the Kannada word, *ukku* or Tamil *urukku*), an alloy of iron and carbon that earned the approbation of being 'the wonder material of the orient'. With its carbon content in the range of about 1.3 to 1.8 per cent, the rest being pure iron, Wootz steel was a highly prized commodity for fabricating the famous Damascus blades, swords, daggers, and even armour. In the Middle Eastern countries, it was also known as Deccani Steel, particularly in Persian markets. Damascus was an important trading centre in the Middle Ages, and European dealers who purchased the sharp blades made from Wootz steel called them the Damascened blades or swords. For, on their polished surface, were beautiful wavy patterns, called *damask*, etched with a vegetable (organic) acid—this made them distinguishable from the other types of blades or swords. The damascened blades were sharp with fine edges, and it is well known that the swords and helmets used by European crusaders proved ineffective against those made of Indian steel, which were being used by Islamic warriors. It would appear that the swords used by these fighters (made of Indian or Wootz steel) were so sharp that not only would they cut the enemy into pieces even if he was protected with armour, but they also retained their lethal capacity, even after continuous use. Isfahan in Persia was one of the major centres for making swords of high quality with the Wootz steel imported from South India. Iron-smiths of Persia and some West Asian countries attempted to produce this type of steel, since they had some knowledge of the South Indian methods, but they did not meet with success. India thus had practically the monopoly of producing high-quality steel in the form of pellets of crucible steel till the middle of the nineteenth century.

Production of Wootz Steel: Francis Buchanan-Hamilton, an employee of the British East India Company, has provided a vivid account of the method of producing Wootz steel, which he witnessed during his travel in the southern region. Titled *A Journey through Madras, Mysore, Canara and Malabar* (1807), his travelogue is an important document of the technique of producing this kind of steel, which was in practice during the first decade of the nineteenth century. His description of the furnace is given below:

> 'The furnace is constructed in a hut, and consists of a horizontal pit and a vertical fireplace, sunk below the ground level. The ash pit is about three-fourths of a cubit in width and height, and conducts from the lower part of the fireplace to the outer side of the hut, where it ends in a square pit, used for drawing ashes. The fireplace is a circular pit that descends from the surface of the ground to the bottom of the ash-pit. Its mouth is a little dilated. Parallel to the ash pit, and a little distance from the mouth of the fireplace, a mud

> wall is erected, in order to shield the workman from the sparks and the glare of the fire. The mud wall is about five feet high. Through the bottom of this passes an earthen tube, which conducts into the fireplace the wind of two bellows, each made of bullocks' hide. The crucibles, made in a conical form, are of unbaked clay.'

According to some archaeo-metallurigsts, Wootz steel was produced efficiently either by (i) carburization of pure or wrought iron in crucibles; or (ii) decarburization of cast iron (which has a higher carbon content) by co-fusion of wrought iron and cast iron. The first method was adopted in the Hyderabad-Karnataka-Tamil Nadu region, while the second was in use mainly in Konasamudram and its neighbourhood in the Hyderabad region. According to Buchanan-Hamilton, the carburization of wrought iron was accomplished as follows: In the specially prepared crucibles, pieces of wrought iron together with 5 to 10 per cent of their weight of charcoal of the plant *Cassia auriculata*, and 1 to 2 per cent of leaves of another plant *Calatropis gigantia*, were placed, and the whole material was covered with one or two leaves of other plants. The mouths of the crucibles were then sealed. The number of crucibles in a single batch of operation varied from six to fifty-nine. Generally, twenty to twenty-five such crucibles were heated in a big hearth fired by charcoal and bellows to reach the necessary temperature. The crucibles were placed with their pointed end downwards in the furnace in a flat area and across a circular shaped pit (about 45 cm in diameter and over 100 cm in depth). Before placing the crucibles, the pit was stacked with straw. Charcoal was replenished from time to time both above and below the crucibles. The bellows which supplied the blast throughout, kept the vigorous charcoal fire active. According to Buchanan, one of the crucibles opposite the blow tube was kept empty, so that it could be removed

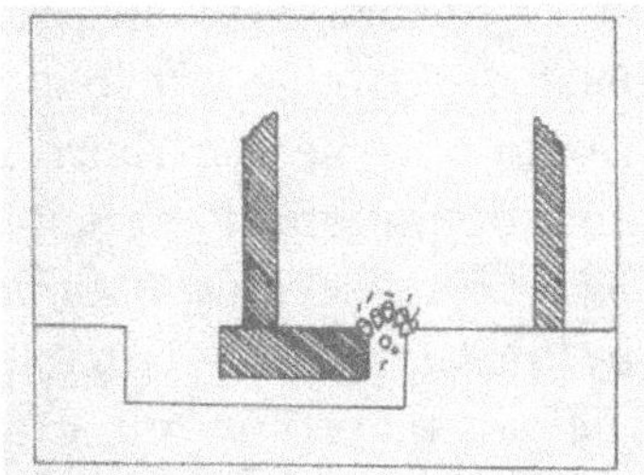

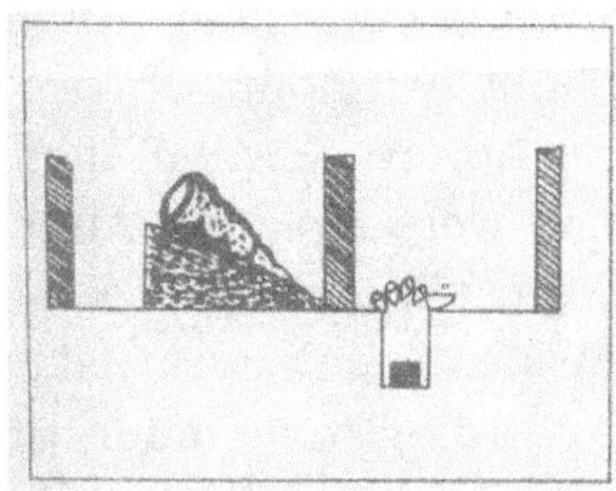

Sections of the steel furnace

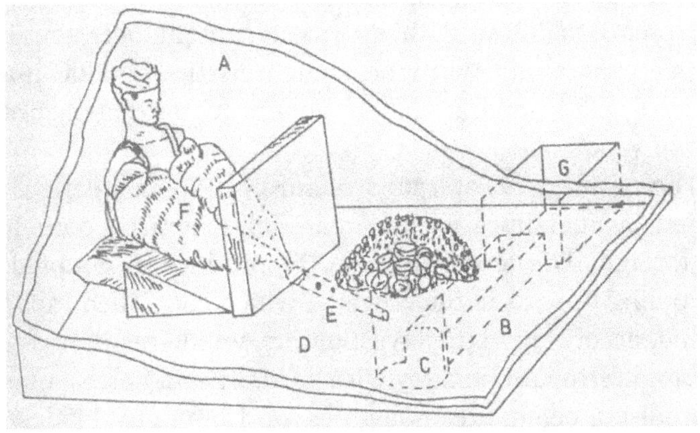

Sketch of a furnace for producing steel

A: Walls; B: Horizontal ash pit; C: Fireplace; D: Ground level; E: Tuyere; F: Bellows; G: Ash collection pit
(After Francis Buchanan-Hamilton's *A Journey from Madras through Mysore, Canara and Malabar*, 1807; II. ch. 7. p.19)
(Courtesy: Project of History of Indian Science Philosophy and Culture, Centre for Studies in Civilizations)

periodically for the replenishment of charcoal below the crucible in the fireplace. The heating operation was continued for about six hours. During this operation, the crucibles were gradually shifted to the hottest zone of the furnace, and this enabled the carburized iron or steel to melt and separate itself from the slag that would float on its surface. Possibly, after about two hours of continuous blast, the twelve crucibles in the hottest centre of the furnace would have been carburized, and were replaced. After the heating operation, the crucibles were allowed to cool, and the Wootz steel (now obtained inside them) was taken out by breaking the crucibles. According to metallurgists, the crucibles played an important role in achieving the desired properties of Wootz steel. The very process of cooling, too, had its expected effect on the end product. The slow cooling of the molten metal would produce a ferro-carbon alloy, with a carbon content between 1.3 to 1.8 per cent along with the characteristic metallic internal structure, and separate itself generally into hemispherical discs of Wootz steel.

The nature and internal structure of Wootz steel has been studied by metallurgists and material scientists, both Indian and foreign. The technique used to produce it, coupled with the cooling process in the crucible, which controlled the rate of cooling, according to the investigators, would result in a hyper-eutectoid ferro-carbon alloy with partially and heterogeneously spherodised cementites. Samples of Damascus blades have been examined, and the microstructure that such a sharp blade displays in the form of a wavy pattern, has been studied. It has been confirmed that the iron-carbon equilibrium and the parameters like heat treatment, mechanical working and casting temperature, all cumulatively contribute to the notable characteristics of Wootz steel. The Science and Humanities Research Institute of Iowa State University, USA, and the Department of Science and Engineering at the Ames Laboratory of the same university, recently carried out experimental work simulating and even replicating the ancient Indian practice of making Wootz steel. They investigated the microstructures and properties of this extraordinary metal. Some metallurgists at Stanford University, USA, also conducted research and developed what they called an 'Ultra High Carbon' (UHC) steel, with its carbon content ranging from 1.3 to 1.7 per cent, which is the same as that of Wootz steel. They have found this steel to have super plastic properties along with the required hardness for use in the production of some automobile and aircraft components.

Cannons: The forge-welding technique in which Indian blacksmiths had excelled in the formation of huge pillars and large beams, was also applicable to the fabrication of cannons, which played an important, though destructive, role in medieval warfare. But the technology of fabricating cannons was not indigenous to India; it was possibly borrowed from Islamic regions or Europe. It may be noted that gunpowder began to be used in India no earlier than the middle of the fifteenth century, and the application of gunpowder was necessary for the cannons to fire a missile or a metallic ball effectively. Several medieval kingdoms, Hindu and Muslim alike, both in the north and in the south (especially the Muslim kings) used cannons in their battles. Cannon technology received a fresh fillip with the arrival of the Portuguese as well as of Babur, the founder of the Mughal dynasty in India, because gunpowder was used in artillery in their field warfare.

Iron Cannons: It would appear that the technique of making cannons of wrought iron was borrowed from Europeans, who were using a kind of forge-welded iron cannon in their ships in the sixteenth century. But it was not long before the iron cannons were discarded in preference of the bronze ones, because the latter could be cast by a simpler technique than forge-welding. Moreover, the front bore or the opening (for the ejection of a cannon ball) could be shaped more easily by this process than in the iron cannon. Bronze cannons could also bear inscriptions which would help in dating them. However, a few forge-welded iron cannons have survived, one of which is at Haidar Burj in Bijapur (Karnataka), which has been dated to 1582. Two more forge-welded wrought-iron cannons exist in the public gardens at Khandwa in Madhya Pradesh, with inscriptions indicating their manufacture in 1585 and 1589. There were also some small-sized cannons belonging to an earlier period, like the one belonging to Sher Shah Suri's reign, dated to 1542-43 based on their Persian inscriptions. An iron cannon of larger dimensions, too, was in use in eastern India, with a length of over 900 cm and a weight of 47 tonnes. Another one at Dacca (lost probably in floods) weighed about 30 tonnes, while the iron ball in it weighed about 465 pounds!

An impressive and massive forge-welded cannon, which was made during the reign of Raghunatha Nayak (1600-45) and is now resting on three concrete supports at Thanjavur in Tamil Nadu, is more than 750 cm in length from end to mouth, including a 32 cm projection at the end of the barrel. Its total weight is estimated to be over 22 tonnes. The outside and the inner diameters of the gun barrel are 93 cm and 63 cm respectively. At Murshidabad in West Bengal lies a large iron cannon called *Jahankosha,* which means the 'conqueror of the world'. It is 525 cm long with a circumference of about 158 cm and bears nine brass plates with inscriptions that indicate its date as 1637 (during the reign of Mughal Emperor Shāhjahān). Possibly, this cannon as well as the others were manufactured by Hindu metal-smiths under the orders of Muhammadan rulers. There is also another forge-welded iron cannon at the garden of the Hazardurai palace museum in Murshidabad, called the Bachhawali: length, 461 cm (of which the length of the powder chamber is 110 cm; the bore diameter, 46 cm). This cannon has several rings for handling it as and when required. The other important iron cannons include those at Bishnupur (384 cm in length; about 8 tonnes in weight); Jhansi (two cannons, one of them belonging to the Maratha period and the other associated with Rani Laximibai of the First War of Independence, 1857); Patna Museum and Asiatic Society, Kolkata (belonging to Assamese King Jayadhwaja Simha, 1648-63); Gulbarga, Bidar and Bijapur, and Basava Kalyan (Karnataka); and Hyderabad.

The forge-welding technique for the production of cannons was effective in itself, hence Indian metal-smiths did not adopt cast iron technology, like the European rulers, which would have yielded better artillery. Nevertheless, the Indian technique of producing composite cannons by casting bronze over wrought iron barrels to strengthen them, and the light cannons that could be used on camelback, were appreciated by Europeans. Several composite cannons belonging to the reign of Aurangzeb speak of the effective indigenous technique of their fabrication. Systematic studies are necessary, however, particularly of the huge cannons that are preserved in forts and other places, to understand their artillery efficacy in warfare. Even as products of macro-iron technology, they need to be examined both scientifically for the nature of their metallic content, and technologically, for the process of their manufacture in sections and rings.

Bronze Cannons: Bronze cannons were preferred to the iron ones, but their production was more expensive. Since they bear the inscriptions by which they can be dated, they were also prized objects in the artillery of a monarch. Turks were noted for their technical skills in the production of bronze cannons. Since they were known also as *rumis*, an expert fabricator of cannon artillery was given the title *Rumikhan*. Among the cannons of larger dimensions, special mention needs to be made of the *Great Gun* of Agra (made of brass; weight: over 53 tonnes; length: 4.2 m and bore: 55 cm) and Malik-i-Maidan (bronze; 4.5 m in length: 70 cm at its bore; about 55 tonnes). The latter was probably cast in Ahmednagar, but it is now perched on top of the Sherzah Burj tower of Bijapur. The Mughal rulers were keen on developing artillery and the use of gunpowder for their warfare. Akbar took several measures in this direction and one of the important personalities in his Court, Fatullah Shirazi, was an inventive genius who developed a machine for cleansing several gun barrels.

Cannons were both small and large. It would appear that the smaller ones ejected balls of about 25 kg in weight, while the larger ones discharged a ball of about ten times this weight. The smaller cannons, in the production of which Indian smiths were proficient, were lightweight due to their composition of both brass and iron. It is interesting to note that when Tipu Sultan, the ruler of Mysore, was defeated in Srirangapatnam (1799), the British captured more than 900 cannons; Tipu's own foundries, located in Srirangpatnam (two) and Bangalore (one), produced about 400 small, elegant cannons with an attractive finish, testifying to the capabilities of Indian metal-workers who operated in the foundries during Tipu's reign.

COSMETICS AND PERFUMERY

Indian traditional methods of preparing a wide variety of cosmetics and perfumes were empirically perfected chemical practices. They enjoyed a special prestige because of their extensive use in worship as well as their appeal to the royalty, elite, and the laity, then as now. It is a human instinct to feel exhilarated through the use of enchanting fragrances, be they natural or artificial. It would appear that even primitive man was fond of fragrant flowers and grasses, some of which were thought to have magical powers. Later, in advanced settlements, *khol*, (collyrium or mineral containing compounds of antimony) was one of the most popular cosmetics, used as it was to emphasize and darken eyelashes and eyelids. In the Indus Valley Civilization, too, *khol* containers have been unearthed along with some of its sticks.

Cosmetics and perfumes not merely enriched lifestyles materially, but also received religious sanction and were used during the performance of rites and rituals, as well as in celebration of marriages and similar joyous social functions. The Vedic gods, Rudra and Agni, were known by the appellation of *sugandha* (pleasantly scented), and Śrī, the goddess of wealth, was believed to have had the gates of her abode made of scented wood. Some Vedic texts speak of the makers of collyrium, the bathing of brides with fragrant water, scented flowers, and incense. Later texts like the epics and even the Buddhist works, like *Lalitavistara* and *Divyāvadāna,* mention the use of incense and other types of fragrances. The Buddhist *Jātaka* stories, too, refer to the use of perfumes by *bhikkus* (Buddhist monks) and *bhikkunīs* (Buddhist nuns). As noted before, from about the sixth century BCE, India experienced a second phase of urbanization (the first one being that of the Indus Valley Civilization), particularly in the Indo-Gangetic plains, which provided a fillip to the cosmetics and perfume industry to meet the demands of the people of the expanding settlements. In fact, during that time, the trade in perfumes had acquired the status of being the best of all commercial activities, including that of jewellery.

The *Arthaśāstra* of Kauṭilya, an important text of the flourishing Mauryan period, mentions the extensive use of perfumes in the royal household. The medical classics, the *Caraka* and the *Suśruta Saṃhitās,* too, have emphasized the role of cosmetics and scents for a healthy life. The former states that the use of fragrances and flower garlands is aphrodisiac, leads to longevity, charm, strength, and pleasing manners; and they even ward off evil effects. The *Suśruta Saṃhitā* has laid down rules or guidelines for the use of cosmetics and scents after getting up from bed, while bathing and also after bathing. The *Kāmasūtra* of Vātsyāyana (fifth century) naturally advocated the use of perfumes, like scents, sandal paste, collyrium and fragrant fumigation of apparel, for an exciting, romantic life.

The earliest detailed account of perfumery can be found in the seventy-seventh chapter titled *Gandhayukti* of the encyclopaedic work, the *Bṛhatsaṃhitā* of Varāhamihira (sixth century). He lists about forty aromatic ingredients and plants, several of which have not been identified properly from the standpoint of modern botanical nomenclature. The text has even indulged in an exercise, through progressive permutations and combinations, of the possibilities of preparing as many as 1,820 varieties of perfumes from sixteen ingredients (names in Sanskrit) as shown in the following table:

Ingredients of perfumes in the *Bṛhatsaṃhitā*

aguru	2	*tvak*	8	*turuṣka*	5	*śaileya*	8
priyaṅgu	5	*patra*	3	*rasa*	2	*kesari*	3
spṛkka	4	*mustā*	8	*tagar*	7	*māṃsī*	4
malaya	7	*nakha*	6	*śrīvāsa*	4	*kunduruka*	1

Since the proportions (column 2) of the basic aromatic substances have been given, it is possible that some experiments could well have been done and verified for their perfumery potential. The chapter title, *Gandhyukti*, is significant because the word, *yukti* has connotation of experimentation and confirmation. The *Bṛhatsaṃhitā* also deals with some scents under the name *sarvatobhadra* (perfume which is good for all purposes), the royal perfume called *Kopacchada* (which means it can mitigate anger), a perfume for increasing sexual urges, toilet powder for perfuming clothing, fragrant tooth-sticks and brushes, perfumes for the mouth and hair, fragrant water for use by the royal family, hair oil smelling of the flower *campaka* (*Michaelia champaka*) and black dyes for colouring white or grey hair—all for embellishing a material and pleasurable life.

The *arca* tradition of worshipping a broad spectrum of gods and goddesses with flowers and other accessories became more pronounced in the Classical Age of India (fourth to eighth centuries) and thereafter, continued with equal vigour over the succeeding centuries. The offering of incense and burning of camphor have been the most popular forms of worship of divine icons. The *Viṣṇudharmottara Purāṇa* lists as many as twenty-one aromatic substances for the preparation of incenses. The *Vāmaṇa*

Purāṇa refers to a blended incense from *guggulu* (*Balsamodendron mukul*), benzoin, and *aguru* (*Aquilaria agallocha*) as Viṣṇu's favourite incense. *Guggulu* incense was supposed to be dear to all goddesses, as was *agaru*, particularly in the lunar month of Śrāvaṇa. There are special types of incenses prescribed for offerings, especially in lunar months, to propitiate a divinity to seek boons or the desired benefits. According to the *Agni Purāṇa*, if a devotee burnt the incense in the lunar month of Bhādrapada, he would attain heaven after death. There was even a tacit belief in the potentiality of specially prepared incenses for acquiring a girl in marriage or winning in scholarly debates. Such were the influences of cosmetics and perfumery, then as now, on the social life of Hindus.

The *Mānasollāsa* of Chalukyan King Someśvara III (twelfth century), highlights the importance of incense of three types: (i) small boll form (*piṇḍadhūpa*) (*dhūpabhoga*); (ii) a powdered form (*cūrṇadhūpa*); and (iii) wicks (*varttidhūpa*). The author was a connoisseur of royal pleasures, one of which was the extensive use of perfumes. The *Kālikā Purāṇa* provides a list of several plants or trees, with parts that could be used as sources for the preparation of incense. But it has emphasized that no animal product like fat or bone marrow should be added to such preparations. This text also describes six varieties of collyrium which are dear to various goddesses. Of all the perfumes, according to this *Purāṇa*, sandal paste (*gandha*) or sandal oil is the most pleasing to gods and goddesses, a tradition which continues even now in orthodox homes and temples as part of worship. Perfumery in the Indian way of living is of a two-in-one dimension—mundane and divinc. It is for this reason that the preparation of perfumes, an intricate operation, had to be carried out only by a trained and efficient perfume-maker. He was not allowed to take up any other work and was enjoined to attain a high quality of production.

Methods of Perfumery: The earliest text that describes the operations for the preparation of perfumes is the *Viṣṇudharmottara Purāṇa* which, in a succinct manner, deals with purification (*śodhana*); emission (*vamaṇa*); purging or taking out the essence (*virecana*); infusion of fragment powders (*bhāvana*); curing (*pāka*); fumigation with desired aromatic substance (*dhūpana*); and scenting (*vāsana*). Another text, *Gandhasāra* by Gaṅgādhara (twelfth century), has described six processes as follows: *bhāvana; pācana* (curing)*; bodha* (tempering or revitalising the perfume); *vedha* (further intensification); *dhūpana* and *vāsana*. According to this text, the powder of an aromatic substance is to be mixed with another fragrant material, made into a scented slurry, stirred well and then filtered five or six times. It is then wrapped with leaves of other chosen aromatic substances, coated with mud and heated over cowdung fire (*pāka* or *pācana*) for the requisite time, which would vary depending upon the perfume. In all of these processes, 'digestion' or careful heating of the material, is to be carried out in a closed vessel, and the fire is to be stoked from above (*gartāpāka*). A variant of this is called *veṇupāka* by which the aromatic material is 'digested' in a bamboo tube, but subjected to heat by steam or mild radiated heat.

In its impressive repertoire, the *Gandhasāra* presents substantial details of many a cosmetic and perfume under several heads: scented water (*gandhodaka*); those for making the mouth fragrant (*mukhavāsa*); perfumed areca nuts and the like; scented oil; unguents; perfumed water for bathing; fragrant water for drinking; production of artificial musks; incense sticks and fragrant wicks for lamps; perfumed powder; artificial scents; extraction of essences from sweet-smelling flowers; and blending of perfumes. More importantly, the text has given seven tables and many aromatic substances which, by appropriate permutations and combinations, could yield a wide variety of enchanting perfumes, their numbers varying from 1,820 to as many as (perhaps a hypothetical proposition) 174,720! Not all the fragrant materials listed in the tables, however, have been identified botanically. But there is no denying that the traditional perfume-makers were knowledgeable about most of them and their uses.

This is evidenced by their classification of aromatic materials in terms of eight categories (*vargas*): (i) leaves (*patra varga*); (ii) flowers (*puṣpa varga*); (iii) fruits (*phala varga*); (iv) bark (*tvak varga*); (v) wood (*kāṣṭha varga*); (vi) roots (*mūla varga*); (vii) exudations (*niryāsa varga*); and (viii) organic (*jīva varga*). Various fragrant plants and their parts have been categorized appropriately under these heads. This methodical approach bears testimony to the knowledge of the perfume-makers about not only the characteristics but also the levels of fragrance of many aromatic raw materials. It seems that individual perfumers or communities of perfumers who perfected their own formulae maintained a proprietary secrecy and passed on that knowledge only to their descendants. No wonder, then, that the author of the *Gandhasāra*, which is a mine of information on perfumes, was unable to provide more details either about the exact proportions of the aromatic ingredients to be used, or the precise details of the methods of their preparations. Nevertheless, the art and technique of perfume production have been presented in sufficient detail to provide at least a glimpse into the processes involved. The canvas of these processes is indeed wide, encompassing details of preparations like fragment pastes, scented sesame and other oils, a broad band of incenses, talc, toilet powders, hair dyes, tooth-powder as well as sticks and artificial perfumes, besides the qualitative aspects of blending, curing, and augmenting the quality of perfumes.

In the medieval period, the tradition of perfumery attained new heights under the influence of the Muslim monarchs. Abul Fazl of the Royal Court of Akbar, in his magnum opus, titled *Āin-i-Akbarī*, deals with the 'Regulations of the Perfume Office of Akbar'. He has recorded that Akbar was exceedingly fond of perfumes and his court chamber was continually scented with flowers and the fumigation of fragrant substances which were burnt in incense vessels made of gold and silver! Akbar himself, according to Abul Fazl, perfumed his body, head, and hair. *Āin-i-Akbarī* has also described natural perfumes under fifteen heads including those from China, Syria and Java, as well as several types of flowers and flowering plants. Abul Fazl states: 'It would be impossible for one, ignorant as I am, to describe every flower of this country (Hindustan). A few have been mentioned by way of specimen. Here are also found abundant flowers transplanted from Persia and Tartary, such as roses, violets, jasmine, etc. The plants, flowers, roots and others of this country, that are used for food, or applied to medicinal purposes are innumerable.'

Āin-i-Akbarī is a storehouse of several kinds of information. Its account of fruits and flowers, especially roses, is fascinating. There is no reference whatsoever either to rose or rose water in early Sanskrit literature, although India had contacts in the ancient period with rose-growing countries like Persia, Babylon, Egypt, Greece and Rome for nearly two millennia. Over the past four centuries or so, the rose has earned a privileged position for itself as an important flower of worship, particularly of female deities. Rose water is also used for bathing divine icons like Subramhanya or Murugan. Rose distilleries are actively productive in some places in North India, and the *attar* of the rose has been recognized as a fascinating natural perfume.

The preparation of perfumes from the rose became popular even though the rose plant was not an indigenous one, transplanted as it was from the region of Iran. It was only from the Mughal period that some varieties of roses began to be cultivated in India, but very soon the rose emerged as the flower *par excellence* with its enchanting varieties, hues and exhilarating fragrance. The Arabic or Persian word, *gulāb*, generally means rose-water and the word, *itr,* means a perfume. The *itr* of roses is believed to have been first extracted by Empress Nurjahan on the day she married Mughal Emperor Jahangir. It is said that a canal in the palace garden was filled with rose water in celebration of the event, and when the royal princess observed a scum on the surface of the water, they got it collected, and found it to have admirable fragrance; it was rightly called *itr-i-jahangiri*.

PYROTECHNICS

The nomenclature, *pyrotechny,* connotes the method of making fireworks and displaying their spectacular light, sparks, and sound. The origin of fireworks is generally traced to China, from where its techniques spread to south, southeast and western Asia, and from there to Europe. Fireworks were associated with the celebration of victory in battles and festivals, then as now. Venetian traveller Marco Polo (thirteenth century) reported the enchanting show of fireworks in China for amusement and celebrations alike. The basic ingredients of gunpowder—namely, sulphur, saltpetre and charcoal—also constitute the components of mixtures used in fireworks. In India, guns and gunpowder were in use at the beginning of the Mughal period (sixteenth century). A Sanskrit text probably belonging to this period or a little earlier, the *Śukranīti*, attributed to sage Śukrācārya, describes the preparation of gunpowder (*agnicūrṇa*) using sulphur, saltpetre, and charcoal in different ratios for use in different types of guns. Probably a century or two earlier, compositions for specific fireworks could have been developed in India.

Literary Sources: Most of the texts concerning pyrotechnics belong to the fifteenth-sixteenth centuries. Among them are *Kautukacintāmaṇi* by Gajapati Pratāparudradeva (1497-1539) and *Ākāśabhairavakalpa* (by an unknown author), both in Sanskrit; *Rukmiṇīsvayamvara* by Ekanātha (sixteenth century) and works of Rāmadāsa, in Marathi; *Bānaśāttiram* and *Bhogarsūtiram*, in Tamil; *Veṭṭikampavidhi*, in Malayalam; a treatise on fireworks in Persian by a Muslim King of Kashmir, Zain-ul-Ābidin (1420-70), and the *Mānasa Mangala*, a Bengali work which contains material information on eleven types of fireworks.

The *Kautukacintāmaṇi* includes a list of twenty-five different substances and accessories which, in certain proportions and combinations, lead to several formulae for the preparation of eight types of fireworks, mostly small and medium-sized rockets (*bāṇas*). Interestingly, a majority of the formulae are more or less similar to the pyrotechnic compositions found in a Chinese text, *Wu-Pei-Chih* (seventh century), and also the text in Tamil by Bhogar, *Bāṇasāttiram*. The basic ingredients include saltpetre, sulphur, and charcoal, as well as yellow-orpiment (arsenic sulphide), red ochre (iron oxide), iron powder, copper oxide, and even magnetic oxide of iron to produce fireworks in a range of colours. Mercury as well as crushed castor seeds and bamboo tubes as containers, are some of the accessories used.

The *Ākāśabhairavakalpa* provides some useful details about four types of fireworks, but intended only for royal entertainment, as follows: (i) fireworks in the form of a wheel ignited by a pyrotechnic composition which explode with the brightness of daylight; (ii) a form of small tree with rockets hung from its branches; (iii) rockets which emit sparks of fire after soaring tangentially; and (iv) rockets which, when ignited, explode with a loud noise. The *Rukmiṇīsvyamvara* mentions (i) *agniyantra* (a small rocket); (ii) *havai* (a rocket going up in air with a stream of fire); (iii) *sumana-mālā* (fireworks producing 'garlands of flowers' as it were; (iv) *bhūnala* (fireworks on the ground emitting a stream of sparks); (v) *cincuṃdarī* (a rocket moving up with hissing sound); (vi) *hātnalā* (fireworks to be held in hand, which when ignited, produce a flow of fiery sparks; (vii) *bāṇa* (a rocket); and (viii) *candrajyoti* (a device that would emit fire, producing a moonlight effect).

Bāṇaśāttiram, the Tamil text, presents detailed formulae for the production of over ninety types of fireworks. A notable fact is that these formulae use metallic dusts of copper, iron, lead, brass, or zinc for such devices, which then sparkle with different colours. Obviously, such techniques would have been arrived at by empirical methods coupled with prolonged observation.

The display of attractive fireworks is common in all countries when special events are celebrated. In India, perhaps from the eighteenth century onwards, fireworks have been associated with some festivals, particularly Dīpāvalī, which is celebrated with éclat and gaiety all over India, as it is believed to signify or symbolize the triumph of the good over evil, and the emergence into light from darkness. The term *dīpāvālī* means illumination from a row of lamps. It is also a special occasion for the worship of goddess Lakṣmī. There are other specific days on which fireworks are displayed. The Persian ambassador, Abdur Razzaq (1445) who witnessed the Mahānavamī celebrations in the court of the Vijayanagar kings recorded his observations on the various kinds of colourful pyrotechnics. His records were full of praise for the dexterity with which the fireworks were produced and displayed. Such fireworks-extravaganzas are still a living tradition in India.

DYES AND PIGMENTS

The use of dyes and pigments goes back to remote antiquity in India as well as in other ancient cultures. Materials of different hues or colours have appealed to the human mind even in pre-historic times. In the Indian context, the ancient rock-paintings in places like Bhimbetka in Madhya Pradesh and in the hills of Mirzapur in Uttar Pradesh (10,000-8000 BCE), reveal the judicious use of natural colours like red, black, and white to embellish paintings. Various colours were also used in the Indus Valley Civilization on ceramics and other objects.

According to Marshall (1931), who was associated with the excavations at Mohenjo-daro, a piece of coloured cotton cloth was found to date as far back as the third millennium BCE. There are references to colourful dresses and costumes in the two great epics, the *Rāmāyaṇa* and the *Mahābhārata*, and in later literary works. Nevertheless, no piece of dyed textile material prior to the medieval period has been found, the reason being that a material like cotton fabric would not survive the destroying effect of the hot, humid climate of India. Even so, that the dyeing of cloth was an accomplished skill in India is amply borne out by the accounts found in *The Periplus of the Erythrean Sea* (first century CE; unknown author), which is a record of the voyage between India and the Roman area. It states, among others, that Ujjain was an important centre for the production of dyed cotton fabrics being exported then. Gujarat was another region from which dyed calicoes were exported. The range and style of printed and coloured cotton costumes are shown in the famous paintings of Ajanta (second century BCE to sixth century CE) and other paintings and murals in different parts of India, including Rajasthan.

It needs to be recognized, however, that in India, colours of different shades as well as attractive designs are inseparably related to or associated with socio-religious customs. Hindus generally regard red as an auspicious colour, but black colour as inauspicious, while the colour of saffron (*kesaria*) is sacred for Jainas. Hindu widows are not expected to wear coloured sarees, only white being acceptable. Muslims consider green to be auspicious, and black apparel is worn on occasions of mourning by Muslims and Christians alike. These social customs are very old, but are changing to some extent of late.

Natural Dyes: Several varieties of natural dyes were used as colouring agents of cotton, silk and other fabrics, as well as in paintings. The application of henna (*mehendi*) by women to colour and decorate their hands and feet is well known. The cultivation of this plant, the leaves of which were used, then as now, for the preparation of the dye, was of economic importance till a few centuries ago. Alta, a lac dye derived from the lac insect (*Kerria lacca*) was an ingredient of paints till the advent of synthetic dyes. The yellow dye obtained from the peels of pomegranate fruit was another natural dye, and saffron was used to impart colour to fabrics, confectionary, and cosmetics. More importantly, turmeric was extensively used not only for its yellow colour but also for its antiseptic as well as depilatory properties.

Being a tropical country, India is blessed with a wide variety of flora. Apart from those used as food, plants have also yielded colouring substances or vegetable dyes. Over the centuries, safflower, madder, *lodhra* (*Simplocos racemosa*), turmeric and indigo have been used as dyestuffs. Red ochre, carbon black or lamp suit, copper sulphate, and antimony sulphide have also been sources of red, jet black, blue and brownish-black colours, respectively. Animal products that impart the colour red are *lākṣā* (insect), *krimirāga* (*Kermococcus*), and *indragopa* (*Coccus cacti*), while yellow matter is prepared from cow's urine (called *rocanā*), as well as from a flowering tree called 'Flame of the Forest' or palāśa (*Butea frondosa*).

The madder plant (*Rubia cordifolia*) has been the most sought after source of red dye for cotton cloth; initially yellow, it turns red when it comes into contact with alum. The real colouring matter in madder goes by the name alizarine which is polygeneric, that is, it produces different colours with different mordants. With iron salt as a mordant, it gives shades from purple to black, depending upon the strength and nature of thc iron salt. Alizarine is also obtained from a type of tree (belonging to the genus *Morinda*). Gulal, the red powder used during Holi celebration, used to be produced from what is called *pathang* wood or the bark of the *sappan* or Brazil redwood tree (Lie sappan L.). A pink or rose-pink or crimson colour is obtained from safflower or the florets of *Carthamus tinctorius*, while the extracts of the root and bark of the *Ratan jote* (*Glabrous recemes*) tree are used for imparting respectively grey and red colour. A beautiful red dye is obtained from the bark and root of a large wood climber called the red creeper *Raktavallī* (*Ventilago madraspatna*).

An extract from the seeds of the shrub or tree *Bixa orellana* is a pigment called *annato*, which yields a bright yellow colour used for dyeing silk and cotton. The bark of the Indian mulberry or *Dāru haridrā* (*Morinda augustifolia*) yields a powder which, with different mordants, gives different shades of yellow. The yellow material extracted from the wood of a jackfruit tree is blended to produce different shades of yellow colour; similarly, the yellow substance obtained from the mango tree is used for dyeing cotton, silk, and even wool. The myrobalan (*Terminalia chebula*) has been a versatile natural source of dyes, since it produces grey if mixed with iron salts, and yellow if alum is added to it.

Pigments: In the past, pigments were usually obtained from minerals, some of which were also ores used for the extraction of metals. Vermilion or cinnabar, which is a compound of mercury and sulphur, is a naturally occurring mineral. If crushed and ground to a powdery form, it serves as a bright red pigment. The *Viṣṇudharmottara Purāṇa* and the *Mānasollāsa* refer to this pigment. However, vermilion does not seem to have been used in the paintings found in the Ajanta, Ellora, and Bagh caves. It is only in some paintings in Rajasthan and Kulu, which are dated to eighteenth and nineteenth centuries, that this pigment is present. Red ochre, an oxide of iron and also known as *geru*, is a pigment which has been used extensively in Indian paintings over the centuries, even though it is not as bright as cinnabar. The reason is not far to seek. Iron ore or mineral occurs abundantly in India and, in fact, it is also known as red-ochre or Indian red. This natural mineral is in both hydrous ($Fe_2O_3.3H_2O$) and anhydrous (Fe_2O_3) forms, the colour of the hydrous form ranging from red to dull yellow, while the anhydrous one is red. The use of this pigment has been confirmed by the chemical analysis of paintings at Ajanta, Bādami, Bāgh as well as in other ancient paintings, besides the subsequent ones in Rajasthan and Kerala. Red lead (sindūr), though mentioned in the *Viṣṇudharmottara purāṇa* as a pigment, has not been used as confirmed by chemical analysis of paintings at Ajanta and other places; but it has been used extensively in the wall paintings of Rajasthan.

Of the yellow pigments, realgar (*manaḥśila*), a mineral (sulphide of arsenic) as well as orpiment (*haridala*, another sulphide of arsenic), yellow ochre—called *Ram Raj* (which is a hydrated variety of iron oxide)—and litharge or massicot (which is an oxide of lead), obtained by heating lead carbonate yielding an yellow powder, have been used in Indian art works. Among the green pigments is *terra-verte* or green earth, which is a mixture of minerals like gluconite or hydrous silicates of iron, magnesium, aluminum and potassium. Depending upon the place or site from which it is obtained, the shade of this natural mineral varies from yellow-green to greenish-grey. *Terra-verte* has also been extensively used in almost all the paintings from the time of Ajanta to the wall paintings of Rajasthan and Kerala. The source of another green paint was malachite (a basic carbonate of copper), which occurs in association with the ores of copper. This pigment began to be used in Indian paintings only from the eighteenth century. Verdigris (the basic acetate of copper), did not appear to have been used in Indian wall paintings, but was a common pigment that found favour with artists of the Mughal period and later miniature painters.

Indigo, a vegetable dye, was the most important blue pigment used for painting (this blue mineral is also referred to in the *Viṣṇudharmottara Purāṇa*). The blue ultramarine obtained from the mineral *lapis lazuli* has been used in miniature paintings. But azurite, another blue pigment, also a basic acetate of copper, does not seem to have been used by Indian artists, although it was extensively used by Chinese artists. The white pigments include chalk (calcium carbonate), kaolin (China clay), gypsum (calcium sulphate), white lead (basic carbonate of lead), zinc white (zinc oxide), talc (powdered soft or soap stone), barium white (barium sulphate) and titanium white (titanium oxide). The chemical analyses of various paintings indicate that only the first four were used by Indian painters.

TECHNIQUE OF PAINTINGS

Paintings on the walls of caves, palaces, temples, or homes were greatly valued for their religious message or decorative and aesthetic appeal. There were also some injunctions against the depictions of themes; those associated with certain deities and sages (*ṛṣis*), for example, were forbidden depictions in homes, but were permitted in temples and palaces. The art of painting, like sculpting, was regarded as a meritorius one, and the painter was expected to have sublime qualities and noble intentions, according to the *Viṣṇudharmottara Purāṇa*. The *Citra sūtra* (the painting manual or rules) of this *Purāṇa* describes three types of painting: (i) on walls (*bhitti saṃskāra*); (ii) on a board (*phalaka*); and (iii) on canvas (*paṭa*). This manual gives details of the method of executing a painting, manipulation of colour tones or tints of different kinds, and related practices. The primary colours recognized are five: white, red, yellow, black, and green, each one of them to be processed to yield the desired shades or tones. This manual also sheds light on pigments like orpiment (*haritāla*), lime, lac (*lākṣā*), vermilion (*hiṅgula*), indigo (*nīl*), lapis lazuli or ultramarine (*rājāvarta*), red lead (*sindūr*), and mica. Other important textual sources dealing with these aspects are the *Harṣacaritā* of Bāṇa; the *Samarāṅgaṇa sūtradhāra* of Bhoja; and *Mānasollāsa* or *Abhilaṣitārthacintāmaṇi* of Someśvara III. The *Śīlparatna* refers to the colours to be used for paintings that are intended to depict religious stories from the *Vedas*, *Purāṇas*, and *Āgamas*. What is known as the 'eye-opening' (*cakṣu unmīlana*) of a figure, to make it look alive with glow of life, is an important but last step in the art of painting. This is accomplished through the painting of the pupils of the two eyes, an extraordinary sensitive skill in which Indian painters excelled.

As regards the techniques of wall paintings, the *Samarāṅgana sūtradhāra* (eleventh century CE) has provided elaborate details of the eight stages of painting: (i) drawing a sketch of the design on plaster with a piece of colour; (ii) preparation and application of the ground (or base) on the plaster; (iii) incorporation of the design; (iv) drawing details of design; (v) addition of further details; (vi) bringing out the effect of light and shade; (vii) preparation of the final form; and (viii) finishing. The *Citrasūtra* gives details about the preparation of the surface as follows:

> 'Brick powder of three varieties (smooth, middle and coarse) should be mixed with clay, a third of it in proportion. To this is to be added fragrant gum, resin, beeswax, honey, *kandara* (a type of grass), molasses, safflower soaked in oil—all in equal proportions. With the addition of some fruit pulp, lime and the like, the whole material is to be soaked in water in a pot, and kept for a month. Thereafter, it is to be applied on the dried wall as a coating, which should be smooth, even, firm and free from uneven patches. When it dries, the surface is to be smoothened by the application of oil, milk and some salt solution followed by careful rubbing.' (Section 40, verses 1-10)

The text adds that 'on an auspicious day, a painter dressed in immaculate white, pure in mind and body, having adored Vedic seers and chanted hymnal prayers and facing the East, should concentrate on the deity to be depicted, and start his work of drawing the form.' This method also appears to have been followed in the Rajasthani paintings. Although, the *Citrasūtra* speaks of lime plaster, such a plaster does not seem to have been adopted in the Ajanta cave paintings. On the other hand, ferruginous earth reinforced with rock powder or sand and fibrous vegetable material was used, as revealed by chemical analysis. However, over this primary or base plaster, a thin white layer of lime or gypsum or kaolin has been coated. Lime plaster is found in the paintings at Bāgh, Sittannavasal, and other places from the seventh century CE onwards. The wall paintings over the centuries are not only a mirror of the socio-religious ideas and practices of the times, but also throw light on the types of coloured costumes and designs that embellished the lives of both the elite and the laity. The artistic painter's expertise also encompassed his practical, but standardized, chemical knowledge of the pigments of different hues and the expertise needed for their aesthetic presentation.

OTHER CHEMICAL PRACTICES

Kauṭilya's *Arthaśāstra* is a mine of information about empirically well understood chemical ideas and practices. The section on mines and factories of this early text (fourth century BCE) provides adequate details about the nature of the ores of gold, silver, copper, lead, tin and iron. It makes a scientifically accurate statement that where there is increase in heaviness, there is increase in metallic content. Its description of the physical appearances of the ores, too, is noteworthy. Among the duties of the chief of mines, was the establishment of factories for the production of metals like copper, iron, lead and tin, and alloys like brass, bronze and steel. Another section of the *Arthaśāstra* is concerned with the examination of pearls, rubies, beryl and other precious stones in a systematic way. This informative text has also described the qualities of pure gold and silver, purification of impure forms of these metals and the skills needed for their setting, stringing, and making of hollow or solid jewellery. As regards minting of coins, the *Arthaśāstra* states that silver coins should be made of four parts of copper, eleven parts of silver, and one part of either iron, tin or lead, while copper coins could be made of four parts of silver, eleven parts of copper, and one part of iron or any other metal.

The production of fermented juices was an intricate one which needed special processing. The process of fermentation demanded the use of a ferment, and the *Arthaśāstra* speaks of *kiṇva* (ferment) to be used depending upon the composition of the fermenting mixture. One such ferment consisted of one *droṇa* (a measure) of mashed *māsa* (black gram) grains, raw or cooked with one-third part of rice grains. The brewers of Kauṭilya's time seem to have been experts in the technique of preparing several types of intoxicating liquors such as *medaka* (from rice grains), *prasanna* (rice flour with particular bark and fruits), *āsava* (from wood-apple and honey), *maireya* (jaggery and the *meṣaśṛṅgī*), *madhu* (from grape juice) and *mahāsurā* (from mango juice and some other ingredients).

The medical classics, the *Caraka* and the *Suśruta Saṃhitās,* have thrown light on a wide repertoire of chemical substances as part of medicinal chemistry, such as minerals, metals and metallic compounds, organic acids as well as alkalies (*kṣāra*) of three types (mild, caustic and those of average strength). Generally, the alkalis were obtained by the lixiviation (that is, separating soluble and insoluble substances) of ashes of certain plants, and the *Suśruta Saṃhitā* furnishes details of the chemical procedure involved. Three other types of alkaline substances, namely, potash (*yavakṣāra*), natron or trona (*sarjikakṣāra*), and borax (*ṭaṅkaṇa*) have been dealt with. This text also describes the characteristics of six types of salts: (i) nitre or potassium nitrate (*sauvarcala*); (ii) rock salt (sodium chloride, *saindhava*); (iii) a reddish-brown granular salt called *viḍa* (a mixture of several compounds); (iv) 'fossil' salt (*audbhida*; a kind of saline deposit); (v) sea salt (*sāmudra*, sodium chloride with traces of magnesium salts); and (vi) a type of saline earth (*romaka*). Apart from alkalis and salts, there is substantial literary evidence to show that some acids were also known, which were by and large organic and obtained from plant sources. Though there is no explicit reference to mineral acids, some of the compositions including nitre, alum and other ingredients would have constituted the ingredients necessary to give rise to mineral acids. The *Rasapradīpa*, a later iatrochemical text (sixteenth century), presents a detailed process for the preparation of a mineral acid called *śaṅkhadrāvaka* (possibly nitric acid).

The use of metals and minerals, metallurgical skills, and the knowledge of many a chemical process like fermentation, distillation, and calcination, were no less pronounced in India than in any other contemporaneous cultures. It needs to be realized that in the West, it was only from the last two or three decades of the eighteenth century that chemistry emerged from an empirical milieu to attain a scientific status. This became possible because of the quantitative experiments that were carried out on the nature of gases, such as air, carbon dioxide, hydrogen, chlorine and oxygen. Joseph Black (1728-90), Henry Cavendish (1713-1810), Joseph Priestley (1733-1804), Carl Wilhelm Scheele (1742-86), and A.L. Lavoisier (1743-94) were the pioneers who investigated the chemistry of gases. The discovery of oxygen by Lavoisier and Priestley added a new dimension to the understanding of several chemical processes. Greek atomism was revived by such noted men of science as Pierre Gassendi, Robert Boyle and Isaac Newton in the seventeenth century. Certain laws like the law of the conservation of mass, the law of definite proportions, the concept of elements, their physical as well as chemical characteristics and their atomic nature, which led to an atomic theory by John Dalton (1766-1844) early in the nineteenth century, provided such a strong foundation to chemistry that it began to make remarkable progress from the nineteenth century onwards. Modern chemistry was included as a subject of study in the Indian universities, the first three of which were established in 1857. Towards the end of the nineteenth century and before Independence, Indian chemists had made significant chemical investigations. Interestingly, in some ways traditional chemical knowledge and skills have been coexisting with modern chemistry in India.

ALCHEMY AND ASSOCIATED PRACTICES

The word 'alchemy' is Arabic, probably owing its origin to the Egyptian *khemit* (black) or the Greek *chyma* (black molten metal). Possibly, the Arabs knew about the Egyptian practices of artificially 'colouring' of inferior metals (Egypt being then known as the country of black soil) so that they looked like gold or seemed to be transmuted into gold. In the Graeco-Roman world, too, such practices were common.

In its twin aspects, namely, *aurefaction* (transmutation of the so-called inferior metals into the noblest metal, gold) and the *elixir of life* (for attaining immortality or 'deathlessness') along with its mercury-sulphur centred ideas and practices, alchemy was not indigenous to India. Its seminal ideas owed their origin to Chinese alchemy, its *yin-yang* postulates and others concerned with what were designated as 'holy immortals'. These were assimilated *mutatis mutandis* in Indian tāntric milieu in the early centuries of Christian Era.

The two Ayurvedic classics, the *Caraka* and the *Suśruta Saṃhitās,* have no doubt dealt with what is called the *rasāyana* or the elixir of longevity. In fact *rasāyana* is one of the eight limbs (*aṣṭāṅga*) of Ayurveda. However, like other systems of medicine, Ayurveda accepts that the body (*śarīra*) is a constantly disintegrating one leading eventually to its death. In other words, it does not believe in immortality, even though it has dealt with certain elixirs including *soma* and gold for increasing longevity or prolonging the life span.

The doctrine of five elements, as noted earlier, is one of the important foundational ideas of Ayurveda and has a wider application in this medicinal system; but it never addressed itself to the problem of change or transformation. On the other hand, in the Greek or Alexandrian tradition, the Empedoclean postulate of four elements—earth, water, fire and air; and the two sets of opposite qualities, namely, hot and cold; dry and moist—provided a theoretical framework for an explanation of the problem of *change* or *transmutation*. It also envisioned the possibilities of transforming one metal into another by effecting changes in the primary qualities of the 'elements'. This concept led to the emergence of Alexandrian alchemy. However, it could not become a source of inspiration for Indian *rasavādins* or alchemists. For, the Indian 'elements' had a much broader canvas or wider connotation than merely an interlinked Greek four 'elements' with four primary opposing qualities. Nor did the Indian five 'elements' have the potentiality of giving rise to such ideas of transmutation among metals. Their metaphysical foundations in the *Upaniṣads* and their physical attributes as part of the *Sāṃkhya,* the *Vaiśeṣika,* and other systems did not permit any esoteric approach. The inspirational source of Indian alchemy lay elsewhere. It was the male-female imagery of Taoism, the powers of immortality attributed to cinnabar as well as gold-elixirs of Chinese alchemists that attracted the attention of *rasavādins* in the tāntric ambiance that was conducive to their assimilation, in an Indian mould.

Origin of Indian Alchemy

Chinese alchemical practices were very old, perhaps going back to about the third century CE. One of the seed-ideas of Chinese alchemy concerning longevity and even immortality relates to cinnabar. This naturally occurring mineral was supposed to be an effective elixir of life. A Chinese text of the fourth century CE, *Pao Phu Tzu* says:

> 'The fact is that the least of the Cinnabar is far better than the best things of the plant world. When submitted to the action of fire, all plant substances are burnt into ashes,

> but cinnabar produces liquid silver (mercury), and this after a whole series of operations can revert to cinnabar again. That far exceeds what plants can do; hence such chemicals can confer longevity and immortality and holy immortals were aware of this.' (Needham, Joseph: *Science and Civilization in China,* vol. V-2., Cambridge University Press, 1974, pp. 126-128)

Significantly, in the early phase of Indian alchemy, cinnabar occupied a distinct place. It was referred to as *rasaliṅga*. The *Mātṛkābheda tantram*, a text belonging to the early centuries of the Christian Era, gives details of its preparation from mercury and sulphur, eulogizing them at the same time as the creative principles of Śiva and his consort, respectively, in a male-female imagery. It also emphasized the divine male-female union that the *rasaliṅga* would represent. In the Chinese culture-area, the Taoist male-female symbolism on the one hand and the concept of immortality associated with the use of cinnabar and other elixirs of life on the other, were intertwined with intense meditation, regulated breathing, a dietary regimen, sexual techniques, and helio-therapeutics and exercises. It was tacitly believed that through these and other esoteric practices, a Taoist would be able to elevate himself into *hsien*—a perfected celestial or earthly immortal. It is worthy of note that the Sanskrit word corresponding to *hsien* is *Siddha,* meaning, the 'perfected', or the one who had attained special powers, called *siddhis*, through intense yogic practices and deep meditation. These were integral to what came to be known as tāntrism.

The tāntric *siddhi* had several pathways for its attainment. One of them was also the intake of some elixirs made of mercury, sulphur, and some minerals besides the gold elixir. It was indeed mystical in many ways and more or less similar to what a Taoist would follow—meditation, sexual techniques, prolonged rhythmic breathing, and the like. More importantly, away from the rigidity of orthodox thinking as well as practices, tāntrism admitted all into its fold, irrespective of sex, caste or creed. It opened itself to the assimilation of new ideas and practices conducive to its overall objective of the attainment of *siddhis*. In the environment of open-mindedness that tāntrism brought forth, Chinese alchemical seed-ideas of mercury and sulphur as well as the concept of immortality were well received and assimilated. They could do so first through the involvement of some Buddhist monks, both Indian and Chinese, to whom esoteric alchemy had a special appeal. This happened in the early centuries of the Christian Era, around the third-fourth century, and continued in the succeeding three or four centuries during the Tang period when cultural contacts, and to some extent commercial intercourse, between China and India were noticeable. During this period, celebrated Chinese monks like Faxian, Xuan Zhuang (Hiuen Tsang) and Yi-Xing (I-Tsing) visited India, and the last two stayed in India for some time. Nāgārjuna, the famous Buddhist monk, went to Mahācīna as an important place for attaining *mahāsiddhi*. The tāntric goddess, Nīlasarasvatī, was stated to be of Chinese origin in her form and attributes. Another goddess, Tārā, was believed to have emanated from the region now comprising Tibet, which was influenced by Taoist thoughts and practices. Mercury was referred to as *amṛtadhātu* by Buddhists of the Mahāyāna persuasion; and one of the early Mahāyāna texts, the *Gaṇḍavyūha sūtra*, which was rendered into Chinese and included in the Chinese *Tripiṭaka*, contains alchemical ideas similar to those of Taoists. The *Mātṛkābheda tantra* refers to *Cīnā* tantra obviously in the context of its alchemical practices.

Alchemical Texts: Alchemical texts in Sanskrit are designated as those belonging to *Rasaśāstra*, the word *rasa*, generally meaning mercury, the alchemical supremo. Texts on this subject in Tamil have no generic name, but are attributed to several authors who are known as *Siddhas*, similar to the Chinese *hsiu*. Of them, Bhogar, an important Siddha according to Tamilian tradition, was a Chinese. One of the alkalis (naturally occurring) used in certain preparations of alchemical significance is called *cīnakkāram* (sodium or potassium carbonate) in Tamilian texts.

The male-female imagery, so characteristic of Chinese alchemy, found its place both in the *Rasaśāstra* and Tamil texts. Several *Rasaśāstra* texts are unfolded in the form of a dialogue between Śiva (male) and Śakti (female). For example, one of the early texts, the *Rasārṇavakalpa* (which is part of a larger tāntric text, the *Rudrayāmaḷatantra*) is in the nature of a dialogue between the Bhairava form of Śiva and Devī (his consort). Mercury was also called by various names (such as *harabīja, haravīrya, and sūta,* which connote the seminal essence of Śiva). Likewise, sulphur denotes the creative principle of Devī, his consort: the combination, mercuric sulphide, signifying the divine union. Indian alchemists, like their Chinese counterparts, had firm faith in cinnabar or mercuric sulphide as the one that had extraordinary properties for the realization of the twin goals—namely, the transmutation of metals and the attainment of immortality.

Though Chinese alchemical seed-ideas were absorbed in the tāntric environment, the *rasavādins* developed a wide alchemical armamentarium that made Indian alchemy stand on its own with certain original attainments. One of them was the processing of mercury itself. Mercury obtained by the distillation of cinnabar was considered to be of good quality and the first eight of the eighteen processes (briefly indicated below) were not deemed necessary for it to exhibit its supreme powers of both transmutation and immortality. One also notices some of these processes in Chinese alchemy, but not in such a detailed manner as it is in the texts like *Rasahṛdaya* and the *Rasaprakāśasudhākara*.

Eighteen Processes of Mercury: The eighteen processes (*saṃskāras*) for mercury are: steaming (*svedanam*); grinding or trituration (*mardanam*); 'swooning' or making mercury lose its form (*mūrchanam*); revival of its form (*utthāpanam*); sublimation or distillation (*pātanam*); potentiation (*rodhanam* or *bodhanam*); restraining (*niyamanam*); stimulation or kindling (*saṃdīpanam*); consumption of the 'essence' of mica (*gagana grāsa*); amalgamation (*cāraṇam*); internal 'liquefaction' (*garbhadruti*); external 'liquefaction' (*bāhyadruti*); digestion or assimilation (*jāraṇam*); colouration (*rañjanam*); blending or preparing it for the potency of transformation (*sāraṇam*); acquisition of the power of transmutation (*saṃkrāmaṇam*); testing its ability of actual transmutation of base metal into gold (*vedhavidhi*); and if the test is positive, its internal use (*śarīrayoga*) for attaining longevity or immortality. In many of these processes, several plants, mica, sulphur, and arsenical substances are used. But it is indeed difficult to understand these processes in terms of modern chemistry. What is important to note, however, is that only when the elaborately processed mercury answers the test of transmutation, should it be taken internally for the transformation of the body into one of 'deathlessness'. For, it was believed that if the mercurial elixir could convert the perishable inferior metals into an imperishable gold of everlasting lustre, it could also transmute the perishable body into an immortal one with the glow of life. Such was the two-in-one goal ever in sight, but never reached either by *rasavādins* or alchemists of other cultures.

Alchemical Materials: The *Rasaśāstra* texts are a repository of valuable information on a large number of chemical substances, namely, mercury, gold, silver and other metals, sulphur, a variety of minerals including arsenic substances, gems, salts, and even some toxic materials. In addition, more than 200 names of plants have been mentioned along with the explanations of how to use them in different alchemical or iatro (medicine) chemical processes. A noticeable characteristic of the *Rasaśāstra* texts is the systematically classified presentation of chemicals used by them, in terms of *mahārasas* (superior substances), *uparasas* (auxiliary or inferior), *dhātus* (metals), *viṣas* (toxic ones), *ratnas* (gems) and *yantras* (apparatus). Mercury is naturally extolled as the most important and beneficial *mahārasa*, the other ones being mica, pyrites, chalco-pyrites, kimberlite, copper sulphate, zinc carbonate, and bitumen (the names in Sanskrit, respectively, are: *abhra, mākṣika, viṃala, vaikrānta, sasyaka, capala* and *adrija*). There are some variations in this list, and a few texts include *darada* (cinnabar) and *kānta* (loadstone) among the *mahārasas*. The auxiliary ones (*uparasas*) are sulphur, red ochre or haematite, iron carbonate or sulphate, alum, orpiment or arsenic trisulphide, stibinite or antimony compound, and another substance called *kaṅkuṣṭa* (not identified yet chemically), which is not listed in some texts. The metals include gold, silver, copper and iron, which are regarded as pure metals, while white lead and tin compounds are considered as odorous ones (*pūti*). The alloys—brass, bronze or bell-metal, those of tin or zinc—are also used. All the known major gems, too, are employed in the chemical processes, the exact nature of which is hard to understand.

The use of several plants, their roots, leaves, seeds or extracted juices, is an integral component of alchemical processes. One of the early texts, the *Rasārṇavakalpa* which, as noted earlier, is part of the tāntric text, *Rudrayāmaḷa tāntra*, describes over forty different types of transmutation processes, and an almost equal number of longevity compositions involving a significant number of (about 100) plants. These are broadly categorized under (i) heavenly medicinal plants (*divyauṣadhi*); and (ii) non-heavenly, or earthly (*tṛnauṣadhi* or *adivyā*). It is the former type of plants that are extolled for bestowing alchemical properties on mercury and its compounds. The *Rasārṇavakalpa* (consisting of more than 800 verses) is a blend of the formulae or recipes marked by alchemical overtones and esoteric practices alike. Of its twenty-nine *kalpas* (a *kalpa* meaning that which is capable of extraordinary performance by its own internal potency), two deal with the properties of sulphur and arsenic sulphide, while the remaining twenty-seven are devoted to the praise and the special efficacies of nearly one hundred plants. One of them, *jyotiṣmatī* (*Cardiospermum helicarbum*, Linn), is hailed for the miraculous power of its oil, namely, the transmutation of base metals into gold, and the attainment of longevity like that of Brahmā (the creator), endowed with a youthful body. The *Rasārṇavakalpa*, more than any other *Rasaśāstra* text, attaches great importance to the 'herbs of immortality' as well as transmutation processes involving them. Probably, the time-tested and proven curative properties of herbs used in the pharmaceutics of Ayurvedic drugs would have influenced the alchemical practitioners in their use of plant products for the preparation of elixirs, although this was possible only after getting rid of the deleterious effects of minerals and metals through processing.

The *rasavādins* were dexterous experimentalists. Nevertheless, their approach to the very process of experimentation was rather obscure and bizarre; yet their multilevel skills and devotion to experimentation, even when it involved many weeks, as in the case of what is called *puṭapāka* (see below), were exemplary. They were motivated by intense faith in alchemy, honesty of purpose, and self-control without any greed. The mediocre and the dishonest, the charlatans and the greedy, were prohibited from practising alchemy.

In the Indian alchemical tradition, there were twenty-seven celebrated *rasavādins*, as recorded in the well-known text, the *Rasaratnasamuccaya* (fourteenth century). They were Ādima, Candrasena, Laṅkeśa, Viśārada, Kapāli, Matta, Maṇḍavya, Bhāskara, Surasenaka, Ratnakośa, Śaṃbhu, Sātvika, Naravāhana, Indrāda, Gomukha, Kaṃbali, Vyāḍi, Nāgārjuna, Surānanda, Nāgabodḥi, Yaśodhana, Khaṇḍa, Kāpālika, Brahmā, Govinda, Lampaka and Hari. But, apart from Nāgarjuna and Govinda, whose texts have come down to us, there is little or no material information about the others. The *rasavidyā* was to be imparted by the preceptor to the disciple through initiation, and the initiated were expected to maintain it as a closely guarded secret, keeping away the uninitiated from the dedicated alchemical experimentation.

The Laboratory

The place in which alchemical and other processes including experiments, were carried out was called the *rasaśālā* (the House of Mercury and other essences). The *Rasaratnasamuccaya* has described it clearly as follows:

> 'The laboratory is to be set up in a place rich in medicinal herbs. It should be spacious, furnished with four doors, and decorated with the portraits of divine beings. It should have several types of apparatus (see below) and contrivances. The phallus of mercury (*rasaliṅga*) in the East, furnaces in the South-East, apparatus and other contrivances in the North-West, facilities for washing and cleaning in the West; and those to be used for drying operations in the South-West. The other devices included *koṣṭhi* (an apparatus for the extraction of essences), pair of bellows, pestle and mortar, sieves of varying degrees of fineness, earthen material for crucibles of the desired shapes, dried cow-dung cakes for heating purposes, retorts of glass, pots and iron-pans.' (VII, verses 1-21)

According to this text, all of these should be installed ceremoniously with sacred chanting.

It was enjoined that the intended chemical operations were to be carried out under the benign influence of the *rasaliṅga*, an esoteric symbol which was made of either a gold amalgam (three parts of gold and nine parts of mercury) or a compound of mercury and sulphur (mercuric sulphide) shaped into a *liṅga* form. In fact *liṅgam* is also the technical word used for mercuric sulphide or cinnabar, particularly in the Tamil Siddha texts on alchemy and medicine. It has tāntric nuances, too, as detailed in the *Mātṛkābhedatantram*. The *rasaliṅga* was, therefore, a source of divine inspiration, and viewed as a tāntric male-female polarity or the non-dual supreme, which was sought to be accomplished through mysterious alchemical practices.

Apparatus: The major texts on the *Rasaśāstra* have a separate chapter devoted to the apparatus (*yantra*) of diverse description that were used for specific purposes. The *rasavādins* had the proven expertise in such processes as the upward, downward, or sideways distillation, condensation, sublimation, trituration, steaming, extraction, and incineration (prolonged heating). The wide variety of apparatus used by them encompassed *mūsa yantrā* (crucibles of several shapes); *koṣṭhi, dolā*, and *sevdanī* (steaming); *ūrdhva, adaḥ*, and *tiryak pātana* (for distillation upwards, downwards, and sideways); *dīpikā, dheki, jāraṇa, garbha, haṁsapāka, vidyādhara, kacchapa, somānala, vālukā, lavaṇa, nalikā, bhūdhara, pālikā, ghaṭa, iṣtikā, nābhī, ḍamarukākhya, dhūpa, sthālī, kaṇḍuka, antarālika* and *khalva*. The names of some of them are suggestive of either their shapes, the material used, or the purpose for which they are to be used.

The crucible, an important contrivance, was made with great care. Generally earthen, it was prepared with the earth of an anthill or white clay, rice husk, iron dust and chalk powder, which were compounded and kneaded in goat's milk to obtain a dough-like mass. It was then shaped into desired forms and sizes, and finally sun-dried. For extracting zinc from calamine, a crucible of the shape of a brinjal (*Solanum melongena*) to which a tubular contraption was attached, was used in the extraction of the 'essence' of zinc from its mineral, calamine. A striking aspect of the important *Rasaśāstra* texts is their detailed and methodical descriptions of ingredients and the crucible. For example, the *Rasārṇava kalpa*, a text probably of the eighth or ninth century, provides some details of the apparatus which consists of two crucibles, each of twelve *aṅgulas*, one of them having a narrow orifice. In this is to be taken some quantity of sulphur; and in the other, an appropriate amount of mercury. Both mercury and sulphur are to be moistened with garlic juice. The sulphur-containing crucible is to be inserted into the one having mercury, and this is to be carefully lowered into an earthen pot over which another pot is to be placed, and their rims luted with clay and cloth. The whole apparatus is heated from outside by fire of cow dung for three days to obtain mercuric sulphide of the desired quality. Crucibles in several shapes like that of a cow's udder, round, or oval ones were used, depending upon the type and purpose of chemical actions.

The *Koṣṭhi* apparatus comprised two pots of suitable sizes (one of them being bigger), measuring sixteen *aṅgulas* in width and forty-eight *aṅgulas* in length, both with rims. The material, such as processed mercury, is placed in the bigger vessel along with other ingredients, and heated from outside by the application of cow-dung fire. In the *dolāyantram*, the substance (mineral or metallic compound) is tied in a piece of cloth and kept suspended by means of a rod which is positioned across a pot half-filled with liquid in such a way that the tied substance is kept immersed in the liquid. The liquid is then heated from outside for several hours so that the chemical interaction between the substance and the material in the liquid takes place to the needed degree. *Svedanī yantram*, as the name indicates, is employed for steaming substances either for purification or for extraction.

The *pātana* contrivances used for distillation are of three types: (i) vertical (ii) downwards and (iii) sideways. For this purpose two well-baked pots of suitable dimensions (one big and the other small) are the vessels used; water is filled in one, while the big one is kept inverted over this pot. The rim or the junction of the two necks of the pots is closed tight by the application of a dough made of lime, raw sugar, and buffalo's milk. The inside bottom of the upper vessel is smeared with the desired substance, and is heated from outside by burning cow-dung cakes over it. When the substance vaporizes, the vapours fall into the water kept in the lower pot, which means that the distillation is carried out downwards (*adaḥpātana*). The opposite operation (sublimation upwards or *ūrdhvapātana*) would also be carried out when needed. A variant of this procedure is the distillation sideways (*tiryakpātana*), which is accomplished by connecting one pot (in which the desired substance is kept), provided with a long tube fitted in an inclined position, and connected with another pot which serves as the receiver. The substance kept in the former, when heated, attains a vaporous form, goes through the inclined tube and collects in the receiver which is kept cooled by water from outside.

In the *dheki* apparatus, a hole is made slightly below the neck of a pot, and one end of a bamboo tube is inserted into it tightly, the other end fitted into a brass vessel filled with water. Mercury, which is mixed with other auxiliaries, is subjected to distillation in this contrivance. The extraction of mercury from cinnabar (mercuric sulphide) is carried out in a device called the *vidyādhara yantram*, which has two pots luted together, one below the other, mercury being kept in the lower pot which is heated. Mercury vapourizes and collects at the bottom of the upper pot. The *vālukā* apparatus, as the name indicates, is a sand-bath for heating substances uniformly and over a long time. If salt is used instead of a sand-bath, it becomes *lavaṇa yantram*. *Dhūpa yantram* is employed for fumigation, particularly when a substance is to be interacted with the vapours of sulphur or arsenic substances.

Several apparatus of the *rasavādins* were simple, purposeful, and generally made of earthen materials; but sometimes glass contrivances like a bottle were also in use for sublimation reactions. Generally, the source of heat was large cow-dung cakes or wood, which would not yield a very high temperature. But the *rasavādins* were experts in attaining the maximum effect by prolonged heating over a long period from days to weeks, and this method was known as *puṭapāka*. There were several types of this method. A cubical pit called the *gajapuṭa* (one *hasta* or twenty-four *aṅgulas* in length, breadth, and height) is dug out and the crucible which contains the minerals or metallic substance, is placed upon a heap of cow-dung cakes. Sometimes as many as five hundred pieces of cow-dung cakes of uniform sizes immerse the crucible. The cakes are fired and the crucible is heated for a long time. In what is known as *mahāpuṭa,* even a thousand cow-dung cakes are packed, but the crucible is sometimes placed almost at the top of the heap of the cakes, and above this, another five hundred cakes are packed, and set fire to. An eighteen-*aṅgula* cubical pit is known as a *varāhapuṭam,* while a fifteen-*aṅgula* one is called a *kukkuṭapuṭa*. The end product of a *puṭapāka*, a metallic compound, now called *bhasma*, is supposed to have attained extraordinary qualities, both physical and chemical, by this heating method. The quality of a *bhasma* is ascertained by two tests: first, when it is pressed between the thumb and the index finger, it should not stick to the fingers; and secondly, when it is thrown in water, it should float for a considerable time, indicating that it has high degree of fineness which will ensure its absorption in the human body when administered with an appropriate adjuvant.

Bhasmas: Several types of *bhasmas,* which are prepared as indicated above and in a finally powdered form (hence the name), are generally oxides or some compounds, depending upon the minerals or metals used. Their preparation, for which thc *rasavādins* laid a solid foundation, is still a living tradition. Though Ayurveda does not fully accept that the *Rasaśāstra* is integral to its concepts and practices, some of the preparations of the latter—particularly the *bhasmas*—have found their way into Ayurvedic therapeutic practices. The most important *bhasmas* are those of mica (*abhraka*) and of metals like gold, silver, and copper. Although they are used in extremely small quantities along with other prescriptions or recipes, they are believed to possess the special properties of prolonging life or preventing the onset of ageing prematurely, and also rejuvenating the human body. In their preparation, however, one or the other plant extract is used with or without mercury or its compounds.

The *Rasaratnasmuccaya* (fourteenth century) states that a gold *bhasma* can be prepared by coating very thin leaves or foils of gold with a paste made of mercuric sulphide and the juice of lime, and incinerating them in a *puṭapāka*. The process is to be repeated ten times, after which the refined *bhasma* becomes fit for use in medicine. The process of preparing a silver *bhasma* is more elaborate: silver foils coated on both sides with black mercuric sulphide which is ground fine in lime juice. These foils are dried, and powdered sulphur is spread over them. They are then placed between two earthen plates, and heated over a strong fire in a sand-bath for a day. After it cools, the product is mixed with pyrites in equal quantities, and again ground well with lime juice. This mixture is subjected to *puṭapāka*, the process repeated twelve times, and the resultant product is reduced to the *bhasma* form. Likewise, details of processes for the other *bhasmas* have been provided by important texts of the *Rasaśāstra*, which shed light on the experimental skills as well as the knowledge of chemicals used by the *rasavādins*.

Transmutation Processes: Experimental techniques of *rasavādins* scaled new heights perhaps in the so-called transmutation process, a complicated one which is hard to understand scientifically. There are tall claims of not only converting inferior metals like copper and lead into gold or silver, but even of quantities of gold emerging from the use of small amounts of processed mercurial preparations as transmuting agents. In general, five types of 'transmutation' methods have been mentioned and, under each, some simple and several complex processes are described. The five methods are: (i) smearing (*lepana*); (ii) throwing (*kṣepa*); (iii) pouring (*kunṭavedha*); (iv) fumigation (*dhūma vedha*); and (v) touching (*sparśa vedha*). While a considerable number of processes involve the use of mercury and its compounds with their presumed potential for transmutation, there are others which adopt different procedures. Even for the same method, like smearing or fumigation, there are variations of the ingredients used as per some texts. A substantial number of 'transmutation' processes have been described in the *Rasaśāstra* texts, but is hardly possible either to present an account of all of them, or to offer plausible chemical explanations of them, even of the simple ones. A few are reproduced here by way of illustration.

Smearing Method: Perhaps the most common method adopted for the so-called transmutation process was the smearing technique. The *Rasārṇvakalpa's* description (verses 357-358) of this method is as follows:

> 'Mercury which is ground well with sulphur and orpiment (arsenic sulphide), and rubbed with small quantity of gold as a nucleus (*bīja*) is to be deposited inside a pit for one mouth. Silver foils, when smeared with this processed material and heated in a closed crucible, will turn into gold! According to *Mānasollāsa* (a text of the eleventh century), if sulphur is macerated seven times in the oil extracted from the seeds of the white variety of a plant called *brahmataru* and, if the resulting mass is coated on to copper foils, and heated in the *puṭa* contrivance, the copper foils become gold!'

Another text gives an elaborate method of transmuting lead into gold with a paste made from the leaves of peepal tree (*Ficus religiosa*) and other plants. A number of smearing methods are elaborated in the *Rasaśastra* texts. Perhaps by the smearing process, since sulphur or sulphides and certain plant juices are used, a thin surface film of golden colour was produced. In the same manner, a silvery white colour was sought to be obtained on brass sheets, as the *Rasaratnapradīpa* (ms. folio 19; lines 4-8) describes in the following way:

> 'Four parts of natron and an equal weight of potash along with borax, and two parts of white mica, two parts of orpiment and two alkaline salts—all of these are to be made into a paste by rubbing them thoroughly with the juice of *Kākamāci* (*Solanum nigrum*). This paste is to be smeared on brass foils, and allowed for some time so that they absorb the paste. Then they are to be heated in a closed crucible when the brass foils turn white. If mixed with equal weight of silver and heated again and again, silver of eleven-sixteen parts purity could be obtained!'

Throwing Method: A somewhat complex process, with extraordinary claims of transmutation, throwing has been described as follows:

> 'To a round-shaped crucible (*vajra-mūṣa*) a tube in the shape of cow's udder is to be attached to its forepart. Mercury, which has been digested with ten times its weight of sulphur, and one *karṣa* (a weight) of gold dust, are to be placed in the crucible. The oil of a plant called *Kanguni* is to be added in such quantity that mercury could be easily immersed in it. The crucible with its contents is to be heated. The same amount of gold dust is to be melted in another crucible. This molten gold is to be thrown on to mercury in the crucible. After it cools down, the solid mass thus produced is capable of transmuting *one thousand* times its weight of gold! One *gaḍyāna* (a weight) of this solid mass, when *thrown* separately on each of the three metals—silver, copper

> and lead in their molten state, transmutes them into gold of excellent quality!' (*Kaṅkalīrasa* ms. folio 15)

Pouring Technique: An example of this method is provided by the *Rasārṇava* (XIV. 86-90) as follows:

> 'One *pala* (weight) of heated mercury and an equal weight of lead are heated in *puṭayantram*, resulting in a *bhasma* of red colour like that of vermillion. This is taken in a copper vessel and made into a paste with cow's urine by constantly stirring the two with an iron rod. When heated, it becomes an amalgam of mercury and lead. One part of this incinerated product and an equal quantity of gold are to be mixed thoroughly, and made into a paste with such ingredients as have the properties of making mercury penetrate into metals. The product now obtained becomes so empowered that it can "transmute" a hundred times its weight of processed silver into gold of high quality!'

The *Rasārṇavakalpa* (verses 78-81) makes even more fantastic claims. It speaks of an amalgam of mercury, incinerated gold, and diamond, which is capable of 'transmuting' *ten million times* its weight of base metals into gold!

Fumigation and Touching Methods: According to the *Rasārṇavakalpa*, (verses 482-90) mercury, copper, the extract from a plant called *nāginī* and the urine of goats, are mixed well and calcined together several times extending over ten weeks. On the eleventh week, the smoke of the resulting mercurial preparation would have the power of transforming base metals into gold or silver. If mercury, after its deleterious properties are removed using plant juices, is digested with gold six times of its own weight, it acquires the special power of transmutation by its 'touch' only!

The foregoing examples reveal the types of chemical process described in the *Rasaśāstra* texts and the exaggerated claims of transmutation. It is indeed difficult to understand the full import of such processes and the mind-set of *rasavādins* who tacitly believed in the transmutation of metals like copper and lead into silver or gold. One wonders whether all of these processes were of a fictitious nature, without actually being conducted! However, the texts contain details of certain chemical processes that need to be experimented upon for their scientific validity, which is by no means an easy task, since the texts do not provide exact quantities or ratios of the chemicals.

Chinese alchemical texts, however, contain some detail about such processes, which were of five types but of a different nature. They are: (i) the development of uniform substratum alloys of a golden-yellow colour; (ii) enrichment of the surface layer; (iii) a special mode of surface-layer enrichment; (iv) formation of coloured surface films; and (v) others for producing similar effects. The golden colour of the end product would give the impression that the creation of golden colour itself was the ultimate goal. For, in the absence of any other definitive chemical tests known then for confirming whether it was gold or not, the attractive colour like that of gold served as a positive indicator, confirming the 'formation' of gold. Towards this end, alchemists used mercury, sulphur, arsenic substances, metals, salts and plant juices. The exact chemical reaction that might occur during the complex process engendered by heating cannot even be surmised until a process of this type is reproduced in a chemical laboratory, and its progress followed and analyzed from modern chemical point of view. The texts provide some sporadic information about the weights and measures of the ingredients used, but they are neither systematic, nor adequate for examining any of the textual processes in a laboratory. Possibly, the compulsions of keeping their method secret might have led the *rasavādins* to conceal certain important details of their own experimentation.

A noteworthy aspect of the transmutation processes, however, was the use of what the *rasavādins* call the '*bīja*' (seeding) to nucleate the process of 'gold-making' or aurifaction. It is conceivable that they also knew the difference between the real gold and the aurifacted or 'transmuted' gold. In the *Rasaśāstra* texts, there does not appear to be any designated name for the aurifacted gold. On the other hand, in Tamilian alchemy, such a distinction was made (*hemam* for artificial gold, and *tangam* for pure gold) and tested by heating both of them, that is, both were subjected to calcinations. According to a Tamil text on alchemy, *Amudakalaijñānam*, attributed to the celebrated Siddha, Agastyar, the aurifacted gold in the process of heating would lose its usual lustre, while the real gold would remain unblemished and continue to glitter.

Compounds of Mercury

Some *Rasaśāstra* texts describe the methods of preparing two compounds of mercury: (i) mercuric sulphide, which is pinkish red (*rasasindūra*) as well as black (*kajjali*); (ii) and mercuric chloride (*rasakarpūra* or *rasapuṣpa*). The details of such methods, as given in the texts, indicate that both of them were accomplished. In addition, the combination of mercuric sulphide with certain stimulants, fine gold dust, etc., known as *makaradhvaja*, was prepared with great skill using mainly purified mercury and sulphur since extraordinary rejuvenating properties were attributed to this compound. Purified mercury was generally obtained by heating cinnabar. Sulphur used to be purified by using milk and some plant juices to vaporize and condense it. Such operations were repeated several times to ensure the high purity of sulphur.

Rasasindūra, the pink-red variety of mercuric sulphide, was prepared by rubbing thoroughly an intimate mixture of pure mercury and sulphur in equal quantities. The blend was then heated in a glass bottle, the outside of which was wrapped with a piece of cloth plastered thick with finely powdered chalk.

Heating was continued in a controlled manner in a sand-bath (*vālukāyantram*) for several hours. In this process, the excess sulphur would undergo combustion and leave the neck of the bottle as a pale blue flame. When this reaction stopped, the mercuric sulphide would be deposited inside the upper part of the bottle, and was scraped off. *Kajjali*, the black variety of mercuric sulphide, was obtained by mixing three parts of pure mercury and one part of pure sulphur, and heated. Sometimes *kajjali* was prepared by the addition of lead also.

Rasakarpūra, also called *Rasapuṣpa* or *Rasakusuma,* was prepared by mixing mercury intimately with an equal quantity of salt (sodium or potassium chloride), and rubbed with the juice of a plant (generally *śnuhī* or *Euphorbia neriifolia*) along with some quantity of iron sulphate, if needed. The mixture was then placed in a glass bottle, the outside of which was plastered with mud. The bottle was placed on a salt-bed (*lavaṇa yantra*) and heated. When the fumes started emanating from the bottle, its mouth was sealed by a piece of chalk soaked in molasses. The heating was continued for six hours or more. *Rasakarpūra*, the bi-chloride of mercury, also called corrosive sublimate, would collect at the inside neck of the bottle as a white substance. The bottle was then allowed to cool, and the material scraped off.

The *rasavādins* knew that metals in their pristine state could not be used for any medicinal composition that was to be taken internally; nor for their so-called alchemical operation. They adopted two pathways for the removal of the deleterious effects of raw metals as such. One of them goes by the name '*māraṇa*' (literally, killing). For this purpose, some plant extracts, copper or iron pyrites, salts and sometimes even sulphur were mixed with the metal and heated. Understandably, the nature of the metal itself would undergo a change into a compound, sulphide or chloride, thus losing the toxicity which is generally associated with the untreated metals. The second pathway, which is rather difficult to explain, goes by the name *sattvapātana* (literally, extracting the 'essence' of metal). Here again the metal (or even a mineral) is mixed with certain plant extracts, salts, and some minerals and subjected to distillation. The distillate is regarded as the 'essence' or the *sattva* of the metal or mineral concerned. The process appears to be a micro-technique of treating the substance to cast off its deleterious impurities.

Rasaśāstra in its earlier stages (sixth to twelfth centuries) had pronounced alchemical overtones, not only with the deification of mercury and sulphur, but was also embedded in esoteric practices, such as tāntric ones. Later, processes leading to mineral medicines with certain curative powers, were developed. Known as *rasacikitsā,* such methods relied mostly on inorganic medicines: however, a number of medicinal plants were also used in their preparation. The *kavirājas,* the medical practitioners who generally follow the Ayurvedic system, appear to rely more on the inorganic medicines of the *rasacikitsā* than on the *materia medica* of Ayurveda. This preference may be due to the easy storage, simplicity of prescription, administration of small dosages, and the rejuvenating potential of some of the inorganic preparations like *makaradhvaja*. Ayurveda and the *Rasaśāstra* have been supplementary to each other in some ways. In any case, the *rasacikitsā* found new dimensions in the Siddha system of medicine (see Chapter 7) which continues to be an alternative healing practice in Tamil Nadu, as well as among Tamilians who live outside of Tamil Nadu all over India and elsewhere.

6

AYURVEDA: A HOLISTIC MEDICAL SYSTEM

Uniquely, India seems to be the only country in the world that can be proud of what may be called medical pluralism; for, coexisting with modern medicine are the traditional systems of Ayurveda, Siddha and Unāni, besides homeopathy and folk medicine. Of the traditional medicines, Ayurveda has been an endogenously developed system, since its foundational ideas are based on the Indian philosophical systems of *Sāṃkhya* and the *Vaiśesika* (fifth century BCE). The Siddha is mostly prevalent in Tamil Nadu and popular among Tamil-speaking people in India as well as in neighbouring countries. It derived its inspiration from the fundamental ideas of Chinese Taoism and alchemical practices, but its medicinal chemistry and pharmacopoeia developed indigenously from about the sixth century CE onwards. The practice of Unāni, which is Graeco-Arabic medicine, entered India around the thirteenth century, and was particularly encouraged during the Mughal rule. In the present, the three traditional systems have been receiving support from the government, both of the centre and the states, and Ayurveda has been recognized by the World Health Organization as an alternative system of medicine.

AYURVEDA: ITS PHILOSOPHICAL FOUNDATION

During the post-Upaniṣadic period from the sixth to the third century BCE, there was an intellectual efflorescence which began to take shape in the form of certain philosophical systems. New ideas about man and nature; man and the observed world of matter, qualities and motion; the methodology needed for the validity of knowledge of the phenomenal world, and the elevation of the human mind with a stronger control over the senses emerged. These were meaningfully delineated in the form of the *Sāṃkhya, Vaiśeṣika, Nyāya* and *Yoga* systems. Alongside, two other systems—the *Pūrva*-and *Uttara-Mīmāṃsā* (relating to higher self)—also came up. It was in this new mood of understanding and systematization that the ideas and practices of medicine too were standardized, leading to the overarching nomenclature of Ayurveda (literally, the science of a long and healthy life). In this process, the earlier Atharvavedic magico-religious healing practices that were superstitiously popular for over a millennium began to recede to the background to some extent. Even so, some aspects of the Atharvavedic religious healing methods were sought to be adopted for curing diseases, the causes for which were not known. Though not strictly true, Ayurveda is regarded as having its roots in the *Atharvaveda*, and in fact, one of the branches of Ayurveda, called the *Bhūta Vidyā* (psychiatry and demonology), is similar to Atharvavedic medicinal practices. But the rationality and the ramifications of Ayurveda are embedded in its philosophic foundational ideas, based on the *Sāṃkhya* as well as the *Vaiśeṣika*, with the epistemological support of the *Nyāya*.

Sāṃkhya, which is considered to be one of the oldest branches of Indian philosophical systems, is primarily dualist, encompassing man and nature (*puruṣa* and *prakṛti*). However, it envisages the plurality of *puruṣas* and thus has a pluralistic dimension too. This system is by and large atheist; although later it became somewhat theistic. Delineating the distinction between nature and the active, sentient being, *Sāṃkhya* regards nature as being in an equipoise or equilibrium of its three *guṇas*—namely, *sattva* (bright and light); *rajas* (energetic and stimulating); and *tamas* (heavy and inert). The interaction by the sentient being is said to disturb the equilibrium, leading to an evolution into intellect (*buddhi*), ego (*ahaṅkāra*) and mind (*manas*) as the psychic aspect of the evolution. The others in

Ādya Sāṃkhya in *Agniveśa tantra* (*Caraka Saṃhitā*)

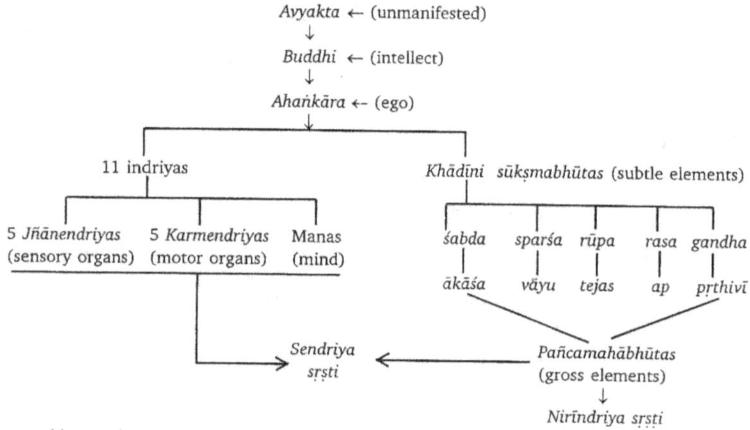

Sāṃkhya as described in the *Suśruta saṃhitā*

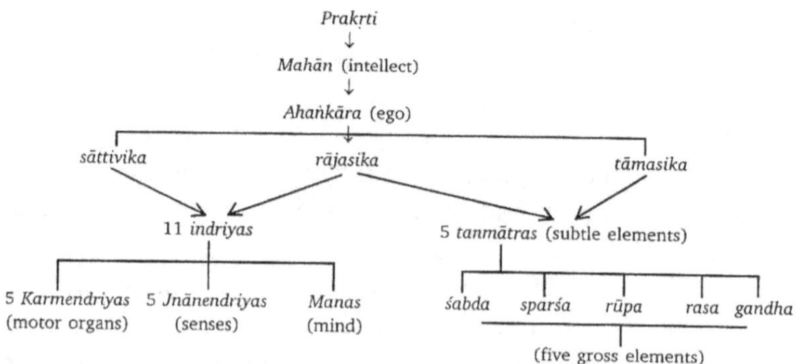

Source: B.G. Gopinath, 'The Foundational Ideas of Ayurveda', in B.V. Subbarayappa's (ed.), *Medicine and Life Sciences*, p. 95

the evolutionary sequence comprise the five sense organs, the five motor organs, the five subtle states called *tanmātras* and the five elements (*mahābhūtas*). The philosophical postulates of the Sāṃkhya have influenced the basic as well as the psychosomatic ideas of Ayurveda.

The *Vaiśeṣika*, with its pluralistic approach to the elucidation of the reals, deals with six *padārthas* or categories: substance, attributes, action, generality, similarity and inherence, besides including what is called *abhava* as the seventh one. But, as it happens, the term *Vaiśeṣika* is not found in any of the Ayurveda classical texts; nor is its the word *padārtha*. However, an exposition of them is found in the first chapter of *Sūtrasthāna,* of the *Caraka Saṃhitā,* which deals with *kāraṇas* or causes of action, leading to the maintenance of the equilibrium or harmony of all bodily elements and processes. The *Caraka Saṃhitā* has devoted substantial attention at the outset, to the cause and effect theory, adopting *mutatis mutandis* the six categories of the *Vaiśeṣika*. In several ways, the Ayurveda approach to the substances, their properties and actions, in the general concomitance as well as in particularity, is in essence an applied *Vaiśeṣika* philosophy.

The *Nyāya* methods of acquiring knowledge too find extensive applications in Ayurveda. Ayurveda has adopted four means or instruments of acquiring knowledge, namely (i) scriptural testimony; (ii) direct perception; (iii) inference; and (iv) experimental confirmation, or investigation and analysis of multiple causative factors, called *yukti.* Scriptural testimony is also called *āpta vākya* (or *āptopadeśa*). This means the acceptance of thc words of an erudite and enlightened person who by virtue of his meditation and experience, has acquired definitive knowledge without any doubt and with an unimpaired memory. Ayurveda's definition of direct perception is not merely in terms of the senses but is also related to the cognition arising from the conjunction or synergy of the soul, the mind, the senses, and the sense-objects, Direct perception has been employed extensively in Ayurveda in the examination of a patient, his disease as well as the identification of raw herbs and the preparation of drugs. Inference, an important method of examination, is of three types: (i) inference of the cause from the effect; (ii) inference of the effect from the cause; and (iii) inference in general. There is, however, one irrational aspect of inference inasmuch as Ayurveda subscribes to the *karma* theory, also by inference. One of the classifications of diseases is that some of them are held to be due to one's past deeds (*karmaja*). Since such detrimental deeds cannot be directly perceived, Ayurveda approaches this problem through scriptural authority and inference.

The foregoing three methods apart, the fourth one or *yukti* is an important rational method of Ayurveda. Caraka is believed to be the originator of this method, and he defines it as an intellectual perception or incisive knowledge of an entity, process, or an event associated with certain causative factors. Significantly, it is arrived at through experimentation and observation. The *Nyāya* methods of debates were also adopted by Ayurvedic practitioners. It was mandatory for a *vaidya* (Ayurvedic physician) to be proficient in logical debates with clarity of expression, pleasant conduct, and expertise in the use of technical words. Participation in professional debates, preceded by intense studies under an able preceptor, was one of the essential qualifications for an Ayurvedic practitioner. For, it was enjoined that such a method would alone enable him to have an exchange of ideas with others, thus sharpening his intellect and eliminating his own doubts.The Ayurvedic classic, *Caraka Saṃhitā*, has laid down even the principles of debate, including the types of discussion to be involved in or avoided, and the quality of the assembly of scholars. It lists as many as forty-four logical terms to be employed as canons of exposition. In no other contemporary medical system, including Greek or Hippocratic medicine, does one find a mental training of this type within the framework of a medical system. Even though the entire intellectual exercise is based on the *Nyāya* philosophy, the Ayurvedic exposition and methodology are superior to it.

While the *Sāṃkhya, Vaiśeṣika* and *Nyāya* have had their perceptible impress on the ideational structure of Ayurveda, the *Yoga* system appeared to have a rather limited influence except on psycho-therapy, which was known as *Sattvāvajaya*, as well as in the accomplishment of holistic healthy living. The term '*yoga*' is used in Ayurvedic medicinal preparations like *cūrṇa yoga* and *taila yoga*, which means a combination of one medicine with another and also a proper therapeutic application. Ayurveda is also concerned with the four-fold goal of life, namely, *dharma* (righteousness); *artha* (material means); *kāma* (desire but not greed); and *mokṣa* (liberation or emancipation from bondage). The *Caraka Saṃhitā* states that these were discussed at an assembly of sage-physicians who were deeply concerned with the need for standardized medical care, and thus endeavoured to evolve an efficacious and holistic medical system, leading to a long life (*dīrghāyu*) with sound health in order to achieve these goals.

According to traditional lore, Brahmā was supposed to have imparted a holistic system of medicine called Ayurveda to Prajāpati (a Vedic deity associated with creation) who, in turn, imparted it to Aśvinī kumāras, the twin divine physicians, who handed it over to Indra, the king of the cosmic gods as well as the god of war, rainfall, and storms. The assembly of sage-physicians, therefore, deputed the sage Bharadvāja to Indra to learn Ayurveda and bring it with him. This is neither real history nor the strength of Ayurveda, but an interesting legend narrated to explain the origin of Ayurveda.

Of significance is the exposition that the sage-physicians were deeply disturbed by the quackery that was rampant in the name of religious medicine. Superstition and healing practices went side by side with no attempt at standardization of medical practices with a sound theoretical basis. According to the legend, the sage-physicians were well versed in the philosophical ideas germane to an understanding of the human body and mind, the ailments and their cure. In tune with the Indian ethos, a divine origin and medium of medicine had to be conceived, and thus Indra, the venerated divinity, was associated with Ayurveda. The sage-physician, Bharadvāja to whom Indra handed down the medical knowledge he received, imparted it to one Ātreya Punarvasu who had six disciples. Agniveśa was the most brilliant of these disciples and it was his compendium, known as *Agniveśa śaṃhitā* or *Agniveśa tantra,* that was later redacted with additions by Caraka under the title *Caraka Saṃhitā* (second century CE). This was further refined by an erudite medical man called Dṛdhbala who lived in the fourth or fifth century. There is also a lineage starting with one of the forms of Viṣṇu, Dhanvanṭari, leading to the compendium by Suśruta which has come down to us as the *Suśruta Saṃhitā* (fourth century CE). These two are the great Ayurvedic classics, based on which a third one, with some improvements, appeared around the seventh century under the title *Aṣṭāṅgahṛdaya,* a treatise by Vāgbhaṭa. These three are recognized as the *Bṛhattrayī* (the great trinity) of Ayurveda. In the course of time, Ayurveda underwent certain improvements in the diagnosis of diseases, identification of new ailments, and *materia medica.* But its foundational ideas remained practically the same. From the ninth century onwards, several texts dealing with Ayurveda appeared. Among them are *Mādhava Nidāna* (ninth century), *Śāṅgadhara Saṃhitā* (thirteenth century), and *Bhāva prākaśa* of Bhāvamiśra (sixteenth century), a triad of texts clubbed under the title *Laghutrayī.*

STRUCTURE OF AYURVEDA

The eight branches of Ayurveda are akin to the specialties of modern medicine: *kāya cikitsā* (internal medicine); *śalya cikitsā* (surgery); *śālākya* (ophthalmology; ear, nose, and throat; dentistry); *kaumarabhṛtya* (paediatrics, obstetrics, and gynaecology); *agada tantra* (toxicology); *rasāyana* (rejuvenation:

geriatrics and nutrition); *vājīkaraṇa* (virility); and *bhutavidyā* (psychiatry and demonology). However, the two Ayurvedic classics deal with these in their own way under different chapters. The innate strength of Ayurveda is its holistic vision of the body, mind, soul, the environment, and the universe. The word *vaidya* for an Ayurvedic physician is significant. For, it literally means a person who has profound knowledge not merely of diseases and their cure, but also of the material and spiritual life, of the harmony between the individual and universe.

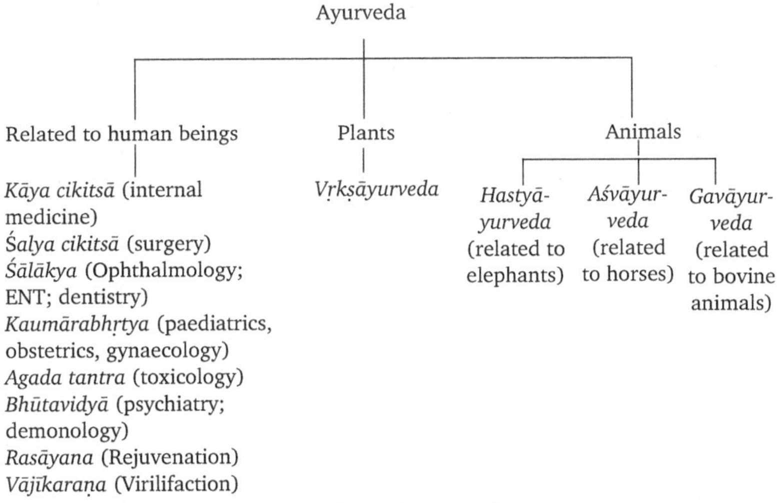

Branches of Ayurveda

As noted before, the *Caraka Saṃhitā*, which deals predominantly with internal medicine (*kāya cikitsā*), is known as *Agniveśa* tāntra of the Ātreya tradition. It has in all 120 chapters under eight heads: *Sūtra* (thirty chapters) or foundational ideas; *Nidāna* (etiology: eight chapters); *Vimāna* (evaluation/assessment of various factors by dietetics, methods of examination, professional debates, etc.; eight chapters); *Śārīra* (anatomy, physiology, genetics, embryology, women and child care, etc.; eight chapters); *Indriya* (features and characteristics of a person, environment, etc., indicating near death of a person, etc.; twelve chapters); *Cikitsā* (rejuvenation and virilification, diagnostic and management aspects, and therapeutic descriptions; thirty chapters); *Kalpa* (drugs and their preparations used in eliminating processes; twelve chapters); *Siddhi* (*pañcakarma* processes and *tāntrayukti*; twelve chapters).

The *Suśruta Saṃhitā*, which accords a distinct position to surgery in addition, is said to belong to the Dhanvantari tradition. The philosophical or fundamental ideas of Ayurveda are much less in this compendium than in the *Caraka Saṃhitā*. This text also has 120 chapters, but under six heads: *Sūtra* (fundamental principles of surgery and food articles; forty-six chapters); *Nidāna* (diagnosis of major disorders, regarding surgical or para-surgical interventions; sixteen chapters); *Śārīra* (origin of the universe, embryology, anatomy, physiology, mother and child care; ten chapters); *Cikitsā* (management of specific diseases, surgical and general therapeutic measures; forty chapters); *Kalpa* (toxicology and management of associated diseases; eight chapters); *Uttara Tantra* (paediatrics, demonology, and *tantrayukti*; diagnosis and management of disorders of eye, ears, nose, throat and head).

One of the basic concepts of Ayurveda is its recognition of the consonance of the macrocosm and microcosm, that is, an individual human being is the miniature replica of the universe. It rightly envisages that an individual (*puruṣa*) and the universe (*loka*) are in continuous interaction with each other, and the relationship between them is one of a donor-recipient in everyday life, whether it is air, water, food articles, or environment. As long as this interaction is wholesome or harmonious, the individual is bound to be healthy; but if there is a disruptive interaction, it would result in a diseased state. This is known as the *Loka-Puruṣa-Sāmya vāda*. According to Ātreya, the preceptor of Agniveśa, one who sees the entire universe in his own self is in possession of profound knowledge by which he attains ultimate peace and wisdom. (*CS*. *Śār*. 5.20)—an inseparable relationship between the self and the universe.

The Five Elements and Ayurveda

The holistic philosophical base of Ayurveda is composed of the theory of five elements: *pṛthivī* (earth), *ap* (water), *tejas* (fire), *vāyu* (air), and *ākāśa* (an ubiquitous non-material element) which, as noted earlier, are integral to the *Sāṃkhya* and the *Nyāya-Vaiśeṣika*. Ayurveda stresses that all substances, inorganic or organic, are made of five elements in varying quantities (if one of them is predominant in a particular substance, the other four will also be present in small proportions).

Ayurveda deals with six tastes or *ṣadrasas* which are composed of the elements as follows:

1. *Madhura* (sweet) : *pṛthivī* + *ap*
2. *Āmla* (sour) : *pṛthivī* + *tejas*
3. *Lavaṇa* (salty) : *ap* + *tejas*
4. *Tikta* (bitter) : *ākāśa* + *vāyu*
5. *Kaṭu* (pungent) : *tejas* + *vāyu*
6. *Kaṣāya* (astringent) : *pṛthivī* + *vāyu*

Metabolism: Ingested food and internal energetic processes

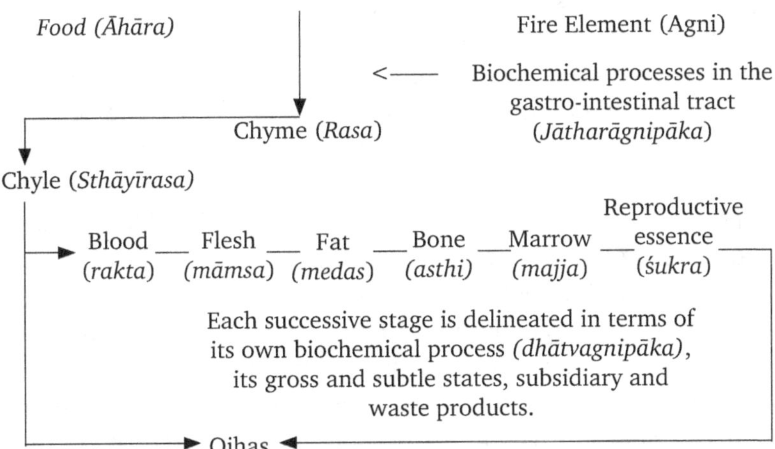

Even in the embryonic stage of development of the foetus, the five elements are stated to have their definitive functional roles: *vāyu* helps division; *tejas*, metabolism; *ap*, percolation; *pṛthivī*, union; and *ākāśa*, the expansion of the foetus. The subtle essences of the elements (*tanmātras*) are believed to accompany the soul, but the soul is not thought of in terms of five elements. Various bodily elements, the colour of the skin, excreta, digestion, and metabolism have been explained in the two Āyurvedic classics in relation to the five elements.

Three Humours (*Tridoṣa-Tridhātus*)

An important concept of Ayurveda is that of three humours (*tridoṣa)*, which is fundamental to all explanations of health and disease and which plays a pivotal role in Ayurvedic diagnostic as well as curative practices. The five elements are the physical constituents that are transformed into the three major biological components of a living body—*vāta, pitta* and *kapha*, called the *tridoṣas*—on the basis of which Ayurveda also tries to explain cogently the whole gamut of bodily processes. It needs to be emphasized that *tridoṣas* connote far more meaning than what is generally understood by its English equivalent, 'three humours'. The word, *doṣa* (vitiation*)*, is significant because of the inherent proneness of each or all of them together to get vitiated and also to vitiate other bodily elements. *Vāta* is regarded as the biological product of the predominance of the two elements, *vāyu* and *ākāśa*; *pitta*, of the element *agni* or *tejas*; and *kapha*, of *pṛthivī* and *ap*. Ayurveda recognizes that (i) the entire human body is composed of a solid sub-stratum; (ii) there is in it an unceasing interplay or interaction of varied biochemical activities; and (iii) they are engendered by energetic influences and several types of motion or movements. These three aspects, which are integrated with life processes, function in a coordinated manner for the health of the entire organic system. Each of the three humours has its material nature too (*vāta*: gas in bowels; *pitta*: bile; and *kapha*: expectorated sputum).

According to some Ayurvedic professors, Ayurveda's major concept is *tridoṣa-tridhātu* (three humours), a three-in-one approach, consisting of *vāta* (regulative bio-energy); *pitta* (catabolic and thermo-genic bio-energy); and *kapha* (anabolic and thermo-toxic bio-energy). Health is the state of their harmony or equilibrium in the body; and disease is the state of their disharmony or disequilibrium. Ayurveda's fundamental concern is the maintenance of the harmony or equilibrium of three humours.

Ayurveda accepts that there would be genetically determined relative proportions of the three vitiators within normal limits. Known as the *doṣaprakṛti* or natural condition, this factor is by and large responsible for the determination of the whole personality or the totality of one's physique (external), physiology (inside), and psyche (the mind-set). Ayurveda also believes that, depending upon the environment and several other factors, an internal matrix, as it were, of the relative preponderance of one or the other of the three humours develops in an individual's genetically determined constitution. Hence, in Ayurveda this concept has a special place and its consideration is essential for a proper understanding of the life processes, health, disease, susceptibility to disease, preventive measures for warding off diseases as well as health-promotive care.

The idea of Nature (*Prakṛti*) in Ayurveda owes its conceptual origin to the *Sāṃkhya* philosophy's *prakṛti*, which is but a harmony or equilibrium of the intertwined three qualities (*guṇas*) of righteousness (*sattva*), energetic (*rajas*), and inertia (*tamas*). It changes in ways determined by the dominance of the quality. These are also reflected in the human psyche, which, according to Ayurveda, is also composed of the five elements: *ākāśa* is represented in *sattva*; *vāyu* and *tejas* in *rajas*; and *pṛthivī* and *ap* in *tamas*. It is explained that the mind-set or the psychic make-up of an individual is determined by the preponderance of one of the three qualities. An apt definition of Ayurveda, in this context, is provided by Suśruta when he says that total health (*svāsthya*) is the state of an individual in whom there is a biological balance or equilibrium of the three humours, the various kinds of digestive agents (*agnis*), the seven supportive tissues (*dhātus*) and waste products (*malas*), besides a harmonious state of sensorial, mental, emotional, and spiritual well-being—a definition which the World Health Organization has also adopted.

In order to achieve or attain such a healthy state, Ayurvedic classics have prescribed an elaborate schedule for a health regimen and code of conduct during the day (*dincaryā*), night (*rātricaryā*), and in the environment of the different seasons (*ṛtucaryā*). These are to be accomplished by proper diet, nutrition, exercise, personal and social hygiene (*sadvṛtta*), along with the needed bodily purificatory processes from time to time.

The concept of body-mind-*prakṛti* apart, Ayurveda expounds a special theory called the *Svabhāvoparamavāda,* that is, natural self-cure or healing on its own. It emphasizes the fact that there is an inherent power or potential in the body itself that would act in self-defence and regeneration, resulting in the natural cure of disease or injury. The role of the administered medicine is to strengthen this natural process of healing. Ayurveda firmly believes that there is always some residual health in a person afflicted with a disease, and the former needs to be fortified to supplement the total cure leading to the mind-body harmony.

As noted earlier, another important foundational idea of Ayurveda is known as *Loka-Puruṣa-Sāmya* (macrocosm-microcosm continuum) or a recognition that an individual is a miniature replica of the universe. The interaction or exchange between the two needs to be wholesome for sound health; otherwise, a disease-state appears.

DISEASES AND THEIR CHARACTERIZATION

There are three major categories of diseases, according to Suśruta: (i) those caused by traumatic situation or infliction by weapons, wild animals, etc., (*ādhibhautika*); (ii) those caused by seasonal and supernatural factors (ādhidaivika); and (iii) those due to psychosomatic, heredity factors, etc. (*ādhyātmika*). The *Suśruta Saṃhitā* has clearly described six distinct stages of pathogenesis or the sequence during the evolution of a disease. They are: (i) accumulation of vitiators (*doṣas);* (ii) their vitiation and aggravation; (iii) their spreading; (iv) their localization; (v) manifestation of the disease; and (vi) further progress of the disease with complications. The first is stated to occur mainly because of slow digestive power, which leads to the formation of what is called the *āmāvasthā,* in which the various channels of the body are blocked in varying degrees. This corresponds in some way to the modern ideas of the affliction of the membrane-biological processes. The second results when this situation is aggravated by the quantitative intensification of the vitiators *(doṣas)*. At this stage, if a proper curative method is not adopted, the vitiators begin to spread to different parts of the body. If unchecked both quantitatively and qualitatively, they lead to the third stage. The next state would be the localization of the vitiating factor in a place that is vulnerable or weak biologically. This is a stage of morbidity that sets off the manifestation of disease which, at this state, may not be in its fully developed form. The penultimate stage is when the disease is in its full virulent form, and in the last or sixth stage, it is in an advanced state with several complications that need to be diagnosed in detail, and remedial measures taken.

Diagnostic Methods

Ayurvedic diagnostic methods and procedures are indeed rational and elaborate alike. The *Caraka Saṃhitā* avers (VI. 4.19) that an Ayurvedic physician, in addition to his training and acquisition of expertise in medical concepts and *materia medica,* needs to be humane and sincere. He should establish a close rapport with his patient, entering the innermost self of his patient. Apart from his knowledge and wisdom, a *vaidya* should be revered for his exalted ethical conduct and service-oriented mind-set, without any desire for material gains. *Caraka Saṃhitā* has stressed this aspect in a subtle way: 'Not for self, not for the fulfillment of any earthly desire or gain, but solely for the betterment of those who are suffering from diseases, should you (*vaidya*) treat your patients and excel in all. Those who sell the treatment of diseases as merchandise collect dust and neglect gold.' (*CS*. Vi. I. 4. 58-59)

Clinical Methodology: In several ways, Ayurveda is a methodological compendium. Its clinical methodology is succinct and effective and several pathways of diagnosis are delineated, the fundamental approach being (i) to know the strength of the patient; and (ii) to know the strength of the disease.

The Ayurvedic clinical examination is purposeful and an elaborate one, though it is largely qualitative; for, the diagnosis is not confined only to the identification of a disease, but takes cognizance of the *whole personality* of the patient and the roots of aetiopathogenesis. The two-in-one diagnostic approach of Ayurveda, therefore, is the examination of the patient and his disease (*rogi-roga-parīkṣā*) together. Thus an Ayurvedic physician examines the human factors in the patient as well as his residual or remaining health even in the diseased state. Therefore, methodically he evaluates (i) the man and his constitution; (ii) the residual health in him even in the afflicted state; and (iii) the nature and intensity of the affliction of the disease itself.

Examination of a patient is comprehensive, and comprises four methods: (i) direct observation; (ii) inference; (iii) the textual testimony; and (iv) *yukti*, by which an Ayurvedic physician adopts experimental investigations to gather evidence, which might either corroborate or disprove the conclusion he would have arrived at. These four methods are known as *Caturvidha parīksā*. There is the six-fold method in which a physician utilizes his own five senses to examine the body of the patient, which would be his direct perception and his inference based on the interrogation of the patient and writing down his case-history is also part of the diagnosis.

The most comprehensive diagnostic method goes by the name *daśavidha parīkṣā* or ten-point clinical examination. This has been clearly described in the *Caraka Saṃhitā* in relation to the examination of a patient and in terms of (i) constitution (*prakṛti*) or the nature of the body and the psycho-physical status, whether it is in a balanced or imbalanced state; (ii) susceptibility to disease (*vikṛti*); (iii) tissues and their qualities (*sāra*); (iv) compactness of the body (*saṃharana*); (v) anthropometry (*pramāṇa*); (vi) adaptability (*sātmya*); (vii) mental stamina (*sattva*); (viii) digestive power or appetite (*āharaśakti*); (ix) exercise endurance (*vyāyama-śakti*); and (x) age and the rate of ageing condition (*vaya*). In the ten-point examination the first place has been rightly accorded to the constitution or *prakṛti* of a patient since it provides several clues leading to a proper understanding of the twin aspects of the patient and his disease.

Ayurveda emphasizes that the constitution or the nature of a patient encompasses both his physical and mental nature in the context of the *tridoṣa* or the vitiated *vāta, pitta* and *kapha* as well as the three *guṇas*: *sattva, rajas* and *tamas*, and their relative proportions. A patient's susceptibility to the disease could be ascertained by the history of his past or the present illness. The status of tissues could be understood through the relative qualities of the chyle (white bodily fluid), blood, flesh, fat, bones, marrow and the reproductive essence which, together with his mental state, could lead to an evaluation of the patient's psycho-physiological health and his overall make-up.

Now in practice, Ayurveda's diagnosis generally comprises the examination of eight points of the body through: (i) pulse-reading; (ii) urine examination; (iii) stool examination; (iv) tongue examination; (v) voice and speech investigation; (vi) skin or tactile sensation; (vii) condition of the eyes; and (viii) study of the face. These are in the nature of indices for gaining an insight of the disorder in the body as a whole. The classical texts, the *Caraka* and the *Suśruta Saṃhitās* do not mention the first two, namely, the pulse and urine examination. These two methods might have been adopted by the Ayurvedic physicians as a result of their working with Unāni hakims in hospitals established by Muslim rulers, or were influenced by Siddha medicine, which employs these two methods for diagnosis.

Pulse and Urine Examination: The investigation of the pulse and urine, which are described in later medieval Ayurvedic texts like the *Bhāvaprakāśa* and *Yogaratnākara*, are intended to find out the nature and extent as well as the intensity of the disease. An accurate reading of the pulse, which depends upon the dexterity of an Ayurvedic physician, is meant to provide vital clinical information to him. For this purpose, both the physician and the patient should sit in a relaxed manner, the elbow and the wrist of the patient need to be slightly flexed, and the physician should place his three fingers (index, middle, and the ring ones) on the radial pulse of the patient at the base of the thumb in such a way that the physician's index finger is nearer to the patient's thumb, while the ring finger is pointed downwards. The pulse is palpated by the physician with his three fingers separately, and also in combination. He should find out from his own experience and intuition, the state of the three *doṣas* (*vāta, pitta* and *kapha*), with the index, middle, and the ring finger, respectively. The principle that lies behind the pulse examination is that its rate and rhythm are controlled by *vāta*, while its amplitude is governed by *pitta*, and the volume of the pulse is determined by *kapha*. The whole examination is a delicate one, and needs an extraordinarily trained physician who, in turn, will be able to arrive at his conclusion about the type of disease, whether it is predominantly due to either *vāta, pitta, kapha* or their combination. His acumen, supplemented by his experience, is displayed in the evaluation of the relative preponderance of the three vitiating factors. The general conclusion (differences in opinion apart) is: The pulse of a patient who has a predominantly *vāta* disease has a high rate and a relatively low volume and amplitude, while that of *pitta,* has a high amplitude and average volume and rate. On the other hand, the *kapha*-predominant pulse has a high volume, low rate, and average amplitude. The mixed counts will, of course, have a mixed presentation.

The examination of urine is also intended to find out the nature and extent of the manifestations of the three vitiated factors. Here the volume, colour, consistency, smell, taste and frequency are investigated in detail. Sometimes the data of urine analysis by modern methods are also taken into consideration, but interpreted in terms of the vitiating elements. One of the traditional methods of urine examination, called the oil drop examination (*taila-bindu-parīkṣā*) is carried out as follows: A patient's first morning-urine is kept in a vessel and mustard or sesame oil is dropped into it in a controlled ambience of temperature, humidity, air circulation and light. The movement and behaviour of the oil drop on the surface of the urine are observed. On the basis of the observation of how the oil drop dips in urine, moves in a particular direction either slowly, fast, or very fast, or whether it continues to remain as a drop, or breaks into parts, an Ayurvedic physician tries to determine the prognosis of a disease in terms of the three humours.

Other Medical Investigations: Examination of stools is carried out by a water-floating technique (*jala-tāraṇa-vidhi*) by placing a certain amount of faecal matter on water in a jar half-filled with water. Faeces that float on water is considered to be what is called the *nirāma* (less morbid) state, while if it sinks quickly and disintegrates, it is regarded as being in the *āma* state, indicative of less digestive power, lack of appetite, nausea, constipation and related states of toxicity. Through the examination of the stool, an Ayurvedic physician, with his own insights, will be able to judge the deleterious factors that are causing the particular ailment.

In Ayurvedic diagnostics, tongue examination, too, has a special place, since the tongue is considered a mirror of the entire body, particularly of the digestive system and its associated physiological processes. Likewise the patient's speech and voice, the appearance of the skin, eyes and face, though external, reveal a good deal of clinical information. Anaemia and jaundice are perceptibly reflected in the eyes, and the disorders or distresses within the body are expressed in the face. Through the adoption of the ten points of examination and a careful analysis of them by his own knowledge and experience, an Ayurvedic physician can identify the nature as well as the intensity of affliction in terms of the vitiating factors, their strength, location of disease, and the disharmony of the *tridoṣas*, enabling him to charter his course of treatment according to these guidelines.

AYURVEDIC APPROACH TO TREATMENT

The word *cikitsā* includes Ayurvedic curative procedures or remedial measures. It is classified under two heads: (i) *naiṣṭhikī*; and (ii) *laukikī*. The approach of *naiṣthikī* is not mundane but spiritual, inasmuch as it is intended for the augmentation of the concerned individual's overall well-being, free from pain and misery of any sort, which would eventually lead to liberation (*mokṣa*). Intense meditation and yogic disciplines are the pathways of this treatment. The *laukikī* approach is concerned with the cure of ailments that afflict an individual both physically and mentally as a result of the disequilibrium in the psychosomatic interlinked make-up and the disharmony of *tridoṣas*. The aim is to restore eventually the equilibrium or harmony by administering appropriate medicines coupled with a dietary regimen and routines for a disciplined daily life. For, according to Ayurveda, total health is the harmony or balance among the *doṣas*, *dhātus*, and the *triguṇas* (*sattva, rajas* and *tamas*).

Three main pathways of curing ailments (*laukikī cikitsā)* have been enunciated in the Ayurvedic classics. They are: (i) Divine therapy; (ii) Rational therapy; and (iii) Psychotherapy. The first type of treatment relates to the cure of peculiar ailments, the causes of which cannot be diagnosed exactly, and they are believed to be the results of one's own deeds (*karma*) in one's past life leading to sufferings in this life in the form of ailments. Ayurveda, thus, believes in the concept of rebirth (*punarjanma*) which is a corollary of the *Sāṃkhya* philosophy which, as noted before, has provided some foundational ideas for Ayurveda. The diseases found to be incurable are believed to require divine intervention through divine therapy. It consists of actions such as offering prayers, incantations, performing rituals with oblations, wearing prescribed gemstones, and daily worship of some divine forms. This type of therapy understandably draws its sustenance from astrological factors, too, and nowadays it is the astrologer more than the Ayurvedic physician who is involved in this type of cure. In any case, divine therapy is difficult to explain scientifically, and appears to be in the nature of a faith-cure, strengthening the psychical status of the afflicted person and his belief that his disease would be cured by divine dispensation.

The rational therapy of Ayurveda is based on its fundamental concepts about the body, physiological processes, the psychosomatic relationship of the mind and body, and the restoration of equilibrium or harmony of the three humours. This therapeutic approach is, therefore, governed by the theory of the five elements, and the concept of the three humours, besides the interrelationship of metabolic processes, diet or drug-dietary regimen, and the intake of a variety of herbal remedies. The curative treatment is not merely for the cure of the disease but also for the uplift of the patient's psychosomatic health, which needs to be integrated with both purificatory and palliative measures.

The purificatory procedure, in general, goes by the name of *saṃsodhana pañcakarma* or five therapeutic measures for attaining the desired level of purification. Purification of the body internally is an essential prerequisite for Ayurvedic curative measures; for, according to Ayurveda, if the human body is biologically purified, not only will it have the innate potential to combat the disease but also, any medication would flow more easily and efficiently towards the site of the disease, enabling it to eliminate the root cause of the disease, a state that would facilitate the restoration of the balance of the three humours and their qualities. Palliative measures are accomplished by the use of specific drugs, dietetics and daily routines, including exercises.

In tandem with its foundational concept that all objects including herbs, food articles, bodily parts, tissues, and even the *tridoṣas* are essentially made of five elements, each with its specific attribute, Ayurveda accepts that when a substance of similar property is added to similar ones, it increases the latters' character or quality, while dissimilar ones would result in their decrease, resulting in a homologous or heterologous effect. Hence Ayurvedic drugs are classified on the basis of substances that are categorized under the five elements, each being predominant like *pārthiva dravya*, (predominantly consisting of the earth element), *āpya dravya* (mainly the water element), and so on. The prescription or application of drugs, therefore, is to increase or deplete similar or dissimilar biochemical situations as diagnosed by the Ayurvedic physician. In other words, if the physician feels that a particular biological component is to be increased, he will select a similar or homologous drug and also the appropriate dietary regimen; likewise, if he finds that some factor is excessive beyond the normal limits, he will administer dissimilar or heterologous drugs to deplete it, along with the needed dietary restrictions. Thus, Ayurvedic therapy is not in the nature of chemotherapy, or eliminating only the micro-organisms or other factors that cause the disease; it has a much larger effect on the bodily mechanism, its nutrition, and feeling of well-being. The diet and the drug go together in restoring the total health of the afflicted person by bringing about an equilibrium of the *tridoṣas* and *triguṇas*.

The third type—psychotherapy—is adopted in the case of mental disorders like insanity and epilepsy. Apart from characterizing them in terms of perversion of mind and intellect, loss of memory and other symptoms in the case of insanity, and occasional unconsciousness and abnormality in the case of epilepsy, some physical causes are also evaluated for such mental states. The impairment of physical activities, memory and intellect is called *prajñaparadha* which, in turn, can be a causative factor of several somatic disorders. In addition to psychotherapy, drug therapy is also advocated in recognition of the Ayurvedic concept of the mind as a substance itself, the disorders of which can be cured by administering medication that would nourish it. The Ayurvedic classic texts, however, have not provided adequate details about psychotherapy which, by and large, appears to be similar to divine therapy.

SURGERY

'It was in surgery above all that the ancient Hindus excelled,' observed D. Guthrie, a medical historian, and added: 'Suśruta described more than a hundred instruments. This was their greatest contribution to the art of healing, and the work was bold and distinctive. It is not unlikely, though difficult to prove, that some of it was of Greek origin. Some, indeed, state that the Greeks drew much of their knowledge from the Hindus.' (D. Guthrie, *History of Medicine*, London, 1920; p. 19) Another noted historian of medicine, Max Neuberger mentions: 'The outstanding feats of the ancient Indian surgery related to laparotomy, lithotomy and plastic operations.' (Max Neuberger, *Geschite der Medizin* (English tr. by E. Playfair), London, 1925; pp. 57-58)

Surgery (*śalya*) is one of the eight medical specialities of Ayurveda, and the *Suśruta Saṃhitā* has provided extensive details about surgery in its first section (called *Sūtrasthāna*) and in the fourth one (*Cikitsā sthāna*). Possibly some proto-forms of minor surface surgery would have been in practice even earlier than the time of Suśruta, but it is in his work that a systematic record of surgical procedures including pre- and post-operative treatments is set out. He is, therefore, rightly called the 'Father of Indian Surgery'. It appears that there was an earlier work under the title *Sauśruta tāntra*, probably of the fifth century BCE, which has not come down to us. It belonged to the Dhanvantari tradition and, as recorded in the introductory portions of the now available *Suśruta Saṃhitā*, Suśruta was a disciple of Kāśirāja Divodāsa Dhanvantari, an eminent surgeon of his time. The extant text is a redacted version by one Nāgārjuna (eighth century CE) with some modifications. This text is an Ayurvedic classic not only in its treatment of subjects like gynaecology, paediatrics, ophthalmology, anatomy, physiology and geriatrics, but also of various surgical operations, surgical instruments, and the practical training to be imparted to a surgeon—all explained in a methodical manner with appropriate medical terminology.

Description and Management of Wounds

Wounds and ulcers, the nature of their inflammation, repair, and management find greater attention and space in the *Suśruta Saṃhitā*. It describes as many as sixty types of management of wounds with elaborate details of dressing of bandages including their indications and contraindications, stitching a traumatic wound, protection from maggot formation, and associated procedures. Suśruta's concern for preventing infection is evident from his descriptions relating to wounds, although neither bacteriology nor the idea of asepsis and antisepsis was scientifically known during his time. He has also systematically dealt with the management of burns and scalds, fractures, dislocations and skeletal injuries, simple and compound fractures. His prognosis of a skeletal injury and its possible complications in advanced states hold good even today.

Reconstructive Surgery

In ancient times it was not uncommon that, as a punishment, an offender's earlobes or nose would be cut off and this would need reconstruction, or there could be situations like a deformed split-lip. All of these required reconstructive or plastic surgery, and the surgeons of that time in India rose to the occasion. The *Suśruta Saṃhitā* provides detailed procedures for plastic surgery. Indeed, this was an outstanding contribution of Suśruta, since it was not before the eighteenth century that plastic surgery was practised in the West.

Suśruta has described the repairing of fifteen varieties of torn or defective earlobes. More importantly, he has explained the principles of rhinoplasty, which are in no way different from modern pedicle grafting. In rhinoplasty, according to Suśruta, a portion or patch of living flesh along with skin, its subcutaneous tissue and circulatory mechanism, is carefully taken out from the adjacent portion of the body of the patient, and dexterously transferred to the sacrificed portion of his nose. Two pipes are inserted into the nostrils to facilitate breathing. The pipes are to be removed after the grafted portion gets healed and is successfully grafted in the nose. A similar technique is to be adopted for labioplasty for severed lips, in which case no pipes need to be introduced. Further, there are also details about cosmetic plastic surgery.

Surgical Ideas and Methods

The *Suśruta Saṃhitā*, in the fifth and twenty-fifth chapters of the *Sūtrasthāna* section, has provided details of the main surgical procedures or operative measures under eight types: (i) *chedanam* (excision); (ii) *bhedhanam* (incision); (iii) *lekhanam* (scraping or curettage); (v) *vyadhanam* (puncturing or tapping); (v) *viśrāvaṇam* (drainage); (vi) *eṣanam* (probing); (vii) *āharaṇam* (extraction by sharp hooks, etc.); and (viii) *sīvanam* (suturing). Vāgbhata in his *Aṣṭāngahṛdaya* (seventh century CE) added five more procedures: (i) *utpātanam* (paring); (ii) *kuṭṭanam* (pricking); (iii) *manthanam* (drilling); (iv) *grahaṇam* (apposing); and (v) *dahanam* (cauterization). More than one of these methods would be adopted in surgical procedures, depending upon the necessity and the nature of operation.

Suśruta has, in his own assessment, thought of the indications that were pointers to which particular procedure could be adopted. Excision was indicated in the surgical treatment of anal fistulas, cysts, moles, tumours, piles, enlarged tonsils and warts. On the other hand, incision needed to be employed in the case of supportive lesions such as several types of abscesses, carbuncles, afflicted mammary glands, bladder stones, closed wounds, etc. Scraping was to be done in leucodermic patches, dental plaques, diseases of the eyelids, certain types of skin diseases, non-healing ulcers, muscular out-growths, etc., all to be carried out with dexterity. Puncturing and drainage were to be adopted in cases of ascites, localized inflammation, skin lesions, hydrocele, elephantiasis, some types of eye diseases, adenoids, etc. Probing was to be resorted for curing sinuses with or without the foreign body inside, while extraction was to be conducted for the removal of tartar from teeth, gravel from the feet and urethra, earwax, a dead foetus, and impediments in the anal canal. As for suturing, Suśruta has recommended that it should be done for fresh wounds after their intensive but careful cleaning. It could be of the continuous, interrupted, or winding types using three types of needles—straight, curved or triangular—and fine threads, except for intestinal sutures for which an organic thread type was to be employed. Suturing also required intensive training.

An Ayurvedic surgeon was required to acquire the needed skills in various surgical procedures before he was expected to practise surgery on human beings. Suśruta could be regarded as the first to have conceived of practical training on experimental models. Since animal surgery *per se* did not exist at that time, some innovations were necessary to achieve proficiency and dexterity in performing the eight basic surgical feats, which were to be adopted after proper diagnosis. Proficiency in incision was attained by practising it on leather bags or the urinary bladders of dead animals filled with mud or some other soft material. Excision was conducted on fruits which were hollow inside; scraping on hides of animals; puncturing and blood-letting on the veins of dead animals or the hollow stem of the lotus; probing, on hollow pieces of dry wood or bamboo with holes; extraction of the teeth of dead animals or jack fruit seeds was the rehearsal for work on human beings; and suturing on cloth (cotton or silk) and leather.

Suśruta has emphasized the importance of keen observation as an inseparable adjunct of practical training. It was essential, then as now, for a medical student to observe with his own eyes the various organs of the body and their exact locations. Suśruta's method of dissection of the human body consisted of placing the cadaver (dead body) in clean flowing water for six to seven days, after which it would disintegrate so that it was possible to scrape it off by a brush made of a particular type of grass (*kuṣa*). Once the cadaver was carefully scraped, a student could observe all the organs and bones of the body.

Suśruta has given detailed descriptions of all the anatomical parts, their shapes, numbers, and arrangement. A surgeon was to have an exact knowledge of the vital points (*marmas*) because during surgery, any injury caused to these spots could lead to shock and trauma that could lead even to the patient's demise. Suśruta has identified as many as 107 vital points and even classified them as follows: nineteen of them (if affected) would prove instantaneously fatal or fatal up to seven days; thirty-three could cause death after about fifteen to thirty days; three would lead to death after the foreign body is removed; forty-four may not lead to death, but could cause incapacity; and eight could result only in pain. Descriptions such as these are hardly found in any contemporary text. Suśruta also thought of theory and practice as being two wings of surgery. He denounced those who practised surgery without intense practical training, because persons with half-knowledge who carried out these procedures could only cause harm to the patient.

Surgical Instruments

The various surgical operations and their success would depend not only on the dexterity of the surgeon but also on the surgical

ŚASTRAS

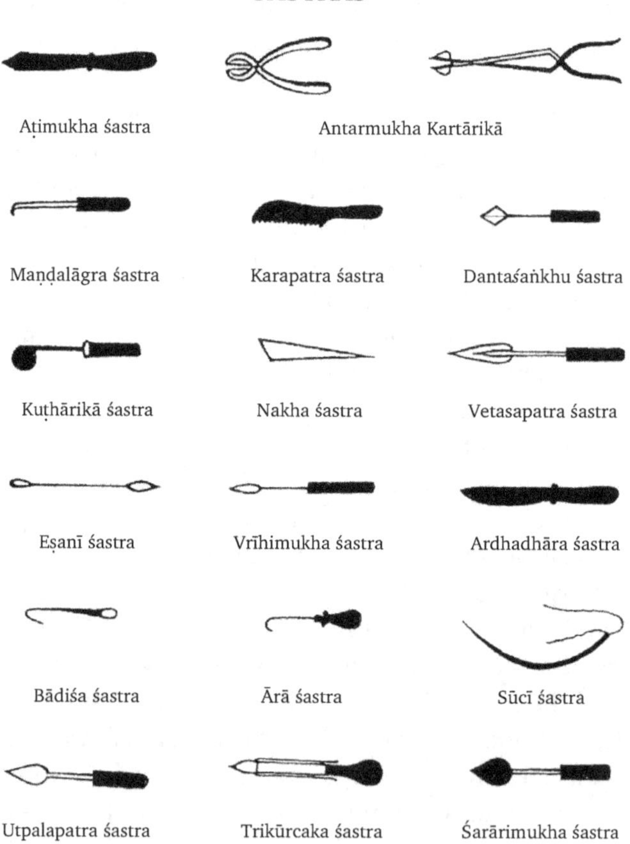

Some of the *Śastras* (Sharp Instruments) as described in the *Suśruta Saṃhitā* and by its commentators (Courtesy: Project of History of Indian Science Philosophy and Culture, Centre for Studies in Civilizations)

instruments and their efficacy. Suśruta has described twenty types of sharp instruments under what are called *śastras* and 101 types of blunt and accessory ones as *yantras*. The *śastras* comprise circular knives (*maṇḍalāgra*); bone-saw (*karapatra*); scalpels (*vṛddhipatra*); nail-parers (*nakha śastra*); finger knives (*mudrikā śastra*); lancets (*utpala patra*); single-edged or half-edged knives (*ardhadhāra*), axe-shaped knives (*kuṭhārikā*); trocars (*vrīhimukha*); awls (*ārā*); probers (*eṣaṇī śastra*); needles (*sūcī*); curved bistouries (*antarmukha*); another type of lancet (*atimukha*); tooth-scalers (*danta-śaṅku*); an instrument with three sharp bristles (*trikūrcaka*); drainage scissors (*śarārimukha*); and special puncturing contrivances.

Suśruta's concern for meticulous surgical performance is evidenced by the details he has provided even for holding the instruments in a particular manner. For example, he advocates that the scalpel and similar instruments should be held firmly at the junction of its handle and blade while performing incisions. If the instrument is to be used for scraping, the surgeon's hand should be raised at an angle. The trocar needs to be held in such a manner that its handle is under the surgeon's palm, with his thumb and index finger holding the instrument.

Suśruta's approach to surgery was humanitarian and innovative alike. If sharp instruments were not available or could not be used, for example, in the case of small children, he advocated that certain substitutes could be effectively used. Called the *anuśastras*, such contrivances included bamboo-bark, rock-crystal, glass pieces, thorns of particular plants, caustics, cauterizers and leeches. Blood-letting or removal of vitiated blood, which otherwise could give rise to several diseases, was an accomplished healing practice. Suśruta has prescribed five methods of removing vitiated blood including venipuncture, incision by instruments, and the use of leeches. He has also given extensive details of the para-surgical practice involving leeches of non-poisonous varieties and even identified their types, besides providing details of the needed pre- and post-operative procedures.

The blunt instruments (*yantras*), 101 in number, which would facilitate surgery have been grouped under five heads: (i) Cruciform instruments (*svastika*) have a fulcrum at the centre with cross-action blades intended to be capable of extracting foreign bodies from the surface or deeper tissues particularly

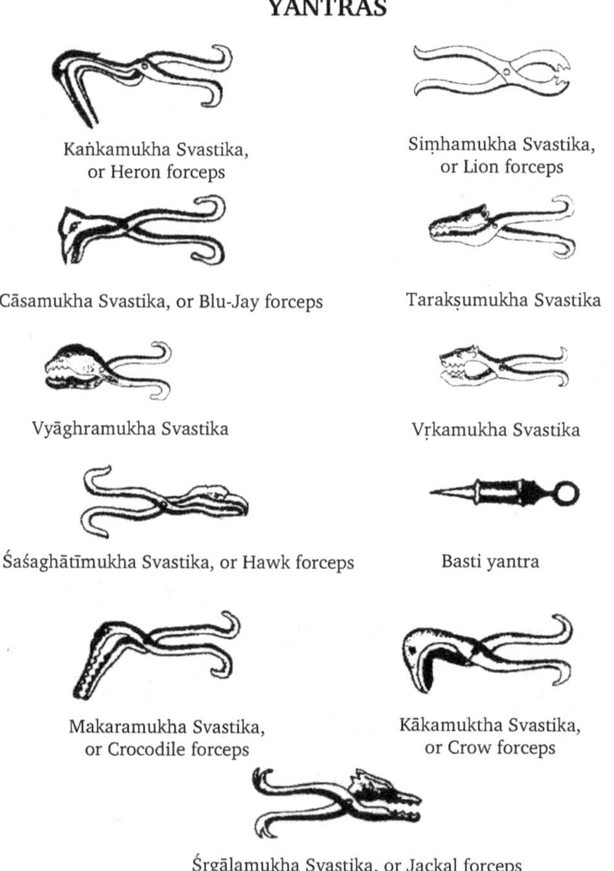

Some of the *Yantras* (diagrammatic) as described in the *Suśruta Saṃhitā*. (Courtesy: Project of History of Indian Science Philosophy and Culture, Centre for Studies in Civilizations)

from bone. They have been named after the mouths of some animals or birds, like *kaṅkamukha svastika* (heron forceps), *siṃhamukha svastika* (lion-mouth shaped forceps), *vyāgramukha* (tiger-mouth forceps), and *vṛkamukha* (jackal or leopard-mouth type of forceps). The dental forceps, bone forceps, and Ferguson's lion jaw forceps, which are used in modern surgery, fall under this category. (ii) Gripping instruments (*sandaṃśa*) are generally in the shape of pincers with or without handles, and used for the extraction of an embedded foreign body in soft tissues. Vāgbhaṭa added a few more contrivances of this type, like those with teeth as well as those with a catch or without it, which were used during dissection. (iii) Scoops and disc-shaped instruments (*tāla*) are utilized for removing extraneous bodies from ear and nose. (iv) Tubular instruments (*nāḍī*) are hollow inside and used for different purposes. Male and female catheters in modern surgery, for the evacuation of urinary bladder belong to this category. (v) Rod-shaped instruments (śalāka) are neither hollow inside nor sharp; these have different shapes, and Suśruta has provided details of eleven types of them while Vāgbhaṭa has added four more.

The aforesaid contrivances apart, both Suśruta and Vāgbhaṭa have described several accessory contrivances (*anuyantra* or *upayantra*) which include many types of bandages, cloth, thread, magnet, medicinal caustics and medicaments. The medicaments comprise pastes, paints, and plasters (*ālepa*). Special therapeutic measures have been recommended both for the pre-and post-operative stages. The *pañcakarma* purification of the body is normally adopted along with medicines that would mitigate the vitiated factors as a pre-operative measure. In the post-operative period, the treatment is one of increasing the strength and vitality of the patient through nutritious food and tonics.

Suśruta's account of surgery, which deals with about forty types of surgical and para-surgical operations, is not only comprehensive but also fascinating. His concept of preparing a patient for operation with *pañcakarma* (which is adopted even today in Ayurvedic hospitals) and other procedures like softening by anointing, sudation, internal administration of medicated oil or *ghee*, and a dietary regimen, were indeed remarkable and scientific; for, they could prevent the possible complications that could arise during or after the surgery. Surgical instruments were sterilized by dipping them in hot oil. The quality of sharpness of the twenty described instruments was ascertained by a process of cutting across a hair, that is, if the instrument could cut the hair, it was deemed to have the desired sharpness. The application of anaesthesia was not known then, but the administration of intoxicating beverages appeared to have been practised to mitigate the pain of surgery. The performance of surgical operations required, then as now, an extraordinary dexterity coupled with profound anatomical knowledge. The manner in which the art and science of surgery have been dealt with in the *Suśruta Saṃhitā*, conferred on it the distinction of being the earliest comprehensive text on surgery.

RASĀYANA AND VĀJIKARAṆA

Apart from internal medicine and surgery, two of the eight branches of Ayurveda, the other six branches, such as ENT, paediatrics, toxicology, geriatrics and rejuvenation, virility, and psychiatry, have also received adequate attention in the Ayurvedic classics. The word used for the body (human or animal) is *śarīra*, in which all kinds of physiological processes are continuously taking place, and which is subject to decay and eventual death. But, to live long in perpetual youth and to experience the joy of harmony in this very life have an irrestible appeal to every human being. Ayurveda as the science of life is not oblivious to this human inclination; accordingly, in its two branches—*rasāyana* (rejuvenation) and *vājīkaraṇa* (virility)—it has expounded the practicalities of attaining them.

Ayurvedic texts have described two types of *rasāyana*: (i) *kuṭi praveśikā*, which is to be undertaken in a specially built resort, a three-chambered concentric house; and (ii) *vātatāpikā* by which a person is exposed to sunlight and air. A sub-type of the first one is called *droṇī praveśikā rasāyana*, in which a person stays in a boat made of *palāśa* (*Butea frondosa*) and consumes only a diet based on milk. The texts have given details of the procedure to be adopted as well as several types of rejuvenating drugs to be administered along with the dietary regimen. All of these, if followed as prescribed, would eventually result in the rejuvenation of the body, preservation of exhilarating health, improvement in intelligence, and protection from diseases. But, it is enjoined that during and after the *rasāyana* therapy, the individual should be in a balanced state of mind and observe right conduct without greed, hatred or anger.

As for the stage in one's span of life for the administration of the *rasāyana* treatment, Caraka seems to be silent on this matter, while Suśruta and Vāghbhaṭa state that childhood and middle age, and even up to seventy years of age would be suitable, while the spring, pre-rainy and autumn seasons are recommended. Nevertheless, a person with mental instability, or who lives a sinful life and is a disbeliever in the *rasāyana* treatment should not undergo rejuvenation therapy. Even the physician who administers *rasāyana* should be of high calibre, free from ego and irritation; he should be truthful and full of compassion. Ayurveda thus considers the *rasāyana* treatment in all its details, but with certain restrictions regarding the individual to be treated and also the physician who treats him.

Similar is the case with *vājīkaraṇa* or virilification, which is advised to be undertaken after the *rasāyana* therapy. Several types of virilificatory drugs have been described in the Ayurvedic texts along with the manner of treatment. These comprise drugs that possess the efficacy of increasing sexual desire and performance, promoting spermato-genesis as well as improving the quality of the sperm and the ovum, leading to healthy progeny. Ayurveda has also identified several plants known to have health-promoting, rejuvenating, and vigorous qualities by which robust health could be preserved, and immunity against diseases fortified. But it has been emphasized in unequivocal terms that both *rasāyana* and *vājīkaraṇa* procedures would produce their desired effect in ample measure only in persons who have a sound mind and body as well as selfcontrol, which are essential for the realization of the fourfold goal of life culminating in *mokṣa*. This branch of Ayurveda has also addressed itself to the solution of the problem of both male and female infertility.

TRANSMISSION OF AYURVEDA

By virtue of its geographical position as a subcontinent of Asia, India in the past had played a key role in the transmission of scientific ideas and techniques. The endogenously developed Ayurveda had attracted the attention of the contemporaneous cultures of West and Central Asia, besides those of Southeast Asian regions. Medicine by and large knows no barriers of region or religion; for, the healing practices recognize the patient as a human being without his regional or religious label.

Greece and India: Among the ancient civilizations, Greece and India had developed medicinal ideas and practices in a methodical way, engendered by ideas of harmony and concordance between man and nature. However, it is rather difficult to trace their mutual influence, even though Alexander the Great and his men invaded India between 326-327 BCE, and some of the Greeks stayed back in India. Around 300 BCE, Megasthenes officiated as Greek ambassador at the court of Mauryan ruler Chandragupta. A Greek physician at the Persian court, who had come to India, recorded that Indian physicians were skilful in curing snakebites, while Alexander's army physicians were ignorant of such cures. Jean Filliozat, a well-known French Indologist, has adduced some evidence which shows that the Hippocratic treatise titled *On Breath*, deals more or less in the same way with its idea of *pneuma* as the Ayurvedic concept of *vāyu* or *prāṇa*. According to him, Plato's discussion on pathology in his *Timaeos* is almost an echo of the Ayurvedic *tridoṣas*. Indian medical knowledge must have seeped through the Parthian Empire which oversaw parts of North India and Greece along with the trade routes of this extensive region. It is significant to note that Megasthenes had described the diseases of elephants and their cures, which were borrowed from *Hastyāyurveda (*Ayurveda for elephants). The *materia medica* of Discorides (first century CE) an army surgeon under Nero, included several Ayurvedic materials like pepper and ginger.

Iran: There is some definitive evidence about the transmissions between India, Persia, and West Asia. Even before Iran was conquered by the Arabs, Burzuya, physician to Iran's King Chores I (sixth century CE), had come to India and carried back the *Pañcatantra* stories; but he is more known for his autobiography, which *inter alia* contained a summary of embryology as known to Indian physicians. Then, there was a famous school of medicine at Gundeshapur in Susiana (Iran) established in the fifth century CE by the Nestorians after they were expelled from the Byzantine Empire, and it was through this Christian sect that Ayurvedic practices were introduced to Iranian physicians. Significantly, the *Caraka Saṃhitā* mentions Kāṅkāyana, a physician from Bāhlika, in an assembly of sage-physicians gathered round Ātreya (the preceptor of Agniveśa, whose work, as noted earlier, constitutes *Caraka Saṃhitā*). Bāhlika was located in Bactria on the other side of the Hindu Kush mountains, which was part of the dominion of Chores I. Around the fourth or fifth century, as noted before, the now famous manuscript called the Bower Manuscript (discovered by Lieutenant H. Bower in the Kuchi region or modern Kucha in the eastern part of Central Asia) is a compendium that contains seven Sanskrit texts, three of which are medical.

Tibet and Southeast Asia: Ayurveda found a congenial home in Buddhist Tibet. The Ayurvedic concept of *tridoṣa* and the ideas on metabolism as well as the Ayurvedic *materia medica* influenced Tibetan medicine a great deal. In the South Asian region, the impact of Ayurveda was evident. One of the inscriptions (ninth century CE) of Cambodia (ancient Kambūja) refers to the medical treatise of Suśruta. The great king of Kambūja, Jayavarman VII (twelfth century) established over a hundred hospitals (called *Ārogyaśālā*) in his kingdom, as authenticated by an inscription (in Sanskrit verses) which also gives details about diet and medicinal herbs supplied to these hospitals. Burma (now Myanmar), too, adopted the Ayurvedic system to a large extent, and it is worthy of note that several Burmese medical terms were actually derived from Sanskrit. Ceylon (now Sri Laṅka) was, and continues to be, a home of Ayurveda, which was introduced there by Buddhist missionaries as early as the first century, although the medical literature in the local language appeared much later.

West Asia: The transmission of Ayurvedic knowledge and practices was more pronounced in West Asia than elsewhere. The scientific centre at the Iranian Gundeshapur along with the medical school was shifted to Baghdad after the Islamic conquest. Between the eighth and twelfth centuries, Baghdad played a pivotal role in the assimilation and transmission of scientific knowledge both from the orient and from the Graeco-Roman world. Arabic emerged as the versatile language of science at the Abbasid Caliphate in Baghdad and, as noted earlier, the Caliphs invited Hindu scholars to Baghdad and promoted translations of astronomical-cum-mathematical texts from Sanskrit to Arabic. The renowned Abbasid Caliph, Harun al-Rashid, patronized learning and actively encouraged the rendering of Ayurvedic classics into Arabic. Even an emigrant Indian scholar well-versed in Ayurveda became the physician-in-chief at the Caliphate's state dispensary. He is also stated to have cured the royal patient of a chronic stomach trouble using Ayurvedic drugs. Besides, he translated the *Suśruta Saṃhitā* into Arabic under the title *Kitāb-Samural-hind-i-susrud.* Ali Ibn Zain, the noted scholar in Arabic and medicine, rendered the *Caraka Saṃhitā* into Arabic as the treatise of '*Sarag*'. Subsequently, the *Nidāna* (on etiology) of Mādhava, and the *Aṣṭāṅgahṛdaya* of Vāgbhaṭa were also rendered into Arabic. Through these translations, Ayurveda became well known in the Arabic world, and spread to the other countries which were under the dominion of the Arabs. Later, in the ninth century, Al-Rāzi, the renowned medical man of that time, wrote a compendium titled *Kitāb-al-Hawi* in which he incorporated several Indian medical practices including some aspects of Ayurvedic *materia medica.* His book was also translated into Latin in the thirteenth century and became the standard medical text in medieval Europe.

Earlier in the tenth century, the Central Asian Islamic physician, Abu-Manṣūr Muvaffak Ali Khiravi, produced a book under the title *Kitāb-al-abniya an-Haqā'iq-al-Adviya* (The Book of Main Properties of Drugs) in which he also adopted the principles of Ayurveda. More importantly, he pointed out that Indians used many drugs which were more efficient, and exclaimed that several of them could not be seen in any parts of the world, except India! Other works too arose from Central Asian Islamic medical men, like *Kitāb-Hidayat al-Mutallimin fil-Ṭib* (The Book of Guidance for Learning Medicine) by Abubakar Rabi ibn Ahmad Ahavaini al-Bukhāri (961-981) and *Donishnamoi Maisari* (The Book of Knowledge of Maisara), which was written in the form of poetry, by Khorasanian Hakim Maisara. Ibn Sīna or Avicenna (980-1037), as he is known in the West, was an acknowledged authority on Unāni medicine (see next chapter) in India. In his scholarly book, *Al-Qānūn* (The Canon of Medicine), he has described about thirty Indian drugs and also referred to Caraka. The chapter on 'Diet' in this book contains Indian dietary practices along with their medicinal implications. In the eleventh century, the encyclopaedic scholar, al-Bīrūnī (973-1048) sojourned in India for more than a decade. Apart from his primary interest in Hindu astronomy, he wrote a book, *Kitāb-al-Saindana-phiṭ-ib,* on pharmacology, in which he described more than 4,000 names of plants, animals and mineral drugs, including 350 Indian ones.

PRESENT STATUS OF AYURVEDA

Despite its strong and viable philosophical foundation with an appreciable epistemological matrix, Ayurveda showed signs of decline over the years, especially under colonial rule. The patronage of Ayurveda was generally in the hands of jāgirdars (landowners) and chieftains in different parts of India. Indigenous medical practice continued as a preceptor-disciple tradition even in the colonial dispensation, and there was hardly any governmental patronage. Yet Ayurveda and folk medicine played a major role in bringing succour to the afflicted, particularly in the vast rural areas. Even in urban and semi-urban areas, during British rule, hospitals dispensing Western medicine were few and far between. There also ensued a divide between those who preferred Western medicine and those who were protagonists of Ayurveda and other forms of traditional medicine. Ayurveda, however, did not lose its sheen and relevance; but it needed revitalization. As a first step in this direction in 1907, an All-India Ayurvedic Congress came into being as a result of the efforts of some leaders of the freedom movement and Ayurvedic experts. This Congress initiated a teaching wing, namely, the All India Ayurveda Vidyapeetha, which formulated certain courses for teaching Ayurveda on sound lines. The Indian National Congress, at its sessions in 1920 and 1938, adopted resolutions for broad-basing and strengthening Ayurvedic education and practices.

It was the great nationalist, educationist, and visionary, Pandit Madan Mohan Malaviya, the founder of Banaras Hindu University (1919), who established an Ayurvedic College in the campus of the university. Two years later, the enlightened Unāni physician and nationalist, Hakim Ajmal Khan, set up the Ayurvedic and Tibbia College in Delhi. A year later (1922), an Ayurvedic School came into being at Gurukul Kangri (near Hardwar) as a result of the efforts of Svami Shraddhanadji. In the next two decades, these initiatives led to the establishments of several Ayurvedic institutions in different parts of India, for teaching and practical training along with attached hospitals, pharmacies, and herbal gardens, even before Independence.

On the eve of Independence, two Committees were set up: (i) the Chopra Committee (1946) headed by Ramanath Chopra; and (ii) the Bhore Committee (1947). The former recommended that (i) Western and indigenous systems of medicines should be harmonized; (ii) the syllabi of integrated courses should be properly worked out, and the curriculum should be uniform for the whole country; (iii) chairs of the history of Indian medicine be established in modern medical colleges; (iv) experts of both systems (modern and traditional) should undertake research in Indian medicine; and (v) drug standardization needs to be accomplished, for which purpose a proper survey and the cultivation of identified herbs should be undertaken, and an Ayurvedic pharmacopoeia compiled. The Bhore Committee was more or less lukewarm towards the promotion of Indian medicine, although it recommended that modern medical institutions should study the indigenous system with a view to investigating their potentiality and the extent to which they can contribute to the advancement of medical knowledge and practice in national interest.

The Central Institute for Research in Indigenous Systems of Medicine was set up in 1952 at Jamnagar. In 1956, a Postgraduate Training Centre in Ayurveda was also set up there. In 1958, the Ministry of Health of the Government of India set up a Committee to evaluate the then existing training and research facilities for Ayurveda. This Committee recommended, among others, the establishment of a Central Council for Indian

Medicine, integration of the old and new medical practices, standardization of the Ayurvedic pharmacopoeia as well as the manufacture of Ayurvedic drugs, and the treatment of Ayurvedic practitioners on par with modern medical practitioners in pay and other respects. The Banaras Hindu University established in 1963 a postgraduate Institute for Teaching and Research in Ayurveda, and emerged as an important nucleus for Ayurvedic research. In 1967, the now well-known Gujarat Ayurvedic University came up, and in 1976, the National Institute of Ayurveda started working at Jaipur. Several states began to open Ayurvedic colleges with hospitals attached to them. In 1988, the Rashtreeya Ayurvedic Vidyapeeth came into being in Delhi. The Government of India set up a separate Department of Indian Systems of Medicine (Ayurveda, Unāni and Siddha) and Homeopathy (AYUSH) in 1995, in the Ministry of Health and Family Welfare, which has been actively promoting these systems, besides establishing directorates of Indian systems of medicine.

Realizing the importance of research in Indian systems of medicine by the application of modern scientific methods, the Government of India has set up separate research councils for clinical investigations, surveys, and identification of medicinal plants and related areas. The need for the standardization of indigenous drugs was recognized by the Pharmacopoeia Committee of Ayurveda, which published the first volume, under the title *Ayurvedic Formulary,* consisting of about 650 formulations.

Ayurveda has been recognized by the World Health Organization (WHO) as one of the alternate systems of medicine. For, the Ayurvedic concept of *Svāsthya* eminently fits into the WHO's definition of health: 'Health is a state of physical, mental, social and spiritual well being.' Over the decades, Ayurveda has found its followers in several countries, both in the West and in the East. Maharṣi Mahesh Yogi, whose name is synonymous with Transcendental Meditation, popularized Ayurveda, particularly in USA and Europe. Deepak Chopra, Vasant Lad and others have been well known as effective emissaries of Ayurveda, and their books are among the best selling in the world, pioneering a movement known as the New Age Ayurveda.

Over the past five decades, institutions for Ayurvedic studies and practices have sprung up in several countries: USA, Canada, Germany, the Netherlands, UK, Switzerland, Italy, Brazil, Australia, Japan, South and Southeast Asian countries (like Sri Lanka, Bhutan, Nepal, Burma, Thailand, and Indonesia) and Mauritius. One of the enduring foundational ideas of Ayurveda is its approach to mind-body balance or harmony, which has attracted the attention of the whole world. From its beginnings over 2,500 years ago, Ayurveda has traversed a long way and passed through several vicissitudes. But its basic ideas and practices fortified by its holistic approach to man *in* nature, or man *in harmony* with nature are its forte; its unitary conception of mind-body-food-drug is its triumph; and its philosophy of total health has been its inner strength.

7

OTHER MEDICAL SYSTEMS: SIDDHA AND UNĀNI

SIDDHA MEDICINE

The nomenclature 'Siddha medicine', owes its origin to the medical thoughts and practices of the Siddhas. The term *siddha* means an 'accomplished' or 'perfected' human being who has mastered the eight *siddhis* (accomplishments), namely, capacity to become as tiny as an atom; or very large; extremely light and able to float in air; capable of expanding; possessing the power to obtain any desired object, and the irresistible *will* to perform; lordship over everything and the power of subduing. These are superhuman powers which could be obtained, according to the tāntric tradition, by intense meditation and yogic practices. Yoga, therefore, is an important component of the life of a Siddha, and a Siddha physician is expected to practise yoga daily to ensure that his expertise as a medical person meets with the desired success.

It has been noted already that the theoretical basis for Ayurveda has been provided by the three Indian religio-philosophical systems: the *Sāṃkhya*, the *Vaiśeṣika* and the *Nyāya*. The doctrine of five elements has by and large structured the various Ayurvedic concepts and practices. On the other hand, the doctrine of five elements did not have any role in the conceptual scheme of Siddha medicine in its formative stages. Ayurveda accepts the degeneration and eventual death of the human or animal body, but the Siddha medical thought-structure firmly believes that the human body can be perfected by certain practices to attain the state of 'deathlessness'. The Siddha medical system has been especially popular in Tamil Nadu and among Tamils who live elsewhere. The Tamilian tradition affirms that the Siddhas who have attained immortality are still alive with their superhuman powers, and their blessings are being sought by the Siddha physicians in their profession.

According to one tradition, there are eighteen Siddhas: Nandīsar, Agastyar, Tirumūlar, Bhogar, Punnākīsar, Pūnaikaṇṇer, Rāmadevar, Idaikādar, Pulakaisar, Karuvūrār, Konganavar, Aluganṇi, Chettai munī, Kalaṅgi Nāthar, Theraier, Pāmbāṭisiddhar, Pulastyar and Sundarāndar. Some contemporary Siddha practitioners speak of their own list with minor variations in names. However, Agastyar, Nandīsar, Bhogar, Rāmadevar and one or two others have been held in high esteem for their contributions to Siddha medicine, and several texts have been extant under their names. Like Ayurveda, a divine origin has also been attributed to the Siddha system; the origin of Ayurveda is believed to be the Creator (Brahmā), while the origin of the Siddha system is traced to Śiva (who is supposed to be the conqueror of death) since the main aim of the Siddha system is the attainment of immortality. A further belief is that Śiva imparted this knowledge to Nandi (the divine bull associated with Śiva) who, in turn, taught it to the sage Agastyar. No wonder, then, that a large number of texts in Tamil on Siddha medicine are in the name of Agastyar or Agattyar as the Tamils generally call him. The dates of the Siddhas are uncertain. Though the Tamil tradition speaks about their remote antiquity, it would appear that Agastyar, Tirumūlar, and Bhogar as well as Nandīsar might have lived in the area now comprising Tamil Nadu around the fifth-sixth century CE, while the other Siddhas were of a later date, extending even to the fifteenth and sixteenth centuries. Of the four revered Siddhas, Tamils admit that Bhogar was probably Chinese—and adept in alchemy and medicinal chemistry—who came to Tamil Nadu and settled there.

Alchemical Origin

A proverb in Tamil saying that 'a physician is the son of an alchemist' is reflected in Siddha medicine; for, its roots can be found in the alchemical two-in-one concept of the 'elixir of life' for attaining immortality, and the transmutation of base metals into gold (the everlasting shining noble metal). It is generally accepted by historians of science that China was the original home of alchemy (see chapter 5). It centred round, to begin with, cinnabar or mercuric sulphide, of which mercury and sulphur were representative of the female-male or *yin-yang* principles of Taoism. Their union, as found in cinnabar, was believed to have the power of conferring immortality. The so-called elixir of gold and its enchanting lustre were also associated with this power. The alchemical rationale, as noted earlier, was that if an inferior metal like lead, tin or copper could be transmuted to the everlasting gold, the perishable human body, too, could be transformed into an imperishable or immortal body. In India, there was an additional dimension to this view—that if the soul is considered eternal, it should reside in a body that should also be immortal. Such thoughts as these, coupled with the ideas of eight superhuman powers, lay behind the origin of Siddha medicine.

In the Chinese repertoire, cinnabar, sulphur and the two sulphides of arsenic were extolled as 'semi-magical substances'. Cinnabar elixirs were sought after for attaining longevity. But to attain it, certain pathways had to be followed, like meditation, controlled respiration, a dietary regimen, heliotherapy, and even some sexual techniques. Through them a devout Taoist (a follower of the Chinese mystic exponent Lao Tze) was expected to elevate himself to the status of an immortal, either Heavenly (*tien sien*) or Earthly (*tih sien*). The methods of a Taoist were more or less similar to those of tāntricism in India. Thus it was in the tāntric milieu in the early centuries of the Christian Era that the seed ideas of Chinese alchemy found support, and they went on to constitute a nucleus for the origin of the Siddha medicine. One of the then celebrated Siddhas, Tirumūlar, in his seminal text *Tirumantiram* (3,000 stanzas), has included in his definition of medicine a statement: '(medicine is that) which ensures a preventive against mortality.' Ayurveda accepts the decay and mortality of the physical body, although it believes, based on the *Sāṃkhya* concept of the subtle body (bereft of the gross five elements), that it would pass on from one birth to another as a continuum. But this is not in any sense a concept of the 'deathlessness' of the body even in its subtle form. The Siddha system, however, envisages and pursues the attainment of the 'deathless state' of the physical body—a vision wholly alien to Ayurveda.

Though the initial source of inspiration could be traced to certain ideas of Chinese Taoism, Siddha medicine really grew in the Indian context. A yogi or a person who is devotedly immersed in intense meditation has a respectable position in Indian society—and Siddhas were highly revered yogis. Though they were nonconformists in their initial stages, they subscribed in a way to the then prevalent Śaiva and Śākta religio-philosophy. Mercury and Sulphur were regarded by them as the divine male and female principles. Their combination—mercuric sulphide—was called *liṅgam* (an appellation for Śiva) by them, in which the (phallic) form of Śiva is worshipped. Such esoteric ideas notwithstanding, Siddhas were skilled experimentalists and developed several medicinal formulations involving mercury, sulphur, arsenic substances, minerals and salts. These involved elaborate and meticulous processing with a wide variety of medicinal plants and also the use of a number of functional apparatus devised by them, leading to what is now called Siddha medicine.

The Human Body

The Siddha concept of the human body is basically a tāntric one in terms of tāntric *cakras* or plexes, evidenced by the tāntric terminology maintained in the Siddha medical parlance. Regarded as psycho-physical nerve centres, the *cakras* are to be stimulated or roused by yogic practice so that the latent energy in the body and mind would become manifest—in much the same way as *Kuṇḍalinī* is roused in tāntric practice to ascend the six *cakras* (*mūlādhāra, svādhiṣṭhāna, maṇipūra, anāhata, viśuddhi* and *ājñā*). Siddha practitioners believe that *Kuṇḍalinī*, the potential energy, would uncoil and reach the blissful state of a thousand-petalled lotus! At this stage, Siddhas believe that a divine union of Śiva and Śakti takes place, and the so-called '*nitai*' (nectar) drops into the bloodstream. This nectar leads to the 'deathlessness' of a Siddha who would thus attain the state of immortality. Central to all of these is intense yogic practice with disciplined and regulated breathing. Yūgimunivar, one of the Siddhas, says that 21,600 deep breaths a day (works out to one respiration during four seconds) in a disciplined manner, would be the secret of remaining young forever!

The Siddha system's unquestionable faith in the tāntric human body has excluded any description of human anatomy based on observation; the only mention of it in the Siddha texts is about 72,000 veins and nerves, seven vital nerve centres, ten vital airs (*prāṇa*) and ten vital pulses (*nāḍī*). Of the vital pulses, three—*iḍā, piṅgaḷai* and *suzhumnai* (in Sanskrit, *suṣumnā*)—belong to tāntric parlance. The number of respirations, according to Siddha tradition, for each of the seven vital centres is: *mūlādhāra* (6,000); *svādhiṣṭhāna* (6,000); *maṇipūragam* (600); *anāhatam* (6,000); *viśuddhi* (1,000); *ājña* (1,000); and the blissful last state, *ājñā* (1,000), amounting to 21,600 respirations a day. Nevertheless at a later stage, the Siddha system also accepted the concept of five elements and related them to these centres as follows: *nela* (*pṛthivī*) is the base for *mūlādhāra*; *nīr* (*ap*)—*maṇipūraka*; *thi* (*tejas*)—*anāhata*; *vali* (*vāyu*)—*viśuddhi*; *viśumbu* (*ākāśa*)—*ājñā* (the words in brackets are given in Sanskrit as equivalents). Like Ayurveda or influenced by Ayurveda, the Siddha system describes five physiological locations (*āśayas*): the stomach (*āmāśaya*), small intestines (*pakvāśaya*), kidney (*jalāśaya*), large intestines (*malāśaya*) and genitals (*śukrāśaya*). However, an explanation of the functions of these locations in terms of digestive and metabolic activities is but scanty. It is significant that Siddha terminology in its several expositions comprises mostly Sanskrit words.

MAIN CHARACTERISTICS OF SIDDHA MEDICINE

Diagnostic Methods

The success of any medical system is dependent on its diagnostic principles and methods as well as its therapeutic procedures. The Siddha system in recent times has adopted the Ayurvedic *tridoṣa* concept for diagnosing a disease as being due to *vāta*, *pitta* and *kapha* or a mixture of any two or all of them. But, the Ayurvedic *triguṇa* (*sattva*, *rajas* and *tamas*) and the *doṣa-prakṛti* or *manas-prakṛti,* that is, the disequilibrium of the body and mind, do not seem to have been adopted to a significant extent by the practitioners of the Siddha system; likewise, the Ayurvedic theory of self-cure or spontaneous healing and also that of fortifying residual health in order to activate the innate immunity of the body. Moreover, the details of diagnostic methods of Ayurveda like the five, eight and ten points of examination, have not been delineated in the Siddha medical texts. However, one finds in these texts extensive details of pulse and urine examinations as important diagnostic tools, besides some associated astrological considerations.

Pulse Examination: The word used for pulse is *nāḍī* (which is a Sanskrit term), meaning the beating or throbbing of the radiant artery felt by three fingers (index, middle and ring fingers) suitably positioned by an experienced Siddha physician. The rate and character of the pulse of a patient are expected to indicate a person's condition of disease or health. They are related, as in Ayurveda, to the *tridoṣas*, although the original texts on Siddha medicine do not throw light on the *tridhātus-tridoṣas*. According to Siddha texts, if the pulse movement as felt by the index finger is similar to that of a fowl, the disease is identified as being due to *vāta*; if it is akin to that of a frog (felt by the middle finger), the disease is recognized as being due to *pitta*; on the other hand, if it resembles the movement of a snake (felt by the ring finger), the ailment is understood as being due to *kapha*.

There are, however, differences in such identifications among the Siddha practitioners. Theraiyer, a later Siddha, suggests that the movement of the pulse in the vitiated states of *vāta, pitta* and *kapha* would be that of a swan or peacock, hen or ant, and fly or vulture, respectively. In any case, it is the acumen, intuition and experience of a Siddha practitioner (similar to those of an Ayuvedic *vaidya*) that would be of immense value for an effective pulse-reading. It is also enjoined that a Siddha practitioner should possess foresight and spiritual experience gained by constant yogic practices which, it is believed, would enable him to examine the pulse accurately, and thus help him in prognosis. Tirumūlar, who is recognized as a renowned authority on Siddha medicine, states that the interrelatedness and actions of the ten vital airs, the six *cakras*, the three nerve-channels (*iḍā, piṅgaḷā* and *suṣumnā*), and three regions of the body which, according to him, represent the sun, the moon, and Agni (the fire god), are also to be visualized by a Siddha practitioner through his spiritual insight, while examining the pulsations of a patient. Thus a Siddha physician is to be a yogi, tāntric and astrologer rolled into one. For, he has to take into consideration—apart from his yogic insights and tāntric esoteric mindset—certain astrological aspects like the time of arrival of the patient at the physician's house as well as the planetary configurations at that time, for determining the intensity of the disease vis-à-vis the planetary influences. This would also enable him to come to a conclusion as to whether the disease is curable or incurable, or if the cure would take a long time.

Urine Examination: The curability or otherwise of a disease is judged by a Siddha physician after an examination of the patient's urine, especially of its colour, and this is an important Siddha diagnostic method. If the urine of a patient is honey-coloured or yellow, the disease is diagnosed as a curable one. On the other hand, if it is ruby-coloured or milky white, the disease is believed to be incurable. The general method of urine examination consists of placing a drop of sesamum oil in a sample of the patient's urine from a height of 8-9 cm and observing how it spreads. If it spreads in a longitudinal fashion, the disease is said to be due to *vāta*; if the oil drop assumes a circular form, the ailment is understood as being due to *pitta;* if there is no spreading, it would mean that the disease is due to *kapha;* and if it elongates and spreads in coils, the general conclusion is that the affliction is due to both *vāta* and *pitta*. If, on the other hand, the oil drop sinks, the conclusion is that the patient is suffering from *sannipāta* (a dangerous fever). In addition to the pulse and urine examinations, Siddha practitioners have been employing the method of questioning a patient, ascertaining the history of his disease, examining his tongue, eyes and faecal matter just as an Ayurvedic physician does.

Therapeutic Principles

Adjuvant: In the Siddha curative procedures, the adjuvant and the dietary regimen are regarded as crucially important. The words being used for these procedures are the Sanskrit ones, *anupānam* and *pathyam*, which are also given due recognition in Ayurvedic therapy. A Siddha physician claims that he can cure several diseases using a single drug, by changing its adjuvant with an appropriate dietary regimen which would play an effective role in the curative process. For example, the *Agastyar Vaidya Kāṇḍam*, a text on drug administration, speaks of a medicine called *kauri cintāmani cendūram* (consisting of specially processed and purified mercury, sulphur and borax) for curing a variety of diseases by changing the adjuvants: if the adjuvant is the powder of three myrobalans, this drug could be efficacious against chronic fevers, rheumatism, hepatitis and bronchitis. With cloves, nutmeg and mace as adjuvants, it can cure stomach ailments like diarrhoea, while with some other adjuvants, the drug can cure even piles and fistula, all with a strict dietary regimen as the concurrent therapy. The Siddha system thinks of the adjuvant as a supporting drug, resulting in synergic action. An adjuvant, according to this system, is to be changed in tune with seasonal variations. Such an approach to the pharmaco-dynamics of adjuvants is also to be found in Ayurvedic therapy.

There is a wide range of adjuvants in Siddha therapy: decoctions, syrups, *ghee*, honey, betel-leaf juice, hot water, cow's milk and certain herbal extracts. However, one of the Siddhas, Theraier, warns that a Siddha physician needs to be extraordinarily careful in selecting an adjuvant in terms of its compatibility with the drug administered. There is a popular saying in Tamil, which means that 'diet itself is medicine and medicine itself is diet'. The Siddha texts are more or less similar to the Ayurvedic ones in thinking of the diet (food) and drug as complementary to each other. Dietary restrictions are specific for certain diseases like those of the eyes, jaundice and diarrhoea.

Siddha Drugs

The main difference between Ayurvedic and Siddha medicines lies in the latter's pharmacopoeia of carefully processed mineral drugs using mercury, sulphur, arsenic sulphides, metals (gold, silver, copper, lead and iron) and their compounds, borax, sal ammoniac and other salts, mica, alkalis, gems and certain animal products. Ayurvedic medicines are by and large herbal. However, in the Siddha preparations, various herbs are also used for extensive processing of minerals, metals and their compounds. Nearly a hundred of such herbs have been identified botanically as a result of the efforts of the Central Council of Research in Siddha Medicine. According to Siddha practitioners, one has to search for these herbs with great care in the hills of Podugai (Tirunelveli, Tamil Nadu), the Koli Hills (Salem) and Seven Hills (Tīrupati). The efficacy of mineral drugs, a Siddha physician believes, is dependent not only upon the type and quality of the plants or herbs used but also on the processes adopted. The processes comprise, among others, distillation, extraction and prolonged incineration or heat treatment carried out in specially designed apparatus and contrivances. These are more or less similar to those employed by the *rasavādins*. In any case, the mineral drugs are to be administered in very small quantities along with the compatible adjuvants.

The Ministry of Health and Family Welfare of the Government of India, through its Research Council for Siddha Medicine, has published a *Siddha Pharmacopoeia,* which contains the details of about 240 drug compositions. Alongside, the Indian Medical Practitioners Cooperative Pharmacy and Stores (Chennai), has published a *Formulary of Siddha Medicines*, relating to approximately 220 preparations. These endeavours are in the nature of a standardization of the Siddha drugs. Nevertheless, traditional Siddha practitioners often prefer to follow their own methods, which are imparted to them by their preceptors.

Siddha drugs can be broadly categorized as follows (Tamil words in brackets): (i) Certain plant products and mineral preparations in fine powder or tablet form (*cūrṇam*); (ii) Incinerated fine powders of metallic compounds and certain salts (*centhūram*); (iii) Calcined powders of metallic derivatives, calcium compounds, saltpetre with or without marine shells (*cunnam*); (iv) preparations containing mercurial as well as some metallic compounds (*kalaṅgu*); (v) Rejuvenating elixirs (*karpam*); (vi) Black-coloured mineral compositions (*karuppu*) generally prepared from mercuric sulphide and other ingredients; (viii) Pills (*māthrai, kuligai*) containing specific plant products with or without some minerals or mercuric compounds; (viii) Linctuses (*lehyam*); (ix) Semi-solid or liquid forms (*melagu* or *kulambu*) prepared from some plant products, minerals with or without processed mercury or sulphur compositions; (x) Medicated syrups (*mānappāgu*); (xi) Fine powders (*parpam*) obtained by prolonged heating of compositions of chosen plant products, shells and salts of mercury with chosen metallic compounds; (xii) Sublimated products (*pātaṅgam*); (xiii) Medicated oils (*tailam*); and (xiv) Distilled compounds (*tirāvakam* or *drāvakam*). It may be noted that some Tamil words have Sanskrit nuances. The word *rasa* is used for mercury as in the *Rasaśāstra* texts in Sanskrit; likewise, *gandhaka* for sulphur. The nomenclatures of other minerals and metals are also akin to their names in Sanskrit.

Muppu

An important difference between Ayurveda and Siddha, however, lies in the concept and practical application of what the Siddha physicians call *muppu* (literally, the union of three salts). This word does not find even a mention in any of the Ayurvedic texts, although certain salts are dealt with by them. *Muppu* is regarded as *prima materia* and is extolled as the 'universal salt', with extraordinary properties. It is believed to be capable of (i) transmuting base metals into gold (*vāda muppu*); (ii) rejuvenating cellular organisms, besides enhancing the potentiality of medicines, when added to them (*vaidya muppu*); (iii) augmenting yogic concentration and increasing mental powers (*yoga muppu*); and (iv) leading to an enlightened state (*jñāna muppu*). *Muppu* is believed to play a key role in the revitalization of the body (*Kāya-kalpa*) and, for this reason, its preparation is an elaborate one. However, the method is kept a closely guarded secret, each Siddha practitioner being unaware of the process adopted by the others for the preparation of *muppu*. According to Siddha tradition, the first prayer of a Siddha is to invoke the divine (*Śiva-Śakti*) blessings for the bestowal of the knowledge of *muppu*. More than forty processes for *muppu* have been described in Siddha medical literature, but the descriptions are enigmatic and mysterious.

There is, however, a general consensus among Siddha practitioners that *muppu* consists of three salts: (i) *pūnīru* (a natural exudate from the soil near sea, that yields Fuller's earth); (ii) *kalluppu* (either rock salt or sea salt); and (iii) *veḍiyuppu* (naturally available potassium nitrate). Their exact processing, individually or together, and the duration of such processing are not available in the texts; nor is a Siddha physician inclined to reveal them fully. Nevertheless, it would appear that the three salts are first purified using small quantities of water from rivers which are considered sacred, along with juices of certain plants, and subjected to prolonged heating. Small quantities of processed mercury (or its chlorides), sulphur, mica, arsenic sulphides, and a few other minerals are also subjected to calcination, separately, and then powdered. This powdery composition is added to the aforesaid three salts to obtain *muppu*, which is thus a conglomerate of several ingredients, but with the three salts present in a greater proportion. *Muppu* in combination with specially processed gold powder or thin gold foils is hailed as the 'elixir of life', with extraordinary potential for attaining longevity. Along with the *Kāya-kalpa* process and yogic practices, it is believed that one can reach what is called the state of 'deathlessness'.

An important constituent of *muppu* is *pūnīr*, which is a mutation of the Sanskrit word, *bhūnīr* (literally, watery exudation from the earth). It is collected by Siddha physicians in certain places of Tamil Nadu, particularly on four full-moon nights, towards early morning when the moon is low on the western sky, in the months of January-April (lunar months: *Puṣya, Māgha, Phālguṇa* and *Caitra*). Since its role is seminal in the preparation of *muppu,* the collection process is accompanied by rituals or worship. The belief is that through such a devoted action *muppu* would be endowed with divine power. *Pūnīr* itself is supposed to be in the nature of male-female union, which, according to traditional explanation, is formed by the action of the rays of the moon (male) on the earth (female) at certain places, and the earth is said to give 'birth' to it on full-moon nights. There are over thirty places in South India where *pūnīr* of good quality can be collected. Kālahasti, Thirualankāḍu (near Chennai), Shivaganga, Pashumalai, Vada Madurai, Dindigal, Thillaivanam, Tenkasi and Kanyākumārī are regarded by Siddha practitioners as the places where *pūnīr* of the best quality can be obtained.

Classification

The Tamil texts on Siddha medicine present a kind of classification of substances, which are to be processed and used for the preparation of drugs. Though this is somewhat akin to that of the Sanskritic *Rasāśastra,* it is not as detailed as in the *Rasaśāstra* texts. The six classifications are: (i) Salts (*uppu*) that are water-soluble inorganic compounds which, when heated, will give out water of crystallization (if any), or decompose when strongly heated; (ii) Toxic substances (*pāṣāṇam*) which are not water-soluble, and do not decompose on ignition; (iii) Auxiliary substances (*uparasam*) which are inorganic, such as metallic compounds, some minerals and silicates; (iv) Metals and alloys (*loham*); (v) Mercury (*rasam*) and its compounds, which sublimate on heating; and (vi) Sulphur (*gandhakam*), which is insoluble in water, but burns when ignited.

Mercury and four types of its compounds are used in Siddha medicine: (i) processed mercury (*rasam*); (ii) red oxide of mercury (*rasacendūram*); (iii) red sulphide of mercury (*liṅgam*); (iv) mercurous chloride (*pūram*); and (v) corrosive sublimate or mercuric chloride (*vīram*). The process of 'binding' mercury and sulphur to obtain mercuric sulphide is called *liṅgakaṭṭu* and is considered to be very important for curing certain diseases. The Siddha texts have described as many as twenty-five kinds of salts, 120 auxiliary substances, sixty-four toxic ones, and a large number of medicinal plants. In addition, some medicinal preparations involving animal products, like cow's urine, dung, cowrie shells, peacock feathers, conch shell and musk have been described. Nine types of gems or precious stones are also used in some nourishing preparations. They are: *zircon* (*gomedhikaṃ*), sapphire *(neelamaṇi)*, coral (*pavalaṃ*), topaz (*puṣparāgaṃ*), emerald (*marakataṃ*), ruby (*māṇikyam*), pearl (*muttu*), lapis lazuli (*vaiḍūryaṃ*) and diamond (*vairaṃ* or *vajram*). The names of various substances used in Siddha medicine owe their origin to Sanskrit equivalents.

Another broad classification that is peculiar to the Siddha system is that certain inorganic drugs have been divided into *nāda* and *bindu*, which have tāntric nuances. These are believed to have a natural affinity to each other, and their combination is presumed to have extraordinary curative potential and to be more powerful than individual drugs. Some substances have also been classified as enemy (*śatru*) and friendly (*mitra*), perhaps because of their mutual incompatibility or compatibility.

In the Siddha iatrochemical or medicinal dispensation, mercury, sulphur and salts play an important role. They also appear to stand conceptually for liquidity, volatility and fixity respectively, governed by their natural characteristics. It is interesting to note the growth of iatrochemistry (medicinal) in the West, led by Paracelsus (1490-1541)—the German-Swiss physician and alchemist who was a crusader amid the craze for esoteric alchemy that had gripped Europe then. He expounded that the alchemical twins, mercury and sulphur, should be viewed in a different light, including salt as the third substance. Paracelsus advocated that the human body should be recognized as a chemical system comprising mercury, sulphur and salt—what he called the *tria prima*. He exhorted that alchemy should transform itself into medicinal chemistry, and set aside the four-humoural theory of the Greeks, of renowned Greek physician Galen (131-201) and Ibn Sīna or Avicenna (980-1037), the established authorities of Graeco-Roman and Graeco-Arabic medicine respectively. Paracelsus was so antagonistic to these medical systems that he burnt the books of both Galen and Avicenna in public. The ideas and demonstrations of Paracelsus led to the acceptance of the view that the human body is a chemical system which, when afflicted with diseases, can be cured by using mineral or chemical substances as medicines. The Siddha system of medicine with its core ideas, albeit being initially alchemical about mercury, sulphur and salt, along with several metallic compounds and minerals, had developed a range of mineral medicines even before Paracelsus thought of them. The pathways of the Siddhas were no doubt different, but they made contributions in their own way to the progress of medicinal chemistry.

Kāya-kalpa (Revitalization of the Body)

Integral to the main objective of Siddha thinking on the material immortality of the body is the intricate process of *Kāya-kalpa*. It is essentially concerned with the rejuvenation and revitalization of body and mind, and to maintain them in a youthful state over a long span of life. Towards achieving this objective, the Siddha system advocates: (i) *prāṇāyāma:* a yogic breathing exercise which leads to the preservation of vital energy within the body and its bloodstream; (ii) conservation of sperm or reproductive essence by continence and yogic concentration; (iii) use of *muppu;* (iv) intake of meticulously processed preparations like calcined powders of metals and minerals, mercury and its compounds, sulphur, gold, copper or mica; and (v) intake of the products of certain chosen herbs in a regimented manner. According to the Siddha physicians, the first three, which can be adopted by disciplined individuals like *yogis*, have the potential to make the body imperishable, if a dietary regimen is scrupulously followed. The other two could be employed for persons with normal health, but not for anyone beyond seventy years of age. It is claimed that if a person undergoes *Kāya-kalpa* at the age of thirty, he will be rejuvenated with the energy of a young boy of sixteen. Even a person of seventy years of age would feel as if he is a young man of twenty years!

Ayurvedic treatments for achieving similar results under the *rasāyana* and *vājīkaraṇa* methods are different from the *Kāya-kalpa* ones inasmuch as the former employ mainly herbal treatments, while the latter utilize mercurial and other mineral compositions in addition to yogic *prāṇāyāma* exercises as well as tāntric elements like rousing the *Kuṇḍalinī*.

Kāya-kalpa is indeed an elaborate procedure, the details of which have been provided in some major Siddha texts. Briefly stated, it consists of (i) cleansing the mouth, throat and bowels by administering specified herbal applications and decoctions; (ii) intake of black pepper every day: five grains to begin with, along with an adjuvant, and the intake to be increased by five in the succeeding days till one hundred grains are consumed in a day; (iii) removal of the deleterious effect of the vitiated *doṣas* by the application of certain plant extracts; (v) invigoration of the nervous system and connective tissues by repeated treatment with a decoction made from the leaves of a neem tree; (vi) toning up the muscular system as well as the removal of morbidity in the stomach, and phlegm in the respiratory system through the application of certain herbal preparations; (vii) strengthening the three *dhātus* (*vāta, pitta* and *kapha*) and expelling the vitiated elements by the consumption of chosen plant extracts and preparations; (viii) increasing appetite, digestive capacity as well as sexual power, and to stimulate other physiological activities through the administration of fine powders of eight selected herbs with appropriate adjuvants; and (ix) invigoration of the genital-endocrine system by the intake of half of a black fowl's egg a day, after it is subjected to some processes, for about twenty days.

An oil bath with medicated oils is a part of *Kāya-kalpa*. The texts have given the names of specific plants for the purpose of medicating the oils, although there are some variations in the names of such plants. *Kāya-kalpa* drugs include several metallic and mineral compounds; specially processed mercury and mercuric sulphide in particular, have pride of place as the most effective for rejuvenation of the body and warding off degenerative diseases. The diet in the course of *Kāya-kalpa* is to be strictly without common salt, sour articles like tamarind, fish or meat, and pungent substances. In addition, a disciplined daily routine as prescribed by the Siddha physician is to be followed. Even now, it is not uncommon that quite an appreciable number of middle-aged (fifty to sixty years) persons go in for *Kāya-kalpa* in a detailed manner. Siddha physicians also prepare a package of *Kāya-kalpa* drugs, including very small quantities of *makaradhvaja* (mercuric sulphide) and powders of some plants like *asvagandha* (*Withenia somnifera*) and *śatāvarī* (*Asparagus racemosus*) to be taken with milk for forty-one days. However, each Siddha physician has his own approach to the *Kāya-kalpa* medicaments.

Present Status: The Government of Tamil Nadu, where the Siddha medicines have a special appeal to common people, and the Government of India, in pursuance of its avowed policy of encouraging traditional systems of medicine, have been promoting the Siddha system to a great extent. A standardized *Siddha Formulary* (over 240 drugs and 100 plants) has been prepared, and this has been a guide for new entrants in colleges. Nearly 475 Siddha pharmacies have been engaged in producing and marketing standard drugs both at home and abroad where Tamilians live in large numbers. Nevertheless, not a few private pharmacies adopt what they call the traditional methods for preparing health-promotive and rejuvenating drugs, medicated oils and the like. The Central Council for Research in Ayurveda and Siddha has been promoting research and clinical trials of some select Siddha drugs like those for curing peptic ulcers and amoebic dysentery, as described in the Siddha texts. It would, however, appear that such scientific evaluations are few and far between. More systematic investigations and clinical trials on a large scale would be necessary for Siddha drugs to join the mainstream of medicines with scientific acceptability. Even so, some of the Siddha drugs, especially those of *Kāya-kalpa* and health-promoters, have become popular. There are several common or home remedies in the Siddha system and these have been brought to light by the Central Council for Research in Indian Medicine in the form of a text titled: *Handbook* (1987). They include simple methods of preparations using readily available ingredients, for curing general ailments like headaches, coughs, ordinary fever, pain, diarrhoea and sore throat. With these and its formulated medicines, the Siddha system has become an important wing of traditional Indian medical systems.

UNĀNI MEDICINE

The word *unāni* is an outlandish nomenclature for *Ionia* of the Greek Archipelago in the same way as the term *Yavana* is used for Greeks in Indian literature. Unāni Ṭibb means Graeco-Arabic medicine; hence, it would be desirable, indeed necessary, to have some idea of Greek medicine in a historical perspective, and its assimilation or transformation in Islamic culture. In the sixth century BCE, it was among the Ionian thinkers that a rational attitude towards the universe without invoking any type of divine origin first arose, led by Thales (624-565 BCE), the renowned philosopher of the Ionian city of Miletus. Later, Anaximander, Anaximenes, Heracletus, Pythagoras and Empedocles were among the other noted pre-Socratic Greek thinkers who expounded philosophical ideas on nature, discarding any type of mythological or religious element. New ideas on medicine came up too. In the fifth century BCE, there were two Greek schools of medicine: the older one at the city of Cnidus, and the later one at Cos, which Hippocrates (about 460 BCE), who displayed a rational attitude to medicine, belonged to. This approach set aside the superstitious and temple-healing practices for which Aesculapius was venerated as the Greek god of medicine (his staff and coiling serpents even now compose the recognized emblem of modern medicine). The new attitude laid emphasis on observation and recording the symptoms of diseases, their natural causes and natural cures. Known as the Hippocratic method, it heralded the dawn of 'rational' medicine displacing the earlier 'religious' medicine. This method was foundationally strengthened by Aristotle (384-322 BCE), the renowned Greek thinker and pupil of Plato (427-347 BCE), who made significant contributions to anatomy and physiology. He coined the word *psyche*, which added a new dimension to the body-mind relationship in medicine. When Alexander the Great died (323 BCE), his empire disintegrated and one of his generals, Ptolemy (not to be confused with Ptolemy, the astronomer-astrologer), ruled over Egypt. Alexandria in Egypt emerged then as an important centre of scientific thinking, including medicine, till about 200 CE when it lost its glory during Roman rule.

Hippocratic medicinal ideas and Aristotle's biological views found an able exponent in Galen of Pergamum (131-201 CE) who worked in Rome, but had already visited Alexandria and imbibed the spirit of natural studies there. Like Hippocrates or Aristotle, Galen believed in the four 'humours'—blood, phlegm, yellow bile and black bile, correlating them with the four Empedoclean elements—earth, fire, water and air and the two sets of opposing qualities: hot and cold; dry and moist. But, his greatest contribution was the concept of *pneuma* as the basic principle of life and his explanation of blood circulation in terms of natural spirit, vital spirit and animal spirit. Like Aristotle, Galen was a teleologist and believed that the inner physiological processes in the human or animal body were part of an intelligent design by the Creator. He was also a practical physiologist, and carried out numerous dissections and experiments on living animals. He emphasized the relationship between the *psyche* and the body, and explained that all temperaments were mixtures, in varying degrees, of the four basic humours. Galen's ideas and works continued to be an authority on medicine in the succeeding centuries.

From the eighth to about the fourteenth century, the scene of scientific activity shifted from the Roman world to the Islamic area. Islamic savants were well-versed in Greek scientific knowledge through Arabic translations of the works in Greek, including those of Aristotle, Galen and Hippocrates. Baghdad rose as a great centre of learning and witnessed translations of classical works on astronomy and medicine. As noted before, Indian medical works too were rendered into Arabic. Hunayn ibn Ishaq (809-77), an erudite Nestorian who was very active in the regular school of translation created by the Abbasid Caliph, Al-Mamūm (813-33) in Baghdad, translated the entire corpus (as many as sixteen books) of Galenic writings and seven of the ten Hippocratic works into Syriac. One of his pupils rendered them into Arabic. The city of Harran provided another group of translators like Thabit b. Ibrahim and Qusta b. Luqa. Owing to their efforts, Greek medicine came to be accepted in the Islamic world and also by the Arabic writers on medicine. In fact, some treatises of Greek medical writers, which are not available now in Greek, were preserved in their Arabic translations.

It would be naïve to think that Islamic medical men indulged only in translations or in transmitting Greek medicine in their regions. On the other hand, they made original contributions through their experiments, observations and perceptions to the growth of Greek medicine, which came under the general nomenclature of Graeco-Arabic medicine. The oldest original medical writer in Arabic was Ali-ibn Rabban Ṭabari, who produced four books, of which *Firdaus al-Hikma* is significant.

Abu Bakar Muḥammad ibn Zakariya al-Razi (865-925),

known to the Latin West as Rhazes, was a Persian who studied in Baghdad under a disciple of Honian ibn Ishaq (809-77), an erudite Nestorian. Well acquainted with Greek, Persian and Indian medicine, he rose to be a great physician for centuries and produced his magnum opus in Arabic—*Kitāb al-Ḥawi,* and his innumerable works on medicine. He was the first to describe the nature and spread of smallpox and measles and, as a physician-in-chief at Baghdad, made original contributions to the medicare of patients. Ali ibn Ābbas al-Majusi (d. 994) was another Arabic physician who flourished in the later half of the tenth century. His original treatise *Al-Maliki* or *Kamil al-San' al-Ṭibbiya* was an admirable text on the principles and practice of medicine. This voluminous text, divided into twenty discourses, each being subdivided into several chapters, sheds light on the theoretical, anatomical, physiological and surgical knowledge of his time. Majusi's own original contribution was his accurate description of the capillary circulations, and the clear directions for the catheterization of the urinary bladder in male patients to remove cysts and other infected glands. His *Al-Maliki* enjoyed great popularity till the appearance of *Al-Qānūn* (The Canon of Medicine) by Ibn Sīna, who was also known as Avicenna.

Ibn Sīna's *Al-Qānūn*

It is Ibn Sīna's great medical compendium that has been the recognized authority of Unāni Tibb in India. During his time he was known as 'Chief Master' (*al-Shaykh al-Rais*) or the 'Second Teacher' (*al-Muāllim al-Thani*), after Aristotle. Born in Bukhara (Central Asia), Ibn Sīnā was revered as a philosopher, poet, man of public affairs and learned physician. At the age of eighteen, he was requisitioned to attend on the ailing ruler of that region and won royal acclaim for the services rendered by him as a physician. His masterpiece, *Al-Qānūn* is an encyclopaedic, systematically arranged and authentic medical treatise, which was used as a textbook for a long time even in Europe through its translations into Latin. It is divided into five books: the *first* discusses in detail the general principles of medicine; the *second* deals with simple drugs arranged alphabetically; the *third* sheds light on anatomical aspects from head to feet; the *fourth* describes several diseases, including those that tend to spread to the other body parts; and the *fifth* presents details of compound drugs. In *Al-Qanūn,* Ibn Sīna has meticulously enunciated the basic principles of Greek medicine to reconcile the medical concepts of Aristotle and Galen. He also wrote a treatise on cardiac drugs, which was translated into several languages: Latin, French, English, Persian and Urdu.

Though *Al-Qānūn* held a special position in medical literature throughout the Islamic world, superseding the earlier medical treatises of Al-Rāzi and Al-Majusi, it did not differ very much from these treatises in basic ideas. There were other medical works, too, that had not lost their appeal, like *Ṭaḍkira al-Kuhḥālin* of Ali ibn Isa, on ophthalmology. Abu Ali al-Hassan ibn Haithām (965-1038) had blazed new trails by explaining that in vision, rays passed from the object to the eyes, and not from the eye to the object, an erroneous proposition put forward by the Greeks. In the western Islamic region of Spain, the well-known surgeon of the Arabic world, Abul al-Qāsim al-Zahrāwi (d. 1013) had produced his surgical work, *Al-Ṭaṣrīf* with illustrations of many surgical instruments used by Arab surgeons. Ibn Zuhr or Avenzoar (1090-1162), who was born in Seville in Spain and whose disciple was the great thinker Ibn Rushd (or Averroes as he was known in the West), was an expert clinician. His work *Kitab al-Teisir* is a repository of several case-histories meticulously collected by him (later rendered into Urdu in India). Averroes, though a philosopher of repute, was also a noted physician who wrote a book under the title *Colliget*, which was regarded as an important book by philosophers and physicians alike. He also wrote a book on the foundational ideas of medicine titled *Kitāb al-Kulliyat*. Ibn Maimum (1135-1204), known as Maimonides, was born in Cardova in Spain and established himself as a philosopher and physician in Morocco and Egypt. He prepared a treatise titled *Fusul Mūsa* on medical principles and wrote on preventive medicine as well as on hygiene. Yet another medical treatise of encyclopaedic expanse was *Dhakhiraḥ-i-Khwarazm Shāhī* by Zain al-Din Ism'al Jurjāni (early twelfth century). There were other important medical treatises too. Unāni Tibb had established itself as a sound medical system by the thirteenth century both in the western and eastern Islamic regions.

Greek medicine underwent noticeable improvements at the hands of Arab physicians because of their contributions to medicinal chemistry and the associated experimental methods, as well as the production of drugs with great care and paying attention to the compositions and standards required. Arab physicians also appeared to have experimented with their drugs on animals before they were administered to human beings. More importantly, Islamic rulers spared no efforts in the establishment of hospitals. One of the hospitals, the Manṣūr Hospital established in Cairo, attracted patients from far and wide. It is on record that it had wards with side fountains—'chanting waters', which had a soothing effect on the patients! There were several other hospital-centres, such as Damascus, Baghdad and some cities in Persia. Graeco-Arabic medicine had made substantial strides in several areas—in symptomatology, pathology, epidemology, ophthalmology, toxicology and dietetics as well as midwifery. It had evolved pharmaceutical techniques leading to the production and use of many drugs that were practically unknown to Greeks and Graeco-Romans. This Graeo-Arabic medical system or Unāni Ṭibb entered India through the hakims who accompanied Muslim rulers and it was assiduously nurtured, especially by the Mughal rulers during the sixteenth to eighteenth centuries.

GENERAL PRINCIPLES OF UNĀNI ṬIBB

Ibn Sīna calls the medical system *tab'iat* (physic), which is derived from the word, *tab'iyaat* (physic or physical). What it denotes is that the system takes physical laws into consideration and rejects blind beliefs and superstitious practices in the name of healing art. It also emphasizes that health and diseases, though different, are related states of the human body, that is, an integrated approach to health and its restoration whenever it is vitiated. Unāni, like Ayurveda, is a holistic system which thinks of both the human body and mind in a sound state, accepting the mutuality of both physiology and pathology. The aim of Unāni medicine is to support and strengthen the recuperative potential that is natural to the human body and also to eradicate diseases. It recognizes the nature of the body and the maintenance of health under seven components: (1) *Al-Arkān* (the elements); (ii) *Al-Mizāj* (temperament); (iii) *Al-Akhlāṭ* (the humours or body fluids); (iv) *Al-A'ḍa* (organs); (v) *Al-Arwāḥ* (pneuma-vital spirit); (vi) *Al-Quwa* (faculties or powers); and (vii) *Al-Afāl* (functions or actions). Unāni believes that the loss of any one of these components could result even in the death of a person.

The Elements: The Empedoclean (Greek) four elements—earth, water, air and fire—are also recognized as the constituents of the human body, just as Ayurveda has conceived of it in terms of five elements (including *ākāśa*). In the Unāni system, these elements represent certain states of matter in the body, namely: the earth is representative of solids; water, the liquids or fluids; air, the gases; and fire, the heat or warmth of the body. They are also involved in the biochemical and mechanical processes that take place in the body all the time. Ibn Sīna has characterized four elements as *arkān* (pillars of the body). He also used the word, *unsūr* (elements), but since it was likely to be mistaken for modern chemical elements, Unāni practitioners do not use this word at present.

As noted earlier, there are four primary qualities associated with these four elements—hot and cold; moist and dry—which exhibit affinity and opposition. Two of them, hot and cold, are believed to be active and energetic while the other two, moist and dry, are considered to be passive and material. It is further believed that the two members of each pair are antagonistic to each other and hence would not be coexistent. The four combinations in the elements are also thought of thus: (i) hot and moist (air); (ii) cold and moist (water); (iii) hot and dry (fire); and (iv) cold and dry (earth). These considerations are in marked contrast with the corresponding Ayurvedic concepts derived from the five elements (*pañcabhūtas)*. But the foundational ideas of Unāni, relating to the constitution or the 'pillars' of the body in terms of 'elements', have some similarity to those of Ayurveda, though the Aristotelian fifth element, *ether*, which in no way corresponds to Indian *ākāśa*, does not figure in Unāni's reckoning of elements. It is important to note that both Unāni and Ayurveda recognize, each in its own way, the role of inorganic elements in organic life processes.

Temperament

The Unāni concept of temperament is related to the elements. The explanation is that when the elements interact and mix with one another intimately through their minutest forms or particles, their qualities also interact with one another. Since the positives or actives are modified by the negatives and passives (because of their opposing natures), the human body naturally possesses a uniform balance of these qualities. This state of a uniform body is called its temperament, a product or a result of the four primary qualities. This may in a way correspond to *prakṛti* of Ayurveda—but it is different in several ways from the latter, since Ayurveda thinks of the nature of the dominant elements at the time of the union of the sperm and the ovary in the womb (*doṣa prakṛti*), and also the psychic state of an individual in terms of the predominance of any of the three qualities (*sattva, rajas* and *tamas*) and their variations (*manas-prakṛti*).

Unāni accepts that in man and other beings, there would be no perfectly balanced temperament, nor anything even close to it, and there are bound to be variations depending upon the birth and genetic aspects. Such imbalances may be due to the excess of one of the four primary qualities (hence four varieties), as well as excess of the dual qualities like hot and moist; hot and dry; cold and moist, and cold and dry in all eight varieties. The first four varieties are known as the simple ones, the second four varieties, the compounded ones. Both varieties can be caused or manifested with morbidity or without morbidity, and thus, counting these, Unāni recognizes sixteen varieties of imbalanced temperaments.

A special feature of Unāni is its acceptance and explanation even of the *temperament* of drugs, and of organs in terms of the primary qualities. It envisages drugs as being hot or cold. This does not, however, connote that a drug is either hotter or colder than the human body; but it would indicate that they would generate a greater degree of heat or cold in the body, while an evenly constituted drug would have the same temperament as the human body. The organs are also supposed to have balanced temperament in accordance with their nature and functions. The hottest parts of the body, according to Unānī, are what it calls (after Galen) the vital force and heart, followed by muscles, spleen, kidneys, breast, etc. On the other hand the coldest are phlegm, followed by hair, bones, cartilage, ligaments, nerves, spinal cord, brain, etc. Likewise, the dry and moist parts of the body have been delineated in the context of temperament. There is also a categorization in relation to age and sex based on the four qualities.

Humours

The Greek four-humoural theory is an intertwined part of the Unāni system. Hippocrates had explained the state of the human body in terms of blood, phlegm, yellow bile and black bile—the four humours that would cause either health or disease depending upon the correct proportions of each other, their strength as well as quantity. Like Ayurveda, Unāni believes that their derangement, mutual separation, and either one or all of them being in excess or deficient, would impair health, resulting in distress, pain and affliction of diseases. Ibn Sīna in his *Al-Qānūn* (Canon of Medicine), has elucidated the humoural physiology and pathology further. Explaining that humour is a fluid substance as the initial product of the metabolized food taken in, he has classified it into two types—normal and abnormal. The normal one, either by itself or in combination with the other humours, is capable of assimilation or absorption into the tissues; the abnormal one is unsuitable for assimilation and the body eliminates it; and it is the normal one that nourishes all parts of the body.

According to Unāni, there are two kinds of body fluids, primary and secondary. The primary ones, both normal and abnormal, are the four humours which in Arabic are known as *dum* (blood), *bulghum* (phlegm), *ṣafra* (bile) and *sauda* (black bile). Blood is regarded as hot and moist in temperament; phlegm, cold and moist; bile, hot and dry; and black bile, cold and dry. Their types have also been described. The secondary fluids, either essential or superfluous, are considered to be also of four types: intra-vascular (*maḥṣoora*), pericellular (*ṭalliya*), intra-cellular (*qareeba*) and cellular (*munwiya*). Since humours are directly related to ingested food, Unāni attaches great importance to a dietary regimen along with appropriate drugs which would correct the imbalances of humours. Conceptually, this is similar to Ayurveda, although it may differ in some details; for example, Unāni physicians take into consideration the humoural *temperament* of a patient and that of the administered drug, while Ayurveda examines the *prakṛti* or the mental state of a patient in terms of the disequilibrium of *sattva*, *rajas* and *tamas*. In both the systems, however, the aim of the treatment is to restore the balance of humours, three in the case of Ayurveda, and four in Unāni.

Faculties

Unāni recognizes three types of faculties and an equal number of functions as follows: (i) nervous (*nafsania*) faculty; (ii) physical (*ṭabbai'yya*) faculty; and (iii) vital (ḥaiwania) faculty. The first one is of two kinds: one, which is responsible for the preservation and maintenance of an individual's health; its seat or centre is in the liver; the second is concerned with the reproductive ability and continuance of progeny, being located in the reproductive organs. The physical faculty is both primary and secondary, the former being involved in the maintenance or preservation of the individual's nutrition, growth, etc., while the latter, in the reproduction and growth of the race. The third or the vital faculty is intimately involved in all types of sensations, movements and brain activities. This faculty is stated to have the heart as its centre, following Galen's concept of the vital spirit. The vital faculty is also responsible for mental states such as emotion, fear and anger. The nervous faculty relates to sensorial experiences of various kinds, cognition, intellect and memory. The Unāni medical texts describe in detail the nature and causes of diseases, the qualities of food and drink, and environmental aspects in the context of health and disease as well as in relation to temperament and the four primary qualities.

Causes

Ibn Sīna has stressed the importance of four types of causes to be kept in view: i) Material causes (*māddi*) are those in which health and diseases are involved, concerning the organs, vital forces, the four humours, and the four elements; (ii) Efficient causes (*f'liya*), which either activate or prevent changes in the human body, like air, food and drinks, retention and evacuation, wakefulness and sleep, sex variations and other habits; (iii) Formal causes (*ṣouriya*) are the temperaments, faculties and structures; and (iv) Final (*tamāmiya*) causes are those which would evaluate or assess the faculties and vital forces. The epistemological aspects of Unāni medicine do not appear to be well structured as those of Ayurveda.

Diagnostics

Temperaments are diagnosed on the basis of the feel of the body, muscularity and fatness, quantity and quality of hair, general complexion and physique, receptivity or sensitivity of organs, sleep and wakefulness, the functional calibre of certain organs, excretions and psychological status.

Examination of pulse, urine and stool plays an important role in Unāni diagnostic methods. As in Ayurveda (later period) and Siddha, pulse examination calls for the special expertise and experience of a Unāni physician. Ten features of the pulse have been taken into consideration, like the size and strength of the pulse, its speed, fullness, consistency, rate, temperature (hot, cold or moderate), constancy, regularity and its rhythm (*wazan*). Ibn Sīna has also provided details about the different types of pulses, including the compound pulses and the normal ones.

Ibn Sīna's prescriptions for urine examination may be summarized thus: urine that is to be examined should be passed in the early morning and in a fasting state. Any food or drink which is coloured yellow or red should not be imbibed the previous night, and the patient should not have undue mental and physical exertions. In any case, the sample of urine to be examined should not have been kept for more than six hours before examination, nor exposed to sunlight and air. Its colour, turbidity, transparency, sediment, density, odour and froth are to be examined. Ibn Sīna has also dealt with the examination of stool in considerable detail.

Unāni treatment of diseases comprises the use of simple and compound drugs, a dietary regimen, and surgical operations, whenever necessary. The disease is treated by drugs that have the capacity to ward off the root causes in an allopathic manner, that is, drugs that are contrary to the nature and pattern of the disease. On the other hand, the preservation of health is sought to be accomplished by the method of prescribing nutritious foods and drinks to bring about a healthy body with a sound mind, as in Ayurvedic practices. The Unāni *materia medica* is extensive, comprising both herbal and animal products. A perceptible way of medicinal care in Unāni relates to various baths—sunbaths, sand-baths, oil-baths, sprinkling of cold water on the face, and most importantly, a bath in a *ḥamām*. It is accomplished in a large building with four apartments, one each for the four seasons, with the needed supply of fresh air and sweet medicated water for baths—both being extremely beneficial for overall bodily fitness and mental invigoration.

GROWTH OF UNĀNI ṬIBB IN INDIA

One of the earliest Unāni physicians in India was Ziauddin Abdal Rafī, the court physician of the last Ghaznavi ruler, Khusrow Malik (1150-86). He also served the succeeding ruler Muḥmmad Ghori. During the Slave dynasty (1206-90), Unāni was in use, but in a subdued manner. Ala'uddin Khalji came to power towards the end of the thirteenth century, and his prosperous rule attracted several scholars including hakims from Iran and other parts of Central Asia. The well-known hakims included Maulāna Badruddin Damishqi, Hisamuddin Marikali and his son, Sadruddin, Matraz, and the surgeon Jaja. Damishqi was regarded as an expert in diagnosis, especially by urine examination, and an able exponent of Ibn Sīna's principles, while Hakim Matraz specialized in pulse reading. Unāni began to develop its foundation in India during Ala'uddin Khalji's rule.

Tughluq Dynasty

During the Tughluq dynastic rule, Unāni began to register considerable progress, the main reason being that the founder of this dynasty, Ghiyasuddin Tughluq Shah (1320-25), was himself well-versed in Unāni medicine and its applications. It is on record that this ruler personally used to treat wounded soldiers of even the enemy! His famous son, Muḥammad bin Tughluq (1325-51) had a profound knowledge of Unāni Ṭibb, and used to cure several patients with dexterity. Under his patronage were able hakims like Khawaja Shamsuddin Mustaufi, 'Zia Mohammad Masud Rashid and Zangi Umar Ghaznavi. An interesting development was the mutual interactions between Unāni and Ayurvedic practitioners. As Unāni Ṭibb was establishing itself on Indian soil, hakims became well aware of Ayurvedic practices. Hakim Mustaufi wrote a text under the title *Majmu ah-i Shamsi,* using Ayurvedic sources in Sanskrit. Hakim Zia Mohammad, the Court physician of the Tughluqs, produced a book *Majmuah-i Diai* (1336) which contained considerable details of Ayurvedic medicines. In this work, several Hindi (Sanskrit) names of diseases and drugs are also found in the place of Arabic and Persian names; in fact, this treatise is a judicious blend of Unāni Ṭibb and Ayurvedic medicines. Such was the spirit of the times. The famous Sūfi physician Sheikh Ziauddin Nakhshabi, who came to India from Baghdad (early fourteenth century), produced two major works: *Tuti Nāma,* which has substantial references to Ayurvedic texts in Sanskrit and *Al-Juziyat wa al-Kulliyat* in which he has discussed the organs of the body, their diseases and the curative procedure.

Firoz Shah Tughluq (1351-88) who succeeded Muḥammad bin Tughluq, was well-versed in Unāni medicine and Islamic astronomy. His famous medical treatise *Tibb-i Firoz Shāhi* provides details of the treatment of some ailments that are not found even in Ibn Sīna's *Al-Qānūn*. He was hailed as an efficient bonesetter and opthalmologist too. In fact, a collyrium, which is a compound of antimony, made by him goes by the name *Surmah-i Firoz Shāhī,* and was used extensively for curing eye diseases. During his time, an erudite Hakim, Ilyas bin Shahab, wrote a comprehensive medical book titled *Rāhat al-Insān.* It was also influenced by Ayurvedic therapeutic measures, Hindi (Sanskrit) names of drugs along with Arabic and Persian equivalents. But his curative methods also included, strangely, certain types of amulets and talismans, thus satisfying the superstitious beliefs prevailing at that time. With vigour, Firoz Shah continued the earlier tradition of establishing hospitals and there were as many as 115 hospitals during his time. In many of them, patients were treated free of charge and with great compassion as ordered by the king. The compassionate king himself used to visit the hospital, examine the patients, and write prescriptions of drugs and diet. Firoz Shah also established several madrasas in which, among others, medicine and astronomy were being taught. During this enlightened king's rule Unāni Ṭibb began to assume new dimensions in the medical care of all people, irrespective of their religion.

Lodhi Dynasty

Sikandar Lodhi, an erudite king, was deeply interested in medicine, both Unāni and Ayurveda. At his instance and encouragement, his minister Bahwa ibn Khawas Khan, who was respected for his learning, wrote an extensive medical compendium including the principles and practices of Ayurveda under the title *Ma'dan al-Shifa'-i Sikander Shāhī* (1512). In its three sections and in all 128 chapters, this elegant treatise, which is voluminous and luminous alike, discusses in great detail the basic principles and practices of Ayurveda, including the three humours. The liberal-minded Bahwa thought that since most of the Unāni drugs were not produced in India in his time and they would not appeal to Indian people by and large, Ayurvedic drugs and treatment deserved their due recognition. He enlisted the assistance of several Ayurvedic *vaidyas*, and himself was conversant with the texts of Caraka, Suśruta, Mādhava, Bhela, and Vāgbhaṭa. Amir Miyan Taha, another minister of Sikandar Lodhi, was an expert both in Islamic sciences and Ayurveda, besides music. He was believed to have learnt by heart many verses on Ayurveda by studying the Ayurvedic classics—a rare feat of a Muslim scholar indeed!

No wonder that even some Hindu Ayurvedic *vaidyas* used to go to him for lessons in Ayurveda.

The complementarity of Ayurveda and Unāni was not confined to the dynasties which ruled northern parts of India. In western India, in Gujarat, during the reign of Muzaffar Shāh (fourteenth century), Shihabuddin Abdul Karim Nagawri wrote three medical works. One of them, *Shifa' al-Khani* (1392) was based on Arabic, Persian and Sanskrit medical sources. It also included medical terms in Sanskrit and Sindhi language. Another book by Nagawri was a dictionary titled *Farhang-i-Shahabi*, which contains names of medicines in Hindi, Arabic and Persian. His third treatise *Tibb-i Shahabi,* also known by its associated title *Shifa'-al-Marad,* describes only those drugs which are tested and describes the nature of the action of certain drugs and foods. During the reign of Sultan Mahmud Shah Begarah, Vāgbhaṭa's *Aṣṭāṅgahṛdaya* was rendered into Persian and titled *Tibb-i Shifa'-i Mahmudi* and the translator was Ali bin Mohammad Isma'il Aswali Asili. The interaction with Ayurveda and the promotion of Unāni thus went hand in hand under Muslim royal patronage.

Deccan

In the Deccan, the Bahamani kingdom under Mohammad Shah encouraged Unāni Ṭibb and there were noted hakims like Nasiruddīn Shirāji and Khwāja 'Amaduddin Maḥmūd Ali Gilāni. The rulers of Golkunda and Hyderabad also patronized this medical system. At the instance of Mohammad Quli Qutb Shah (1580-1611), the founder of Hyderabad city, an imposing and functional hospital *Daral-Shifa* (1595) was built on the banks of the Moosi river. It had enough capacity to treat as many as four hundred patients at a time. Hakims and Ayurvedic physicians worked side by side in this hospital, which also had a teaching centre. The Iranian Mir Momin, an erudite scholar of literature and medicine, won the approbation of Qutb Shah and wrote two medical texts: (i) *Ikhtiarat-i Qutb Shāhi*, a critical account of an earlier medical text by Ali bin Jamaluddin al-Hysayni Ansari, which deals with simple and compound Unāni drugs; and (ii) *Risālah-i Miqdariah*, which contains accounts of weights and measures.The Sultan also encouraged Shamsuddin Ali Husayn al-Jurgani, a renowned ophthalmologist of his time, to render into Persian an Arabic manual on ophthalmology. This was produced by him under the title *Tazkirah al-Kahhalin* (Memoir of Oculists) thus introducing into India this important work on eye diseases and their cure. Hakim 'Abdullah Tabib wrote *Tibbi-i Farid*, which deals with Unāni medicine, dietetics, and hygiene. Saifiuddin Muhammad Tabib Gilani and Shamsuddin ibn Nuruddin Tabid, who was also a court physician, and Taqiuddin Muḥammad ibn Sadruddin were among the other noted hakims in the Hyderabad region during the sixteenth and seventeenth centuries.

The tradition of medicine continued to receive patronage in the Adil-Shahi dynasty of Bijapur which promoted both Unāni and Ayurveda. During this rule, many scholars came to Bijapur from Persia and West Asia. Hakim Muhammad Qasim Hind Shah, generally known as Firishta, authored a book titled *Ikhtiarat-i Qasmi*, better known as *Dastur-al Atibba* (1590). He has recorded in the introduction to this book that he first studied the Unāni Ṭibb in detail and also the Ayurvedic system. He has recorded further that the Ayurvedic theories and practices were sound, and that he was closely associated with the vaidyas. The Deccan, which came under the Muslim rulers, gradually developed into one of the important areas of the Unāni Ṭibb.

Mughal Period

Unāni medicine attained new heights during Mughal rule (1526-1800). The medical literature produced during this period, which is more or less preserved in its entirety, excelled that of the previous three hundred years both in its content and form. Several noted hakims contributed enormously to its internal strength and external popularity. Hence, this period is rightly recognized as the Golden Period of Unāni medicine in India. There were eminent Unāni physicians, who under royal patronage as well as with their own expertise, held the Unāni tradition aloft and brought succour to people from all walks of life—rich and poor alike. Muhammad ibn Yusuf Harawi came to India along with Babur (1526) as his personal physician, and wrote several works on Unāni both in prose and poetry.

Hakim al-Mulk Abul Fath Gilāni had also earned a name for himself as a great Unāni physician during Akbar's regime. Hailing from Gilān (in Iran) along with his two brothers who were also hakims, he produced not only books on medicine but also left behind his tested prescriptions for the cure of various ailments. Akbar's royal patronage was so inspiring that several physicians began to come to India from Iran, which was actually emerging as the training school for India. Hakim Mir Fahullah Shirāzi (from Shiraj in Iran) won the acclaim of Akbar who conferred on him the title of *'Aḍḍud Dawlah*. Shirāji was a multi-faceted intellect in medicine, philosophy, astronomy, mathematics, and even astrology. His fame also rested on his invention of a grinding machine or flour mill and the fabrication of a gun which could fire twelve shots in a round. His pupil, Hakim Ali Gilāni, was a respectable physician during the reign of Akbar and his descendents too, were able Unāni physicians. There were many other established hakims who were encouraged by the Mughal Kings.

FAMILY TRADITIONS

The Sharifi Family

The extensive patronage extended to the Unāni Ṭibb by Mughal monarchs catalyzed it in such a manner that it became a traditional pursuit among some families of physicians. Of them, two families, the Sharifi and the Azizi families merit special consideration. The origin of the Sharifi family goes back to its connection with a Sufi saint, Khwājah Abdullah Ahrar who lived in Samarqand (fifteenth century) during the reign of Tamerlane, and the family medical tradition of his succeeding generations. During the time of Aurangzeb, Hakim Wasil Khan of this tradition was in India, and one of his grandsons, Hakim Sharif Khan, became the torchbearer of this family. He was born in Delhi (1725) and, after becoming proficient in medicine under his father and uncle who were renowned Unāni physicians, Sharif started his own clinic and rose to become a reputed medical practitioner himself. He has left behind several medical works which include *'Ilāj al-Amraz*, which is an Unāni pharmacopoeia; *Talif-i Sharifi*, which deals with the properties and actions of Indian drugs and foods; *Matab-i Sharif Khan*, his medical diary and prescription; *Hashiyah Sharah al-Asbab Wa al-'Alamat* and *Madanik al-Hikam*. His son, Hakim Sadiq Ali Khan, was an able exponent of Unāni Ṭibb and authored several medical treatises. He was devotedly followed by his son, Hakim Ghulam Mahmud Khan, the author of *Ziya' al-Absar*, and the latter's three sons, Hakim Abdul Majid Khan, Hakim Wasli Khan, and the famous Hakim Ajmal Khan.

Hakim Ajmal Khan (1863-1927), who learnt Unāni medicine, its principles, drugs and practices from his father and elder brothers, had studied *Al-Qānūn* of Ibn Sīna in depth. He was not only a skilled and efficient physician but also a prolific author. His medical treatises include *Al-Tuhfat al-Hamidiyal fil'l Sana'at al-Taklisiyal*, *Al-Wajizah*, *Muqaddimat al Lughat al-Tibbiyah* and *Al-Ta'un*. More importantly, he spared no efforts in advancing the cause of Unāni medicine in India through the establishment of educational institutions and research laboratories. This was his great contribution and earned for him a special position in the promotion of Unāni medicine. Ajmal Khan was the personal physician of the Nawab of Rampur and he diligently studied the Arabic and Persian manuscripts on medicine preserved in the library there. According to him, several principles and practices of the Unāni system needed revision and reinterpretation from the modern scientific point of view. He established the institution, Anjuman-i Ṭibbiya (1906) to run the *Ṭibbi* school that had been started earlier by his brother, Hakim Majid Khan. He also established the Unāni School for Women (1909). His brother, Hakim Wasil Khan, had set up a private Unāni and Ayurvedic company for the production of drugs of both systems; in 1907, it was named *Hindustani Dawakhana* and went on to produce and sell drugs of such high standards that its earnings also supported the Ṭibbiya institution. Ajmal Khan's educational zeal and efforts resulted in the establishment of an Ayurvedic and Ṭibbiya College (in Karol Bagh in Delhi), the foundation stone of which was laid by Viceroy Lord Hardinge, and opened by Mahatma Gandhi (1921).

Hakim Ajmal Khan was a nationalist to the core. He pioneered an All India Ayurvedic and Unāni Conference as early as 1910, with a view to promoting both the systems by bringing together the *hakims* and the *vaidyas* on the same platform. This was at the time when modern medical education was spreading, though sporadically, and some modern hospitals were springing up in different parts of the country fully supported by the British colonial government—which was lukewarm towards the indigenous medical systems. But the hakim not only stood against colonial prejudices but also took steps for conducting researches on some indigenous drugs on modern scientific lines. Dr Salimuzzaman Siddiqui, who had established himself as a reputed chemist, was appointed by the hakim as the Director of the Research Institute of Ayurvedic and Unāni Ṭibbiya College. Researches were conducted on an Indian medicinal plant, called *asrowl* in Urdu or *sarpagandha* in Sanskrit (*Rauwolfia serpentina*). Some of the alkaloids which were isolated by scientific experiments were named *Ajmaline, Ajmalinine, Isoajmaline, Neoajmaline,* etc., after the reputed hakim. A great patriot, Hakim Ajmal Khan renounced the title of *Haziq al-Mulk* given to him by the British government (1907) in protest against its atrocities. But his popularity and social service were duly appreciated by the Muslim community which conferred on him the title *Maish al-Mulk*. Ajmal Khan, who had forged a strong friendship with Mahatma Gandhi, was also the President of the Indian National Congress (1921).

The Azizi Family

Like the Sharifī family in Delhi, the Azizi family based in Lucknow produced several hakims of high reputation. They nourished the tradition of Unāni medicine at a time when the patronizing Mughal rule was coming to a close, and the oppressive British colonial rule was on the ascendancy. The family derived its name from the enlightened and erudite Hakim Abdul Aziz, son of Hakim Muḥammad Ismai'l and grandson of the noted Hakim Muhammed Ya'qub in Lucknow. He was the author of a treatise on compound drugs titled *Risalah Tohfah-i Azizi* (1888). He was compassionate and treated his patients, rich or poor, with equal attention. He also left behind his diary of prescriptions of tested drugs. More importantly, he founded a medical school, *Takmil al-Ṭibb*, which is now a college of Unāni Medicine in Lucknow. When the epidemic plague broke out in 1902 into a virulent form in Lucknow, this institution provided succour to the afflicted people, and the relief measures were led by Hakim Aziz himself along with his two sons, Abḍ al-Rashid and Abd al-Hamid. The curriculum of the College and the training in Unāni Ṭibb had been of a high order and won the appreciation of Unāni hakims throughout the country.

Hakim al-Hamid, the second son of Aziz, paid greater attention to the teaching of Unāni medicine in Takmil al-Ṭibb, established by his father. In 1926, when the Government of India set up a Board of Indian Medicine, he was one of its nominated members. Arising out of the deliberations of this Board, three institutions were established: the Ayurvedic School at Hardwar, the Ṭibbi School at Lucknow and the Ṭibbia College at Aligarh Muslim University. Hakim Hamid was of the view that the structure of traditional medical systems like Unāni and Ayurveda, should be kept intact and given the necessary freedom to progress within their own conceptual frameworks. He was also of the firm view that colleges of mixed education in traditional and modern medicine would not be beneficial to either of them. Hakim Hamid was the recipient of several honours, both national and international, including the Padma Bhushan (1991). He was also the chancellor of the Aligarh Muslim University.

Hakim Abdul Hameed (1908-2000) was another humanist to the core. His tireless efforts and outstanding contributions to the progress of Unāni Ṭibb as well as education have earned for him a permanent place in the history of Unāni medicine in contemporary India. A noble personality, his consultation was free, and he treated the poor with great compassion. He established the Institute for the History of Medicine and nurtured it personally. It was not long before it led to the birth of Jamia Hamdard, New Delhi, which is now a university.

Unāni Ṭibb Today

During colonial rule, Indian states like Rampur, Tonk, Patiala, Bhopal and Hyderabad patronized Unāni medicine. Dai'rat-al-Ma'arif, an institution in Hyderabad was very active in publishing several works on medicine in Arabic and Persian including Al-Razi's *Kitāb al-Hawi* in twenty-three volumes. The Nizamia Ṭibbi College (Hyderabad); the Ayurvedic and Ṭibbia College (Delhi); the Takmil al-Ṭibb College (Lucknow) and the Ajmal Khan Tibbia College (New Delhi) were established during the British reign. After Independence, more than a dozen new colleges for the promotion of Unāni Ṭibb have come up. In 1979, the Central Council for Research in Indian Medicine of the Government of India, was divided into separate councils and, as a result, the Central Council for Research in Unāni Medicine came into existence. It has been engaged in several medicinal activities like clinical research, drug standardization, cultivation of medicinal plants and literary research and publications. *Hamdard Dawākhana*, in Delhi, is the largest pharmaceutical company today. About a decade ago, a National Institute of Unāni Medicine was set up in Bangalore for postgraduate education and research in Unāni Ṭibb.

Though it entered India from West and Central Asia with medicinal concepts and practices of its own, Unāni Ṭibb has been an integral part of traditional Indian systems of medicine, along with Ayurveda and Siddha. Its role in the medical pluralism of India and the health care of the vast masses has been considerable.

8

THE NEW ENTRANT: MODERN SCIENCE IN INDIA

Modern science as we know it today—versatile and methodology-centred—did not originate in India, but came from the West (as has been noted already). For this reason, it is essential to have some idea of its origin and growth in the West, particularly in Europe.

The origin of modern science can be found in the new intellectual efflorescence leading to what is generally called the Renaissance in Science, which emerged in the wake of the Renaissance in the Arts and Letters in Europe in the fourteenth and fifteenth centuries. In the emergent new mood, traditional ideas about matter and motion, space and time, as well as the model of the universe, especially of the solar system, began to be questioned and even revised. Greek philosopher Aristotle (427-347 BCE), an illustrious pupil of Plato, had thought of the universe as being finite and composed of a series of spheres (the form of a sphere being perfect) arranged around the earth as a central entity (geocentric). His model of the universe was: the central earth, concentric spheres (in succession) of water, air, fire, planets and fixed stars, and above all of them, was the sphere of what was called the *Primum Mobile* or the Unmoved Mover. Aristotle projected the idea that planets would move with uniform velocity and in circles around the earth. He thought that circular movements were perfect, since the circle is the perfect figure. His geocentric concept, which was geometrically formulated by Hellenistic astronomers such as Ptolemy (second century CE) was absorbed in the theological frame of the Middle Ages in Europe, including the Church of Rome. Aristotle and the Pythagoreans (followers of Ionian Greek philosopher and mathematician Pythagoras) had also emphasized the distinction between the motion of celestial bodies and that of terrestrial bodies—a speculative duality that suited Christian theological interpretations. The Aristotelian geocentric model of the universe, supported by the Church, reigned supreme till the sixteenth century, a long period of over 1,800 years!

Downfall of Aristotelian Ideas

In 1543, a seminal book *De Revolutionibus Orbium Coelestium* by Nicolas Copernicus (1473-1543) appeared, enunciating a theory that was to change man's concept of the solar system. A Polish astronomer and orthodox religious person, Copernicus put forward the hypothesis that the sun was the centre around which the planets including the earth move—a heliocentric postulate instead of the Aristotelian-Ptolemaic geocentric model. Even in the Copernican postulate, several ancient ideas were retained. The Copernican model still assumed that the universe was spherical and finite, terminating in the sphere of fixed stars. It also accepted that the movements of the celestial bodies were always circular and with uniform velocities. It still invoked epicyclic and eccentric models for understanding the true positions of planets like the Ptolemaic one. Copernicus had, however, argued that in the heliocentric approach the calculations regarding planetary motions and positions would be simpler. Hence the central position enjoyed by the earth and exhorted by the Christian orthodoxy became disturbed and suspect alike. Towards the end of the sixteenth century, a Danish observational astronomer, Tycho Brahe (1547-1601) put forward a model in which the planetary movements (in circles) were heliocentric and the sun, along with the planets as well as the starry sphere, moved around the earth (geocentric), thus reaffirming the concept of the circular motion of the celestial bodies, which still reigned supreme.

However, the downfall of Aristotelian ideas was imminent. In 1584, an Italian of turbulent spirit, Giordono Bruno (1547-1600), published some booklets in London, proclaiming the doctrine of the infinite universe in contradiction to the Aristotelian view of a finite universe. A century earlier, Nicolas of Cusa (1401-64) had also propounded the theory that the earth had motion and that the universe was infinite. Since creation is fundamental to Christianity, which expounds that God created a finite universe, Bruno incurred the wrath of Christian religious leaders who opposed his views of an infinite universe in space and time as well as his other ideas, especially his theory that there are worlds other than ours. It was for these heretical expositions that Bruno had to suffer and he was burnt alive in 1600. However, the spirit of the Renaissance in science continued to gather momentum. In the same year, there appeared in London a refreshing book (in Latin) titled *On the Magnet, Magnetic Bodies, and Concerning the Great Magnet of the Earth: A New Physiology* by English physician and scientist William Gilbert (1546-1603). Besides recording a series of experiments on the properties of the magnet, the last section of this book expounded a system of the universe similar to that of Bruno.

Galileo Galilei

Galileo Galilei (1564-1642), an Italian intellectual of indomitable courage, began systematic experimental investigations (1585) of the mechanical concepts of Aristotle. Soon he countered Aristotelian physics and demonstrated that the Aristotelian idea of falling bodies, that heavier bodies would fall to the ground faster than lighter ones, was wrong. Galileo's famous experiment (1591) from the Leaning Tower of Pisa demonstrated that weights of one pound and of a hundred pounds dropped from the top of the Tower, reached the earth together. Although there is no satisfactory evidence for this story, Galileo had by then demonstrated the importance of experimentation and verification. This led him to prove conclusively that the Aristotelian idea on motion, that bodies fall with velocities proportional to their weights and inversely proportional to the densities of the medium through which they fall, was erroneous. With these new demonstrations and other aspects of motion, a solid foundation for modern dynamics was laid by him. Galileo wrote a treatise titled *De Motu* (On Motion) which, however, was not published during his time.

The years 1609 and 1610 were a watershed in his intellectual, but flamboyant, life. For, in 1609 Galileo devised an instrument with one convex lens and one concave lens that gave a magnification of nine times. Early in 1610 he wrote *Messenger of the Heavens*, recording his observations of the moon's surface, four satellites of Jupiter, sunspots and others. Later, he observed the phases of Venus and the rings of Saturn. Soon he became an ardent exponent of the Copernican heliocentric system. Expectedly, the religious Church was antagonistic to his exposition of Copernican heliocentric ideas. In 1630 Galileo completed his book *The Dialogue Concerning the Two Chief World Systems*, which was published two years later and in which he expounded the Copernican model. Thereafter, he had to face Papal inquisition and imprisonment, and all his works were proscribed. But Galileo had already heralded the dawn of a new spirit and a new attitude to Nature. His dictum was that the *Book of Nature* is in the language of mathematics, that is, the application of mathematics to the understanding of physical problems of matter and motion, space and time. He threw fresh light on motion and its measurement. In fact, his concepts and experiments had in them the seeds of Newton's first and second laws of motion. Galileo thus became a pioneer in the development of the science of mechanices. Nevertheless, though an exponent of the Copernican system, Galileo in a way had not challenged the Aristotelian idea of the circular uniform motion of celestial bodies.

Johannes Kepler

A contemporary of Galileo was the German protestant Johannes Kepler (1571-1630), who was perhaps the first to apply the principles of mathematics for an understanding of the laws of celestial motion. Kepler's was indeed a many-sided personality—mystic, astrologer, Pythagorean and Platonic, but also had a rational mind. Adopting the Copernican view of the universe and, as part of it, the solar system with its planetary orbits, Kepler began to search for a law of motion that was binding together the members of the solar system. However, in 1596 he sought to fit in the five Platonic bodies (that is, geometrical figures with equal sides and equal angles) between the five intervals of the six planets, so that each polyhedron was inscribed in the same sphere about which the next one was circumscribed.

Kepler tried to explain the structure of the solar system by this conceptual arrangement, even though it corresponded to reality in no manner. In a way, he was a mystic Pythagorean, with a strong faith in the so-called 'Platonic bodies'. But the rational attitude in him came to the fore, relegating his mysticism, when he joined Tycho Brahe in Copenhagen and worked in Brahe's astronomical observatory there. In 1601, when Brahe died, Kepler became his legatee, and in 1609 came out his seminal publication *Astronomia Nova* (or *New Astronomy, Based upon Causes, or Celestial Physics, Treated by Means of Commentaries on the Motions of Mars*), in which he indicated that the earth attracts a stone just as the stone seeks the earth; in other words, two bodies near each other will attract each other if they are beyond the third body's influence.

More importantly, this publication dealt with the first two planetary laws which bear his name: (i) Planets move around the sun, not in circles but in ellipses, the sun being one of the foci; (ii) a planet moves not uniformly but in such a way that a line drawn from it to the sun sweeps equal areas of the ellipse in equal times. Kepler's third law appeared a few years later (1618) in his *Epitome Astronomiae,* which states that the squares of the period of revolution round the sun are proportional to the cubes of their distances. He thus undermined the Aristotelian foundation of planetary circular and uniform movements.

The dismissal of the presumed circular motion of celestial bodies by Kepler destroyed Aristotelian authority to a great extent. It also gave credence to such geometrical forms as the ellipse (which existed only in the minds of geometers till that time). It even recognized the reality of the ellipse in the realm of matter and motion. Both Galileo and Kepler conceived of and explained a mathematical universe. In fact, Galileo became the prime votary of what may be called 'Science is measurement'. This positioned science along the path of verifiability and reproducibility, casting off speculative concepts and irrational beliefs. However, even during the time of Galileo and Kepler, the irrational Aristotelian divide of the terrestrial and celestial motions had not been totally vanquished. Both of them had more or less banished the earth-centred universe, but the conception of a sun-centred universe needed additional factual support.

Isaac Newton

In 1687, a path-breaking publication in Latin titled *Philosophiae Naturalis Principia Mathematica* (Mathematical Principles of Natural Philosophy or *Principia* as it is usually called) transformed the course of modern science. Its author was the great seventeenth-century genius, Isaac Newton (1642-1727) of Cambridge in England. He was born in the year in which Galileo had died. Apart from the three laws of motion which go by his name (for which Galileo had provided the seed ideas), he enunciated the law of universal gravitation and proved that the force that causes a stone to fall to the earth is the same as the one that keeps the planets in their path. In other words, Newton mathematically showed that there was a relationship between the so-called celestial motion and terrestrial motion; and that the law of universal gravitation operates equally in the heavens and on the earth. With this Newtonian exposition, the Aristotelian cosmology which was till then accepted and authorized by the Church, was shattered. The mathematics of the heavens, fortified by the Newtonian formulation, was revolutionary in character. For, the planetary motions appeared to be mechanical, capable of measurement, and positional prediction. Newton's mathematical explanation of the motion of celestial bodies was based on their mutual attraction and properties of gravitation, leading him to the conception and quantitative understanding of universal gravitation. Newton demonstrated mathematically the incongruity of the Aristotelian differential approach to celestial and terrestrial motion. More importantly, this also led to a mechanical concept of the universe, which had the potentiality of foreseeable determinism in a causal framework. The science of mechanics, with its verifiability and reproducibility, emerged as an important edifice for understanding the motions of all visible and tangible objects, celestial and terrestrial alike.

The seventeenth century was one of revolutionary ideas. Apart from the law of universal gravitation, Newton conducted experiments on optics and put forward a theory of the propagation of light, which goes by the name of the corpuscular theory of light in the form of a continuous stream of finest particles. On the other hand, the Dutch Christian Huygens (1629-95) expounded a wave-theory of light propagation. A mathematical genius, Huygens worked in the Royal Observatory at Paris (1671-73) and produced his famous work *Horologium Oscillatorium,* on the mathematical analysis of the principles of dynamics. This treatise was rated high for its scientific importance, perhaps next only to Newton's *Principia*. Huygens also measured the acceleration due to gravity by experiments with a seconds pendulum.

One of the most influential thinkers of the seventeenth century was René Descartes (1596-1650), who not only emphasized the nature and the aim of the scientific method but also put forward his bold views on cosmology (infinite and devoid of any empty space). However, his 'Vortex theory' failed to explain the planetary motions, including the theory of Kepler. Francis Bacon (1561-1626), a powerful exponent of the scientific process, was known for his famous four 'Idols' (of the tribe, the cave, the marketplace, and the theatre) or false notions, prejudices and erroneous ways of understanding natural phenomena.

In England, Robert Boyle (1627-91) and Robert Hooke (1635-1703), who were contemporaries of Newton, made notable contributions to the foundations of chemistry and physics. Boyle, known for his law (Boyle's law) which states that the volume of a gas varies inversely as the pressure upon it at a constant temperature, wrote *Sceptical Chymist* (1661) which relegated the then still reigning Greek doctrine of the four elements (earth, fire, water and air) to the background. Moreover, Boyle tried to revive the atomic philosophy of Greek thinkers, Leucippus and Democritus. In this direction, he drew upon the expositions of French philosopher, Pierre Gassendi (1592-1655). Newton too accepted the Greek atomic views, as did the English philosopher, John Locke (1632-1704). The revival of Greek atomism, in contradiction to the Aristotelian views on matter and motion, enlarged the scientific horizon of both chemistry and physics in the nineteenth century.

Notable Trends in Scientific Investigations

Physical and Chemical Investigations: In the eighteenth century, physics began to make rapid progress. The Newtonian concept of mass, namely, the weight of a body divided by gravity, his laws of motion, and other experimental observations accelerated the rise of measurable methodology pertaining to matter and motion, space and time. Several investigations of far-reaching importance were conducted on basic physical phenomena like heat, light, electricity and magnetism as well as chemical reactions—including those involving gases. The most notable among them were methods of temperature measurement and the development of the thermometer; the discovery of latent heat by Joseph Black (1728-99); application of this phenomenon to the improvement of the steam engine by James Watt (1736-1819); the nature of heat, its transformation and conversion into mechanical force; static as well as current electricity, and the latter's production by Alessandro Volta (1745-1827), which showed the relationship between chemical changes and electricity; electro-magnetism and its quantitative aspects determined by Hans Oersted (1777-1851), Francois Arago (1786-1853) and André-Marie Ampere (1775-1836); Michael Faraday's (1791-1867) production of current electricity as power; experiments with gases by Joseph Black (1728-99), Henry Cavendish (1731-1810) who discovered hydrogen, Joseph Priestley (1733-1804) and William Scheele's (1742-86) chemical discoveries encompassing gases and inorganic and organic acids; the discovery of oxygen by A.L. Lavoisier (1743-94) and Priestley. More importantly, it was Lavoisier who scientifically conceptualized chemical 'elements' in the sense they are understood in modern chemistry. He also enunciated in general terms the law of the conservation of mass in a chemical reaction. With him, chemistry began to assume the status of a scientific discipline, transcending its mere empirical nuances and discarding alchemical endeavours.

Astronomy: With the advent of the telescope, observational astronomy was pursued scientifically both in England and on the continent. In the Paris observatory, a micrometer, a telescopic device for measuring smaller angular distances was devised by Huygens who also measured the acceleration due to gravity by experiments with the seconds pendulum. In England, John Flamstead (1640-1719), the Astronomer Royal at the then newly erected observatory at Greenwich (near London), produced a star catalogue. He was succeeded by Edmund Halley (1656-1742), of Halley's Comet fame, who had brought to light Newton's acumen in mechanics and was a noted astronomer. James Bradley (1693-1762) and William Herschel (1738-1822) strengthened observational astronomy and postulated the possible form of the observable universe (lens-theory). Astrophysics, too, registered substantial progress. The use of the telescope added new dimensions to observational astronomy to such an extent in the eighteenth century that the earlier Hellenistic models receded to the background, and a new vision and structure of the universe began to emerge.

Biological Investigations: Biological science, especially physiology, did not lag behind. Starting with the adaptation of the thermometer for clinical purposes by the Italian inventor Santorio (1561-1636) and the conception of blood circulation by William Harvey (1578-1657), physiology began to scale new heights. Bodily muscle actions in terms of mechanical principles were explained by Giovanni Alfonso Borelli (1608-79) who was also an eminent astronomer and a friend of Galileo, and of the famous Italian microscopist, Marcello Malphigi (1628-94). Antony Van Leeuwenhoek (1632-1723) conducted incisive microscopic investigations on the structure and behaviour of living beings; both Malphigi and Leeuvenhoek revealed the nature and extent of capillary vessels, corpuscles of the blood, secretory functions of the glands, and the fibrillary character of muscles. Just as the telescope was a gateway to new astronomical knowledge, the microscope opened up the portals to minute cells and organisms. It thus laid bare the innermost details of plant and animal life. As knowledge of the numbers of living organisms rapidly increased, a scientific way of their codification or classification became essential. Carl Linnaeus

(1707-78) was a Swedish pioneer in this endeavour and his system of classification was expounded in his *Philosophia botanica* (1751) in which *species* were grouped into *genera,* the *genera* into *orders*, and the *orders* into *classes*, the nomenclature being mostly in Latin. He is well known for the 'binomial nomenclature', a system of identifying every known living thing by two Latin names, the first being that of its genus and the second that of its species.

In the Linnaean classification of plants, the classes and orders were based on the number and arrangement of the parts in the flower, the number of stamens (male parts) being a determinant of the class, while the number of 'styles' (female parts) of the flower would be the criterion for the order. His book, *Systema Naturae* was first drafted in 1735 and its tenth edition (1758) is recognized to be the most authoritative source for botanical classification. The French naturalist, Georges Cuvier (1769-1832) was another noted authority, particularly of the animal kingdom, who laid more emphasis on the inner parts rather than the external characteristics. Robert Brown (1773-1858), W.C. Williamson (1816-1895) as well as Johannes Müller (1801-58) and Richard Owen (1804-92), besides several others, placed biology on a solid, enduring foundation.

Nineteenth-century Science: A Bird's Eye View

The nineteenth century witnessed far-reaching developments both in the physical and biological sciences. In the beginning of the century, John Dalton (1766-1844) enunciated his atomic theory (1808) in his classic publication *New System of Chemical Philosophy*, in which he explained chemical combinations. He also stressed that atoms were indivisible and indestructible. Joseph Louis Proust (1755-1826), Claude Louis Berthollet (1748-1822), and others threw light on the then emerging law of definite proportions. French chemist Joseph Gay-Lussac (1778-1850) formulated the relationship of the expansion of all gases with an increase in temperature. In Italy, Amedeo

Avagadro (1776-1856) put forward his famous hypothesis (1811) that equal volumes of all gases, at the same temperature and pressure, contain an equal number of molecules. Several comprehensive chemical investigations were also carried out with regard to gases, atomic and molecular weights as well as classifications based on them. The study of chemistry entered a new phase with the emergence of the Periodic Table (1870) by Dimitri Mendeleev (1834-1907) and Lothar Meyer (1830-95). This enabled chemists to foretell as well as investigate the existence of elements unknown at that time.

Physics forged ahead in the nineteenth century and also experienced an attempt at a synthesis of certain physical phenomena. The doctrine of energy was an important synthetic strand that quantitatively explained the relationship or transformations of heat, light, electricity, magnetism and sound. J.P. Joule (1818-89), a pupil of John Dalton, published a seminal paper, 'On the Calorific Effects of Magneto-Electricity and the Mechanical Value of Heat' (1843), in which he clearly brought out the relation between heat and work. In 1847, another paper by Joule, 'On Matter, Living Force and Heat', expounded that *living force* was one of the most important qualities with which matter is endowed, and it cannot be destroyed in any of the transformations or conversions. This was nothing but *energy,* the term employed by William Thomson, later Lord Kelvin, (1824-1907). The law of conservation of energy (1847), which owed its formulation to Hermann Helmholtz (1821-94), became a revolutionary turning point in the progress of physics.

In biology, too, new ideas emerged: the physiology of plants; cell theory, the nature and structure of organs and tissues; protoplasm and protoplasmic activity, including enzyme action; human bodily processes as a chemico-physical mechanism; circulation of blood; digestive metabolism; and the supremacy of the nervous system and neuron physiology. All of these shed fresh light on the very mechanism of living as well as the development of life. More importantly, the evolution of life and the animal kingdom leading to the emergence of man, were studied in a systematic manner. In the middle of the nineteenth century, Charles Darwin (1809-82) put forward his Theory of Evolution, and his classic, *Origin of Species,* appeared in 1859.

The last quarter of the nineteenth century witnessed scientific investigations of seminal importance, notable among them being the discovery of radioactivity, X-rays and the electron as well as the important work of Gregor Mendel on inheritance. But it was during the first four decades of the twentieth century that there was a revolutionary expansion of the scientific horizon—Quantum theory; Special and General theory of Relativity; the discovery of proton and neutron; the wave-nature of particles; Uncertainty Principle; atomic structure, matter-energy relationship—which gave rise to new (Quantum) physics while the nineteenth century one began to take on the appellation of Classical physics. Fundamental ideas like DNA, RNA, Double Helix and the like added a new dimension to biology. Astronomy and astrophysics too attained significant heights in the twentieth century; likewise space-technology, atomic energy, electronics and information technology; even cosmology, which was rather in the background in the nineteenth century, was pursued in the twentieth century with new theories about the origin of the universe. All of these were in the nature of an enchanting registry of the triumph of science and its method.

The word *science*, as noted before, began to be used from the middle of the nineteenth century as an overarching nomenclature for such methodical investigations primarily based on experimentation, observation and inference, but supported by processes of induction and deduction. It was called 'Western Science' then, and began to be introduced in the lands, including India, 'discovered' by European conquerors. Even earlier, in the latter half of the eighteenth century, such an introduction was in the nature of scientific surveys of the natural wealth of lands under colonial rule.

SCIENTIFIC SURVEYS IN INDIA

The beginning of the seventeenth century witnessed the establishment of two East India companies: one in Britain under the name 'The Governor and Company of Merchants of London Trading into the East Indies', which was granted a charter by Queen Elizabeth I on the last day of 1600; the other, 'The United East India Company of the Netherlands' came up in 1602. Six decades later (1664), the French East India Company was registered in France. European administrators, missionaries, naturalists, engineers and medical men who were associated with these trading companies in India were instrumental in formally introducing the methods of Western science in India.

Early in the sixteenth century, the Portuguese had set foot on the western coast of India (Goa). The Portuguese physician and botanist, Garcia da Orta, had scientifically studied the flora of the region and produced a book on the medicinal and other plants of India. A few Jesuit missionaries were also active in geographical and astronomical investigations. Some employees of the Dutch East India Company had already demonstrated their ability as collectors and classifiers of Indian flora. The most notable among them was Heinrich Van Rheede tot Drakenstein (1637-92), the Governor of Dutch possessions on the Malabar coast to whom the flora of the region looked enchanting. He studied a wide variety of plants, flowers and seeds with the help of his European colleagues as well as a few traditional medical practitioners of Malabar, since medical people had an intimate knowledge of the plants and herbs of the region. With the help of local assistants and artists, Drakenstein produced a monumental compilation under the title *Hortus Malabaricus*, which was published in Amsterdam in as many as twelve volumes with 794 plates (1686-1703). This great work was of inestimable value to the Swedish botanist Carl Linnaeus, in the nomenclature of Indian plants in his renowned work *Species Plantarum*. John Gerard Koenig, a Dane, who had worked with

Linnaeus earlier, was in South India (Madras coast) at that time and began to collect plants and study them scientifically. He was also with the Dutch Mission on the Malabar coast before he was appointed the Natural Historian of the British East India Company in the Madras Presidency. Koenig organized a small group of botanically interested persons under the name 'United Brothers', who were involved in the collection of a large variety of plants, exchanging specimens as well as their knowledge of botanical nomenclature and the associated descriptions. Gradually, the small group expanded, and some of them took up studies in the flora of Bengal Presidency too. To Koenig belongs the credit of having introduced the Linnaean binomial system of nomenclature of plants in India. He also sent some of his collections to the University of Lund in Sweden, thus making Indian flora known to European botanists.

It was, however, the British East India Company that was involved to a great extent in the implantation of Western science in India. But its calculated efforts were largely governed by its exploitative commercial interests after it won the Battle of Plassey in 1757. The traders soon emerged as rulers of India as well.

As a first step in this direction, a thorough knowledge of the geographical features of the new land was necessary and several European surveyors who were employed by the Company set about the task in a scientific manner. In 1761, Plainstead conducted a survey of the coastal region of Chittagong, while Hugh Cameron surveyed the 24 Parganas. Three years later in 1764, James Rennell was active in conducting the survey of the river basin of the Ganges since that would help in determining the waterway for carrying goods from Calcutta. His efficient survey work won accolades from the Governor-General, Robert Clive, who appointed Rennell as the Surveyor-General of India (1767). The survey work continued unabated in the Bengal and Bihar regions as well as in northeastern parts of Assam. To Rennell belongs the credit of preparing the first *Map of Hindoostan* (1783), which was far more presentable and reliable than the *Carte de l'Inde* (Map of India) produced by the well-known French geographer, Bourgignon d'Anville in 1752, the *Carte de l'Inde* being more or less based on the limited geographical data of Jesuit missionaries like Father Bouchet (1719).

Survey of India

As the political ambitions of the East India Company extended towards South India, a survey of this part of India also loomed large. Sustained attempts were thus in place especially for surveying the strategically important regions. Early survey attempts were made by Robert Kelly who, in a planned manner, tried to cover the southern peninsula. It was, however, Michael Topping who played a pioneering role in establishing the Madras Observatory (1792) and a Surveying School (1794). He implemented successfully his novel idea of a continuous series of triangles that would encompass the whole of India in the course of time. For this purpose, he determined as a first step, a nearly 500 km line of triangles along the coast from Madras to Palk Strait, a bold endeavour at that time. Topping appeared to have favoured a trigonometrical survey of India which, of course, began to take rapid strides early in the nineteenth century. In the meantime, Collin Mackenzie (who belonged to the Madras Engineers Corps) nucleated the topographical survey. This led to a purposeful delineation of the natural and other characteristic features of the newly surveyed area. Mackenzie prudently examined the geographical details of Deccan area, including Mysore, then a princely state. The second half of the eighteenth century witnessed the survey work being undertaken in northern India as well as in eastern India, including Odisha.

However, it was in the nineteenth century that the survey work registered substantial progress. This was aided by the use of a number of instruments like the theodolite, telescope, chronometer, compass, plane tables and sextants. The defeat of Tipu Sultan in the Mysore War towards the close of the eighteenth century gave added strength to the British colonial masters. Through their political and economic sagacity, they consolidated the territory that was already under their control as well as the new territories coming under their jurisdiction. In such an expanding situation, the survey of India on new lines, especially the trigonometrical survey, was pursued with great vigour. The East India Company had employed competent men known for their expertise in trigonometrical calculations and geodesy as well as geographical matters; these led to the scientific mapping of India with its plains, hills, mountains and other geographical features.

In 1800, a trigonometrical survey was undertaken in southern India and the first area to be covered was obviously the newly subjugated Mysore region. William Lambton (1738-1823), who was well versed in the geodetic methods of determining the figure or shape of a terrain as well as its gravity measurements, commenced (1800) what came to be known as the Great Trigonometrical Survey of India, with the measurement of a base near Bangalore. During the next eighteen months he determined a preliminary triangulation of the Mysore area. For the first time the inhabitants of Mysore region were exposed to the scientific instruments he used: a theodolite, a circular graduated instrument, zenith sector, telescope, thermometer, chronometer and steel chains. These instruments and the survey method employed by him were to result in a new impact on the minds of the local people.

Lambton pursued the survey work so meticulously for nearly fifteen years that by 1815 he had compiled a fairly accurate map of the southern region. He even determined the largest arc ever measured in any other country at that time and so close to the equator. The entire peninsula, up to a latitude of 15°N, was covered with a network of triangles and the geographical positions of several prominent points were fixed. Lambton's sustained scientific survey activities were hailed as being of extraordinary importance to geodetic science. In fact, his explanations and methodical presentations of the geodetic results were relied upon by other surveyors in India and elsewhere in the succeeding decades.

There was another side of Lambton, and this was his humanism, his loving care of his assistants, both British and Indian. The Indians, braving the vagaries of weather, sickness and accidents as well as being away from their homes and families, took care of his instruments, besides looking after his daily material comforts. Lambton's conduct with the native employees was so exemplary that the other surveyors could not but treat them with humanistic concern and dignity.

In 1815, Collin Mackenzie became the Surveyor-General, but he did not have any authority over the survey work being done under Lambton. However, his work, which had included the areas beyond the Madras Presidency, came under the overall jurisdiction of the colonial government under the name of the Great Trigonometrical Survey of India as well, since it continued in other parts of India, including the lower Himalayan region and areas of the Bombay Presidency. In 1822 an atlas of the whole country was prepared on the quarter-inch scale based on the survey of different regions, and this became the standard map of India throughout the nineteenth century.

Lambton's able assistant was George Everest, who rose to become an outstanding Surveyor-General (1830) as well as Superintendent of the Great Trigonometrical Survey. His tenure of thirteen years proved to be a remarkable period for high-quality geodetic exercises in India. His planned work had the full support of both the colonial administration in India and the British government in London. He laid emphasis on the use of accurate instruments which he had either obtained from England or attempted to produce in India, helped by Henry Barrow, a skilled mechanic and efficient instrument-maker he brought from London. He also spotted an Indian mechanic, Mohsin Hussain of Arcot in South India, who was demonstrably proficient in mechanical repairs and the reconstruction of old instruments. He greatly encouraged Hussain and, when Barrow left for England, appointed him a mathematical instrument-maker at Calcutta, offering him an additional personal allowance of Rs 150 per month, a gesture which Hussain enjoyed even after Everest departed from India. Everest's instruments included a large theodolite with an azimuth circle of 36″ diameter, two double vertical circles, heliotropes, and compensation bars that he used instead of chains. A theodolite assembled by Hussain (in the 1850s) is still preserved in the Victoria Memorial Museum at Calcutta as a shining example of the capabilities of Indian technicians even in the middle of the nineteenth century. His credibility and inventive talents were such that Everest would even take him on his fieldwork tours. In the scientific survey of India, the accomplishments of Everest, namely, the Great Arc covering a distance of about 1500 km from Bidar in the then Nizam's territory (now in Karnataka) to a place called Banog in the Himalayas, and the three subordinate meridonal arcs determined by him, merit special attention.

In his endeavours, Everest was ably assisted by Andrew Waugh who became the Surveyor-General after his retirement and extended the work of his mentor. Between 1845 and 1850 he proceeded with the triangulation of the northeastern Himalayan series. This was recognized as the largest series undertaken in any part of the world then, covering over 2500 km from the base at Dehradun to that at Somakhoda in the Purnea district of West Bengal. During this extensive survey, the heights of as many as seventy-nine Himalayan peaks were determined, using theodolites. Of them, the highest, called the Fifteenth Peak, was 29,002 feet above sea level as determined by Waugh—and he named it Mount Everest in honour of his mentor, George Everest.

The survey work during the tenure of Waugh was extended to the Punjab as well as towards Sind, besides the Kashmir region. On the retirement of Waugh in 1861, the government bifurcated the offices of the Superintendent of the Great Trigonometrical Survey and of the Surveyor-General. Colonel Walker, who was appointed Superintendent of the Great Trigonometrical Survey, and Colonel Thuillier who became the new Surveyor-General, continued their work with great zeal. The result was that the survey of India towards the close of the nineteenth century attained remarkable success, owed not a little to the contributions of Indian associates.

The conspicuous among Indian assistants was Radhanath Sikdar (1813-70), a distinguished alumnus of Hindu College. Known for his mathematical capabilities even at the young age of eighteen, he joined the Great Trigonometrical Survey of India and worked with George Everest and Andrew Waugh. He ably prepared what were known as the *Auxi Tables,* of value for the computations of the survey work. His intellect and devotion to work impressed George Everest so much that he wrote about Sikdar as follows:

> 'Of the qualifications of Radhanath I cannot speak too highly; in his mathematical attainments there are *few* in India—European or native—that can at all compete with him; and it is my impression that even in Europe these attainments would rank very high. As a computer he is indefatigable, and there is no person in my department so thoroughly skillful in the application of the variants formulae... Eventually he will furnish a convincing proof that the aptitude of your countrymen (Indians) for the practical as well as the theoretical parts of mathematics, is in no way inferior to that of Europeans.' (Philimore, *Historical Records of the Survey of India*, Dehra Dun, IV. 340-41; 461)

Appreciative words such as these indicated the response of bright Indians and their contributions to the progress of what was referred to as *Western Science*. An astute computer, Radhanath compiled the meteorological observations too, which appeared in the *Journal of the Asiatic Society of Bengal* from 1858 to 1864. He was also a member of the Physical Sciences Committee of the Asiatic Society for several years. More importantly, in 1864, Radhanath was elected a Corresponding-Member of the Society of Natural History (Bavaria), a rare honour conferred on an Indian by the internationally reputed German Society.

It needs to be emphasized that the surveyed area of India was then more than one-fourth of the total triangulated area of the world, and the largest that had been undertaken so far by a single survey organization. The related astronomical observations had to be fitted into the position of this area on the globe. By 1912, India had been fairly well covered by a series of triangulation, both primary and secondary. A large number of topographical map sheets of contoured surveys within the borders of India and Burma, were produced on scales between $^1/_8$ inch and one inch to a mile, and also on scales of 16 and 32 miles to an inch for public use. By 1940, more than two million map-sheets of different districts of India were made available. In 1923, an air-survey branch was set up for aerial mapping and photography, using the then efficient methods. By 1935, the air-survey library emerged as a repository of aerial photographs of about 12,000 sq. miles.

Botanical Investigations

The survey work and the associated geographical details were necessary for the East India Company to augment its political ambitions. Alongside, some employees of the Company were also interested in the flora of the new land and were active in botanical investigations motivated by the works of European naturalists at home.

Royal Botanic Garden: The year 1787 was a landmark in the history of botanical investigations in India. In that year came into existence what was then called the Royal Botanic Garden (now

India Botanic Garden) at Sibpur, Calcutta, on the west bank of the river Hooghly. It had a fourfold objective: (i) promoting economic benefits to the region by growing commercially important plants; (ii) augmenting food resources and raw materials of plant origin; (iii) importing and acclimatizing or cultivating economically useful plants from other counties; and (iv) fostering experimental studies in plant sciences.

One of the compulsions of establishing a botanic garden then was that teak wood for building the freight vessels at the Calcutta dockyard had to be imported from Burma. When there was some difficulty in getting the Burma teak, it was realized that local efforts would have to be made to grow teak trees in a garden in Calcutta, and the trial yielded the desired results. The British official who mooted this suggestion was Robert Kyd, who had established a name for his horticultural interests. He became the first Honorary Superintendent of the Royal Botanic Garden which then covered an area of over 300 acres. His devoted efforts led to the growing of a large number of species, particularly those of economic importance, including the commercially beneficial cinnamon. During his tenure of six years, Robert Kyd also laid a solid foundation for sustained botanical investigations as well as for proper nurturing of commercially important exotic plants. His successor, William Roxburgh not only expanded the Garden but also initiated and implemented the associated botanical investigations. An astute researcher as he was, Roxburgh had already won high acclaim when he was the East India Company's naturalist in the Madras Presidency through his discovery of indigenous pepper plants at Samulcotta in the Godavari district (now in Andhra Pradesh). He systematically catalogued all the species which he had cultivated and nurtured in the Garden, and produced a catalogue of about 3,500 specimens with relevant details and elegant illustrations for which he employed local artists. His work was published in 1814 under the name *Hortus Bengalensis,* in two parts, by his intimate friend and botanist William Carey, who himself had established and nurtured an enchanting botanic garden at Serampore near Calcutta. Carey also edited and published Roxburgh's well-known botanical compilation *Flora Indica* (excluding that of the Himalayas), which came out in three volumes (1820,1824 and 1832) after his death. Another compendium of Roxburgh was *Plantae Coromandelianae* (three volumes; 1795, 1802 and 1819).

Carey also meticulously prepared over 2,300 coloured illustrations of various plants encompassing as many as thirty-five volumes! These were of great value and a tremendous help to J.D. Hooker, the renowned British botanist of the Kew Garden in London, for his monumental publication tilted *Flora of British India*. Roxburgh's volumes, which are on the Linnaean system, have been recognized as shining examples of systematic work and a source of inspiration for subsequent botanical investigations in India. He has been rightly called the 'Father of Indian Botany'. Roxburgh was succeeded by Buchanan Hamilton, another noted botanist (also a zoologist), as the Superintendent of the Royal Botanic Garden and he produced an extensive commentary on the voluminous *Hortus Malabaricus* of Van Rheede, besides sending a large number (about 1,500) of specimens and drawings (400) to his fellow-botanists in Europe.

The Royal Botanic Garden at Calcutta continued to play a pivotal role in the scientific investigations not only of Indian flora but also that of the neighbouring regions of Nepal and even Singapore where the East India Company had its suzerainty, both political and economic. The Dane, Nathaniel Wallich, who was Professor at the Medical College of Calcutta, succeeded Buchanan Hamilton in 1815. Since he had a special interest in botany, he soon began to explore and collect the flora of Nepal, Assam, Penang in Malaysia, Singapore, and other nearby regions. He systematically classified his rich collections, but donated a major part of his specimens to the Linnean Society established in England, which the European botanists found extremely useful for their own investigations and classifications. The three volumes of *Plantae Asiaticae Rariores* with 300 coloured plates by Wallich were published by the East India Company in 1832. Another book of his, *Tentamen Florae Nepalensis,* was an important book on the flora of the Nepal region.

The Royal Botanic Garden with a large number (over 15,000) of trees and shrubs as well as an amazing collection of herbaceous species was, and continues to be, an exquisite storehouse of great value to varied botanical studies. Its Herbarium has extensive dried plant materials of the entire Indian subcontinent as well as the other parts of Asia and far off regions of Australia and Europe. Recognized as one of the most important herbaria of the world, its contribution to global botanical knowledge has been immense, fortified by constant exchanges of specimens with other herbaria particularly with the Kew Garden in Europe. With its more than two million botanical specimens and periodical interchanges with other leading herbaria in the world like the British Museum in London, the Jardin des Plants in Paris, gardens of St Petersburg, Java and Sri Lanka, the Botanic Garden at Calcutta emerged as a significant scientific enterprise of the Indian colonial administration.

Two cyclones—in 1864 and 1867—on the Bengal coastal region caused extensive destruction to the Royal Botanic Garden. The 1864 cyclone brought in its wake six to seven feet of sea (saline) water and submerged practically the whole Garden. The situation was worsened by the devastating second cyclone within three years. In the meantime the colonial government had appointed a new superintendent of the Garden, George King. Under his direction, the Garden was renovated and reorganized with a phytographical arrangement. A medical man from the University of Aberdeen, George King was a dynamic and imaginative person. He had earlier functioned as the superintendent of a botanic garden at Saharanpur

(United Provinces) and also as Assistant Forest Conservator at Dehradun. With his rich experience and versatality, he was able to bring back the Royal Botanic Garden to its pristine glory. It was George King who initiated steps for the ushering in of the Botanical Survey of India in 1890. In recognition of his sustained efforts, the Government appointed him *ex-officio* Director of Botanical Survey of India. Besides being engaged in the collection of the flora of the Malay Peninsula, King started the well-known and useful publication titled *Annals of the Royal Botanic Garden*. In 1897 he left India to take up the position of Director of the Botanic Garden at Kew in London.

The Botanical Survey of India played a pivotal role in the augmentation of several botanical studies. David Prain, who succeeded George King and had already made a name for himself as an able curator of the Herbarium, initiated taxonomic studies relating to *Papaveraceae, Leguminosae, Scrophulariaceae* and the like. He published a series of volumes under the title *Noviciae Indica*, besides his own book titled *Bengal Plants.* The importance of botanical investigations was also recognized by the departments in the Presidencies. Several botanists were active in their researches and noted among them were: Theodore Cooke (Principal of Science College, Poona and Director of Botany, Bombay region); J.F. Duthie (northwestern India in Saharanpur); and M.A. Lawson (southern region of Madras). Several valuable publications like *Flora of the Bombay Presidency* (Theodore Cooke, in 2 vols; 1901-08); *Flora of the Upper Gangetic Plain* (J.F. Duthie) and the *Flora of Madras* (M.A. Lawson) enriched global botanical knowledge. The botanic garden at Saharanpur was equally active. Noted investigations were by George Govan, John Forbes Royle and Thomas Thomson, who shed light on the natural history of the Himalayas. Thomson collaborated with John Dalton Hooker (in London) in the publication of the first volume of *Flora Indica* (1855). The concerted efforts of the Botanical Survey of India and of several eminent botanists contributed greatly to the scientific understanding of the floral wealth of India in the global context.

By the end of the nineteenth century, the botanical investigations in India by and large conformed to systematic profiles and studies in economic botany. J.D. Hooker's monumental work *Flora of British India* was completed in seven volumes in 1897. The publication of the flora of the various provinces was also undertaken. In 1896, George Watt's *Dictionary of the Economic Products of India* also came out. Charles Baron Clarke (1832-1908) who was associated with the Garden at Saharanpur, though a mathematician by training, produced a monograph on the ferns of northern India.

In the first decade of the twentieth century, florist investigations in different parts of India continued with renewed vigour. With the establishments of the imperial and provincial Departments of Agriculture, the mycologists and economic botanists of these departments were engaged in applied researches. The Imperial Agricultural Research Institute came up in Pusa (Bihar) in 1903 and botanical investigations continued as part of agricultural research. In the first decade of the twentieth century, there did not appear to be much research in botany in any of the universities or colleges attached to them, these institutions being more concerned with teaching botany than research. However, after the First World War, the Departments of Botany of the universities were active in researches in morphology, anatomy, embryology, plant physiology, systematic botany, applied botany, palaeo-botany and ecology to which several Indian botanists made substantial contributions.

Meteorology

A thorough knowledge of the general climate, rainfall and temperature variations from season to season of the new land was essential for the Europeans—seamen, engineers, missionaries, medical men, and even administrators who would often move from one place to another to follow their vocations. Rainfall, humidity and the range of temperatures began to be recorded using barometers, hygrometers and thermometers, respectively. T.D. Pearse, an employee of the Company, meticulously made such recordings for over three years (1785-88) at Calcutta. Rainfall measurements were also recorded in a few other places, including Ujjain in Central India. But it was in the nineteenth century that steps were initiated for systematic recordings over specified periods in different parts of the country that were under colonial rule. Among them were: tidal observations near Calcutta by James Kyd for the years 1805 to 1828; preparation of weather-temperature registers by G.T. Hardwick at Dum Dum covering a period of eight years from 1816 to 1823; similar recordings by James Princep at Banaras; also registering the rainfall and draught conditions for eight seasons from 1831 to 1838 by R. Everest. These provided the needed incentive for further progress in meteorological studies in a planned manner. In 1848, for the first time, the daily maxima and minima temperatures were recorded and in 1856, hourly observations too were started.

Storms of devastating proportions occurred at intervals, prompting methodical studies of their periodicity or regularity. H. Piddington, who was the curator of the Museum of Economic Geology and president of the Marine Court of Enquiry, carefully observed such storms, and compiled as many as twenty-three of them, which were published in his *Memoirs* on the *Law of Storms*. It was he who coined the term 'Cyclone', (for the curling type of storms, derived from the Greek word *Kuklos,* meaning the coil of a snake). He also compiled an account of major cyclones that occurred in the Bay of Bengal between 1829 and 1857.

The interest evinced by colonial administration and its technically qualified employees in meteorology, stimulated some princely states to take up meteorological studies. In 1836, the Raja of Travancore established an observatory for both astronomical and meteorological recordings in

Trivandrum (now Thiruvanantapuram). (John Allan Broun worked in this observatory as an astronomer from 1852 onwards). A branch observatory was also set up in 1855 at a hilly place known as Agastya-mūla, a peak of 6,200 ft on the Western Ghats. The hourly observations made there related to atmospheric pressure, temperature and humidity. The colonial administration gradually set up a number of meteorological stations in different parts of the Indian subcontinent, which led to a somewhat general understanding of its meteorological profile. However, it was felt that the observations being made needed to be intensified with a uniform plan and, therefore, a comprehensive plan was prepared in 1863. In 1864 a devastating cyclone lashed the upper coast of the Bay of Bengal, which resulted in destructive consequences on a large scale in Calcutta and its neighbourhood—80,000 human beings and a large number of cattle died, and several ships were damaged.

Two years later, in 1866, a catastrophic famine occurred and the Bengal region reeled under its scourge. All these disasters emphasized the importance and necessity of augmenting meteorological observations. Between 1866 and 1871, several observatories were established and assigned with the task of recording the atmospheric data and issuance of cyclone warnings and tidal effects, besides the desired astronomical and magnetic studies through efficient scientific instruments like self-registering anemometers, standard barometers, thermometers, pluviometers, hygrometers and tidal gauges.

Several meteorological observatories were set up at an all-India level. In 1875 the India Meteorological Department was established, with H .F. Blanford as Meteorological Reporter to the Government of India. The main functions of the department comprised the preparation of daily weather charts and issuing weather summaries, collection of seismological data, terrestrial magnetic studies, and observation of solar phenomena. Blanford energized the whole range of activities of the all-India department and the provincial departments alike. By 1885 there were as many as 128 observatories, and an additional twenty-two in the Bengal Presidency. There was also one in Dehradun for the collection of data concerning the forest weather established at the Forest Research Laboratory there, on the initiative of Blanford.

An able administrator, H.F. Blanford was a devoted experimentalist, meticulous compiler and analyzer of the observational data. He prepared a catalogue of cyclones that occurred in the Bay of Bengal up to the end of 1876. With his colleague John Eliot, he made notable contributions to the understanding of the nature and extent of monsoon rainfall and tornadoes. Eliot, who succeeded Blanford in 1889 as the Meteorological Reporter to the Government of India, introduced a system of daily reports on the weather at an all-India level. It was only in 1884 that an Indian trained in physical science could get employed in the department and he was Lala Ruchi Ram Sahni, who was associated with the preparation and issuance of daily weather reports. It is significant that even today, Blanford's *India Meteorologists Vade-Mecum* and *The Climate of India, Burma and Ceylon* as well as Eliot's *Handbook of Cyclonic Storms in the Bay of Bengal* and the *Climatological Atlas*, constitute the very foundation of our knowledge of weather and climate in India as they were at the end of the nineteenth century.

The India Meteorological Department continued to play a major role in the forecasting of seasonal rainfall, which was beneficial, then as now, to our predominantly agricultural country. Gilbert Walker succeeded Eliot and he was elevated as the Director-General of Observatories. During 1906-07 he introduced new correlation-coefficients for the monsoon forecasts. This was improved further by C.W.B. Normand (who succeeded Walker), leading to a reliable forecasting of monsoons. Before 1914, daily weather prediction in India was based on surface data. Since then, what was called upper-air data has also been utilized to arrive at better results. Such data comprised the velocity and direction of wind at various heights above the ground, these being collected from a network of as many as forty or more stations in different parts of India. Several Indian meteorologists pioneered investigations on the upper atmosphere and ionosphere, winning national recognition for their contributions to Indian meteorology, even in the colonial dispensation of the time.

Modern Astronomy: Investigations in India

The colonial government was interested in promoting astronomical studies in India for its own benefit. In 1792, the Governor of Madras, Charles Oakeley, established what was then known as the Madras Observatory for 'promoting the knowledge of astronomy, geography and navigation in India'. This was facilitated by the earlier (1787) private efforts of William Petrie, a Member of the Government of the Madras Presidency, who, on his own, had set up an observatory with some instruments and these were transferred to the new observatory. John Goldingham, a Fellow of the Royal Society of London, was the first noted astronomer who, along with his deputy, H. Warren, made important astronomical observations. Goldingham's successor, Thomas Glanville Taylor, placed the new institution on solid foundation by the middle of the nineteenth century, with new instruments like telescopes, a circular measurer and astronomical clocks. He endeavoured to prepare a Star Catalogue (about 11,000 stars) which was published in 1844.

N.R. Pogson, an able astronomer who equipped the Observatory with instruments of precision including a transit circle and an equatorial, also prepared a catalogue of about 5,000 stars. He worked in this Observatory for 30 years (1861-1891) and discovered six minor planets between the orbits of Mars and Jupiter, as well as six variable stars. His able Indian associate, C. Raghoonathachary, made important astronomical observations particularly those relating to the light variation in variable stars.

Pogson had also thought of setting up a branch of the Madras Observatory in Palni or Nilgiri Hills to record specially the photographic and spectroscopic observations of the Sun and stars. But it did not materialize during his tenure. In 1891, when Mitchie Smith became the first Director of the Observatory, the project began to take shape. As a result, in 1900, the Solar Physics Observatory came up in Kodaikanal on the upper Palni Hills, but administratively under the control of the India Meteorological Department. Investigations on the sunspot spectra, solar prominences and their hydrogen content, night-sky spectra and the like, besides the meteorological and seismic studies, were undertaken.

Earlier in 1836, the Rajah of Travancore had established an observatory at Trivandrum (now Thiruvananthapuram). John Alan Braun, who was Astronomer to the Rajah, worked there. Poona (now Pune) had an observatory called the Maharajah Takht Singhji Observatory (established by the Maharajah) which possessed the largest telescope (20 in. Grubbs Reflector) in the country then, where a diligent astronomer, K.D. Naegemwala made notable observations, particularly of eclipses including the solar corona of an eclipse which occurred in 1898.

Early in the twentieth century, the Nizamia Observatory came into existence (1908) near Hyderabad due to the patronage of the Nizam. It took part in the international photographic mapping of the sky between 17 and 24 degrees North assigned to it. In Kodaikanal, Evershed who succeeded as Director of the Solar Physics Observatory, studied the solar prominences, spectra of sunspots and discovered the radial motion in sunspots, which is now known as the 'Evershed Effect'.

Astronomical investigations apart, India participated in the global study of the earth's magnetism, which was organized by the Gottingen Magnetic Union to record simultaneous magnetic observations at fifty stations (1834-41) including three in India (Madras, Shimla and Trivandrum). As a sequel to it, the Magnetic

Observatory was established in Colaba, Bombay (now Mumbai). This Observatory was shifted to Alibag near Bombay in 1904.

Another important discovery related to geophysics known as the Theory of Isostatic compensation (a condition in the earth's crust whereby equal earth masses underlie equal areas down to an assumed level of compensation) was formulated by A. Pratt, who was a mathematician.

Geological Studies and Explorations

European administrators and naturalists, engineers and surveyors were fascinated by the variegated geological formations of India. Apart from their pursuits in their own fields, some of them began to explore the mineral wealth of India. In 1820, H.M. Voysey, basically a surveyor but well versed in geology, who was associated with William Lambton in his trigonometrical survey work, published an account of the diamond mines of South India. He also studied the mountainous terrain north of the river Krishna and named the rocks which constituted the mountains the 'Clayslate Formation'. P.M. Benza, a surgeon by profession in Madras, noticed the basaltic dykes while examining the geology of the Nilgiris. He also brought to light the kinds of iron ores available in the southern region. Yet another medical man, T.G. Malcamson studied the formation of the Deccan trap and the Lonar Lake, while J.T. Newbold (in South India), James Franklin, W.S. Sherwill and J. Homfrey (Central and Eastern India), with their expertise in geology, revealed coalfields as well as some mineral deposits such as of manganese in these regions.

Palaeontology: In the Siwalik region of the Himalayas Hugh Falconer pursued palaeontology with vigour. Proby Cautley who was engineer-in-charge of the Jamuna canal project and two other engineers, William Baker and Henry Durrand, worked with Falconer in the investigations of fossil fauna. The discovered fossils have been associated with the names of Falconer and Cautley: *Mastadon perimensis* Falc. and Caut; *Sivatherium giganteum* Falc. and Caut; and *Equus sivalensis* Falc. and Caut. The originals of these fossil bones have been preserved in the British Museum (London); the reproduced ones or the replicas are preserved in the Indian Museum at Calcutta. These discoveries added a new dimension to palaeontology, as reflected in the *Memoirs of Falconer and Cautley* on the fossil species, and a descriptive catalogue of the vertebrate fossils of the Siwalik hills, the Narmada valley and other regions by Falconer himself, which have become classical treatises on palaeontology. The fossils of mammalian fauna of the Siwalik Hills discovered by H. Falconer and P.B. Cautley expanded the horizon of knowledge of the evolution of mammals. For their outstanding work, the prestigious Wallaston Medal in duplicate was awarded (1837) to both Falconer and Cautley. The Geological Society of London, of which the famous geologist Charles Lyell was the president, vested the Society's honours on young, bright Falconer who was then under thirty years! When Falconer died (1865), the presidents of the Royal, Linnaen, Geological, Geographical and Ethnological Societies of England joined together to raise 'Falconer Memorial Fund' and installed his bust in the precincts of the renowned Royal Society of London.

While the palaeontological studies were of scientific importance, the colonial government was naturally more interested in the commercial exploitation of the mineral wealth of India; the coalfields in the eastern parts of India in particular attracted the attention of the East India Company. In 1836, Governor-General Auckland set up a committee for the investigation of coal and other mineral resources of India with John McClelland of the Medical Services as its secretary. A year before, in 1835, the Geological Survey of Great Britain had been established by the British government to further geological investigations of its own land. The British Association for the Advancement of Science was also interested in the exploration of the coalfields and other mineral wealth of India. As a result, in 1846, D.H. Williams of the British Geological Survey came to India and explored the coalfields in Raniganj, Jharia, Dhanbad, Taldange and the neighbouring areas. Systematically, he uncovered the nature and extent of the coal reserves. With his dynamism and geological expertise, Williams laid the foundation of the Geological Survey of India, before he died, struck by jungle fever in 1848 in Hazaribagh in Bihar.

Geological Survey of India: The geological investigations became broad-based and were vigorously pursued from 1851 onwards with the arrival of Thomas Oldham, a dedicated and energetic geologist. He was already a Fellow of the Royal Society of London and professor of geology in the University of Dublin. The colonial government too, under Governor-General Lord Canning, was keen on exploiting various minerals, starting with the coalfields in Jharia, Bokaro, the Narmada Valley and other places. For this purpose, the Geological Survey of India was established in 1851 with Thomas Oldham as its first director.

In addition to surveying and examining the several coal-belts in Bengal and adjoining areas, Oldham proposed the name 'Vindhyan' for the great sandstone formation of northern and central India which, he noted, could not be correlated with any of the existing European formations. He showed the identity of the extensive thick layers of sandstone and shales with plant fossils of the Vindhyan areas, comparing them with coal formations in Burdwan, Hazaribagh and Cuttack. He christened the range of thick and ferruginous sandstone formations *Mahadevas.* Oldham was also interested in seismology and, under his stewardship, the first catalogue of Indian earthquakes was prepared, followed by a list of about 300 thermal springs. To Oldham belongs the credit of having recruited into the Geological Survey of India, some bright Indian geologists as apprentices. Ram Singh (1873), Kishen Singh and Hari Lal (1874) were the first Indians to work in the Geological Survey of India.

Oldham had also built up a team of efficient geologists: W. Theobald, H.B. Medlicott, and the Blanford brothers. Medlicott, who had established himself as a meticulous geologist, was professor of natural history and economic geology in the Thomason College of Engineering at Roorkee. He joined the Geological Survey of India and, encouraged by Oldham, studied Himalayan geology (1857-61); he also examined the glacial action and gneissose granite in the Duladhar and Kangra valleys and surveyed the northern side of the Narmada valley, showing that the rocks of the Raipur basin did not belong to the coal-bearing series. More importantly, Medlicott suggested the term *Gondwana*, for the extensive southern continent. He succeeded Oldham (1876), became director of the Geological Survey of India in 1885, and expanded its activities in a planned manner.

Distinguished geologist W.T. Blanford joined the Geological Survey of India in the same year when Medlicott became its director. He had already surveyed the coalfields at Raniganj and other adjoining areas, and had worked with Oldham on the correlation and classification of geological formations of southern peninsular India. His brother, H.F. Blanford, though basically a meteorologist as noted before, was also interested in geological investigations. Along with W. Theobold and his brother, W.T. Blanford surveyed the Talchir boulder-beds in the Cuttack region, and threw fresh light on their formation. He also examined the physical geography of the western desert, and to him goes the credit of publishing for the first time a geological map of India on the scale of one inch to one mile. In recognition of his outstanding geological investigations in India, the Geological Society of London awarded him the prestigious Wollaston Gold Medal. His brother H.F. Blanford, F. Stoliczka and Bruce Foot examined the geology of South India including the Nilgiri Hills, and the first two brought out in four volumes the *Cretaceous fauna* of South India. V. Ball, who was an economic geologist, made significant contributions to the understanding of the geological formation of the Andaman and Nicobar islands in the Bay of Bengal. William King, who succeeded H.B. Medlicott as director of the Geological Survey of India in 1887, explored the coal deposits of Singareni in the then State of Hyderabad. T.H. Holland examined the geology of Madras Province and proposed the name *Charnockite* to the acidic pyroxene-bearing rocks existing there, in honour of Job Charnock, the founder of Calcutta. It was Holland who discovered the high-grade large deposits of iron ore, chromite and magnesia in the southern districts of Madras Province, besides the hard corundum near Mysore.

Of the extensive geological investigations carried out in the Indian subcontinent, special mention needs to be made of the great Vindhyan and the Gondwana systems as well as the Gangetic alluvium as land deposits. The vast southern continent of Gondwanaland of the upper carboniferous glacial epoch encompassed Australia, South Africa, and even South America. Thus, Indian geological explorations enhanced the knowledge of the earth as a whole. The study of earthquakes in India, too, elucidated the main types of earthquake waves and the internal structure of the earth. The Geological Survey of India, though saddled with the exploitation of the mineral resources of India by the colonial government in furtherance of its enlightened self-interests, contributed substantially to global geology.

By the end of the nineteenth century, the Geological Survey of India had as many as fourteen officers out of whom two were Indians. P.N. Bose, a bright and accomplished geologist, was the first Indian to hold a graded position in the Geological Survey of India in 1880, while P.N. Dutta was another able geologist, who discovered the extensive deposits of manganese ore in the Bandwara and Chhindwara river valley in 1888. P.N. Bose examined the geology of the then Central Provinces, the Vindhyan and the igneous rocks of Raipur and Balaghat. He was the first to provide detailed microsections of rocks in his progress reports. A nationalist to the core, P.N. Bose chose to come out of the Geological Survey of India in 1903 when the British T.H. Holland, who was junior to him by eight years, was appointed director of the Geological Survey of India, superseding him. Soon he became the geologist of the state of Mayurbhanj and discovered large deposits of iron ore there. He had a chance meeting with the industrialist Jamshetji Nusserwanji Tata who was contemplating the manufacture of iron and steel—and this encounter led to the establishment of an iron and steel works at a place now called Jamshedpur, in memory and honour of the industrialist.

Indian geologists also made significant contributions to geological investigations during the pre-Independence period, notable among them being the elucidation of the *Gondwana* and the Deccan Trap, as well as the tectonics and stratigraphy of the Himalayas, the Salt Range and the Indo-Gangetic trough. Their explorations and studies enriched the knowledge of economic geology and palaeontology. Geology, as a subject, found its place in the curricula of several universities. The associated fieldwork and investigations augmented the knowledge of Indian geology.

Zoological Investigations

The meteorological, botanical and geological investigations made headway in the nineteenth century as they were beneficial for the colonial government. There was, however, some disinclination on the part of European naturalists towards pursuing zoology in the beginning. Even William Jones, the president of the Asiatic Society, who had an extraordinary interest in natural history, was lukewarm towards zoological studies. Nevertheless, some naturalists led the way. An early distinctive zoological study was by Brian Houghton Hodgson, who in 1829 shed light on a new species of *Buceros*. One of the first animals studied by him was a serow, which he named *Antilope Thar*. He also identified several new species of animals including *Paradoxurus, Cerves, Plecotus, Talpa macruka,* Merceline birds, porcupines and Indian snipes. Hodgson, in addition to his being an exemplary recorder of Indian vertebrates, was well known for his detailed studies on the birds of Nepal, and the kind of animal he called *Takin*. All of these—his meticulous researches and observations on the sheep and goats of the sub-Himalayan region and Tibet—monkeys, musk deer, otters, antelopes, mammals and flying squirrels were catalogued systematically, with impressive drawings (published in 1847).

Another zoologist, S.R. Tickell, who earned a name for himself in Indian ornithology (the study of birds) was one of the best field naturalists. His contributions to the scientific understanding of gibbons, brown flying squirrels and the sloth bear are well known; likewise his investigations on a new species of hornbill and the new genus of the *Gadidae* of Arkan. He produced an authentic and well-illustrated manuscript on Indian mammals, and this served as a source of reference for other naturalists like W.T. Blanford. This manuscript is still preserved in the library of the Zoological Society of London.

E. Blyth, the first curator of the Asiatic Society Museum, was keenly interested in the study of specimens of birds and mammals not only of India but also of Burma, Ceylon (now Sri Lanka), Afghanistan and the Malay Peninsula. His publications *The Catalogues of Birds* (1852) and *Mammalia* (1863), based on the Asiatic collections, are regarded as classics among zoological publications. His own investigations were concerned with wild sheep, *Magaderma, Cuculidae, Phylloscopus*, reindeer and certain kinds of asses. Blyth, who was rightly regarded as the founder of what came to be known as a school of field zoologists in India, was eulogized as one of the finest field zoologists of his time by no less a person than Charles Darwin, the propounder of the Theory of Evolution. Blyth was indeed an exemplary zoologist inasmuch as he kept up an active correspondence with other naturalists both in India and abroad, and produced several memoirs with indefatigable energy.

W.T. Blanford was a dynamic naturalist who, in addition

to his own important contributions to geology, began to collect and describe the various kinds of reptiles, molluscs and birds of different parts of the country. His knowledge of these was so vast and accurate that, when he retired and went to England, he was appointed the editor of a series of volumes on the fauna of British India. He himself brought out two volumes on birds and one on mammals which were published between 1888 and 1891. Herpetology or the study of reptiles attained new heights with the investigations of W. Theobald, H. Benson, T. Hunter and W.T. Blanford. Among the other noted zoologists were G. Navil (who specialized in marine and estuarine mollusca); H. Godwin-Austen (birds of Assam); F. Stoliczka (molluscas, invertebrates and vertebrates); R. Lyddekar (fossil vertebrates of India, antelopes, etc.); McClelland and F. Day (*Fishes of India*); A.L. Adams and Aitchison (mammals); Jerdon (author of *Birds and Mammals of India);* A.O. Hume (birds); G.S. Roden (bears and the Himalayan game animals), and R.C. Wroughton (rodents).

The establishment of the Indian Museum at Calcutta was an inspiring source for the promotion of research, especially in zoology. Over the years, the scientific staff of the Museum contributed a great deal to the zoological knowledge of a wide variety of fauna. A. Alcock's publication titled *Materials for a Carcinological Fauna of India* (in six parts; 1885-1900) was an important source-book on marine crabs of the Indo-Pacific region. The Marine Survey Department set up by the colonial government in 1874 undertook deep-sea biological investigations, particularly in the Andaman and Nicobar region in the Bay of Bengal, the mouth of the Ganges near Calcutta, as well as the Persian Gulf area. The Asiatic Society played a catalytic role in moving the colonial government to undertake deep-sea dredging in Indian waters and also investigate the conditions of life in the sea on the lines of British *Challenger* (ship) expedition. Another survey ship, *Investigator,* initiated studies on marine life, and its work continued for more than three decades (the 1880s to 1914) until the First World War interrupted it. However, another expedition into the Arabian Sea called the John Murray Expedition under the leadership of R.B.S. Sewell, enriched the knowledge of Indian marine fauna. Sewell and N. Annandale also explored the fauna on the banks of the Chilka Lake (a marine lake on the Odisha coast), leading to further studies of the fauna of brackish water.

Before Independence, several Indian zoologists had made important contributions in helminthology, entomology (study of insects) including economic entomology, ichthyology (study of fishes), herpetology (reptiles), ornithology (birds), and mammology (mammals). With the introduction of zoology as a subject in the curricula of the newly established universities, zoological researches in these areas were encouraged to a great extent. Morphology or the study of the structure of animals (Indian type) received adequate attention. Higher education in zoology also received its due recognition by some Indian universities. On the eve of Independence, zoology in India had gained its own status as an important biological discipline.

The field organizations which were engaged in surveys, meteorology, geology, botany and zoology, had to their credit a number of valuable scientific publications and played an important role in fostering a scientific climate in India. That the colonial government encouraged them in its own self-interest cannot be denied, but it does not diminish their contributions towards enlarging the scientific base in India. The fact that these organizations have become an integral part of the promotion of science in India even after Independence reveals their importance, then as now.

INTRODUCTION OF WESTERN EDUCATION

The East India Company, which began to establish political rule from 1757, was disinterested in developing an educational system for the people of the new land. For, even at home in

Britain then, the situation was not inspiring. Before the industrial revolution that began in the middle of the eighteenth century, public education was not pursued as a social necessity. Primary education was by and large in the hands of the religious clergy, and of organizations involved in the promotion of Christian knowledge through what were called Charity Schools. The industrial revolution demanded new methods and skills for which workers had to be trained, making it necessary for a versatile educational system to be in place. In the latter half of the eighteenth century, a substantial number of public schools sprang up, but the introduction of scientific subjects into the school curriculum was still a far cry. There were luminaries of science in the great universities—Oxford and Cambridge—but their work and ideas reflected their personal acumen rather than any effective or purposeful scientific syllabus. Religious studies and philosophic discourses received greater attention. It was only in the beginning of the nineteenth century when new socio-economic changes resulted from the industrial revolution that the British public education system began to give more emphasis to science. In the middle of that century, some attempts were made to introduce scientific subjects in school education. Cambridge University established natural sciences trips in 1848, and Oxford's Scientific School of Physical Sciences came up in 1849. Mathematics was, however, recognized as a seminal subject by both the universities.

The situation was rather different in India. Towards the end of the eighteenth century, European scholars employed by the East India Company developed an interest in studying Indian traditional sciences as well as religious literature in Sanskrit. The newly founded Asiatic Society encouraged studies in Vedic texts, Indian philosophical systems, traditional astronomy, mathematics and other scientific subjects found in Sanskrit manuscripts with the help of local *paṇḍits*. Impressed as they were with the wealth of knowledge contained in them, in mathematics, astronomy and medicine, not a few European scholars were inclined towards the promotion of traditional education. They were called Orientalists, while those in favour of the introduction of Western liberal education with English as the medium of instruction were known as Anglicists.

Warren Hasting, the Governor-General at the time, approved of the establishment of a *madrasa* in Calcutta in 1781, largely due to the pressure of Muhammadan religious leaders, to educate young students in Muhammadan law and what was called the Islamic 'rational sciences'—all in Arabic and Persian. Educational training in this institution underwent some changes; in 1826 an English section also began to function in this institution.

In 1791, some *paṇḍits* of Benaras, who were advocates of Sanskrit learning, prepared a scheme with the support of Jonathan Duncan, the British Resident in Benaras, for the establishment of a college for the teaching of Hindu laws, literature, philosophy, mathematics and medicine—all in Sanskrit. After Duncan relinquished his office in 1799, this college suffered a setback with misrule, and a committee of European gentlemen was set up to administer it. In 1830 a separate English class was also added to this institution in order to teach scientific and arts subjects in English.

Educational Institutions in India before 1857

The first perceptible landmark in the establishment of educational institutions in colonial India was that of the Hindu College or the *Vidyālaya* (also known as *Mahāvidyālaya* as well as Anglo-Indian College), which was established in January 1817 at Calcutta. Early in 1816, several Indian and European thinkers residing in Calcutta assembled to discuss the manner of providing liberal education on the European model to Indian students. David Hare, a watchmaker by profession and a philanthropist, prepared a plan of the proposed institution. Raja Ram Mohun Roy, the Raja of Burdwan, Raja Radha Kanta Dev, David Hare and a few other prominent Hindu and

European gentlemen participated in the meeting, which was held at the residence of Sir Edward Hyde East, Chief Justice of the Supreme Court at Calcutta. That was when they decided to establish the Hindu College and set up a committee consisting of ten Europeans and twenty Indians. The main object of this college was to 'impart education to the sons of respectable Hindus in English, Indian languages, European science and literature'. The experiment of educating the natives in the European—particularly the British—methods of imparting education and learning were adopted successfully. The ideas of eminent British thinkers like Francis Bacon, John Locke, David Hume, Jeremy Bentham and Adam Smith appealed to bright young Indian students, who were also attracted to scientific subjects, and all of them were taught in the medium of English.

In the beginning, the Hindu College had only twenty students, and for a decade the number of students did not reach even a hundred, the average being about seventy. In 1827-28, however, the number of students rose to over 430. The subjects taught included natural and experimental philosophy, chemistry, mathematics, algebra, British and European history, as well as works of William Shakespeare and John Milton. Within two decades the college earned the admiration of people of all walks of life. C.E. Trevelyan in his book *On the Education of the People in India* (Calcutta, 1838), wrote: 'A command of the English language and a familiarity with its literature and science has been acquired (by the students of the *Hindoo college*) to an extent rarely equalled by the schools in Europe.' The General Committee of Public Instruction (1835) set up by the East India Company remarked that 'a taste for English has been widely disseminated and independent schools conducted by young men reared in *Vidyālaya* are springing up in every direction.' But H.H. Wilson, who was profoundly involved in oriental learning and who was the secretary of this General Committee, showed his inclination to encourage oriental learning in preference to European liberal education, including science. The Committee even chose to recommend the establishment of a Sanskrit school in Calcutta. However, the Committee's attempts met with serious disapproval both in India and England.

In India, Raja Ram Mohun Roy, the versatile social reformer, vehemently protested against this move of the General Committee. In a letter* to the Governor-General, Lord Amherst, in 1823 he had exhorted as follows (Sharp. H., *Selections from Educational Records*, I, Calcutta, 1920; pp. 101-2):

> 'Humbly reluctant as the natives of India are to obtrude upon the notice of government the sentiments they entertain on any public measure, there are circumstances when silence would be carrying this respectable feeling to culpable excess. The present rulers of India, coming from a distance of many thousand miles to govern a people whose language, literature, manner, customs and ideas, which are almost entirely new and strange to them, cannot easily become so intimately acquainted with their real circumstances as the natives of the country are themselves. We should, therefore, be guilty of a gross dereliction of duty to ourselves, and afford our rulers just ground to complain at our apathy, did we omit, on occasion of importance like the present, to supply them with such accurate information as might enable them to devise and adopt measures calculated to be beneficial to the country, and thus secured by our local knowledge and experience, their declared benevolent intentions for its improvements.
>
> The establishment of a new Sanskrit School in Calcutta evidences the laudable desire of government to improve the natives of India by education—a blessing for which they must ever be grateful; and

*The letter is reproduced here as it reveals the inner thoughts and aspirations of prominent Indians of that time.

every well-wisher of the human race must be desirous that the efforts made to promote it should be guided by the most enlightened principles, so that the stream of intelligence may flow in the most useful channels.

When this seminary of learning was proposed, we understand that the Government in England had ordered a considerable sum of money to be annually devoted to the instruction of its Indian subjects. We are filled with sanguine hopes that this sum would be laid out in employing European gentlemen of talents and education to instruct the natives of India in mathematics, natural philosophy, chemistry, anatomy and other useful sciences, which the nations of Europe have carried to a degree of perfection that has raised them above the inhabitants of other parts of the world.

While we looked forward with pleasing hope to the dawn of knowledge thus promised to the raising generation, our hearts were filled with mingled feelings of delight and gratitude; we already offered our thanks to Providence for inspiring the most generous and enlightened nations of the West with the glorious ambition of planting in Asia the arts and sciences of modern Europe.

We find that the government are establishing a Sanskrit School under Hindu Pandits, to impart such knowledge as is already current in India... This seminary (similar in character to those which existed in Europe before the time of Lord Bacon) can only load the minds of youth with grammatical niceties and metaphysical distinctions, of little or no practical use to the possessors or to society. The pupils will thereby acquire what was known two thousand years ago, with the addition of vain and empty subtleties since produced by speculative men, such as is already commonly taught in all parts of India.

The Sanskrit language, so difficult that almost a

> lifetime is necessary for its acquisition, is well known to have been, for ages, a lamentable check on the diffusion of knowledge; and the learning, concealed under this almost impervious veil, is far from sufficient to reward the labour of acquiring it...'

In England too, the Court of Directors of the East India Company strongly resented the move of the Orientalists of the General Committee of Public Instruction. Even the members of the Committee in India, were equally divided between the encouragement of oriental learning and Western education, including science.

At that stage Lord Macaulay came to India (1834) as a legislative member of the Supreme Council and was made the president of the General Committee of Public Instruction. In his historic minute dated 2 February 1835, he wholeheartedly supported the Anglicists' wish to promote English education as follows:

> 'We are not fettered by any pledge expressed or implied; that we are free to employ our funds as we choose; that we ought to employ them in teaching what is best worth knowing; that English is better worth knowing then Sanskrit or Arabic; that the natives are desirous to be taught English and not desirous to be taught Sanskrit or Arabic...that it is possible to make natives of this country thoroughly good English scholars, and to this end our efforts ought to be directed... It is impossible for us, with our limited means, to attempt to educate the body of the people. We must at present do our best to form a class of people who may be interpreters between us and the millions we govern—a class of persons, Indian in blood and colour, but English in taste, in opinions, in morals, and in intellect...' (*Selections from the Records of the Madras Government*, vol. I, Madras, 1885, Appendix II)

The establishment of Hindu College as a private enterprise stimulated similar developments elsewhere in the Bengal and Bombay Presidencies. In 1829 a seminary was also set up in Calcutta and this institution was popular with nearly 600 students.

The Bombay Education Society was active in the early decades of the nineteenth century and, on its initiative, an organization under the name Elphinstone Native Education Institution was set up in 1820. It established some English schools and colleges in Bombay and adjoining areas, one of which was the Central School of the Elphinstone Institution in Bombay. The Elphinstone Native Educational Institute was another institution which was modelled on the Hindu College at Calcutta. In the Madras Presidency, there was no significant private activity for promoting English education in the early decades of the nineteenth century. However, the Provincial Government encouraged the establishment of what were known as the Collectorate and Taluk schools in different parts of the Presidency. In upper India, a number of schools sprang up in Benaras, Kanpur, Allahabad, Etawah as well as in Meerut. In Central India, Sagar and Jubbalpore had schools in which English, some scientific subjects and the Persian and Hindustani languages were also taught.

Missionary Institutions

With the direct or indirect patronage of the East India Company, Christian missionaries were devotedly active in establishing several schools not only in metropolitan cities but even in far-off interior regions. Alongside, they were also involved in proselytizing activities. Jesuits and Protestants alike attracted several thousands of young minds. Among the early missionaries whose educational efforts and institutions played a notable role was the one at Serampore near Calcutta, which was a Danish settlement. The name of William Carey stands out in this endeavour. In 1818, perhaps influenced by the Hindu

College which had come up a year before, he and his colleagues successfully began a college in Serampore for the education of what they called Asiatic Christians. The curriculum followed in this college included Sanskrit, Arabic, history, geography as well as European science and literature. The scientific subjects comprised astronomy (the solar system), the laws of motion, chemical experiments, botany, zoology and geology. In several ways, the Serampore College pioneered scientific education along the right lines inasmuch as it had a fairly impressive laboratory, a museum, and even an observatory. The members of the mission published several scientific books in English as well.

Another institution which sprang up in Calcutta nearly a year after the establishment of the Serampore College was the Bishops' College (1819). It was founded by an organization called the Incorporated Society for the Propagation of the Gospel in Foreign Parts. The brain behind it was Dr Middleton, the first Bishop in India of the Church of England. Its main objective was to train the native youth as well as European young men who were in India, but in Christian doctrines and disciplines. Though the studies were by and large limited to theology with Hebrew, Greek and Latin as subsidiaries, the College encouraged instruction in English and other oriental languages, philosophy and mathematics. The Church of Scotland, too, established a college in Calcutta in 1823 on similar lines. Another institution called the Bengal Auxiliary Missionary Society maintained an English school at Kidderpore. Down south, there were two missionary colleges: the Christian College (1837) and St Joseph's College (1844). Wilson College, which came up in Bombay in 1834, emerged as an important educational institution. In the first half of the nineteenth century, hundreds of missionary schools and colleges dotted different parts of India. A refreshing aspect of this devoted but calculated enterprise was the encouragement of female education. In 1819, the Baptist mission opened the first school for girls in Calcutta, and David Hare—of Hindu College fame—founded a women's school at his own expense. A number of what were called *Zenana* schools, staffed by only lady teachers, also emerged. In western India, the American mission was active in the promotion of female education. The Ladies' Society for Native Female Education, supported by some Christian missions, energized the setting up and the superintendence of a large number of schools in Calcutta for female education.

The spread of education through schools and colleges apart, the Calcutta School Book Society (1817) and the one at Madras (1820) were involved in the preparation and publication of books at affordable prices and making them available for students in schools, colleges and seminaries. Books on mathematics, physics, astronomy, geography, chemistry and other subjects were published from time to time. These also included some material taken from Sanskrit sources, particularly in mathematics in which Bhāskarācārya II's *Līlāvatī* also found a place. John Clark Marshman of the Serampore Mission produced a text on astronomy and titled it *Golādhy*āya, a name borrowed from *Siddhāntaśiromaṇi* of Bhāskarācārya II.

Technical Education: Medical (before 1857)

The East India Company, of necessity, had to provide some kind of medical education to Indians so that they could work in hospitals, many of which were intended for the military. A substantial demand for trained Indian doctors (called native doctors) who could be of service to the army, compelled the Company's medical board to prepare a concrete plan of medical education for select natives. In 1822 such a plan was prepared and approved by the Governor-General-in-Council. This led to the establishment of a medical school at Calcutta, called the School of Native Doctors. The initial strength of this School was not to exceed twenty students, and no student who was not capable of reading and writing the Hindustani language in the Nāgari or Persian scripts, and below the age of eighteen or above twenty-six years, was admitted to the School. Certificates from the Provincial Medical Board were issued after candidates were found duly qualified. Surgeon James Anderson was the first Superintendent of the School for Native Doctors (1822). But his untimely death brought in another surgeon, Peter Breton, to succeed him and, during the latter's seven-year term, a comprehensive vocabulary of medical terms in English along with their equivalents in Persian and Hindustani was compiled, and this proved valuable to the students. Some pharmaceutical texts as well as medical and anatomical discourses too were rendered into Persian and Hindustani from their English versions. Breton gave importance to native curative methods and operational procedures. John Tytler, who succeeded Breton, was a trained physician, also interested in oriental medicine. He improvised the medical teaching of anatomy and surgery, offering courses in chemistry and pharmacy too. However, he was unable to introduce dissection of human bodies, although he adopted zootomy or the dissection of animals. Tytler's enduring contribution was his meticulous preparation of informative medical notes and their translations into Persian and Hindustani, which were useful for students.

The School of Native Doctors was a source of inspiration for the Calcutta Sanskrit College as well as the Calcutta *madrasa* to open medical classes in their institutions. Though Ayurveda was taught in the Sanskrit College, the European medical system was not excluded from translations. In 1831, a small hospital was also attached to this College. In the Calcutta *madrasa*, Unāni medical courses were offered, and later, the European system of medicine too was taught there for students who opted for it.

Towards the end of 1833 the colonial government constituted a committee, with J. Grant as its president, to examine and submit a report on the state of medical education in the country. The committee recommended that the government itself should establish a medical college for Indian students, in which various branches of medical science cultivated in Europe and a knowledge of the English language would be a *sine qua non* for the students. For, it was argued that the English language had several advantages over oriental languages in the process of instruction as well as the preparation of textbooks. This recommendation, which was accepted by the government, was well in time as Macaulay's minute in favour of English as the medium of instruction soon followed it (1835). The minute sounded the death knell of not only the School for Native Doctors but also the medical classes of the two Oriental institutions in Calcutta. The result was that in Calcutta, a medical college, and in Madras as well as in Bombay, two medical schools were established in 1835. The first one came up in the very beginning of the year (28 January 1835), giving a kickstart to modern medical education in English in India.

Medical College of Bengal: The first medical college of colonial India, the Bengal Medical College offered intensive medical education and training to native young students between the ages of sixteen to twenty, irrespective of caste and creed, as the promulgated order of the Governor-General stipulated. Initially, the intake of students was limited to fifty and each student was given a monthly stipend. The first batch of pupils was to undergo training and receive education in accordance with the model adopted in European medical institutions for a period of not less than four years, but not exceeding six years. Students successful in the final examination were to receive certificates that would qualify them to enter public service. However, their nomenclature was still 'Native Doctor' and their initial salary was Rs 30 per month but with pension benefits. On the other hand, a European Superintendent of the course was drawing Rs 1,200 per month and his European assistant, Rs 600 per month! Dr M.J. Bramley was the first Superintendent of this Medical College. Two others, H.H. Goodeve (anatomy and medicine), and W.B. O'Shaughnessy (chemistry) joined the college, which started with twenty selected students.

The most impressive feature of this new medical college was that from November 1836 to April 1837 (six months), all the students were continuously engaged in the dissection of the human body, instead of the earlier practice on animal bodies. It is on record that on 10 January 1836, a student called Madhusudan Gupta, holding a scalpel in hand, followed Dr Goodeve into the godown where a dead body lay ready, and dissected the body by making a long and deep incision in the breast. Some controversy about this record apart, it would appear that Madhusudan Gupta and four other courageous pupils too were involved in the performance of the dissection of a human body, which was carried out for the first time by Indian students. The medical college had a museum, a good laboratory managed by O' Shaughnessy, and a small hospital with thirty beds as well as an outdoor patients' dispensary. In 1848 the hospital was developed to a great extent, by which time the College had also attracted students, both Hindus and Christians. Among the Hindu students were Brāhmins, Kayasthas, Vaidyas, and Vaiśyas, Christian students being more in number than Brāhmins.

Madras Medical College: In the middle of 1835, a medical school was established in Madras. Earlier, the small hospital established by the provincial government enabled the students to imbibe some practical knowledge about the application of medicine. Some of them were also appointed to the positions of medical subordinates to assist British surgeons and assistant surgeons. With the establishment of a medical school near the general hospital in Madras, systematic medical education for select students became possible. Between 1846 and 1848, medical teaching was raised to a higher level with the appointment of competent professors in anatomy, physiology, ophthalmology and chemistry. By 1850 the educational courses offered included practically all the important branches of medical science and surgery. In 1851, the Medical School became the Madras Medical College, with European surgeons playing a leading role in its functioning. The spread of modern medical education in Madras, however, was not as easy as it was in Calcutta because of the general apathy and, even antagonism, of the people, towards the European system of medicine. New scales higher than those given to native doctors in Calcutta had to be offered to the incumbent native doctors in Madras as an inducement. In the course of time, the Madras Medical College had, however, earned a good reputation, and the students who came out of it were in great demand both in the army and for the medicare of the general public.

The Grant Medical College of Bombay: In 1835, Robert Grant became the Governor of Bombay Presidency, and he was deeply concerned with the unhygienic drainage system and frequent appearance of diseases which haunted the city. Noticing that medical education had not received its due attention, he established a society of physicians and other experts who were interested in the promotion of scientific and medical education, under the name of the Medical and Physical Society. Encouraged by the success of the Bengal Medical College, this Society recommended that similar efforts be made in Bombay too. Robert Grant's official support to this recommendation resulted in obtaining the approval of Lord Auckland, the then Governor-General-in-Council for starting a medical college. But unfortunately, Grant died (9 July 1838) just before the approval was given. A public meeting held in the Town Hall of Bombay to condole the death of the eminent Governor of Bombay adopted a resolution that, in honour of Robert Grant, the proposed medical college be named the Grant Medical College. For this purpose, a substantial memorial fund was also raised for the construction of the building, which was completed in October 1845 and the Grant Medical College started functioning in November 1845. The first Superintendent of this College was Dr C. Morehead and, to start with, it had twelve selected students of whom five were given free education and seven were granted stipends.

That hospital-training is integral to medical teaching was realized by those who began the new medical colleges. The Bengal Medical College at Calcutta had a general medical hospital as well as a surgical one attached to it. There was, however, inadequate financial assistance for the maintenance of the hospital in Calcutta. On the other hand, the Madras Medical College hospital was in a better position. In Bombay, more than the governmental grants for medical education and the maintenance of the attached hospitals, the local businessmen and merchants subscribed adequate funds for these purposes. Jamsetjee Jeejeebhoy, a wealthy business magnate, offered the government a benefaction of one lakh rupees for the establishment of a large hospital with about three hundred beds, and the government supplemented it in equal measure. The Jamsetjee Jeejeebhoy (JJ) Hospital came up in 1845. Professors of the Grant Medical College were empowered to function as medical officers of the hospital.

In Calcutta, a new hospital catering to about 350 patients at a time was established towards the end of 1852. The medical college at Calcutta received the approbation and recognition of the Royal College of Surgeons of England in 1846. In 1854, the Grant Medical College at Bombay was also recognized, and a year later, so was the Madras Medical College. The Calcutta University introduced three types of medical degrees: (i) Licentiate in Medicine and Surgery (LMS); (ii) Bachelor of Medicine (MB); (iii) Doctor of Medicine (MD). The Madras University, however, did not embark upon the first degree, but offered Bachelor of Medicine and Mastery of Surgery (MS) degrees. The Bombay University also introduced two degrees: the Licentiate and Doctor of Medicine. The regulations and syllabus did not differ much in these three universities in the beginning, but in the course of time there were some modifications to enhance the efficacy of the courses. The Punjab University (1882) had a medical college at Lahore, named King

Edward Medical College (1910), which was upgraded from the earlier medical school. Though intended for the residents of the Punjab state, it also admitted students from the neighbouring provinces, including Baluchistan. Although the Calcutta Medical College lost its status as an independent college after the establishment of the Calcutta University (1857), it continued to play a major role in medical education and attracted students from different provinces. The Grant Medical College in Bombay and the Madras Medical College, too, were active and reputed for their methodical medical education, including surgery.

In order to strengthen auxiliary medical education, a number of medical schools (28) were established in various parts of India in the latter half of the nineteenth century. Among them, in the Bengal Presidency, Patna, Dacca and Cuttack had a medical school each to serve the eastern region besides the one at Dibrugarh; while in the Madras Presidency, medical schools were established at Tanjore, Madurai, Dindigul and Nellore. In the Bombay Presidency, Poona and Indore had a medical school each. An important development was the promotion of medical education for girls and the training of nurses and midwives.

Engineering Education

The survey and the construction works undertaken by the colonial government in different parts of the country demanded the employment of trained native personnel. In Madras, a survey school was already in existence towards the close of the eighteenth century. In Bengal, the Hindu College and the Calcutta madrasa offered teaching courses in surveying, and the trained surveyors were employed in the revenue department. The early decades of the nineteenth century witnessed increasing activity in public works. Major irrigation (canal work) and engineering schemes were also taken up in northern India. James Thomason, the Lieutenant-Governor of the North-West Provinces, initiated the Ganges Canal project and it was decided to build a large workshop at Roorkee, related to irrigational and constructional works. This enterprise naturally called for a supply of qualified engineers, overseers and auxiliary technical personnel. In 1845, a small engineering class had been started in nearby Saharanpur. Two years later, in 1847, with a view to meeting the steady demand for engineers, Thomason submitted a memorandum to the government, making out a strong case for the establishment of an institution in Roorkee for the education of civil engineers. Soon an engineering college was established at Roorkee for training students under three categories: (i) for appointment as sub-assistant executive engineers; (ii) non-commissioned officers or overseers and assistant overseers for public works department; and (iii) a vernacular class for the training of Indian youth who sought appointments in the subordinate branch of engineering and surveying. By 1852, the Roorkee Engineering College had emerged as a dynamic institution supported by a scheme of stipends for students. It was a multidimensional college since it had prepared plans for setting up a wing for mathematical and scientific instruments, a museum of economic geology, and a printing press for the publication of scientific works of proven merit. This college also offered training in civil engineering to army personnel. With the liberal grants from the government for the augmentation of engineering education and since Thomason was the brain behind this, the new college at Roorkee was befittingly named the Thomason College of Engineering.

In Bengal, the Council of Education emphasized the need for technical education particularly in civil engineering, geology and mineralogy. In 1848, Lord Dalhousie, the Governor-General, recommended to the Court of Directors in London that an engineering institution be opened in each of the Presidencies. But the establishment of a separate college of engineering at Calcutta took seven more years. Eventually, it was in 1855 that an engineering college came up, to begin with, in the premises of the Writers' Building there, and Lieutenant E.C.S. Williams, an engineer and Superintendent of the Topographical

Survey joined the college as its principal in 1856. The subjects of study included mathematics, architecture, surveying, geology, astronomy, botany, chemistry, natural philosophy, topographical drawing, physical geography and mapping. The courses were designed to ensure that a student who passed out of this college would be able to prove his efficiency, particularly in the public works department and allied engineering work.

In the Presidency of Bombay (in 1854), a civil engineering class was started to train candidates for the subordinate services of the public works department. In 1857, on the advice of the Government of India, the Bombay government established a civil engineering college at Poona (now Pune) for training young Indian students for the position of officers in the department of public works. Later, with the establishment of a university in Bombay in 1857, this college was named the Poona College of Science, which offered courses in scientific and engineering education, as well as agriculture. It had four departments: (i) civil engineering; (ii) agriculture; (iii) forestry; and (iv) training overseers. The educational training in this college was comprehensive and diverse as it included a wide range of subjects—geometry and integrated calculus, optics and astronomy, mining and metallurgy, architecture, mechanical engineering, analytical chemistry, botany and meteorology. It also offered the lower degree of a licentiate in civil engineering. The Poona College of Science was noted for its flexible courses as well as rigorous examination and emerged as a focal point for engineering education in India at that time. Though the Mysore state government had established an engineering college at Bangalore, it soon abolished it, preferring to send select students to study in the Poona College of Science, and offering them suitable stipends. The Hyderabad state, too, followed this step.

The Bengal Engineering College had come up after the establishment of the Calcutta University (1857). It was active in training assistant engineers, surveyors and overseers for the public works department. However, for various reasons, including the absence of qualified teachers, it was amalgamated in 1864 with the Presidency College. But in 1870, it regained its independence and began to function in the new premises near the Sibpur botanical gardens. Popularly known as Sibpur Engineering College, it was officially designated as The Government Engineering College, Howrah. It laid emphasis on practical work or fieldwork, such as the laying out of railway tracks, trigonometrical surveys and colliery work. In the late 1870s, mechanical engineering was treated as a separate discipline of engineering education. This was reflected in the Calcutta University's award of a licentiate or bachelor's degree in either of the branches, namely, civil engineering or mechanical engineering. Each had a four-year course of instruction and one year practical training. The training of overseers, too, was bifurcated on these lines. Towards the end of the nineteenth century (1898), the Sibpur Engineering College began to provide agricultural education in addition. But it did not last long because in 1904, the government had established the Imperial Agricultural Institute in Pusa in Bihar for agricultural education.

The civil engineering education at Madras (Guindy) had two levels: Collegiate and School. The former came up in 1862 to train students for a university degree, which was the prescribed qualification for a candidate to be appointed as an engineer. The syllabus for this course was more or less similar to that adopted by the Bengal Engineering College. In the 1880s, the civil and mechanical branches were reorganized and the degrees offered accordingly, although civil engineering was more popular among students. The school section of the college ran four separate classes for the training of four categories of technicians: (i) subordinate engineers; (ii) surveyors;(iii) draughtsmen; and (iv) sub-overseers and artisans. An important feature of this School was a system in which anybody, without being its student, could appear for an examination and, on passing it, could obtain certificates of proficiency as draughtsmen or surveyors.

Industrial Art Schools

In 1851, Dr Hunter, a surgeon in Madras, opened what he called the School of Industry for the production of various articles of daily use. A year earlier, he had founded a school of industrial arts for inculcating the humanizing dimensions of the fine arts, both schools beginning as a private enterprise. In 1855, however, they were amalgamated to form the Government School of Industrial Arts in which training in the production of decorative articles, whether of metal, wood, stone or clay, was offered to talented young native students so that they could be gainfully employed.

In 1854, a privately run School of Industrial Arts was also opened at Calcutta by the Society of Industrial Arts, in which many European gentlemen were interested. Hundreds of students were engaged in the production of diverse decorative articles. The sale of these products did not, however, generate adequate funds for the maintenance of the school, which meant that it had to depend upon government support for its existence. Thus the industrial schools in Madras and Calcutta were managed largely by private initiatives, with attendant problems of uncertainty about their financial position. In Bombay, the industrial giant Jamsetjee Jeejeebhoy, who had already made munificent donations to the government for establishing a hospital, also offered one lakh rupees to the government to set up a school for industrial art and manufactures. The school, which came into existence with the approval and support of the government, was given a solid foundation by a British superintendent and drawing master. Its creations, such as sculptures, decorative painting, metal work, wood engraving, ornaments and the like, with their aesthetic appeal, found a ready market and earned an enviable reputation for the school.

In 1888, the Victoria Jubilee Technical Institute came

up in Bombay with donations made by the Ripon Memorial Fund, the Bombay Municipal Council, and the Mill Owners' Association of the city. This institute was established by the industrialist Dinshaw Manackjee Petit, who made a spacious building available to the institute, valued then at three lakh rupees. To begin with, the institute offered courses in electricity, magnetism, sound, light, heat, mechanics and machine-drawing with extensive practical training. At first, arrangements were in place for admitting 120 students, but within a month, the number was doubled to meet the mounting demands, particularly of the industries that needed trained workers. A textile school was attached to it, named after Ripon, and well equipped with modern textile machinery capable of producing exquisite finished cloth from cotton. The Ripon School of Textiles accommodated batches of sixty students who, upon completion of their training, found employment in cotton mills. In 1889, the institute opened evening classes for teaching factory and mill workers who were given practical training in select courses for three years, and certificates issued to successful candidates.

In the closing decade of the nineteenth century, the Madras Presidency had six industrial schools: three in Madras, which worked well, and three in other districts, that did not work satisfactorily. One of the schools in Madras called the Nazareth Industrial School, was for native Christian orphans. It was active in the promotion of a wide variety of objects ranging from woodwork to blacksmithy and clothing, all of which found a ready market. The Bombay Presidency had several industrial schools which were mostly managed privately. The Bengal Presidency, too, had five schools, but they were operating with varying degrees of achievement, the most conspicuous one at Midnapore. On the other hand, in the Punjab and Northwestern Provinces and Oudh, as well as in Central India, the industrial schools, which were few and far between, did not have successful records.

The Government of India set up an Educational Committee in 1882 and, following its recommendations on technical education in October 1884, the government adopted a policy resolution that the native youth should be encouraged to undertake industrial and commercial pursuits. It was recognized that technical education should not be looked upon as separate from general education. Science being the foundation of every branch of technical education, the study of the principles and methods of science must also find a place in the scheme of general education. It was also realized that instructions in the fundamentals of science bearing upon industrial applications should form a necessary part of the curriculum of elementary schools. In England too, then, the Royal Commission on Technical Education had recommended such measures. The colonial government had different but futuristic ideas on technical education, expressing its view as follows:

> 'Technical education proper is the preparation of a man to take part in producing effectively some special article of commercial demand. In India at the present time the application of capital to industry has not been developed to the extent which in European countries has rendered the establishment of technical schools on a large scale. But the extension of railways, mills and factories, exploration of minerals and others, expansion of trade, and the enlarged intercourse with foreign trade ought in time to lead to the same results in India as in other countries. Therefore, the Government of India should support technical education of special character as an extension of general education and it should promote and countenance such technical education as may be applied to the service of existing industries, which will profit by the aid of scientific research, scientific method, and higher manipulative skill.' (Selections from *Educational Records of the Government of India*, vol. 4, p. 59)

Agricultural Education

The East India Company was lukewarm about the education of farmers, though the Company's revenue officials admired their traditional knowledge and skills of farm production and management. This attitude, however, could not last long because of the compulsions experienced by the colonial government in the form of a steady decline in its revenue and the reoccurrence of devastating famines. In the nineteenth century particularly, famines visited the country frequently in one part or the other. The 1860 famine resulted in crop failure of nearly 50,000 square miles in the North-West Province, Agra, and Rajasthan, causing food-scarcity and distress to a large number of people, and this also entailed a serious loss of revenue to the government. Similarly, the Odisha famine (1866-67) also spread to the coastal areas of down south, and even to Calcutta in the northeast. Another famine (1868-69) in the northwestern parts of India, including the Punjab and Rajasthan, had caused enormous devastation. Yet another famine in 1876-78, too, was widespread. These famines emphasized the necessity of increasing food production, improvement of agricultural operations, and better storage practices. There was another compulsion from the exploitative colonial government's point of view, namely, that cash crops needed to be grown on a considerably large scale. Further, the civil war which broke out in the USA around this time, had prevented the British textile mills from getting American cotton and this forced the British government to depend on India to feed the Manchester mills. The production of tea and jute in which the British companies had a stake also needed to be encouraged. Economic compulsions such as these made the colonial government set up specific departments which would look after agriculture and associated commerce. Concomitantly, this initiative was followed by the organization of agricultural departments in the provinces too. Even so, the thrust of the government was more on the generation of revenue rather than on the promotion of scientific agriculture through agricultural education and farm research.

Between 1868 and 1900, however, there were a few attempts to initiate agricultural education. Not only were some agricultural schools set up but, wherever such schools were not possible or feasible, courses in scientific agriculture were also offered in certain schools of general education. In Madras (now Chennai), an agricultural school was opened in Saidapet (a suburb) in 1868, in which the training in cotton production and several exotic plants was given special importance. As a result, cotton cultivation showed an improvement, and experimental farms proved effective in that region. Ten years later in 1878, the school was raised to the status of a college and select students pursued a three-year diploma course, with a short practical training course on experimental farms. Those who passed out of this college found remunerative employment easily.

The Poona College of Science had a department of agriculture which started functioning from 1879 as a sequel to the famine of 1877. It included an agricultural farm to provide practical training to students. During its first ten years, this institution awarded its own certificate for proficiency in agriculture to successful students. During 1889-90, with the approval of the Bombay University, it introduced a three-year course in veterinary sciences and started a veterinary hospital for practical instructions. It also set up a bacteriological laboratory for the study of cattle diseases. The Baroda College had an agricultural department, under the charge of a qualified agricultural specialist from Edinburgh University. The diploma in agriculture awarded by this college to successful students was recognized by the Bombay University.

In the Central Provinces, an Agri-Horticultural Society had been established in the 1860s, and a model agricultural farm had come up in 1871. However, it took more than a decade and a half for the opening of an agricultural course of one year duration attached to this farm, and another decade and a half to elevate the institution to the status of a college (1906) with a three-year diploma course, affiliated to the newly established Nagpur University. Nevertheless, this college was popular among the students of agricultural education. In the United Provinces, an agricultural school was opened in 1893, which attained the status of a college in 1906.

It was rather strange, however, that in Bengal there was no agricultural college or school worth the name, and even the agricultural branch of the Bengal Engineering College at Sibpur did not meet with the desired success until after the beginning of the twentieth century. The general picture was that educational institutions for agriculture were not appreciable in number, nor did they receive adequate support and encouragement. The colonial government even admitted this inadequacy in a resolution on agricultural education, stating that the country possessed no institution capable of imparting a comprehensive agricultural education, and that the existing schools and colleges had not succeeded in providing theoretical or practical training. Education in forestry, however, received a fillip when a forest school was set up in Dehradun as a Central School to cater to the needs of different provinces which had extensive forests. Established in 1878, the first theoretical course on forestry was offered by this school in 1881. It had a comprehensive instructional syllabus taught by a competent faculty, turning out capable forest officers and guards annually.

The importance of agricultural education began to be realized by the government towards the end of the nineteenth century to such an extent that agriculture classes were also introduced in several high schools in Dhulia, Ahmedabad, Belgaum, Nasik, Noida, Sholapur, Surat and Hyderabad. In addition, several village schoolmasters were given a six months' course on agricultural farms with a view to making them useful for the communication of new methods and improved agricultural practices in rural areas.

At the invitation of the colonial government to examine the status of agricultural education, J.A. Voelcker, the consulting chemist to the Royal Agricultural Society in London, who was in India during 1889-91, and in 1893, submitted an extensive report to the government. One of his recommendations was that the universities should recognize the importance of the study of agriculture and grant degrees instead of diplomas. Another recommendation was that agricultural education should also form a part of the general education system as a professional subject. In the meantime, the government had supported the organization in conducting a series of agricultural conferences between 1888 and 1896, which laid emphasis on the production of textbooks and reading material on scientific agriculture as well as offering practical agricultural courses.

In the first decade of the twentieth century, there was some progress in agricultural education, but the First World War intervened (1914-18). During this distressing period not only did agricultural schools and colleges take a back seat but even a college like the Poona College of Agriculture was practically turned into a war hospital for the wounded. The postwar period, however, witnessed some spurt in the growth of agriculture. In 1928, the Royal Commission on Agriculture was constituted by the government to examine the Indian agricultural situation in all its details. The Commission in its report had indicated several guidelines for the improvement of India's agriculture and, as part of it, the promotion of agricultural research and associated education. The establishment of the Imperial Council of Agricultural Research and of the agricultural research institutes and departments in several states meant that there was a demand for qualified scientists and technicians, and hence several agricultural colleges received the much needed support. But they were still a part of the general university system in colonial India, without a separate university for agriculture. The concept of the integration of teaching, research, and field or extension work at the university level, which was developed after Independence, led to the founding of the first Agricultural University at Pantnagar in UP in 1960, signifying the advancement of agricultural education.

ESTABLISHMENT OF UNIVERSITIES

In 1854, Charles Wood (later Lord Halifax) who was then the president of the Board of Control of India, sent an Educational Despatch which was described as the *Magna Carta* of English education in India, intended as it was to provide a basic structure of education to be adopted by the colonial government. Two decades earlier (1834), Lord Macaulay had arrived in India when Lord William Bentinck was the Governor-General-in-Council, and both had laid the foundation for the promotion of English and the liberal Western type of education. Bentinck had, in fact, directed that all governmental funds earmarked for education be expended on imparting to the native population a knowledge of English literature and science through the medium of English.

Wood's *Educational Despatch* was a reaffirmation of this approach, but its merit lay in recommending the establishment of universities in India. Its conceptual thrust was, however, in terms of the diffusion of European knowledge in the colonial land and an aversion to oriental or traditional science and philosophy which, it thought, were replete with serious errors of thinking. Nevertheless, on the question of language, the *Despatch* took a liberal and practical view, since it recognized the important role of both English and vernacular languages in the diffusion of European knowledge. In addition, it envisaged that in the course of time, the vernacular languages themselves could be enriched gradually by the translations of European books into these languages, thus ensuring that European knowledge could reach all classes of people.

Wood found that the time was ripe for the implementation

of the *Despatch's* recommendations regarding the establishment of universities in colonial India. By then, in 1854, the progress of liberal education had witnessed the emergence of bright, qualified native candidates for employment by the government in professional and other services. On the basis of this progress, which was noticeable particularly in the Bengal and Bombay Presidencies, the *Despatch* gave precedence to the cities of Calcutta and Bombay for the foundation of universities, with Madras following. The main objective of the proposed universities was to recognize the educational attainments of students through the award of academic degrees in different branches of the arts and sciences. The University of London was explicitly recommended as the model to be followed. The *Despatch* emphasized that Indian universities would be only affiliating and examining bodies, the actual function of teaching being left to the affiliated colleges and other institutions. Since many of these institutions were not well equipped to teach important subjects, the idea of university professorship was also mooted to make up for such deficiencies. Organizationally, each university would be headed by a chancellor, a vice-chancellor, and a Senate of Fellows (academic).

The *Despatch* further recommended that the existing boards and councils or committees of education should be replaced by departments of education. Among its other recommendations were the adoption of a system of the inspection of educational institutions, the mechanism of grants-in aid, stipends and scholarships, promotion of female education, establishment of normal schools, support for vernacular education and the preparation of textbooks. It was, however, recognized that the task of educating the eligible students of the vast population of India was indeed formidable, the resources of the government being so meagre. The *Despatch*, therefore, suggested what came to be known as the 'downward filtration theory', which envisaged the education of the upper classes that would in turn 'filter it down' to the lower orders of society. The *Despatch* thus subscribed to the notion that it would be impossible to reach the lower strata until the upper strata had been educated. Even though the *Despatch* had recognized that the education of the country's vast rural population should be the primary and direct responsibility of the colonial government, it wanted the universities to provide the benefits of education to the higher classes of India.

The governments in India and England took nearly three years to deliberate upon Wood's *Despatch*, and formulate rules and regulations for the university structures, examinations, the award of degrees, and associated issues. However, the first step was taken to constitute the departments of education in the Presidencies, appointment of Directors of Public Instruction, and inspectors of schools and colleges. In 1857, three universities were established at Calcutta (January), Bombay (July) and Madras (September) in the very year in which the colonial government witnessed what is called the Sepoy Mutiny (now recognized as the First War of Independence).

An entrant to the university system was to be sixteen years of age and should have passed the entrance examination. Thereafter, the student was expected to prosecute studies in one of the affiliated colleges or institutions for three years to earn a BA degree, although the degree would only be conferred four academic years after the entrance examination. The Calcutta University had formulated certain other rules: (i) candidates who, within five years from the date of successfully passing the entrance examination, have been placed in the first of the two divisions, in which all the passed candidates for the BA degree are arranged, could be examined for an Honours degree in any one or more subjects to be approved by the Senate; and (ii) every candidate who had passed the examination in Honours was entitled to the degree of MA without further examination or regular fee.

In the Faculty of Arts, provisions were made for two kinds of degrees: BA and MA. For professional faculties of medicine and civil engineering, the degrees were: Licentiate in Medicine and Surgery (LMS) and Doctor of Medicine (MD); later on the Bachelor of Medicine (M.B) was introduced. Likewise, for civil engineering, they were L.C.E. and B.C.E. (Bachelor in Civil Engineering). The two wings of the university were: the Senate which comprised the chancellor, vice-chancellor and other Fellows (thirty in all). The Syndicate consisted of the vice-chancellor, three Fellows of the Faculty of Arts and one each from the Faculties of Law, Medicine, and Civil Engineering. The universities of Bombay and Madras had a more or less similar structure; the entrance examination of the Bombay and Madras Universities was called 'matriculation'. In none of the three universities, however, was there a separate degree for science. In the Madras Presidency, the colleges or institutions affiliated to the Madras University were many more in number than the other two universities. Physics, chemistry, botany and zoology, along with mathematics and drawing were among the subjects for the BA degree examination of the three universities.

The fourth university was the Punjab University, which was established in 1882 through an Act by the Government and, for this purpose, the existing Punjab University College (est. 1869) was elevated to a university. A significant aspect of this new university was that, from its inception, it had a faculty of science for the degrees of BSc and MSc, in addition to five other faculties, namely: oriental studies, arts, law, medicine and engineering. The science course was mainly built up around three core subjects—mathematics, physics and chemistry, to which one of the biological subjects (botany or zoology) or geology was added so as to make scientific education as comprehensive as possible. Such courses were also adopted by the Bombay University as well as other universities, with certain optional modifications.

Each of these universities had its own emphasis on one or the other of the science subjects. It was rather difficult to assess the popularity or otherwise of the science stream as there was no recognition or grading of success at university examinations separately for the art and science subjects. It would, nevertheless, appear that the art subjects and the professional ones like law, engineering and medicine attracted more students than mere science subjects. According to the figures of 1885-86, out of 712 students of the colleges in the Bengal Presidency who took up university examinations in 1885-86, as many as 538 had taken up literary courses and only 134, science courses; the arts students registered a 60 per cent success while the share of science students was 36 per cent. The situation was more or less similar in the other universities. In the new University of Punjab, there was no student registered (1886) for the science degree offered by it. The arts and the professional subjects of law, medicine and engineering were preferred by students as their degrees were seen as a stepping stone to employment in the government.

The Education Commission set up by the colonial government in 1882-83 had suggested that a new university for the North-Western Provinces and Oudh and the Central Provinces, might be established at Allahabad. Till then several colleges in these regions like the Muir Central College in Allahabad (1872) and the Muhammadan Anglo-Oriental College at Aligarh (1881) were affiliated to the University of Calcutta. There was a growing feeling that the educational needs of the North-Western Provinces were not met with satisfactorily by the Calcutta University.

Early in 1885, the Director of Public Instruction of this region sought the opinion of over twenty distinguished educationalists as well as of the provincial government on the more fundamental question of whether the new university should be both an examining and teaching body, unlike the other four which were only the examining ones. The emergent general opinion was, of course, that the objective of the new university should be the promotion of European learning and science as well as professional subjects like law, engineering and medicine. But opinion was divided on the issue of its being both a teaching and affiliating university. While the Director of Public Instruction himself was strongly in favour of only its teaching functions, there were others who had their reservations about this issue. The Allahabad University was eventually established by an Act passed in September 1887, initially providing for the faculties of the arts and law. Three other faculties of science, medicine and engineering were added later by a notification from the Governor-General. The Act, however, cleverly avoided any reference to the character of the new university—whether it should be a teaching-cum-examining body or merely an examining one.

The 1882 Educational Commission had deliberated upon the problems of female education and, as part of it, science education. The Commission was seriously concerned with the lack of demand by females for such education. The main reasons were: social and other regressive attitudes, particularly of the orthodox Hindu and Muslim communities, and the then prevalent custom of child marriage that necessitated the girls' withdrawal from school at an early age. However, the small minorities of Christians and Jews as well as the progressive section of Hindus, including the followers of the Brahmo Samaj, sent their girls to schools and colleges. But their number was minuscule when compared with that of boys. In 1882, the first lady student (Chandramukhi Bose) earned a BA degree from Calcutta University and two years later, creditably passed the MA examination in English. Nevertheless, medical education attracted women students. Madras Medical College as well as the Grant Medical College at Bombay specially encouraged women students professionally, though the Calcutta Medical College was slower in doing so initially.

By the end of the nineteenth century there were only five universities (Calcutta, Bombay, Madras, Lahore and Allahabad) that were affiliating and examining bodies. They received scant financial support from the government, but sustained themselves largely by the examination fees paid by the students. This situation was not to last long. In December 1898, the new Governor General-in-Council, Lord Curzon, arrived in India and showed a keenness to effect reforms in university education. He appointed a University Commission in 1902, and in 1904, through his efforts, the Indian Universities Act was promulgated—which enabled the universities to pursue both teaching and postgraduate research and to conduct examinations. The Act had a provision for the offer of financial grants to the universities from the government.

The universities were further given powers for granting recognition and inspecting high schools and colleges under their jurisdiction. They were also empowered to make appointments to academic positions. In 1910, education which till then was under the control of the Home Department, was placed under an independent Department of Education, a step that emphasized the importance attached to education by the Government of India. In 1913, the government reviewed its educational policy and adopted a resolution of far-reaching significance. This resolution, among others, (i) advocated the establishment of additional but smaller universities of the teaching type; (ii) recommended an increase in the salaries of teachers and also enhanced grants-in-aid to the universities by the government; (iii) emphasized the necessity of facilities for research, welfare of the staff, and refresher courses for training teachers; and (iv) reaffirmed the policy of reliance on private effort in secondary education. The policies outlined in this resolution were expected to accelerate the educational progress in all the provinces. While some improvement or development was noticeable, the outbreak of the First World War in 1914 had a retarding effect on the universities in the next three to four years. However, between the years 1916 and 1929, as many as thirteen new universities—Benaras (Hindu University), Mysore, Patna, Osmania (Hyderabad), Aligarh Muslim University, Rangoon (Burma), Lucknow, Dacca, Delhi, Nagpur, Andhra (Waltair),

Agra and Annamalai—were established.

The Indian Universities Act of 1904 and the Government of India's increasing concern for the improvement of higher education did have a salutary effect, though not to the desired degree. The Calcutta University began to play a leading role in the promotion of scientific education. In 1917, this university took several steps for the promotion of postgraduate teaching and research. Two enlightened and philanthropic Bengalis—Tarakanth Palit and Rash Behari Ghosh—endowed large trust funds to be administered by the Calcutta University for scientific and technical education as well as research. Postgraduate classes also commenced in physics, chemistry, applied mathematics and experimental psychology. Some more endowments resulted in the creation of more chairs in scientific subjects, and thus the Calcutta University emerged as the nerve centre of advanced scientific teaching and research in the country. Several scientific pioneers like J.C. Bose, P.C. Ray, C.V. Raman, M.N. Saha and S.N. Bose—in one way or the other—were associated with this university. The fertile brain behind these developments was that of an outstanding mathematician and enlightened educationist, Ashutosh Mukherjee. The other universities too did not lag behind though their financial position was not as sound as that of the Calcutta University. Several of them, however, had chairs in physics, chemistry, mathematics, botany, zoology and geology.

In the first half of the twentieth century and more so after the First World War, scientific and technical education even in the West was largely governed by the industrial and technical needs of a country. The development or augmentation of facilities for science-teaching in schools and colleges in India was also far from being satisfactory, largely because of the backwardness of India's industrial development then. Even in such a situation, the importance of scientific education as a way to train the mind and sharpen the intellect to understand Nature and the phenomenal world rationally, was recognized by several Indian thinkers as well as some thoughtful European administrators in India. Nonetheless, the objective of scientific education in schools, which was by and large meant to be the backbone of higher scientific education in colleges and universities, remained a record of unfulfilled objectives. It was mainly due to the fact that education was almost a responsibility of the provinces which had inadequate finances. Scientific laboratories in most of the schools and colleges were ill-equipped. The result was that the basic scientific knowledge of a student entering the portals of universities could not be of a high order. The examination system, too, encouraged memorizing of the textual material rather than testing the fundamental concepts of science assimilated by a student.

The universities, however, endeavoured to attain tolerable standards in scientific education by a judicious blend of the curriculum of scientific subjects with practical classes in the laboratories. The introduction of Honours courses both in the physical and biological sciences as well as geology by some universities, demanded adequate laboratory facilities, which were gradually becoming available. But, there were also differing standards in the conduct of Honours examinations in different universities. Notwithstanding this situation, postgraduate studies and research in some universities gained momentum, emulating the example of the Calcutta University. Though the scientific educational spectrum on the eve of Independence was appreciable, there was a strong feeling in the academic community that (i) proper scientific education needed the reorganization of the secondary school system to lay a sound foundation of the scientific method and the basic principles of science; and (ii) teaching scientific subjects in universities required supplementation with adequate laboratory facilities and experimental training.

During the nineteenth century, though scientific and technical education made some progress, it was designed to sub-serve the interests of the colonial government. From the last quarter of this century there was an increasing realization that the promotion of science and higher education would be necessary as an asset for national development. The momentum of the freedom movement which began to take rapid strides from 1920 onwards catalysed new attempts in the promotion of scientific and technical education. But the forerunners of such thoughts and initiatives were the Indian Association for the Cultivation of Science (Calcutta, 1876) and the Indian Institute of Science (Bangalore, 1909).

NEW INSTITUTIONS

Indian Association for the Cultivation of Science

The last quarter of the nineteenth century witnessed devoted efforts by Indian intellectuals and social reformers towards national regeneration in diverse ways. An early effort in this direction was in Calcutta by Mahendra Lal Sircar (1833-1904), a product of English education and one of the brightest students of Calcutta Medical college. He had a strong grounding in science and its rational methods, since in his time both physical and biological sciences were being adequately taught in medical college. It occurred to Sircar that science, the new intellectual pursuit, would never strike deep roots in this country of vast masses through the process of its introduction in the educational curriculum alone. What was needed, he thought, was an institutionalization that would lead not only to the creation of a mass interest in science but also the training of scientists capable of undertaking original researches, as exemplified by the Royal Institution of Great Britain and the British Association for the Advancement of Science.

Mahendra Lal Sircar

In 1869, he delivered a lecture and also published it in the *Calcutta Journal of Medicine*, titled 'On the Desirability of a National Institution for the Cultivation of Science by the Native Indians'. This paper revealed his perception and inner thoughts about the importance of science and its relevance in the Indian context. He wrote:

> 'We want an institution which shall be for the instruction of masses, where lectures on scientific subjects will be systematically delivered, and not only illustrative experiments performed by the lecturers but the audience should also be invited to perform them by themselves. And we wish that this institution be entirely under native management and control. The kind of knowledge which is best calculated to remove prejudices, and promote the spirit of tolerance in the mind is what one passes by the name of physical sciences...'

To counter fears that this rational approach to nature and life would undermine religion, he stated in a public lecture that science would lead to a firm belief in divinity as well as 'a devout attitude of mind before the First Cause'. He was of firm conviction that the proposed institution should be free from governmental interference. His initiative, though creative and meaningful, encountered two kinds of opposition: one was from the people of the traditional or orthodox section of the society who viewed his endeavour as an attempt to subvert religion and religious practices; the second was from the members of the India League, a middle-class political association that had come up at that time. They advocated that, instead of a scientific research organization, a mechanics institute of the type England had established, should be started for training unemployed youth. Some of them were also averse to the spirit of nationalism associated with scientific pursuit and wanted to restrain it in order to earn the favour of the colonial government. But Sircar had the support of such respected leaders as Surendranath Banerjee and Father La Font of St

Xavier's College, who persuaded the then Lieutenant-Governor, Sir Richard Temple to offer support to him. Father La Font brushed aside the opposition of the India League. Sircar's dream became a reality when several gentlemen of Calcutta offered donations and Richard Temple made rent-free accommodation available for the proposed Indian Association for the Cultivation of Science. A setback came, however, when the bureaucratic British stipulated that the Association should invest Rs 70,000 for the establishment of the new institution.

The Indian Association for the Cultivation of Science was established in 1876, the first landmark of a nationalist endeavour with a clear perspective. Its aim was to enable the natives of India to cultivate science in all its aspects with a view to its advancement by original research. But its main effort during the first thirty years of its existence was towards the promotion of scientific teaching at the collegiate level and the arrangements for popular lectures on scientific subjects. In the very first year, Father La Font and Sircar were appointed honorary lecturers in physics. Father La Font obtained certain apparatus and instruments from France for demonstrating his lectures on electricity, optics and heat. In the next two or three years, several physical instruments and chemical apparatus were procured. Father La Font's discourses, about twenty to thirty a year, on various aspects of physics, particularly optics, acoustics and spectroscopy with practical demonstrations, were extremely informative and popular, and he continued with his lectures for nearly seventeen years—a remarkable and exemplary effort indeed! Sircar himself taught static and dynamic electricity, magnetism and heat. An interesting aspect of his lectures was his enchanting articulation of the subject from a historical point of view. Over the years, the Indian Association for the Cultivation of Science had made its impact on the general public and men of science alike. Father La Font and Sircar, were its pillars of strength and nurtured this infant organization with great care and affection.

Several other brilliant scientists, too, were involved in the activities of this Association. Jagadis Chandra Bose who had established himself as an eminent researcher at the Presidency College, used to arrange practical classes in physics for some time. His scientific demonstrations comprised several types of experiments which included Danish physicist Hans Christian Oerstead's methods, polarizations of currents, measurements of electrical resistance of wires, magnetic properties as well as demonstration of static electricity. In 1887, the Indian Association for the Cultivation of Science had the benefit of the expertise of the outstanding mathematician Ashutosh Mukherjee who delivered excellent lucid lectures on mathematics, mathematical physics and physical optics. Chemistry, too, was taught by Tara Prasanna Roy (inorganic), Ram Chandra Dutt (organic) and Chunilal Bose (industrial chemistry). Similarly, lectures, with demonstrations, were delivered by well-known experts in astronomy, geology, botany and zoology. From a perusal of the spectrum of lectures delivered at the Association, it was evident that they were not organized around the college or university syllabus for BA or MA examinations, but were of great benefit not only to the students who attended them but also to the interested general public. The popularity of lectures was such that, in a decade, they began to attract several hundreds of persons including students from various colleges. It is interesting to note that Acarya Prafulla Chandra Ray, then a student of the Presidency College who later became a famous chemist, attended these lectures. The Calcutta University granted recognition to the Association for sending candidates for FA and BA examinations in scientific subjects. From 1895, some bright students started working in the laboratory of the Association to prepare themselves for the MA degree of the Calcutta University. Most significantly, it was the Indian Association for the Cultivation of Science that attracted into its fold the genius, C.V. Raman whose epoch-making discovery—now called the *Raman*

Effect—on the scattering of light, was made in the laboratory of this institution.

Indian Institute of Science

In contrast to the foundation of the Indian Association for Cultivation of Science which was started by several people led by M.L. Sircar, the establishment of the Indian Institute of Science in Bangalore owed its initiative and sustenance to one dedicated visionary, Jamsetji Nusserwanji Tata (1839-1904), a wealthy industrialist-philanthropist of Bombay (now Mumbai). Born in the Parsee Tata family at Navasari, a small town near Surat in the state of Gujarat, Jamsetji moved to Bombay along with his father and family when he was just thirteen years old. At nineteen, he passed out of Elphinstone College as a 'Green Scholar', a conferment equal to a degree. In the next twenty-five years, his business acumen and wise investments in the textile industry as well as landed property in Bombay and elsewhere had made him one of the wealthiest industrialists.

Jamsetji had a triple-stranded vision in his conceptual approach to national reconstruction: (i) generation of hydro-electric power; (ii) production of iron and steel; and (iii) promotion of higher education, besides a model hotel that would extend the warmth of Indian hospitality. He accomplished the last one in the form of Taj Hotel, which now majestically overlooks the Gateway of India in Mumbai. But he left the first two to his sons, Dorabji Tata and Ratanji Tata to pursue and accomplish them. The promotion of higher education which would pay greater attention to science was dear to his heart. For, he believed that the

Jamsetji Nusserwanji Tata

advancement of any country, particularly India, would depend upon higher education and the development of the faculties of its bright students, who would be able to serve the nation and elevate it economically. To offer benefaction to such a cause was what he called 'constructive philanthropy', which appealed to him most. Since the early 1890s, he had been pondering over the inherent weakness of the educational system in general and scientific education in particular in colonial India. He was also well aware of the limitations of the five universities established by the colonial government, in arousing scientific curiosity and providing intellectual stimuli. Jamsetji's dream was to establish and nurture such an institution of higher learning or a university, which he had determined to sponsor. He was passionately involved in such efforts to make his dream come true till the end of his life, but the pathway was by no means smooth as he had to surmount difficult problems.

In 1889 Lord Reay, the Governor of Bombay and chancellor of the Bombay University, in his convocation address referred to the sickening state of affairs of Indian universities and their stagnant role as merely examining bodies with no academic excellence on their own. He exhorted:

> 'It is only by the combined efforts of the wisest men in England and the wisest men in India, that we can hope to establish in the home of learning, real universities which will give a fresh impulse to learning, to research and, to criticism which will inspire reverence and self-reliance to future generations of our and your country men...'

Lord Reay's exhortation reflected Jamesetji's innermost thoughts. It catalysed the wise man's disposition towards the establishment of an institution of higher learning or university, making available his own munificent contribution towards its implementation. Jamesetji soon began to ponder over a scheme of a 'real university'. At the same time, in 1892, he also set up a committee to select some students of extraordinary merit who, on easy loans from his Trust, could be sent every year for higher studies in England. *Tata scholars*, as they were generally known, returned to India and contributed to India's growth in their own way.

In November 1896, Jamsetji addressed a letter to Lord Reay in which he expressed his genuine interest in establishing a new University and willingness to support it by his benefaction.* Soon he secured the educational expertise of Burjorji Padshah who had served for some time as the vice-principal of a college in Karachi and resigned from that position. Jamsetji sent him to Europe to study the organizational structures of several universities and to interact with distinguished educationists. Towards the end of 1898, Padshah prepared a comprehensive but tentative scheme under the title 'Teaching University of India'. Jamsetji as well as Padshah favoured the Johns Hopkins University in Baltimore, USA, as a model for the proposed university, but Jamestji had to encounter some difficulties to realize his dream. Lord Curzon, the newly appointed Governor-General-in-Council had arrived in India on the last day of 1898. When the university proposal was submitted to him, he had his own doubts about it and his lukewarm attitude persisted for quite some time. In the meantime around 1899, Jamsetji had secured the consent of the Dowager Maharani (Vanivilasa Sannidhānam) of Mysore for a gift of land (about 370 acres) in Bangalore, and some financial support.

Jamsetji patiently pursued his university scheme with the Government of India until it was re-examined by the eminent

*There is a view articulated by IISc administration in its centenary year (2009) that Swami Vivekananda suggested this idea to J.N. Tata. This is erroneous since there is no documentary evidence for it. That the new University scheme was Tata's alone is evident by his letter to Swami Vivekandanda (November 1898).

chemist, William Ramsay who was recommended by the Royal Society of London. He came to India towards the end of 1900 and after studying the scheme carefully, set aside the concept of the university as formulated by Padshah. He suggested the name 'Indian Institute of Research' for the proposed institution and not only advocated that the Institute should play a leading role in the promotion of industries, but even recommended that it be located in Bangalore because hydro-electric power would be available to the proposed institution from a nearby place. But Ramsay's cost-estimates for ushering in the Institute, even after taking into account Jamsetji's benefaction and the Mysore government's financial support, was on the high side, needing additional subvention from the Government of India. Ramsay's proposals, therefore, needed further evaluation. At the instance of the Government of India, Orme Masson of the University of Melbourne and Colonel Clibborn of Roorkee College re-examined the proposal with a realistic approach and suggested that the name of the proposed institution should be *Indian Institute of Science*, a nomenclature that eventually prevailed.

But the Government of India had to surmount another problem—it was the Joint Trust settlement proposed by Jamsetji which, he thought, would ensure the financial stability of the new institution. The Joint Trust envisaged the involvement of his two sons and the new institution in the financial dispensation, but primarily for the benefit of the institution. Since Curzon's government viewed it in a different light, Jamsetji was prepared to withdraw it, but he died in May 1904 before his dream was realized. His sons, Dorabji and Ratanji Tata, however, continued his mission and transferred the immovable properties in Bombay to the Treasurer of the Charitable Endowments in Bombay; the Mysore government gave 370 acres of land to the Government of India with a generous offer of five lakhs of rupees for the construction of a building and an annual grant of Rs 50,000 to the Institute; and the Government of India too agreed to make appropriate funds available both for the construction of building and running expenses. All of these including the annual income from the donated properties of J.N. Tata in Bombay, enabled the Government of India to adopt a suitable resolution and issue the vesting order on 27 May 1909 along with the scheme of administration of the Indian Institute of Science.

In 1906, on the suggestion of Dorabji and Ratanji Tata, the Government of India requested the Secretary of State for India in London to get the advice of the Royal Society on the selection of a suitable director for the Institute. Morris Travers who had earlier worked with Ramsay on the discovery of rare gases, was eventually appointed the first director of the Institute. Though he had to encounter some problems in the beginning, Travers set about his task with great zeal and accomplished the construction of the building and equipment of the laboratories. He also planned the areas of scientific investigations. The Institute opened its doors in 1911 with two departments, the General and Applied Chemistry and Electrotechnics, and the Department of Organic Chemistry came into being a few months later. After Travers sought voluntary retirement in 1914, he was succeeded by A.G. Bourne (chemist) who was then the director of public instruction in the Madras Presidency. Later, in 1922 M.O. Foster, also a chemist, became the Director. Thus the first three Directors of the Institute were foreigners (British) and chemists too.

The applied investigations carried out at the Institute during its first five years led to the establishment of factories, such as the sandalwood oil factories in Bangalore and Mysore; a soap factory (Bangalore); an acetone production factory (Nasik); a thymol factory (Hyderabad-Sind); and a bamboo production centre (Bangalore). Later, a lac factory was also established in Ranchi (now in Jharkand). During the First World War, which broke out shortly after its establishment, the Institute took up several investigations of practical value.

The first Indian director was the Nobel Laureate C.V. Raman who joined the Institute in 1933. During his four years of stewardship, Raman energized scientific activities of the Institute and made the Department of Physics an internationally famous one. J.C. Ghosh, a well-known chemist, succeeded C.V. Raman as director in August 1939, but the Second World War erupted in less than two months after he took over the directorship. With his amiable disposition and scientific eminence, Ghosh began reshaping and expanding the Institute even during the turbulent war period. Conscious of the internal financial constraints of the Institute and with rare perception, Ghosh was successful in tapping external funding sources. For the first time in the history of the Institute after nearly thirty years, he produced a four-year plan for the varied scientific activities and infrastructure of the Institute.

Walchand Hirachand, an industrialist, had set up a factory in Bangalore primarily for automobiles in 1940. But because of the compulsions of the war, it was transformed into Hindustan Aeronautics for the assembly and repair of aircrafts with the financial participation of the Government of India and of the Mysore state. The Government of India also offered a financial grant to the Indian Institute of Science to establish a department of aeronautical engineering for training purposes, and this department came up in 1942, adding a new dimension to the Institute. Ghosh succeeded in arranging some financial support for the award of stipends to six trainees from the Sir Dorabji Tata Trust. The government of Mysore too offered capital grant as well as a recurring annual grant for this new venture. In 1943, the Institute also started a new course in chemical engineering and soon took up a wide range of investigation with the financial support of the Council of Scientific and Industrial Research, which had been established by the Government of India in 1942. A full-fledged department of chemical engineering came into being in 1947. Two more departments—the department of metallurgy and the department of internal combustion engineering—were established in 1945. A department of power engineering was also conceived before Independence. In 1911, the Institute had only three departments; it expanded with the department of biochemistry (1921); physics (1933), metallurgy (1945), and chemical technology and chemical engineering (1947). With such multi-disciplinary departments for teaching and research on the eve of Independence, the Institute had established itself as a premier scientific institution of the country.

LEARNED SOCIETIES AND COORDINATING ORGANIZATIONS

The first learned society, though not strictly an exclusive one for scientific studies, was the Asiatic Society, which took its first tentative steps in Calcutta as early as 1784. European scholars involved in antiquarian studies as well as probing into the natural history of India in different regions, felt the need to meet together periodically to enable them to exchange information and to have scholarly discussions. Thirty European intellectuals assembled in Calcutta on 15 January 1784 under the presidentship of Robert Charles, the second judge of the Supreme Court, and adopted a resolution to establish an association called 'The Asiatick Society' (later known as the Asiatic Society of Bengal; this name was also changed in 1936 to the Royal Asiatic Society of Bengal) and to hold weekly meetings on every Thursday at 7 p.m. The brain behind this move was that of William Jones who had come to India in 1783 as a judge of the Supreme Court at Calcutta. He had already won reputation for his literary attainments in Latin, Greek, Arabic, and Persian. Naturally, Jones was elected the founder-president of the Asiatic Society, a position he held till his death in 1794.

In his opening address, Jones outlined the objectives of the Society in broad terms, which included investigations on mathematics, and astronomy of Asia as well as the ideas about the physical world, skills shown in medicine and surgery, chemical practices, agriculture, music, architecture and the arts as well as religious institutions of the region. With this wide canvas of investigations, the objective of the Society was, as Jones stated, 'If now it be asked what the intended objectives of our enquiries are: we answer "*Man and Nature*—whatever is performed by the one and produced by the other within the geographical limits of Asia."' The Asiatic Society played a key role in bringing to light the scientific investigations of the naturalist members of the Society who were interested in studying the rich fauna and flora as well as minerals and climate of India. Jones himself shed light on the antiquity of the Hindu zodiac and even certain aspects of Indian flora. He also prepared a catalogue of Indian plants with their names both in the Sanskrit and Linnaean generic nomenclature.

The Asiatic Society had no building of its own till 1808, when it moved into its own building on Park Street, Calcutta. It was only in 1829 that a few Indian scholars were elected as members of the Society for the first time. It had two committees: one for natural history, physics, medicine and improvement of the arts; and the other, for literature. The Asiatic Society was a pioneer in the publication of the researches of its members, and it started its own periodical, *Asiatick Researches* in 1789, twenty volumes appearing in the succeeding years. In 1829, it was divided into two parts: one to publish only the scientific papers, and the other for literary communications. In 1832, the Society started publishing the *Journal of the Asiatic Society* (the first number bore the title *Journal of the Asiatic Society of Bengal*). In a period of about fifty years, more than 500 papers in mathematical and physical sciences, 560 in zoology, 320 in geology, and eighty in botany were published in this journal.

The varied investigations carried out by active members of the Asiatic Society on natural history and other fields led to the accumulation of many objects, including plant specimens, animal skeletons, stuffed birds, minerals, coins, and other artefacts. Since these treasures needed preservation, the Asiatic Society resolved in 1814 that a museum was needed to serve as a repository and to house their display. Nathaniel Wallich was chosen as the honorary superintendent of what was designated as the 'Oriental Museum of the Asiatic Society'. To begin with, it was divided into two sections: (i) archaeological, ethnological and technical; and (ii) geological and zoological. In 1835, the Government of India which had taken steps to search for new coalfields and new minerals, thought of setting up a museum of economic geology at Calcutta and offered financial support to the Asiatic Society to preserve the government's collections, including fossils and palaentological specimens. By 1855, the collections were so diverse and vast that the Society appointed two curators, Edward Blyth and H. Piddington, who classified them systematically, making the scientific knowledge latent in them accessible to other scholars and laymen.

The Indian Museum

The time had arrived for the establishment of a well-planned and delightfully displayed museum. The Asiatic Society nucleated necessary steps and effectively negotiated with the Government of India not only to transfer its own collections but also to offer the needed financial support for a new building and its annual maintenance cost. In March 1866, the government promulgated an Act which envisaged a suitable building for the preservation and exhibition of archaeological artefacts, objects of natural history, minerals, and even records of the Geological Survey of India. The proposed museum, named the Indian Museum, was to be set up under a Trust, the members of which were to be nominated by the government. The first curator was John Anderson, who was earlier a professor of natural sciences in Free Church College at Edinburgh. The museum came into being in September 1866; but the new building became ready for occupation only in 1875 and the galleries were thrown open to the public in April 1878. A new wing of the museum was constructed in 1891 since additional room was required to display industrial collections shown in the Great Exhibition held in Calcutta in 1883. The Indian Museum is more or less modelled on the pattern of the well-known British Museum in London. The public galleries are divided into: (i) archaeological; (ii) geological; (iii) industrial; (iv) zoological;(v) anthropological or ethnological; and (vi) art sections. The art section was under the care of the Trustees, while the others were under the respective departments of the Government of India. The zoological and anthropological sections of the museum eventually energized the establishment of the Zoological Survey of India (1917), which among its other objectives, surveys the fauna and endangered species of India, maintains and develops national zoological collections, and publishes the results of its work.

Archaeology

The facilities and encouragement for antiquarian investigations, including epigraphical studies, provided by the Asiatic Society began to take strides mainly due to the efforts of devoted scholars like H.T. Colebrooke and H.H. Wilson. Several other European scholars explored cave paintings like those of Ajanta, Ellora and Amaravati. Archaeologist and historian James Princep made bold attempts to decipher the Kharoṣṭi and Brāhmī scripts. The Rock Edicts of Aśoka received special attention and some of them, including the one at Sarnath, were meticulously studied. As a result, a new insight into the ancient history of India began to emerge, and this stimulated further excursions into field-archaeology which was pursued by European investigators, although in a limited way. Realizing the importance of an apex body like an Archaeological Survey of India with the support of the government, Alexander Cunningham who had by then made valuable contributions to Indian archaeology, persuaded the government to promote field-archaeology in a planned manner. In 1861, Lord Canning, the then Governor-General-in-Council, established the Archaeological Survey of India, and his choice of the first archaeological surveyor naturally fell on Cunningham.

Intensive archaeological explorations were then carried on by Cunningham, Markham Kittoe and others. Epigraphical studies began to be undertaken by Edward Thomas, Bahu Daji and J. Stevenson. Robert Bruce Foot of the Geological Survey of India as well as H.P.L. Meswiser of the East Indian Railway unearthed some pre-historic sites like the palaeoliths at Pallavaram and neoliths in the valley of the Tosney river, respectively. In the early decades of the twentieth century, the Government of India provided necessary encouragement and financial support to archaeological excavations, and these were supplemented by the provincial governments.

Under John Marshall as Director-General of the Archaeological Survey of India in 1902, a new era of archaeological activity began with the unstinted support of Lord Curzon, who was greatly interested in antiquities. Marshall reshaped the Archaeological Survey and initiated a series of explorations. In the 1920s, the ancient Indus Valley Civilization or Harappa culture, came to light and this was announced, as noted before, by Marshall in the *Illustrated London News* in 1924. Marshall himself was engaged in the epoch-making excavations at Takṣaśīla from 1912 to 1934 till his retirement. Pataliputra, Sanchi, Besnagar, Nalanda, Sarnath, and a number of other ancient sites were also excavated. From 1921 onwards, archaeology which till then had received divided attention from the central and provincial governments, became the responsibility of the central government, which increased grants for excavations. The preservation of all monuments, regardless of provincial barriers, also became a central charge. The epigraphy department was also encouraged and its journal, *Epigraphia Indica*, became a valuable document of inscriptions being discovered from time to time. This series and other publications of the Archaeological Survey of India have led to a rational understanding of Indian history from pre-historic to medieval times.

The Indian Science Congress Association

By the first decade of the twentieth century, scientific education and research had made notable progress. The establishment of the Imperial Bacteriological Laboratory at Poona (1890, but moved to Muktesar in 1893), the Haffkine Institute or the Plague Research Laboratory in Bombay (1899), the King Institute of Preventive Medicine at Guindy, Madras (1903), the Imperial Agricultural Research Institute (1903) at Pusa, the Forest Research Institute (1906) at Dehradun, and the Pasteur Institute of Southern India at Coonoor (1907), enlarged the horizon of scientific research in India. Eminent scientists like J.C. Bose, P.C. Ray, and Asutosh Mookerjee were beacons for other researchers, whose number was increasing. In such a scientific atmosphere, there was, however, a growing feeling of isolation among scientific researchers in the absence of specific scientific societies or a common platform for all of them to meet together periodically for scientific discussions and exchange of ideas.

In 1911, two British scientists employed in India, P.L. Simonsen of Madras and P.S. MacMohan of Lucknow, came forward with a proposal which envisaged the formation of an association of scientific researchers along the lines of the British Association for the Advancement of Science established in England about fifty years earlier. Their appeal evoked a favourable response from scientific investigators all over India. In 1912, a committee of seventeen scientists representing different branches of science was set up, and the Asiatic Society was entrusted with the responsibility of organizing the first congregation, called the Indian Science Congress. The first meeting of the Indian Science Congress was held in the premises of the Asiatic Society

Sir Asutosh Mookerjee

on 15-17 January 1914 with Asutosh Mookerjee as its first president. Asutosh's conception of the Science Congress was a far-reaching one—in his presidential address, he emphasized that scientists and science-enthusiasts would be able to bring their objectives and views prominently into public notice by periodical meetings and discussions, and even bring them to the attention of the government. He emphasized that governments would require to be reminded of the claims of science on public funds and their use for societal betterment.

The first session of the Indian Science Congress consisted of six sections—chemistry, physics, geology, botany, zoology and ethnography—at which only thirty-five papers were presented. The annual sessions of the Congress, which lasted one week in the month of January of each year, began to be held in different centres of the country, on the model of the British Association for the Advancement of Science. The second to seventh sessions of the Congress were held, between 1915 and 1920, at Madras, Lucknow, Bangalore, Lahore, Bombay and Nagpur and later, in other chosen places. In 1935, the name was changed to the Indian Science Congress Association.

The year 1938 (January) witnessed the silver jubilee session of the Indian Science Congress Association. By then, the scope of the sections had enlarged and there were as many as thirteen sections—mathematics (which was noted for its absence in the first session in 1914), physics, chemistry, geology, geography and geodesy, botany, zoology, etymology, anthropology, agriculture, medical research, veterinary research, physiology and psychology. The number of scientific papers communicated to different sections increased from thirty-five in 1914 to over 800 in 1938, besides holding ten joint discussions of various sections and twenty-two discussion-meetings under individual sections. The silver jubilee session was jointly organized in collaboration with the British Association for the Advancement of Science and scientists not only from UK but also from other parts of the world attended, a tradition that is still being pursued.

The eminent British scientist, Lord Rutherford, discoverer of the fundamental particle, proton (1920) and Nobel Laureate was to preside over the silver jubilee session of the Science Congress; but his sudden demise a few months before the appointed date prevented Indian scientists and others from listening to his enlightening address (Jan 1939). However, his presidential address was received and published by the Asiatic Society. Lord Rutherford, in his thoughtful address, had thrown light on the scenario of scientific education in India at that time as follows:

> 'The universities in India have in later years made substantial progress both in teaching and research in science; yet it must be borne in mind that still greater responsibilities are likely to fall on them in the near future. This is in a sense a scientific age where there is an ever increasing recognition throughout the world of the importance of science to the national development. It is natural to look to the universities and technical institutions for the selection and training of the scientific men for this development.'

The first president, Asutosh Mookerjee and the jubilee president, Rutherford, had articulated the primal role of science and technology in the development of a nation of vast poverty-stricken masses. In the 1940s, just before Independence, this kind of thinking stimulated the minds of not only scientists but also leaders of the freedom movement.

Scientific Societies

The annual meetings and discussions of specific subjects at the Science Congress led to the emergence of a number of specialist societies. These enabled researchers in specialized scientific fields to come together for more intimate academic interactions. But even earlier, there were some attempts in this direction. The first such society was the Mining and Geological Institute of India established in 1906. A year later, in 1907, the Indian Mathematical Society at Poona came up with the name 'Analytical Club', which was later changed to the Indian Mathematical Club. The Calcutta Mathematical Society was established in 1908, and the Benaras Mathematical Society in 1918. The well-known Institution of Engineers (India) was founded in 1910 with its head office in Calcutta and branches all over India. The other societies which were established included the Indian Botanical Society (1921); the Indian Psycho-Analytical Society at Calcutta (1924) with its branches in other places; the Geological, Mining and Metallurgical Society of India (1924); the Indian Psychological Association (1925); the Society of Biological Chemists, Bangalore (1931); the Geographical Society, Calcutta (1933); the Indian Physical Society (1934) at Calcutta; the Biochemical Society (1934); the Indian Society of Social Science and Indian Physiological Society (1935); and the Indian Anthropological Institute (1936) at Calcutta. The societies which had no fixed headquarters were associated with the places where their secretaries or presidents worked. The emergence of specialized scientific societies contributed in no small measure to the augmentation and encouragement of scientific activities. The various research institutes and specialized scientific societies pointed to the need for establishing a coordinating scientific organization at the all-India level.

Board of Scientific Advice and Others

In 1902, the Government of India constituted a Board of Scientific Advice in which the Royal Society of London was intimately involved. Its object was to monitor and coordinate the various scientific activities that were being pursued by the field-survey organizations like the Geological Survey of India and the Botanical Survey of India as well as other research institutes under the government. The Board decided to produce an annual report and send it to the Royal Society through the Secretary of State for India in London. However, this beneficial method that was in the nature of an index of scientific pursuits at the all-India level, despite its limitations, was for one reason or the other suspended in 1924.

In 1911, the Indian Research Fund Association was established by the Government of India for the promotion of researches and their application for the prevention of diseases. It played an important role in the coordination of diverse medical researches that were underway in different parts of the country, besides advising the provincial governments on public health problems. In 1929, the Imperial Council of Agricultural Research was begun with substantial annual grants from the Government of India, following the recommendations of the Royal Commission on Agriculture. These grants resulted in the expansion of agricultural and veterinary research in several centres; these, in turn, offered scientific inputs of practical value to farmers, thus leading to some increase in agricultural production. It had two auxiliary committees: the Indian Central Cotton Committee set up earlier in 1921 and the Central Jute Committee (1937). They were involved in encouraging intensive research on the higher yields of cotton and jute crops respectively. There were sectoral associations like the Indian Lac Committee, the Indian Coffee Association and the Indian Tea Association, which were engaged in the promotion of research and distribution of quality products, each in its own area.

The Three Academies

Between 1930 and 1935, three science academies were born in India.

The National Academy of Sciences: The first was the Academy of Science of the United Provinces of Agra and Oudh, which came into being in 1930 as a result of a Memorandum of Association signed by seven scientists, including Meghnad

Saha, N.R. Dhar and P.C. MacMohan who, as noted before, had played a significant role in nucleating the Indian Science Congress Association. The Memorandum envisaged two levels of scientists, Members and Fellows, who had distinguished themselves in science and technology. The Academy started with fifty-seven ordinary Members and nineteen Fellows. Soon it began the publication of two journals under the name of *Proceedings of the National Academy of Sciences; Section A: Physical Sciences;* and *Section B: Biological Sciences*. Eminent foreign scientists and technologists, too, were elected as Fellows of the Academy. There was a growing feeling that this Academy, as its name indicated, was regional in character, though it had national canvas. It was, therefore, renamed the 'National Academy of Sciences, India' in 1932. Registered under the Society's Act, its main objectives included the cultivation of science and technology in all its branches, promotion of researches relating to the problems of national importance, and bringing out scientific publications.

The Indian Academy of Sciences: Towards the end of 1930, the Government of India addressed letters to all provincial governments, various scientific departments, learned societies, universities, and the Indian Science Congress Association, seeking their views on the desirability of establishing a National Research Council as an apex body that would also adhere to, and cooperate with, the International Council of Scientific Unions. In 1933, Richard Gregor, the editor of *Nature* in London, was on a visit to India for interactions with the editor of *Current Science* (which had been started by certain scientists at the Indian Institute of Science and Central College, Bangalore) to explore the possibility of ushering in a science academy at the national level. The various opinions favouring the establishment of a national scientific organization, its composition and functioning were consolidated and put up in the form of a resolution to the Indian Science Congress Association at its annual meeting held in January 1934. A year before, C.V. Raman, who had taken over the directorship of the Indian Institute of Science and the editorship of *Current Science,* wrote an editorial emphasizing the important role of research in the economic, social and political development of a nation. He pointed out the imperative need of an all-India scientific organization that would focus enlightened public opinion on the doctrine of science as material and spiritual wealth. The Indian Nobel Laureate was of the view that the Academy should have the necessary authority to advise the government, universities and scientific institutions on the problems that might be referred to it for consideration. Besides, it should also be in a position to negotiate on behalf of Indian scientific workers with similar institutions abroad.

The January 1934 session of the Indian Science Congress Association which was presided over by M.N. Saha, approved the idea of an All-India Academy on the model of the Royal Society of London. It also set up an Academy Committee of some distinguished scientists, including C.V. Raman. However, the attitude and approach of Raman and those of Saha as well as of the other members were at variance. A man of quick decision, Raman immediately resigned from the Academy Committee and soon founded his own academy, the *Indian Academy of Sciences*, with its headquarters at Bangalore. It was registered on 27 April 1934 with about 160 Foundation Fellows, including himself as its founder-president. The Dewan of Mysore, Sir Mirza Ismail, inaugurated the Academy in July 1934 and the government of Mysore offered a recurring annual grant of Rs 3,000, besides donating ten acres of land in Hebbal (not far away from the Indian Institute of Science), for the construction of a permanent building for the Academy.

Raman organized scientific meetings regularly under the auspices of the Academy at the Indian Institute of Science. He started the *Proceedings of the Indian Academy of Sciences*, a monthly. In one year, the twelve issues contained about 2,000 pages and 130 illustrations of research results, and these were personally scrutinized by Raman. His own department of physics at the Indian Institute of Science, which was about a year old by then, contributed a substantial number of research papers on physics. The other departments of the institute, too, did not lag behind. The discussion-meetings, lectures and the publications laid a solid foundation for the Indian Academy of Sciences, which has maintained this scientific tradition to this day in an exemplary manner.

The Academy Committee of the Indian Science Congress, however, continued its deliberations and submitted its report to the Indian Science Congress session held in Calcutta in January 1935. The recommendation of the Committee for the establishment of what was named the National Institute of Sciences of India, as an all-India body of scientists and representing all branches of science, was unanimously adopted.

The National Institute of Sciences of India: The name of this body was changed to the Indian National Science Academy in 1970. Inaugurated in the Senate Hall of the University of Calcutta on 7 January 1935, this organization's first president was L.L. Fremor with S.P. Agharkar as one of its secretaries. It started functioning in the premises of the Asiatic Society of Bengal in Calcutta. In 1945, the Government of India decided to recognize the National Institute of Sciences of India as the premier scientific society representing all branches of science. Its headquarters moved to Delhi in 1946. The Government of India also started providing substantial grants to this all-India institution for its various activities. In addition, a capital grant was also made in 1948 for its own building, the foundation stone of which was laid by Prime Minister Jawaharlal Nehru in 1948. The Institute moved into its own building in 1951 and it was designated in 1968 as the adhering organization in India to the International Council of Scientific Unions (ICSU) and its affiliated bodies under UNESCO.

Among the main objectives of this national science body

are: (i) promotion of scientific knowledge, including its practical applications to problems of national welfare; (ii) coordination among scientific academies, societies, institutions and government scientific departments; (iii) to act as a body of scientists of eminence for promoting and safeguarding the interests of scientists of India and to present internationally the scientific work carried out in India; and (iv) to promote and maintain liaison between science and humanities.

In 1942, the Government of India established the Council of Scientific and Industrial Research, and in 1946, an Atomic Energy Board. These and the Indian Council of Agricultural Research became the leading scientific organizations before Independence while the National Institute of Sciences of India began to play a supportive role as and when the situation demanded, a position that continued even after Independence.

SCIENTIFIC PIONEERS

As noted before, the colonial government, in furtherance of its enlightened self-interests, introduced Western education and as part of it, scientific and technical education. This step was more in the nature of a *formal* introduction rather than *real* promotion of scientific research and augmentation of scientific activities. But towards the closing quarter of the nineteenth century, Indian pioneers, with their extraordinary vision of national regeneration, spearheaded the promotion of science and scientific research. As noted before, Mahendra Lal Sircar and J.N. Tata heralded the dawn of this national regeneration, each in his own way; but they were not laboratory scientists. The first Indian who attained fame as an outstanding researcher was Jagadis Chandra Bose who pioneered scientific experiments and research activities, and became a role model for budding scientists.

Jagadis Chandra Bose (1857-1937)

Born in Mymensingh (now in Bangladesh), Jagadis Chandra Bose spent his boyhood in Faridpur where his father, Bhagawan Chandra Bose, worked as deputy magistrate. The young Jagadis studied in a village *pāṭhaśālā* (school) with rural boys and sons of fisher-folk. He loved nature and common people alike, but had a curious mind, and was always asking questions. At the age of nine, he joined Hare School and later St Xavier's School at Calcutta. The enchanting lectures on physics and experiments demonstrated by Father La Font at St Xavier's College had quite an impact on the young mind of Jagadis. After graduating from Calcutta University at the age of nineteen, he had an urge to go to England and sit for the Indian Civil Service examination, a goal of bright students then, but his father dissuaded him. Nevertheless, in 1880, though he went to England to study medicine in the University of London, he took up natural science at Christ College, Cambridge. Jagadis passed the Tripos examination with distinction, obtained his BA degree in 1884, and a BSc degree the next year. He returned to his motherland in 1885, and was offered the post of officiating professor of physics at the Presidency College, Calcutta, only to face a piquant situation. For, he was offered a salary half of what the British teachers were paid. Jagadis refused to accept the salary as a protest and worked in an honorary capacity for three years with indefatigable energy. He was extremely popular with his students because of his inspiring lectures accompanied by extensive demonstration of experiments. Impressed by his brilliance and capabilities as a teacher, the Presidency College authorities recommended a full salary for him from the date of his joining the college.

The Presidency College at Calcutta had no adequate laboratory facilities when he joined it. Jagadis Chandra Bose spared no efforts in equipping a laboratory, even fabricating certain types of instruments with great ingenuity. He carried out

Jagadis Chandra Bose

various experiments on refraction, diffraction and polarization. His ingenuity led him to invent a device for producing electromagnetic waves and he succeeded in generating electromagnetic waves with smaller wavelengths of a few millimeters from 25 mm to 5 mm. In November 1894, Bose gave the first public demonstration of wireless transmission using electromagnetic waves to ring a bell. His invention won high appreciation in England, and found a place in science textbooks.

It was because of this invention that Bose was regarded as the father of wireless telegraphy, since his invention had preceded that of Guglielmo Marconi. Bose was well aware of the lucrativeness of his findings, but he shunned that temptation. In 1901, in his letter to Rabindranath Tagore, Bose wrote: 'I wish you could see that terrible attachment for gain in this country...that lust for money. Once caught in that trap there would have been no way out for me!' He believed in the universality of science and scientific researches.

The early publication of Bose was titled 'On Polarization of Electric Waves by Double Refracting Crystals', which appeared in the May 1895 issue of the *Journal of the Asiatic Society of Bengal*. His researches also included the determination of refractive indices of several substances, the role of thickness of the air-gap between two dielectric slabs and the rotation of the plane of polarization. In 1897, when he was invited to deliver a lecture with a demonstration at the Royal Institution in London, he had taken with him the instrument (similar to a microwave spectrometer with a transmitter and receiver) constructed by himself. An able instrument-maker as he indeed was (he gave Sanskrit names to his instruments), Bose also constructed devices in which mechanical or light stimuli could produce electrical responses, as he turned his interest to plant physiology. The main focus of his interest in this area was to find out or establish that all the characteristics of response displayed by animal tissues are equally exhibited by plant tissues. In 1900, he read a paper 'On the Similarity in Responses in Inorganic and Organic Matter', at the International Congress in Physics held in Paris, and later at the British Association for the Advancement of Science at Bradford in England. He submitted to the Royal Society of London a preliminary note on the 'Electric Response of Organic Substances' in which he had shown the strong electric response from plants to mechanical stimuli. This paper, however, was not published, since it met with criticism, especially that of Sir John Burdon Sanderson who was a leading electro-physiologist of the time. His attempts to publish two more papers on this subject with experimental proof also encountered the same fate.

Nevertheless, his new interest in physiology necessitated the invention of a special type of apparatus and his inventive genius rose to the occasion when he fabricated an optical lever to magnify and photographically record the minute movements of plants; a resonant recorder to determine with remarkable accuracy within a thousandth part of a second, the period of response of what is known as the touch-me-not plant, *Mimosa pudica*; and the oscillating recorder for minute lateral movements. He also devised a crescograph for recording the micrographic growth-movements of plants. With this he was able to obtain high magnification, which enabled him to record automatically the elongation-growth of plant tissues and their modifications through varied external stimuli. He also perfected a magnetic crescograph of high magnification. In 1907, his book *Comparative Electro-Physiology* was published.

Bose was elected a Fellow of the Royal Society in 1920, in recognition of his outstanding research. Earlier in 1903, he had been conferred the Companionship of the British Empire (CBE) by the British government, and in 1912, he was the recipient of the award of Companionship of the Star of India (CSI); soon he was conferred an honorary DSc by the University of Calcutta; likewise, the Banaras and Dacca Universities conferred honorary degrees on him. Bose retired from the Calcutta University in 1915. But he did not stop there. He had already been contemplating a research institute and in 1917 bequeathed to the nation an institute of excellence now known as the Bose Institute. Significantly, he said that this Institute was *not a laboratory, but a temple*. He died on 23 November 1937, leaving an indelible imprint as a great pioneer of Indian scientific movement.

Prafulla Chandra Ray (1861-1944)

Prafulla Chandra Ray

In the second half of the nineteenth century, a family of agriculturists lived in a village in the district of Khulna (now in Bangladesh) and it was in this family that Prafulla Chandra Ray was born. His father, Harish Chandra Ray, imbued with a spirit of providing education to everyone, had established one school for boys and another for girls. Young Prafulla's interest in education and concern for the poor were in no small measure due to the influence of his father. After the family moved to Calcutta, Prafulla entered the famous Hare School, like Jagdish Chandra Bose. He was deeply interested in language and literature. When he was hardly ten years old, he had already acquired an adequate knowledge of English and Bengali literature as well as Latin and Greek. At the age of thirteen, he joined Albert School which had been established by Keshab Chandra Sen, an ardent follower of the Brahmo Samaj movement. He came out of the school with flying colours, standing first in the examination. He passed the entrance examination in 1879 and joined the Metropolitan Institute (now named Vidyasagar College after Ishwar Chandra

Vidyasagar). Here he came under the influence of great teachers such as Alexander Pedlar, who would deliver inspiring lectures on chemistry at the Presidency College. These lectures led Prafulla to take up chemistry for his BA examination, but his passion for literature remained undiminished.

He secured the Gilchrist Prize scholarship after passing the competitive examination, which enabled him to take up the BSc degree course in the University of Edinburgh in 1885. Two years later, he was also awarded a DSc degree and the Hope Prize scholarship that helped him do research for another year. Back in India in 1888, he could not find a job, except for working for almost a year in the laboratory of Jagadis Chandra Bose, whose friendship he had valued. But, to his good fortune in 1890, he was appointed assistant professor of chemistry in the Presidency College on a salary of Rs 250 per month. There, he was admired by students for his teaching skills, experiments and impressive lectures. Significantly, Meghnad Saha, who later became a brilliant physicist and Shanti Swarup Bhatnagar who also rose to be an outstanding chemist and the architect of national laboratories, were among his numerous students.

P.C. Ray's pioneering research was on mercurous nitrite and its derivatives. He also worked on certain alkaline earths, methyl-ammonium and other alkaline ammonium nitrites. When he joined the University College of Science, Calcutta, he studied the compounds of some metals and sulphur and also guided a number of students who conducted chemical researches. P.C. Ray has been rightly regarded as the founder of the Indian Chemical School. But he was also interested in the ancient chemical practices of India and published *The History of Hindu Chemistry* in two volumes (1902; 1925).

P.C. Ray was instrumental in establishing the Bengal Chemical and Pharmaceutical works in Calcutta, an Indian venture in an industry that was totally under foreign control then. Though initially this chemical and pharmaceutical factory faced severe competition, P.C. Ray's devotion and hard work coupled with the services offered by his students, resulted in the Indian enterprise making its mark. Apart from his own chemical and pharmaceutical concerns, he lent a helping hand to other Indian initiatives in setting up textile mills and other factories. Thus P.C. Ray played an important role in ushering in industrialization that had just started in India in the beginning of the twentieth century. In all of these, he was guided by intense nationalism.

The *khadi* movement and the use of *carkhā* (spinning wheel) started by Gandhiji had a profound impact on P.C. Ray. He used to spin yarn with a *carkhā* at least for an hour every day, and wore only *khadi* or *khaddar* (clothes from homespun yarn). This earned him the sobriquets 'Sir Khaddar' and 'Carkhashri'. He retired in 1921 as professor of chemistry of the Calcutta University Science College with which he continued to be associated in one way or other. Even before retirement, he had donated his salary to the department of chemistry which he headed. A dedicated humanitarian, P.C. Ray engaged himself in social service, and organized a relief committee when gigantic floods swept north Bengal, for which Gandhiji once called him 'a Doctor of Floods'. A bachelor, Ray lived simply, sharing his accommodation with poor students. The life of this eminent chemist, promoter of industries, and humanist came to an end in 1944.

Srinivasa Ramanujan (1887-1920)

A mathematical prodigy was born in Erode of Tamil Nadu a year earlier than C.V. Raman, but lived only for thirty-two years. He was Srinivasa Ramanujan. His father worked as a petty clerk in a small factory. That the young Srinivasa was a prodigy with gifted mathematical talent was evident when, at the age of thirteen, he was engaged in intricate arithmetical and geometrical problems in high school in Kumbakonam. He came across a book, *Synopsis of Elementary Results in Pure and Applied Mathematics* by George Schoobridge Carr, and its formulae, theorems and proofs fascinated the young mind of Ramanujan. Even at his tender age, Ramanujan conceived of several mathematical ideas. He used to solve the problems on loose sheets of paper and write the results in his notebooks. These later became famous as *Ramanujan's Notebooks,* which are still being studied.

Srinivasa Ramanujan

With a first class in mathematics in the matriculation examination, Ramanujan joined the Government College in Kumbakonam in 1904. Two years later, he went to Madras and entered Pachaippa College. But he failed twice in the first year Arts examination because he totally neglected the study of other subjects like history and English literature. Besides, he fell ill and when he appeared for the examination he passed only in mathematics. Illness ravaged him frequently, but even then he used to be engrossed in mathematical problems and scribbling numbers. His parents, apprehensive about his mental health and in the hope that marriage would help him to lead a normal healthy life, married him off to Janaki Ammal, a young girl of about nine years! This added responsibility forced Ramanujan to look for a job; but indifferent personal traits came in his way of getting a suitable one.

He met Ramachandra Rao, collector of Nellore and founder-member of the Indian Mathematical Society, to seek his help in getting a job, saying that he wanted just a small amount to live on so that he could pursue his research. Eventually, it was Francis Spring, director of the Madras Port Trust who gave Ramanujan a clerical job on a salary of Rs 25 a month. In 1913, the mathematical prodigy wrote a letter to G.H. Hardy, the famous mathematician of Trinity College, Cambridge, sending along with it a set of 120 theorems and formulae, besides a key formula in hypergeometric series, which later came to be known after him. Hardy was convinced that Ramanujan was indeed a mathematical genius and, therefore, invited the young prodigy to Cambridge, making arrangements for his travel and stay in Cambridge.

Ramanujan sailed for England in March 1914. But he had to encounter some criticism from his orthodox relatives to whom foreign travel was an anathema. He appeared to have sought divine intervention to help him go abroad. He had infinite faith in the goddess Nāmagiri of Namakkal and it was even believed that he was born due to the blessings of this goddess sought by his mother. As the family story goes, the goddess appeared in a dream and instructed the parents to grant permission for his voyage, so he set out at last. After nearly a month's passage in the ship, he landed in Cambridge where he found himself a stranger in an inhospitably cold climate; and he had to cook his own food, being a vegetarian and Brāhmin. But his mind was always immersed in mathematics. During his five-year stay in Cambridge, he published as many as twenty-one original papers, five of them being in collaboration with Hardy. His mathematical prowess encompassed composite numbers, algebra of inequalities, continued fractions, probability theory and the like. The Hardy-Ramanujan circle method in number theory and Roger-Ramanujan identities in the partition of integers became landmarks in mathematics. When asked to rate top mathematicians of his time on a scale of 100, G.H. Hardy gave himself 25, two other distinguished mathematicians, J.E. Littlewood 30 and D. Hillbert 80, but gave Ramanujam 100 upon 100—although Hardy found that Ramanujan was not a systematic mathematician. Nevertheless, Ramanujan obtained the BA degree (by research) of the Cambridge University in 1916 and was elected a Fellow of the Royal Society of London early in 1918. Later in 1918, he was elected to a Trinity College Fellowship—a rare distinction achieved by an Indian. The Madras University, on the advice of the Government of India, granted Ramanujan 250 pounds a year as an allowance for five years from April 1919 soon after his return from Cambridge.

But he was not destined to enjoy for long the benefits of this grant, nor his international fame. Tuberculosis began to decimate him and he was soon hospitalized. Even on the hospital bed he used to play with numbers and associated mathematical problems. The end of a mathematical prodigy, who lived passionately and breathed only mathematics in and out, came in Madras (now Chennai) on 26 April 1920 when he was just thirty-two.

Chandrasekhara Venkata Raman (1888-1970)

One of the most eminent experimental physicists of the twentieth century and the first Indian physicist to win the Nobel Prize in Physics, Chandrasekhara Venkata Raman was born in Tiruchirapalli in Tamil Nadu, which lies on the banks of the river Kaveri. His father, Chandrashekar Iyer departed from his family tradition of agriculturists and became a teacher in mathematics and physics in a local school. He was also a lover of music and Sanskrit literary tradition. Young Raman's interest in music as well as in the physics of musical instruments owed its inspiration to this ambience even as a boy. His early education was in Vishakhapatnam. After his intermediate examination, he moved to Madras, joined the BA class of the Presidency College and, in 1905, earned the distinction of being the only student who passed with first class, winning a gold medal in physics. Two years later he obtained his masters degree.

C.V. Raman

Even as a student at the Presidency

College, young Raman undertook investigations in acoustics and optics. When he was just eighteen years old, one of his research papers was published in the *Philosophical Magazine* of England, and later another was published in the prestigious British journal, *Nature*. There was, however, no attractive career opportunity in the field of scientific pursuits. He, therefore, passed the Financial Services examination in 1907, standing first in this competitive examination, an achievement that earned him the first lucrative governmental appointment as assistant accountant-general at Calcutta.

But C.V. Raman's heart lay in scientific pursuits and he earnestly searched for an opportunity to pursue scientific research. It was one of those rewarding evenings, as it were, for science in India and Raman in particular, when on returning from his office in a tramcar, his searching eyes saw a signboard, on Bow Bazaar Street, of the Indian Association for the Cultivation of Science (IACS), which, as noted earlier, was established by M.L. Sircar and others. His joy knew no bounds and on an impulse, he got off immediately from the tramcar and went into the building. He met Amritlal Sircar, the son of M.L. Sircar, who was the honorary secretary of IACS, and asked him if he could conduct scientific research there during his spare time. His initiative paid off and once he got Sircar's approval, he shifted to a house next door to the IACS building and devoted his morning and evening to research, attending his office during the day. It was in the sparsely equipped laboratory of IACS that C.V. Raman first carried out his investigations on the physics of vibrations and musical instruments, practically alone except for the assistance of Ashutosh Dey, affectionately called Ashu Babu, a devoted attendant.

Raman's dedicated pursuit of scientific research at IACS had brought him in close contact with Asutosh Mookerjee, the dynamic vice-chancellor of Calcutta University, who had succeeded in establishing the University Science College in 1915, with endowments from enlightened philanthropists of

Calcutta. Asutosh Mookerjee offered the newly created Palit professor of physics post to Raman, but the salary was much lower (Rs 600 a month) than the amount he was earning as the assistant accountant-general (Rs 1,100 per month).

However, Raman's inner urge for scientific research was so compelling that he readily discarded his lucrative position in the Government of India and joined the Calcutta University in 1917. After Amritlal Sircar's death, he was elected Honorary Secretary of IACS in 1919. He thus had the advantage of two laboratories, one at the Science College and the other at IACS. He continued his research in IACS, now focusing on the scattering of light. Though, according to the terms governing the Palit professor of physics post, he could have kept himself free from teaching, Raman also involved himself in giving lectures. His teaching was so inspiring that he attracted many young people, for whom he emerged as a role model. Students came to him from different parts of the country for postgraduate studies and were inspired by his experiments as well as teaching. Raman used to forget to eat sometimes, work late right through the night and even sleep on one of the laboratory tables!

Raman's scientific interests encompassed sound and vibrations, the theory of musical instruments, optical studies, colloids, the molecular scattering of light, X-rays, magnetism and magneto-optics. He was an ardent lover of music and used to say that he should live much longer since he had not heard all the music he wanted to hear! He collected a wide variety of instruments like the *mṛdangam, tabala, veena,* violin and *nāgasvaram*. By about 1918, he had offered a scientific explanation of the complex vibrations of the strings of musical instruments, and of the characteristic tones emitted by *mṛdangam, tabla* and others. In fact, Raman was elected Fellow of the Royal Society in recognition of his outstanding investigations. Interestingly, he even collected a large number of specimens of butterflies and diamonds, the physical properties and structure of which attracted him. But it was his fascination with colours that led to his famous discovery of the scattering of light.

On his way to Oxford by ship to attend a conference there, Raman was spellbound by the startling blue colour of the Mediterranean Sea. On his return, he carried out some experiments at IACS laboratory with water samples collected from different seas. He then arrived at the conclusion that the blue colour of sea water is due to the scattering of light by water molecules. His students, under his guidance, conducted several experiments and threw light on the mode of scattering of light in diverse media. One of the ablest students was K.S. Krishnan and, under instruction and guidance from Raman, he took up experimental investigations on the anomalous scattering in certain liquids and vapours. These painstaking experiments finally led to the discovery of the internationally famous *Raman Effect*. The pathbreaking conclusion was announced by C.V. Raman to the press on 28 February 1928. A joint scientific communication by Raman and Krishnan was also sent on the 16 February 1928 for publication in *Nature*; it was published on 31 March 1928. He delivered a lecture in Bangalore on this discovery under the title 'New Radiation' and on his return to Calcutta, he immediately had his lecture printed at the Calcutta University Press, and mailed its copies to many scientists all over the world.

C.V. Raman's discovery of the new phenomenon was hailed as a remarkable achievement and the spectral lines in the scattered light came to be known as 'Raman Lines'. K.S. Krishnan along with K. Venkateswaran (both were students of Raman) collaborated with Raman, who acknowledged their assistance with affection. Among the accolades, the British government conferred a knighthood on Raman when he was awarded the Nobel Prize in Physics in 1930. During the twelve years after the discovery of the Raman Effect, a large number of research papers were published all over the world on its different aspects and several chemical compounds were studied in relation to it—

such was the great impact that Raman's discovery produced. At the Nobel Prize ceremony in Stockholm, Raman demonstrated the significance of his discovery by using alcohol as the chemical substance; but in the celebratory reception held in the evening, he shunned it!

In 1933, C.V. Raman left Calcutta and joined the Indian Institute of Science in Bangalore, as its first Indian director. Just as he had built up an active school of physics in Calcutta, he also established a flourishing department of physics at the Institute. Another set of notable investigations by C.V. Raman was in the field of ultrasonics, which led to the well-known 'Raman-Nath theory of diffraction of light by high frequency sound waves'. A brilliant mathematical physicist, Nagendra Nath was his collaborator, and this joint discovery was regarded as next in importance to the *Raman Effect*. Nagendra Nath also studied the lattice dynamics of diamonds. At the invitation of Raman, Max Born, an outstanding physicist, worked at the Institute. The Neutrino theory of light expounded by him and Nagendra Nath was another important investigation in the department of physics of the Institute. There were several other experimental investigations which won international acclaim for the department of physics of the Indian Institute of Science under the scientific leadership of C.V. Raman.

As the director of the Institute, however, he did not possess the much-needed administrative tact, and lived in his own world of ideas and scientific investigations, often disregarding heads of the other departments and even the council of the Institute. He was forced to resign in 1937, but allowed to continue as professor and head of the department of physics till he retired at the age of sixty years in 1948. He bequeathed to the nation the now familiar Raman Research Institute and gave all his property and savings to the Institute. In 1954, the Government of India bestowed the *Bharat Ratna* on him, the country's highest civilian honour. But, his devotion to scientific research and commitment to guide research undertaken by his students was such that he did not go to New Delhi to receive the *Bharat Ratna* award, since it clashed with the submission of a thesis of one of his PhD students. Although till his last days he was very active scientifically in his own Institute, he remained aloof from the substantial scientific progress that was taking place in the country. This great genius breathed his last on 21 November 1970 at Bangalore.

Meghnad Saha (1893-1956)

An eminent astrophysicist who won international acclaim for his Theory of Thermal Ionization of Elements that could explain the ordered spectral lines of stars, Meghnad Saha was also an intense nationalist. In his own way, he contributed substantially to the promotion of science in Independent India. He was born in a village in Dacca district (now in Bangladesh) as the fifth child of his parents and his father was a petty grocer with a meagre income, barely managing to feed his large family. Even as a young boy, Saha's intellect was in evidence and his father managed to send him to an English school in a small town some ten kilometres away from the village. A local medical practitioner there let Meghnad stay in his house with free board and lodging. The boy stood first in the middle school examination and was the recipient of a scholarship which enabled him to study in the Collegiate School in Dacca in 1905. The spirit of nationalism was so strong in young Meghnad that he joined the boycott movement during the visit of the British governor in protest against the repressive measures being adopted by the alien government. Besides losing his scholarship for this brave act, he was also debarred later from appearing for the Financial Civil Service examination. He shifted to another school which offered

Meghnad Saha

him free studentship and in 1909, he passed the Entrance Examination, standing first among those who appeared from East Bengal. After passing the Intermediate Examination from Dacca College and winning a scholarship, Meghnad joined the Presidency College in Calcutta. His teachers were J.C. Bose and P.C. Ray. While Satyendra Nath Bose, P.C. Mahalanobis, and Nila Ratan Dhar were his brilliant student-contemporaries. He passed both the BSc examnation in mathematics and MSc in applied mathematics with distinction.

In 1917, the vice-chancellor of Calcutta University, Asutosh Mookerjee, recognized the merit of Meghnad Saha as well as of Satyendra Nath Bose in mathematical physics. Therefore, he invited them to join the department of mathematics. However, the head of the department of mathematics did not have a genial temperament to get along with the brilliant young entrants, compelling both Saha and Bose to join the physics department instead. Saha's original papers on theoretical physics, published in such noted journals as the *Philosophical Magazine* of London and the *Physical Review* of USA, earned him the DSc degree of Calcutta University in 1918.

In the course of his preparation for his lectures from select books on astrophysics, Saha knew that the spectrum of the sun's chromosphere had been noticed by the British astronomer, N. Lockyer and Jansen, as early as in 1868, and that Gustav Robert Kirchoff of Heidelberg had shown that the spectral lines were of the elemental composition of the sun. Nevertheless, the main cause for this phenomenon was still unknown, nor had any theory been advanced till then to explain the presence of such spectral lines. Saha pondered over this problem and developed what is called the Theory of the Thermal Ionization of Elements, which could explain several aspects of this phenomenon. This ionization formula is an important milestone, and hailed as the twelfth major discovery in astrophysics. By this brilliant theoretical formula Saha won international acclaim.

In 1919, Saha went on a two-year study visit to Europe as an awardee of the prestigious Premchand Roychand Studentship. He worked in Alfred Fowler's spectroscopy laboratory at the Imperial College of Science in London, where he met the noted chemist, Shanti Swarup Bhatnagar. Soon they became close friends and both joined hands for the promotion of science in Independent India. Later, when Saha was working in Professor W. Nernst's laboratory in Berlin, Asutosh Mookerjee offered him the Khaira Chair in Physics. At the same time, the Universities of Benaras, Aligarh and Allahabad had also offered professorships to him. He chose Allahabad University, but here too he did not have sufficient grants for scientific research. But the situation changed after he was elected a Fellow of the Royal Society in 1927, and he was able to build an active department of physics at the Allahabad University, raising it to international status. It was during this time that he laid solid foundation for the National Academy of Sciences, India at Allahabad, as well as the Indian National Science Academy. In July 1938, he returned to Calcutta University as the Palit Professor of physics, and established the department of nuclear research at Calcutta University, which in 1948 was rightly named the Saha Institute of Nuclear Physics.

Saha had varied interests. He is remembered also for his systematic studies on the nature and control of floods. He wrote extensive essays on river-management, irrigation, floods and related topics, which helped in the formulation of several river valley projects like Damodar Valley, Bhakra-Nangal, and Hirakud; in fact, Saha is regarded as the chief architect of the Damodar Valley Project in West Bengal. He served as a member of Parliament and also provided valuable suggestions from time to time for the formulation and implementation of our country's Five-Year plans. A humanist, Saha was often involved in flood relief measures and helping the victims of communal disturbances in their rehabilitation. An outstanding physicist, planner and humanist all his life, Saha passed away on 16 February 1956.

Satyendra Nath Bose (1894-1974)

A classmate of Meghnad Saha and an equally brilliant physicist, Satyendra Nath Bose was born in Calcutta as the eldest of seven children and the only son of his parents. His father, Surendra Nath Bose, was an accountant in the East India Railway. After his school education, Satyendra Nath joined the Presidency College (1909) where he came under the intellectual influence of eminent teachers like J.C. Bose as well as P.C. Ray, whose patriotic fervour had a deep impact on young Bose. In 1913, he took his BSc examination and two years later, the MSc examination, achieving the first rank in both. The far sighted vice-chancellor of the Calcutta University, Asutosh Mookerjee, who was adept at spotting and encouraging talented persons, appointed S.N. Bose as a lecturer in the university to teach courses in mathematics and modern physics.

Bose began to acquire proficiency in French and German in order to keep himself well informed of the latest developments in modern physics by reading books and articles in journals which were appearing in these languages. Because of the outbreak of the First World War, books published in Europe were not easily available to him. However, there was one instructor in the Bengal College of Engineering, P.J. Bruhl, who possessed some important textbooks on modern physics, which helped Bose teach postgraduate students such advanced topics as relativity and quantum physics. In 1921, he left Calcutta University and joined the physics department of the Dacca University. Around this time, his friend D.M. Bose, who had done research in particle physics in Germany, had returned to India and brought with him new publications on the quantum theory published in Europe. While studying them, S.N. Bose noticed that Max Planck, the exponent of the quantum theory, had some assumptions and formulated an equation rather approximately. Bose pondered over it meticulously and found out a better method to work out the equation. He produced a four-page research paper titled

'Planck's Law and Light Quantum Hypothesis', and sent it to journals in India and abroad for publication; but all the journals rejected it. Bose was immensely disappointed, but did not lose heart. Mustering courage, he sent his paper in 1924 to the great genius, Albert Einstein, which turned out to be the right course to take; for, Einstein was impressed with Bose's bold concept. Einstein himself translated it into German and sent it for publication to the prestigious German journal, *Zeitschrift für Physik,* which soon published it. Einstein had also praised the paper of Bose as an important step in quantum physics and he even adopted the approach of Bose in his own work, and this particular field of Bose's research came to be known as Bose-Einstein Statistics (now called only Bose Statistics). Elementary particles like photons that obey Bose Statistics are called *Bosons* in honour of S.N. Bose—a recognition of his originality in the area of physics.

Bose regarded Albert Einstein as his *guru* and had the privilege to meet him in Berlin when he was in Europe (1924-27). On his return to India, he was appointed professor of physics in the Dacca University. After serving this university for eighteen years, he became Khaira professor of physics at the University of Calcutta. Bose's desire to meet Einstein again in USA did not materialize for a strange reason, namely, that since he had visited the Soviet Union earlier, Americans thought that he might be a communist and did not grant him a visa!

S.N. Bose

Bose was one of the rare brilliant scientists who searched and pondered beyond science, into literature, art and music. He studied several works of Bengali, French and English literature, and even translated some French short stories into Bengali. He had realized the importance of Indian languages as a vehicle to convey scientific ideas to the common man. He was of the view that if science was to reach the masses, it should be taught in the regional languages. In 1948, he established a science association in Calcutta called *Bangiya Vijnana Parishad* which, since its inception, had been publishing a popular and thoughtful periodical titled *Jnan O'Bijnan*. Bose was also a gifted musician as well as a critic of music, besides creating a few *ragas* and playing on bowstring musical instruments. An enlightened individual and eminent scientist, Bose was elected a Fellow of the Royal Society of London (1958). As vice-chancellor of the famous Vishwabharathi University, he also left behind his imprint on this great institution. When he was about to complete eighty years, Bose suffered a massive heart attack and his end came on 4 February 1974.

Prasanta Chandra Mahalanobis (1893-1972)

The first Indian statistician who won international acclaim, and who brought in statistical methods for core planning processes at the national level, was Prasanta Chandra Mahalanobis. He was born in an affluent family, his father being an established dealer in sports goods, and his mother being the sister of the famous physician, Sir Nilratan Sircar. They belonged to the liberal Brahmo Samaj traditions and had intimate contacts with social reformers and enlightened intellectuals. Young Prasanta, influenced by this environment, had a liberal mind with a rational outlook. In 1908, he passed the matriculation examination from the Brahmo Boys' School in Calcutta, followed by the Intermediate Examination (1910), and BSc (1912) from the Presidency College. Later he studied in King's College in Cambridge, and passed the Natural Sciences Tripos with a first

P.C. Mahalanobis

class in 1915. He also had the opportunity to spend some time with Ramanujan who was then in Cambridge.

Mahalanobis had a liking for physics and wanted to work with renowned physicist C.T.R. Wilson in the famous Cavendish Laboratory in Cambridge. However, he developed an interest in statistics when he began to read the bound volumes of the reputed journal, *Biometrica.* He found that statistics was a new tool of measurement and analysis, capable of wide application even in such fields as meteorology and anthropology. When he was in India on vacation, the First World War was in full spate and he decided not to return to Cambridge. He was appointed a lecturer in physics at the Presidency College, but in his spare time he began to study statistics, an area that was not popular then in India either in the universities or in the governmental organizations relating to scientific or economic subjects. Mahalanobis had to face this vacuum in his mission to promote statistics in India. Fortunately for him, three distinguished persons—Rabindranath Tagore, Nilratan Sircar and Brajendra Nath Seal (who was professor of philosophy at the University of Calcutta and was also a mathematician in his own way) encouraged Mahalanobis and lent their helping hand to the advancement of statistics.

To Mahalanobis, statistics was not merely a mathematical exercise in the theory of probability, nor a collection of data and their routine analysis. It was a new technique for increasing the efficiency of human efforts in the widest sense. With his own originality, he successfully applied statistical principles to problems in areas such as meteorology including rainfall and floods, demography, epidemeology and anthropology. His statistical studies were found to be very useful for the implementation of projects like the Hirakud hydro-electric project and the Damodar Valley multipurpose hydro-electric project. He started a statistical laboratory in his own room and worked on his themes with his young students for some years. This was the nucleus for the emergence of the Indian Statistical Institute in 1932. In 1933, *Sankhya*, the first Indian journal of statistics, was started with Mahalanobis as its editor. One of the earliest problems that engaged the concern of the Indian Statistical Institute was in the field of jute, estimating the acreage and the yield of this crop by random sampling, which was successfully accomplished by the statisticians of the Institute. As a result of his efforts, postgraduate studies in statistics were started in Calcutta University in 1941 with Mahalanobis himself as the professor and head of this department. Within two years after Independence, the Government of India established a Central Statistical Unit which, in 1951, became the now well-known Central Statistical Organization. His name in statistics is enshrined through his contribution known as the *Mahalanobis distance* and *Fractile graphical analysis.*

In the post-Independence period till his death in 1972, Mahalanobis contributed substantially to the formulation of the successive Five-Years Plans with statistical insights, and won the admiration of Prime Minister Jawaharlal Nehru. He was elected a Fellow of the Royal Society in 1945. Intellectually active till the last, Mahalanobis died in 1972 at the age of seventy-nine.

Birbal Sahni (1891-1949)

An outstanding palaeo-botanist, Birbal Sahni was born in Bhera, now in Pakistan, in a family known for its educational attainments. His father, Ruchi Ram Sahni, encouraged the young boy to collect plants and fossils, and carefully study them. After his early education in Lahore, Birbal obtained a BSc degree from the University of Punjab in 1911. He went to England and entered Emmanuel College in Cambridge, and soon he undertook research at the Botany School there under the guidance of the well-known botanist A.C. Seward, whom Birbal regarded as his role model throughout his life. Though he began with the conventional investigations of morphology and anatomy of living plants, Birbal developed an abiding interest in fossil plants. His first research paper which was published

Birbal Sahni

in the journal *New Phytologist* was, 'On the presence of foreign pollen in the ovules of Gingkobiloba, and its significance in the study of fossil plants'. In fact his investigations of fossil plants earned for him the DSc degree of London University in 1919. Ten years later, the University of Cambridge, in a rare gesture, awarded him the ScD degree in 1929 in recognition of his work and in 1936, he was elected a Fellow of the Royal Society of London.

When Sahni was professor of botany at the Lucknow University, he attracted students from far and wide, and the department of botany of the Lucknow University earned an enviable position. He had realized that palaeo-botanical studies would be fruitful if they were carried out with a background of geology, since plant fossils and geological formations had implications for one another. He used to say that 'fossil plants represent the debt that botany owes to geology'. This conviction led to his concerted efforts to start a department of geology in the University of Lucknow in 1943. No one at that time was better suited to head this new department than Sahni himself since he was well versed in stratigraphical geology on which he used to deliver lectures to postgraduate students before they started their morphology course.

Birbal Sahni was a meticulous researcher and his relentless investigations led to the discovery of gymnospermous plants of the Jurassic Age, which he named the *Pentoxyleae*. He did research on the Indian Gondwana fossil plants, because one of the accepted geological theories is, that the Indian subcontinent was originally a part of a super continent named Gondwana, comprising South America, Africa, India, Australia and Antarctica. The most abundant of the fossil plants are of what is known as the *Glossopteris* type, the fossils of which were discovered in India. Sahni examined them and concluded that their presence indicated a cold temperate but not glacial conditions. Exploring the Rajmahal Hills of Bihar in which plant fossils abound, he demonstrated that the Gondwana flora was Jurassic (geological span of time from about 213 to 144 million years ago) without any species that are characteristic of the Cretaceous period (144 to 70 million years ago). Sahni's palaeo-botanical acumen manifested itself in his reconstruction of a cycad-like plant, the fossils of which were discovered near Amrapara in Santhal Paragana district of Bihar. He showed that the stem, leaves and flowers of this fossil plant, which was earlier thought to be of a different group, belonged to the same plant. He named this plant *Williamsonia sewardiana,* in honour of his preceptor, A.C. Seward. With similar sentiment, a new genus and species of a cone-bearing plant with fern-like leaves, has been named *Pentoxylon sahanii* after Sahni.

One of the founders of the Indian Botanical Society, Birbal Sahni was elected as its president in 1924. He was the recipient of the Barclay Medal for Biological Research awarded by the Royal Asiatic Society of Bengal, and the Sir C.R. Reddy National Prize for Natural Science (1947). He initiated steps for the establishment of an institute devoted to palaeo-botany, donating his own funds for this cause. The government of Uttar Pradesh and the Government of India also provided suitable grants. On 3 April 1949, Jawaharlal Nehru laid the foundation stone of the new Institute for Palaeo-botany. However, fate had willed otherwise for Birbal Sahni. For, after a severe heart attack, he breathed his last, just five days after the foundation-laying ceremony. But he had bequeathed to the nation a rare institution that stands as one of the landmarks in the progress of botany in post-Independent India.

Shanti Swarup Bhatnagar (1894-1954)

Shanti Swarup Bhatnagar

A chemist of repute and also known for his leadership in building scientific institutions, Shanti Swarup Bhatnagar was born at Bhera in Shapur district in the Punjab. His father, a graduate of the Punjab University, had embraced the Brahmo Samaj, and, in consequence, lost his share of his ancestral property. He died when Shanti Swarup was just eight months old, leaving the family quite poor. The boy grew up under the influence of his grandfather, an engineer, in Sikandrabad (now in Pakistan). He studied in A.V. High School and also Dayal Singh High School in Lahore under the guidance of its headmaster who was a classmate of his father. His science teacher was Maulvi Talib Ali Paband. After passing the matriculate examination with a first division and securing a university scholarship, Shanti Swarup joined the Dayal Singh College in Lahore in 1911. Here he developed an abiding interest in chemistry and in particular, the fermentation process. He also published a paper in a college journal, titled 'Fermentation Phenomenon of Pomegranate Juice'. Young Shanti Swarup had literary tastes and was an amateur playwright too, in the Urdu language.

A bright student, he passed the intermediate examination securing a high first class division, earning encomiums from his principal as the ablest boy with all-round capabilities. The principal expressed the view that, if given the opportunity of developing talent in some European or American centre of scientific research, the boy would make remarkable strides in research. After his BSc degree, and with the financial support of the Dayal Singh College Trust, Bhatnagar sailed for America in 1919. But, on his way when he found that he was unable to proceed from London to America, he took permission from the Trustees to do postgraduate research in the laboratory of chemistry professor F.G. Donnan of the University College, London. His initial field of research, under Donnan, was on emulsification, a theme of colloidal chemistry, and he obtained his DSc degree for his thesis on the 'Invention of Emulsions'.

In 1921, Bhatnagar returned to India and joined the newly established Banaras Hindu University as professor of physical chemistry where he built up an active school of physico-chemical research. Despite his busy schedule, his literary interests did not recede to the background. He composed what is known as *Kulgeet* (University song) in Sanskrit, which is sung even today in BHU on ceremonial occasions. In 1923, when Bhatnagar was in England attending a session of the British Association for the Advancement of Science held in Liverpool, he came across an advertisement in the widely read journal, *Nature*, for the post of director of University Chemical Laboratories in Lahore. He applied for the post and was appointed to this position, despite the fact that his contestant was none other than his own former professor, a British chemist. In this research institution too, Bhatnagar created an active school of research in colloidal chemistry and photochemistry. With one of his colleagues, he devised an interface balance to measure minute changes in the magnetic susceptibilities of substances in chemical reaction, and this device was licensed to a firm in London. Commissioned by the noted British publisher, Macmillan, Bhatnagar published a valuable book (1935) titled *Physical Principles and Applications of Magneto-Chemistry,* which enhanced his reputation as an important Indian chemist. He also started a department of petroleum research in the Punjab University out of the funds he personally received from a British oil company for solving its drilling problem in saline earth.

Bhatnagar was destined to play a substantial role at the national level. Sir Arcot Ramaswamy Mudaliyar, then a member of the Viceroy's Council, convinced Lord Linlithglow of the need for a Board of Scientific and Industrial Research to meet the scientific and technical needs of the Second World War which had broken out in 1939. The Board was set up in 1940 with Shanti Swarup Bhatnagar as its first director. In 1942, it was elevated to the Council of Scientific and Industrial Research (CSIR). Its laboratories, which were functioning in the Government Test House at Calcutta, were shifted to the Delhi University campus because of the risk of the Japanese invasion. Several products of high importance during the war period came out of these laboratories under the guidance of Bhatnagar. After Independence, with the active support of the scientifically oriented Prime Minister Jawaharlal Nehru, Bhatnagar conceived and planned the establishment of a series of national laboratories under CSIR, which Nobel Laureate C.V. Raman called the 'Nehru-Bhatnagar Effect'. The Urdu poet that Bhatnagar was, he was in fine tune with another Urdu poet and national leader, Maulana Abul Kalam Azad, who was then the minister-in-charge of scientific research. Before Bhatnagar died of a heart attack on the night before the New Year's Day in 1955, twelve of the national laboratories had been established by him as the legacy of a man of exemplary devotion to the promotion of science in India.

Homi Jehangir Bhabha (1909-64)

Homi Jehangir Bhabha

Known as the architect of India's nuclear energy programme, Homi Jehangir Bhabha's name is associated with the main Atomic Research Centre in Trombay, Mumbai. He was born in Bombay (now Mumbai) in an illustrious Parsee family. His father, Jehangir Bhabha, a student of Oxford University, was a renowned advocate, and associated with the business house of the Tatas. His mother, Meherbai, was the granddaughter of the well-known philanthropist and the first Baronet, Sir Dinshaw Petit. His paternal grandfather, H.J. Bhabha, was the inspector-general of education in the state of Mysore, who had helped J.N. Tata in his endeavour of founding the Indian Institute of Science in Bangalore. Young Bhabha's early education was in John Cannon High School. Later, he pursued his education in Elphinstone College and the Royal Institute of Science, Bombay. His grandfather had a large library of his own which contained several latest books on science, paintings and art, and his father too had an impressive collection of music records. Bhabha's mind was influenced by all of them and at the age of fifteen, he had already read books on Einstein's theory of relativity. He was also exposed to discussions on industrial projects being conceived by Sir Dorabji Tata, the first son of J.N. Tata.

Bhabha's innate inclination was towards physics and mathematics. But, acquiescing to the wishes of seniors in his family who wanted him to take up mechanical engineering and enter the Tata enterprise of the Iron and Steel Company at Jamshedpur, Bhabha went to Cambridge in 1927 to study engineering there. His father assured him if he passed the engineering examination creditably, he could then pursue mathematics for which the necessary financial support would be available. Bhabha did pass the mechanical engineering Tripos with a first class in 1930, and joined the Cavendish Laboratory, which transformed his later career. His brilliance and aptitude for theoretical physics won him several scholarships, which also enabled him to go to Europe and work with distinguished physicists like Wolfgang Pauli in Zurich, Enrico Fermi in Rome and Hans Krammers in Utrecht. In 1934, he was awarded the prestigious Isaac Newton Fellowship for his first scientific paper published in 1933, which dealt with the role of electron showers in the absorption of gamma radiation. He came into close contact with some of the leading scientists of the time like Paul Dirac, Niels Bohr, Ernst Rutherford and Walter Heitler. As a result, he became deeply involved in the investigations of high energy physics, and most of his (about fifty) published papers were, therefore, in this field. In 1937, he expounded the 'Cascade Theory of Electron Showers', along with Heitler, which is well known today as the 'Bhabha-Heitler Cascade Theory', which sheds light on the process of electron showers in cosmic rays. He continued to work in Cambridge till the early months of 1939.

In the middle of 1939, Bhabha was on a holiday in India, when the eruption of the Second World War made it impossible for him to return to Cambridge anytime soon. Early in 1940, he accepted the position of a special reader for six months in the department of physics (of which C.V. Raman was professor and head) of the Indian Institute of Science at Bangalore, and delivered a series of lectures on cosmic rays. At that time, Nobel Laureate R.A. Milikan of the California Institute of Technology, USA, was also in the Institute delivering lectures on the experimental methods and results of cosmic ray research. Milikan and his associates also conducted rubber balloon experimental flights at Bangalore, Agra and Peshawar.

In the middle of 1940, Bhabha was offered an honorarium of Rs 1000 per month by the Sir Dorabji Tata Trust of Bombay to conduct his own investigations on cosmic rays. With more financial aid from this Trust, Bhabha succeeded in setting up a Cosmic Ray Research Laboratory at the Indian Institute of Science, of which he was appointed reader-in-charge but with the title of professor as a personal distinction. He gave up the idea of returning to England and devoted himself to the national cause of promoting advanced scientific research in India. He received two offers, a professorship of physics at the University of Allahabad and also at the Indian Association for the Cultivation of Science at Calcutta. Declining both, he chose to remain at the Indian Institute of Science at Bangalore. He created a small group which included V.A. Sarabhai (who later became the architect of India's Space Research Programme) and conducted cosmic ray experiments at a height of about

35,000 feet with the help of US Air Force planes then situated at Bangalore. Such experiments were also carried out by Sarabhai in Kashmir at a height of about 14,000 feet. Bhabha, on his part, worked on certain aspects of theoretical mathematical physics and published several papers. While at the Institute, he was elected a Fellow of the Royal Society of London and was also awarded the coveted Adams Prize of the Cambridge University.

Bhabha's innermost desire was to build a scientific institution of excellence in India where bright young minds could venture into frontier areas of science with full academic freedom. His approach to institution-building was different from that of the others. It centred on accomplished individual scientists as the nucleus for further development of research in their own areas. This proved a success and was in sharp contrast with the general trend of creating first the departments of scientific subjects and then recruiting scientists. In 1944, Bhabha wrote a letter to the Sir Dorabji Tata Trust in which he emphasized that at that time there was no large school of research in India for dealing with fundamental problems of research, and that it was essential for the country to have one such vigorous school. His reasoning was that when nuclear energy was successfully applied to electric power production, India would not have to look abroad for its experts, but would find them readily available in the country itself. J.R.D. Tata, the chairman of the Sir Dorabji Tata Trust, concurred with Bhabha's lofty ideals for promoting advanced research in national interest, and encouraged him to submit a formal proposal for setting up such an institute. With the financial support of the Sir Dorabji Tata Trust and on the initiative of H.J. Bhabha, the Tata Institute of Fundamental Research came up in the middle of 1945, and Bhabha moved from Bangalore to Bombay to nurture this Institute.

Nuclear Physics: C.V. Raman, M.N. Saha, and S.N. Bose had no doubt won international acclaim for their path-breaking investigations in the then frontier areas of physics. The farsighted Raman and Saha had also thought of promoting research in nuclear physics, each in his own institution. R.S. Krishnan, a student of Raman, was deputed by Raman to work in the Cavendish Laboratory, Cambridge. On his return from Cambridge in 1940, R.S. Krishnan submitted a proposal for setting up a nuclear physics laboratory at the Indian Institute of Science's department of physics. M.N. Saha had also sent his student, B.D. Nagchaudhuri to the Lawrence Radiation Laboratory at Berkeley in USA for a doctorate degree. He had made arrangements for a cyclotron to be shipped from there to India in 1942 for its erection and operation in the premises of the Calcutta University to conduct research in nuclear physics. While the magnet and other parts of the cyclotron had arrived, the important high vacuum pumps that were shipped later from Berkeley to Calcutta, could not get across since the freighter carrying them was torpedoed by the Japanese in the Pacific Ocean. Thus Saha's initiative for the promotion of nuclear research did not meet with the desired success.

At the Indian Institute of Science, too, the result was not different, but the story had a different pathway. R.S. Krishnan, under the direction of C.V. Raman, put forward his proposal for research and teaching in nuclear physics, strengthening his proposal for the utilization of the new equipment envisaged for the department of power engineering for research in nuclear physics.

Around this time, early in 1946, the Government of India had constituted an Atomic Energy Board at the national level as well as an Atomic Energy Research Committee under the Council of Scientific and Industrial Research. The latter had eight members including H.J. Bhabha as its chairman, M.N. Saha, S.S. Bhatnagar, and K.S. Krishnan among them. In order to examine the proposal of R.S. Krishnan, the Council of the Indian Institute of Science also set up a special committee under the chairmanship of its director J.C. Ghosh, with H.J. Bhabha, K.S. Krishnan, M.N. Saha, and H.J. Taylor (of Wilson College,

Bombay) as its members. Bhabha, Bhatnagar and Taylor did not favour this proposal, pointing out that all large-scale research in nuclear physics in the near future should be concentrated at one big centre to make use of the rather limited funds of the government in an effective manner. Besides, Bhabha thought that such a centre would provide the environment needed by trained nuclear scientists for their mutual interaction, which would bring out the best from each of them. Bhabha thus prevented the Indian Institute of Science from having a nuclear research facility, a lacuna that still persists in that institution, though it has a wide spectrum of research covering practically all other branches of science and technology.

Bhabha believed firmly that the growth of the Indian economy depended upon the country's electricity-generating capacity. He was convinced that the production of atomic power in a phased manner would go a long way in augmenting the total production of environment-friendly electricity in the country. In 1948, soon after Independence, the Atomic Energy Act was passed by the Parliament, and the Atomic Energy Commission was set up with Bhabha as its first chairman. Right until his untimely death in a plane crash in a snowstorm on Mount Blanc on 24 January 1966, Bhabha dedicated his life to the strengthening of India's atomic energy programme with the unstinted support of Prime Minister Jawaharlal Nehru. The Atomic Research Centre at Trombay, which he had set up and nurtured, was re-named the Bhabha Atomic Research Centre in his honour by Prime Minister Indira Gandhi in January 1967. The recipient of several honours and awards, both Indian and foreign, Bhabha was elected president of the first international conference on the peaceful uses of atomic energy which was organized by the United Nations at Geneva in 1965, a glowing tribute to an Indian atomic scientist with an impressive international reputation, who advocated the peaceful use of atomic energy for human welfare all over the world.

Other Distinguished Scientists: Along with the pioneering scientists who had made outstanding researches and won international acclaim, were other distinguished scientists like K.S. Krishnan, K.R. Ramanathan, Sisir Kumar Mitra, D.N. Wadia. S.P. Agharkar, U.N. Brahmachari and S.L. Bhatia; but the list is by no means exhaustive. K.S. Krishnan, who was associated with C.V. Raman in the discovery of the *Raman Effect*, had made significant contributions in the field of crystals and optical and electric properties of liquids. K.R. Ramanathan who had worked with Raman on the molecular scattering of light, developed an electro-magnetic theory of light scattering. He had also made important studies in meteorology. Sisir Kumar Mitra had to his credit remarkable contributions to the development of broadcasting in India, being a specialist in radio physics and wireless; his book *The Upper Atmosphere* is a classic in this field. D.N. Wadia was a reputed geologist and shed scientific light on the Himalayan geology in particular. S.P. Agharkar, an eminent botanist, was associated with the Indian National Science Academy from its very beginning and worked hard purposefully to place it on a solid foundation. U.N. Brahmachari was well known for his research on the bacteriology of *Vibrio cholerae*, which causes cholera, while S.L. Bhatia was noted for his work in physiology.

Science: The Emergent Picture

Even in the colonial dispensation, the canvas of scientific research in India was a wide one—including physics, chemistry, biochemistry, botany, zoology, medicine, geology, anthropology, archaeology and other fields as well. Scientific departments of the central and provincial governments either set up new research institutes or expanded the activities of the existing ones. A good number of universities instituted courses for Honours degrees and advanced studies in scientific subjects with adequate laboratory facilities, which noticeably augmented the nature and extent of scientific research. Many Indian students and researchers were involved in original scientific investigations. These evoked the remark from the Nobel Laureate, Lord Rutherford, in his Presidential Address to the Indian Science Congress Association (January 1939; read posthumously) that 'India is now taking an honourable part and an ever-increasing share in the advance of knowledge in pure science'.

The socio-political environment that had surfaced from the last quarter of the nineteenth century had an invigorating effect on the progress of science in India. It is not insignificant that in 1876, S.N. Banerjee, a former civil servant, founded the Indian Association which heralded the dawn of organized criticism of the British colonial policies, and in the same year, as noted already, M.L. Sircar and others established the Indian Association for the Cultivation of Science. Since its inception in 1885, the Indian National Congress, which spearheaded the freedom movement, had been seriously concerned with India's industrial progress as well as scientific and technical education. In 1899 it adopted a resolution hailing J.N. Tata's concept of establishing an institution of higher learning and his intended philanthropic contribution for its sustenance. In 1905, the Congress also started the *swadeshi* as well as the national education movement. Even though the educational enterprise did not go on expected lines, it ushered in new vistas of thinking on education and in furtherance of national resurgence.

After the First World War and under dedicated leaders of great vision and exemplary integrity, of whom Mohandas Karamchand Gandhi (1861-1948) and Jawaharlal Nehru (1889-1964) were in the forefront, the freedom movement gathered new momentum. The age-old casteism that had been the divisive element of Indian traditional society, lay dormant and the upliftment of the downtrodden assumed paramount importance. Another divisive factor in the form of regionalism was spurned as the new spirit of nationalism came to the fore. Gandhiji's adherence to and practice of non-violence, truth and simple living as well as his conviction in the purity of means to achieve the ends, had a profound impact on the elite and the laity alike. This new value-based sociopolitical atmosphere had a stimulating effect on those who were involved in the promotion of education and scientific research. It is interesting to note that between 1928 and 1930, *Pūrṇa Swaraj*, the ultimate goal of the freedom movement (complete independence from foreign rule, and self governance), assumed a concrete form in the declaration of Independence Day by the Indian National Congress. During the same time, India's scientific acumen attained great heights with the path-breaking discovery of the *Raman Effect* and the award of the Nobel Prize to Raman. In 1929, the Imperial Council of Agricultural Research was established. Some parallels between certain facets of the freedom struggle and Indian scientific landmarks can also be observed. For example, in 1914, the year in which the Indian Science Congress Association was in place, the ebullient national leader, Bal Gangadhar Tilak was released from imprisonment and this provided a new stimulus to the freedom movement. The 1935 Act of the colonial government of India granted limited provincial autonomy and ushered in the system of an elected legislature which, it was thought then, was in many ways a blueprint for Independence. In the same year the National Institute of Sciences of India (renamed the Indian National Science Academy in 1970) came into existence as a national coordinating scientific institution, which in 1945, received recognition from the Government of India as a premier scientific society representing all branches of science. The year in which the Quit India Movement was launched (1942) also witnessed the emergence of the Council of Scientific and Industrial Research which was to play a major role in the establishment of a chain of national laboratories immediately after Independence.

The broad spectrum of scientific research in various areas displayed not only the high scientific attainments of international relevance and appreciation, but also several others of theoretical and practical importance. These were, however, the results of the interest and expertise of individual scientists. But, not all of them had practical applications leading to societal betterment. There was also no general policy or planning of scientific investigations either by the central or provincial governments. Even so, H.J. Bhabha and S.S. Bhatnagar, each in his own way, were involved in the utilization of science and technology for national progress on the eve of Independence.

Around the same time was one great leader, Jawaharlal Nehru, who not only catalysed the freedom movement but also inspired and encouraged scientific development, since he was convinced of the role of science and technology as effective instruments for elevating the living standards of the vast masses of the people of India.

Gandhiji had a different perception of science and technology; for, he saw religion, not science or technology, as the means to foster the moral strength of an individual and of the society as a whole. Jawaharlal Nehru described him as the 'greatest Indian of the day and the greatest Hindu of the Age...a man of God'. However, Gandhi was not totally opposed to science. He said: 'What I object to is the craze for machinery, not machinery as such.' He was an uncompromising critic of industrial civilization which, he envisioned, would be a surrender of human beings to machinery at the cost of moral principles and humanism—a prophetic vision indeed! His concept of Indian nationhood was the ensemble of harmonious village communities living together with religious tolerance, and a thriving village economy through small cottage industries. His approach was in contradistinction to the modern scientific and technological advancement of the nation.

Nehru, Religion and Science: Jawaharlal Nehru, who paved the way towards modernity in India, had a different viewpoint. Educated in the principles of science in England, he was not a laboratory scientist, but a man of science in the much larger laboratory of life. He was convinced of the rationality of science

Jawaharlal Nehru

and of the substantial role of technology in the uplift of the material standards of life of all people. During his imprisonment by the colonial rulers from August 1942 to September 1944, he pondered over the future of India, and also reflected upon science, religion and philosophy in the Indian as well as global context. About religion he wrote:

> 'Religions have helped greatly in the development of humanity, they have laid down values and standards, and pointed out principles for the guidance of human life. But with all the good they have done, they have also tried to imprison truth in set forms and dogmas, and encouraged ceremonies and practices which soon lose all their meanings, and become mere routine. Instead of encouraging curiosity and thought, religions have preached submission to nature, to established churches, to the prevailing social order. The belief in a supernatural agency which ordains everything has taken the place of reasoned thought and inquiry.' (*Discovery of India*; pp. 21-22; 620-622)

Nehru was not unaware of the limitations of science too. He pointed out that science had 'ignored the ultimate purpose of life, and looked at facts alone'. Yet, he emphasized:

> 'It is the scientific approach, the adventures and the critical temper of science, its search for truth and new knowledge, its refusal to accept anything without testing...its reliance on facts and not preconceived notions, all of these are necessary not merely for the applications of science but for life itself and the solution of its many problems.'

While delivering the convocation address at the Allahabad University in 1946, Nehru expressed his conviction in these words: 'It is Science and Science alone that could solve the problems of hunger and poverty, of insanitation and illiteracy, of superstition and deadening custom and tradition, of vast resources running to waste, of a rich country inhabited by starving people.' Realizing the importance of the scientific attitude or the 'scientific temper' as he used to call it, he emphasized time and again that a dispassionate 'scientific temper' would be essential for the solution of most problems of life and of India.

During the seventeen years of his stewardship as Independent India's first prime minister, Nehru laid a solid foundation for the multi-level growth of science and technology. More importantly, he committed the entire nation to science and technology by moving a Scientific Policy Resolution in the Indian Parliament in March 1958—perhaps the first of its kind in the world.

> 'The aims of this policy are, among others, "to foster, promote and sustain, by all appropriate means, the cultivation of science and scientific research in all its aspects—pure, applied and educational; to ensure an adequate supply, within the country, of research scientists of the highest quality, and to recognize their work as an important component of the strength of the nation; to ensure that the creative talent of men and women is encouraged and finds full scope in scientific activity; to encourage individual initiative for the acquisition and dissemination of knowledge and for the discovery of new knowledge in an atmosphere of academic freedom; and, in general, to secure for the people of the country all the benefits that can accrue from the acquisition and application of scientific knowledge...as also to accord the scientists an honoured

> position by associating them with the formulation of policies."'

Since Independence and over the past six decades, India has registered substantial progress in scientific agriculture, atomic energy, space technology, industrial production, information technology, pharmaceuticals, defence needs and other areas. Alongside, India's traditional sciences have been receiving financial and institutional support from both the central and state governments. Their past continues to be relevant to the present. Traditional astronomy coexists with modern astronomy, the *pañcāṅga* playing its role in the sociocultural life of people. Traditional medicine has been complementary to modern medicine. Even traditional technologies have carved a niche for themselves amidst modern sophisticated technologies. India's past also resides in its present in various ways.

FURTHER READING

[The following is an indicative bibliography of some important published books suggested for further reading. But, interested readers can find extensive bibliographies at the end of these books. They include primary sources and important articles.]

CHAPTER 1: AN INTRODUCTORY OVERVIEW

Basham, A.L., *The Wonder That was India, (*Calcutta University Press, Calcutta, 1918).

Bloomfield, Maurice, *The Religion of the Veda*, (G.B. Putnam Sons, New York, 1908).

Bose, D.M., Sen, S.N., and Subbrayappa, B.V. (eds), *A Concise History of Science in India* (revised ed.), (University Press, Hyderabad, 2008).

Brennand, W., *Hindu Astronomy*, (Caxton Publication, Delhi, 1988).

Dasgupta, S.N and De, S.K., *A History of Sanskrit Literature,* (Calcutta University, Calcutta, 1947).

Hiriyanna, M., *Essentials of Indian Philosophy,* (George Allen and Unwin, London, 1949).

Macdonell, A.A., *India's Past,* (Clarendon Press, Oxford, 1927).

Mazumdar, R.C., *History and Culture of Indian People,* vols I-X, (Bharatiya Vidya Bhavan, Bombay, 1950-59).

Pingree, D., *Census of the Exact Sciences in Sanskrit,* Series 1-4, (American Philosophical Society, Philadelphia, 1970-81).

Rahman, A., Alvi, M.A., Ghori, S.A.K., and Murthy, K.V.S., *Science and Technology in Medieval India: A Bibliography of Source Material in Sanskrit, Arabic and Persian,* (Indian National Science Academy, New Delhi, 1982).

Sankalia, H.D., *The Prehistory and Protohistory of India and Pakistan,* (Deccan College, Poona, 1974).

Sen, S.N., *A Bibliography of Sanskrit Works on Astronomy and Mathematics,* (Indian National Science Academy, New Delhi, 1966).

Sarma, K.V., *A Bibliography of Kerala-based Astronomy,* (Vishvesvaranand Institute, Hoshiarpur, 1972).

Siddiqi, M.Z., *Studies in Arabic and Persian Medical Literature,* (Calcutta University, Calcutta, 1959).

Subbarayappa, B.V. (ed.), *Science in India: Past and Present,* (Nehru Centre, Mumbai, 2007).

Winternitz, M., *A History of Indian Literature,* I-III, (Calcutta University, 1959), (Motilal Banarsidass, Delhi, 1963).

Zimmer, H., *Hindu Medicine,* (Johns Hopkins University, Baltimore, 1948).

CHAPTER 2: EARLY SCIENTIFIC IDEAS AND PRACTICES

Altekar, A., *Education in Ancient India,* (Nanda Kishore and Bros, Benares, 1951).

Bhattacharya, K. C., *Studies in Philosophy,* (Progressive Publishers, Calcutta, 1956).

Bhaduri, S., *Studies in Nyāya-Vaiśeṣika Metaphysics,* (Bhandarkar Oriental Series, Poona, 1948).

Davids, Rhys, *Outlines of Buddhism: A Historical Sketch,* (MacMillan, London, 1934).

Das Gupta, S.N., *The History of Indian Philosophy,* vols I-V, (Cambridge University Press, Cambridge, 1922-25).

Deussen, P., *The Philosophy of Upanishads,* (T & T Clark, Edinburgh, 1919).

Dutta, B., *The Science of the Śulba,* (Calcutta University, 1932).

Griffith, R.T.H., *The Hymns of the Ṛgveda,* (E.J. Lazarus, 1891; 1963).

Hiriyanna, M., *Outlines of Indian Philosophy,* (Allen and Unwin, London, 1932); *The Essentials of Indian Philosophy,* (MacMillan, New York, 1949).

Kane, P.V., *History of Dharmaśāstra,* vol. V, (Bhandarkar Oriental

Research Institute, Poona, 1974).

Kapali Shastry, T.V., *Lights on the Upanişhads,* (Sri Aurobindo Ashram, Pondicherry, 1945).

Keay, F., *Ancient Indian Education,* (Oxford University Press, Oxford, 1918).

Keith, A.B., *The Religion and Philosophy of the Veda and Upanişads,* Harvard Oriental Series, (reprint), (Motilal Banarasidass, Delhi, 1970).

Mackay, E.J.H, *Further Excavations at Mohenjo-daro,* 2 vols, (Archaeological Survey of India, New Delhi, 1938); *Excavations at Chanhu-daro,* (New Delhi, 1936).

Mandal, K.K.A., *A Comparative Study of Space and Time in Indian Thought,* (Chowkhamba Sanskrit Series, Varanasi, 1968).

Majumdar, G.P., *Botany in Ancient India: Past and Present,* (Ramakrishna Centenary Memorial, Calcutta, 1961); *Upavana Vinoda,* (The Indian Research Institute, Calcutta, 1935); *Vanaspati,* (Calcutta University, 1927); 'The History of Botany and Allied Sciences in Ancient India' in *Archives International d' History les Sciences,* (International Academy of History of Science, Paris, 1955).

Majumdar, G.P. and Banerjee, S.C. (eds), *Kṛṣi-Parāśara,* (Asiatic Society, Calcutta, 1963).

Marshall, J. (ed.), *Mohenjo-daro and the Indus Civilization,* 3 vols, (Archaeological Survey of India, New Delhi, 1982).

Mishra, Umesh, *Conception of Matter According to Nyāya-Vaiśeṣika,* (Allahabad University, Allahabad, 1936).

Mookerjee, Radhakumud, *Ancient Indian Education,* 3rd edn (Motilal Banarasidass, Delhi, 1960).

Roy, Mira, 'Vedic Medicine' in B.V. Subbarayappa (ed.), *Medicine and Life Sciences in India,* (Centre for Studies in Civilizations, New Delhi, 2001).

Sarkar, N.N. (ed.) with an Introduction, *Vṛkṣāyurveda of Parāśara,* (*Journal of the Asiatic Society of Bengal,* 16 (1), 1950).

Stcherbatsky, Th., *The Buddhist Logic,* 2 vols, (Leningrad, 1925).

Sen, S.N., 'Mathematics' in *A Concise History of Science in India,* (University Press, Hyderabad, 2009); 'Astronomy', ibid.

Sharma, Chandradhara, *A Critical Survey of Indian Philosophy,* (Nand Kishore and Bros., Benaras, 1952).

Sharma, H.D. and Sardesai, N.G. (eds), *Amarakośa* of Amarasiṃha, (Oriental Book Agency, Bombay, 1941).

Subbarayappa, B.V., *Indian Perspectives on the Physical World,* (Centre for Studies in Civilizations, New Delhi, 2004).

Vats, M.S., *Excavations at Harappa*, 2 vols, (Archaeological Survey of India, Delhi, 1938-40).

Whitney, W.D., *Atharvaveda Saṃhitā*, 2 vols, (Harvard University Press, MSS., Cambridge, USA, 1905; reprint, 1984).

Tatia, Nathumal, *Studies in Jaina Philosophy,* (Jaina Cultural Research Society, Benares, 1958).

Wheeler, M., *The Indus Civilization,* (Supplement to the Cambridge History of India, 1968).

Zysk, G.K., *Religious Healing in the Veda*, (American Philosophical Society, *Transactions*, new series, vol. 75, 1985).

CHAPTERS 3 and 4: ASTRONOMY: ITS COMPUTATIONAL AND OTHER DIMENSIONS and MATHEMATICS: APPROACH AND ACHIEVEMENTS

Bag, A.K., *Mathematics in Ancient and Medieval India*, (Chaukhambha Orientalia, Delhi, 1979).

Balachandra Rao, S., *Indian Mathematics and Astronomy*, (Bharatiya Vidya Bhavan, Bangalore, 2004).

Balagangadharan, R., 'Mathematical Analysis in Medieval Kerala', B.V. Subbarayappa and N. Mukunda (eds), *Science in the West and India*, (Himalaya Publishing House, Bombay, 1995).

Bhanu Murthy, T.S., *A Modern Introduction to Ancient Indian Mathematics*, (Wiley Eastern, New Delhi, 1992).

Cajori, F, *A History of Mathematical Notations*, 2 Vols, (Dover Publications, Chicago, 1928).

Chatterjee, Bina (trans.), *Śiṣyadhīvṛddhidatantra* of Lallācārya, (Indian National Science Academy, New Delhi, 1981).

Crozer, V.S, *A Survey of Mathematics*, (New York, 1968).

Datta, B. and Singh, A.N., *History of Hindu Mathematics*, parts I and II, (Bharatiya Kala Prakashan, Delhi, 2001).

Dixit, S.B., *Bharatiya Jyotish Shastra*, parts I and II, translated from Marathi into English by R.V. Vaidya, (Government of India, 1969 and 1981.

Dvivedi, Sudhakara (ed.), *Siddhāntaśiromaṇi* of Bhāskara II, (Kashi Sanskrit Series, Varanasi, 1929); English translation of *Gaṇita* by Arka Somayaji, (Rashtriya Sanskrit Vidyapeetha, Tirupati, 1980).

Gangooly, Phanindralal (ed.), translation with notes by Ebenezer Burgess, *Sūryasiddhānta,* (reprint: Motilal Banarsidass, Delhi, 2000).

Gaṇitayukti-bhāṣā of Jyeṣṭhadeva, ed. tr. by K.V. Sarma with Explanatory Notes by K. Ramasubramanian, M.D., Srinivas and M.S. Sriram, reprint, (Springer, 2009)

Gupta, R.C., 'Second Order Interpolation in Indian Mathematics up to the Fifteenth Century AD', *Indian Journal of History of Science*, 4 (2), (New Delhi, 1969).

Ifrah, Georges, *From One to Zero: A Universal History of Numbers,* (Viking Penguin Inc., New York, 1985,).

Kim Plofker, *History of Mathematics in India: From 500 BCE to 1800 CE*, (Princeton, 2009).

Kuppanna Sastry (trans. and notes), *Pañcasiddhāntikā* of Varāhamihira, (PPST Foundation, Adyar, Madras, 1993).

Kuppanna Sastry, T.S., and Sarma, K.V. (ed. and trans.), *Vedānga Jyotiṣa* of Lagadha (Indian National Science Academy, New Delhi, 1985).

Labat, Rene, *Manuel D'Epigraphie Akkadienne: Syllabaire, Signes, Ideogrammes*, (Paul Geuthner, Paris, 1986).

Needham. J., *Science and Civilization in China*, (Cambridge University Press, 1959).

Pingree, D., 'History of Mathematical Astronomy in India', *Dictionary of Scientific Biography*, XV, Supplement I, pp. 533-633 (New York, 1978).

Ramasubramanian, K and Sriram, M.S., *Tantrasangraha of Nīlakaṇṭha*, Somayāji, Hindustan Book Agency, New Delhi, 2011.

Sachau, E.C. (trans.), *Al-Bīrūnī's India,* 2 vols, (Low Price Publications, Delhi, 1989).

Sastry, Shama, *The Vedic Calendar* (Indian Antiquary, Calutta, 1912).

Sengupta, P.C., 'Aryabhata, the Father of Indian Epicyclic Astronomy', Journal of Department of Letters, (Calcutta University, 1929).

Sen, S.N. and Shukla, K.S. (eds): *History of Astronomy in India,* (Indian National Science Academy, New Delhi, 1985).

Shukla, K.S. (ed), *Mahābhaskarīyam of Bhāskara I,* with translation and notes, (Lucknow University, 1960).

Shukla, K.S. (ed. and trans.), *Vateśvara Siddhānta,* (Indian National Science Academy, New Delhi, 1985).

Sen, S.N, 'Astronomy' and 'Mathematics', in *A Concise History of Science in India,* (Universities Press, Hyderabad, 2009).

Shukla, K.S., and Sarma, K.V., (trans. and notes), *Āryabhaṭīyam,* (Indian National Academy, New Delhi, 1976); translated into English by Clark, W.E. (University of Chicago Press, Chicago, 1930).

Sriram, M.S., Ramasubramanian, K. and Srinivas, M.D. (eds), *500 Years of Tantrasaṅgraha,* (Indian Institute of Advanced Studies, Shimla, 2002).

Saraswathi, T.A., *Geometry in Ancient and Medieval India,* (Motilal Banarsidass, Delhi, 1979; reprint 1999).

Srinivasa Iengar, C.N., *The History of Ancient Indian Mathematics,* (The World Press Private Ltd, Calcutta, 1967).

Subbarayappa, B.V., *The Tradition of Astronomy in India,* (Centre for Studies in Civilizations, New Delhi, 2008).

Subbarayappa, B.V. and Sarma, K.V., *Indian Astronomy: A Source-Book,* (Nehru Centre, Bombay, 1985).

Somayaji, D.A., *A Critical Study of Ancient Hindu Astronomy in the Language of the Modern,* (Karnataka University, Dharwar, 1971).

CHAPTER 5: CHEMICAL TECHNIQUES AND ALCHEMICAL PRACTICES

Agrawal, D.P. and Chakrabarti, D.K. (eds), *Essays In Indian Prehistory,* (B.R. Publishing Corporation, Delhi, 1979).

Balasubramaniam, R., *Marvels of Indian Iron Through the Ages,* (Rupa & Co., New Delhi, 2008).

Banerjee, N.R., *The Iron Age in India,* (Munshiram Manoharlal, Delhi, 1965).

Bhat, M.R. (trans.), *Bṛhatsaṃhitā* of Varahamihira, (Motilal Banarsidass, Delhi, 1984).

Buchanan-Hamilton, F., *A Journey from Madras through the Countries of Mysore, Canara and Malabar,* vol. II (reprint), (Asian Educational Services, New Delhi, 1988).

Dixit, M.G., *History of Indian Glass,* (University of Bombay, Bombay, 1969).

Fazl, A., *Āin-i-Akbarī,* Blochmann, H. (ed.), 2 vols., (Asiatic Society, Calcutta, 1867-77).

Ghosh, A. (ed.), *An Encyclopaedia of Indian Archaeology,* vol I, (Motilal Banarsidass, Delhi, 1989).

Gopinatha Rao, T.A., *Elements of Hindu Iconography,* 3 vols, (Indological Book House, Varanasi, 1972).

Gururaja Rao, B.K., *Megalithic Culture in South India,* (University of Mysore, Mysore, 1979).

Hegde, K.T.M., 'The Painted Grey Ware of India', *Antiquity,* XLIX, (1975).

Joshi, D., *Rasaratnasamuccaya,* (Indian National Science Academy, New Delhi, 1987).

Kangle, R.P. (trans.), *The Kauṭilya Arthaśāstra,* 3 vols, (Bombay University, Bombay, 1960-65).

Neogi, P., *Iron in Ancient India,* (Indian Association for the Cultivation of Science, Calcutta, 1914); *Copper in Ancient India* (1918).

Rama Sastri, R. and Deveerappa, H. (eds), *Śivatattvaratnākara.,* (Oriental Research Institute, Mysore, 1974).

Ray, P., *History of Chemistry in Ancient and Medieval India,* (Indian Chemical Society, Calcutta,1956).

Ray, P.C., *History of Hindu Chemistry,* 2 vols, (Calcutta, 1902 and 1925).

Read, John, *Through Alchemy to Chemistry,* (G. Bell and Sons, London, 1957).

Shrinogdekar, G.K. (ed.), *Mānasollāsa* or *Abhilaṣitārthacintāmaṇi* of Someśvara, 3 vols. (Oriental Institute, Baroda, 1961).

Subbarayappa, B.V. (ed.), *Chemistry and Chemical Techniques in India,* (Centre for Studies in Civilizations, New Delhi, 1999); 'Chemical

Practices and Alchemy' in *A Concise History of Science in India,* (University Press, Hyderabad, 2008).

Sundararajan, K.V., *India's Religious Art,* (Cosmos Publication, New Delhi, 1983).

Vyas, R.T. (ed.), *Gandhasāra of Gangadhara and an Unknown Author's Gandhavāda,* (Oriental Institute, Vadodara, 1989).

CHAPTERS 6 and 7: AYURVEDA: A HOLISTIC MEDICAL SYSTEM and OTHER MEDICAL SYSTEMS: SIDDHA AND UNĀNI

Abdulah Sahib, P.M. and Thyagarajan, R. (eds), *Agastyar Kāṇḍam* (Tamil), (Palani Temple Siddha Publication, Madras, 1975).

Azmi, A.A., *Basic Concepts of Unani Medicine: A Critical Study,* (Jamia Hamdard, New Delhi, 1995).

Bhagvat, S.J., *A Short History of Aryan Medical Science,* (Rare Reprints, New Delhi 1981).

Dash, B. and Kasyapa, L. (eds and trans.), *Ayurvedasaukhya* of Todarananda, (Concept Publishing, Delhi, 1981); *Fundamentals of Ayurvedic Medicine,* (Bansal and Co., New Delhi, 1978).

Dwarkanath, C., *Introduction to Kāyachikitsā,* (Popular Book Depot, Bombay, 1959).

Filliozat, J., *The Classical Doctrines of Indian Medicine: Its Origins and Its Greek Parallels,* Dev Raj Chanana (trans.), (Munshiram Manoharlal, New Delhi, 1963).

Gruner, O.C., *A Treatise on the Canon of Medicine of Avicenna,* (Luzac and Co., London, 1936).

Jolly, J., *Indian Medicine,* C.G. Kashikar (trans.), (Munshiram Manoharlal, New Delhi, 1977).

Keshwani, N.H. (ed.), *The System of Medicine and Physiological Concepts in Ancient and Medieval India,* (India Publications, New Delhi, 1974).

Kunjalal, B., *Suśruta Saṃhitā* with English translation, 3 vols, (Chowkhamba Sanskrit Series, Varanasi, 1963).

Kuppuswamy, T.N., *Siddha Medicine,* (Tamil Nadu Government Publication, Madras, 1954).

Kutumbaiah, P., *Ancient Indian Medicine*, (Oriental Longman, Calcutta, 1962).

Leslie, C. (ed.), *Asian Medical Systems: A Comparative Study*, (University of California Press, Berkeley, 1976).

Meulenbeld, G.J., *A History of Indian Medical Literature*, 4 vols, (Groninger Oriental Studies, 15, Groninger, Egbert Forsten, 1999-2002).

Meulenbeld, G.J. and Dominik Wujastyk (eds), *Studies on Indian Medical History* (Seminar Papers), (Groninger, Egbert Forsten, 1987).

Mukhopadhaya, G.N., *History of Indian Medicine*, (Calcutta University, Calcutta, 1923).

National Siddha Formulary, (Ministry of Health, Government of India, New Delhi, 1978).

Pillai, N.K., *History of Siddha Medicine*, (Tamil Nadu Government Publication, 1979).

Ray, P. and Gupta, H.N., *Caraka Saṃhitā: A Scientific Synopsis:* (Indian National Science Academy, New Delhi, 1980).

Ray, P., Gupta, H.N. and Mira Ray, *Suśruta Saṃhitā: A Scientific Synopsis*, (Indian National Science Academy, New Delhi, 1980).

Sarma, P.V. (trans.), 4 vols, *Caraka Saṃhitā*, (Chawkhamba Orientalia, Varanasi, 1981);

Shah, M.H., *The General Principles of Avicenna's Canon of Medicine*, (Naveed Clinic, Karachi, 1966).

Siddiqi, M.A., *Unani Medicine,* (Institute of History of Medicine, New Delhi, 1961).

Subbarayappa, B.V. (ed.), *Medicine and Life Science in India*, (Centre for Studies in Civilizations, New Delhi, 2001).

Subramanian, S.V. and Madhavan, V.R. (eds), *Heritage of the Tamils: Siddha Medicine,* (International Institute of Tamil Studies Madras, 1984).

Singhal, G.D. and Patterson, T.J.S., *Synopsis of Ayurveda based on a Translation of the Suśruta Saṃhitā,* (Oxford University Press, New Delhi, 1993).

Sreekantha Murthy, K.R. (trans.), *Aṣṭāṅgahṛdayam* of *Vāgbhaṭa*, 3 vols, (Krishnadas Academy, Varanasi, 1995).

Udupa, K.N. and Singh, R.H. (eds), *Science and Philosophy of Indian Medicine,* (Vaidyanath Ayurveda Bhavan, Varanasi, 1978).

Velan, A.S., *Siddhar's Science of Longevity and Kalpa Medicine of India,* (Sakti Nilayam, Madras, 1963).

Zysk, K., *Medicine in the Veda,* (Motilal Banarsidass, New Delhi, 1977); *Asceticism and Healing in Ancient India: Medicine in the Buddhist Monastery,* (Oxford University Press, New York, 1991).

Zillur Rahman, S., 'Unani Medicine in India: Its Origin and Fundamental Concepts' in Subbarayappa, B.V. (ed.), *Medicine and Life Sciences of India,* (Munshiram Manoharlal, New Delhi 2001).

CHAPTER 8: THE NEW ENTRANT: MODERN SCIENCE IN INDIA

Anderson R.S., *Building Scientific Institutions in India: Saha and Bhabha,* (McGill University, Montreal, 1975).

Bernstein, H.T., *Steamboats on the Ganges: An Exploration in the History of India's Modernisation through Science and Technology,* (Orient Longmans, Bombay, 1960).

Chakrabarti, P., *Western Science in Modern India,* (Permanent Black, New Delhi, 2004).

Crawford, D.G., *A History of the Indian Medical Service 1600-1913,* 2 vols, (London, 1914).

Dasgupta, S., *Jagadis Chandra Bose and the Indian Response to Western Science,* (Oxford University Press, New Delhi, 1999).

Dasgupta, U. (ed.), *Science and Modern India: An Institutional History, c. 1784-1947,* (Centre for Studies in Civilizations, New Delhi, 2010).

Kumar, D., *Science and the Raj,* second edn, (Oxford University Press, New Delhi, 1997).

Markham, C.R., *A Memoir on the Indian Surveys,* (Kegan Paul, London, 1978).

Mazumdar, R.C., *History and Culture of Indian People, vol. 10, Part 2: British Paramountcy and Indian Renaissance,* (Bharatiya Vidya Bhavan, Bombay, 1959).

Mukherji, S.N., *History of Education in India: Modern Period,* (Acharya Book Depot, Baroda, 1955).

Mukherjee, S.N. and Subbarayappa, B.V, *Science in India: A Changing Profile*, (Indian National Science Academy, New Delhi, 1984).

Nehru, J., *Discovery of India*, Centenary edn, (Oxford University Press, New Delhi, 1985).

Phillimore, R.H., *Historical Records of the Survey of India,* 4 vols, (Government of India, Dehradun, 1945-58).

Prashad, B., *The Progress of Science in India During the Past Twenty-five Years*, (Indian Science Congress Association, Calcutta, 1938).

Qaisar, A.J., *The Indian Response to European Technology and Culture,* (Oxford University Press, New Delhi, 1982).

Ranganathan, S.R., *Ramanujan: The Man and the Mathematician,* (Asia Publishing House, Bombay, 1967).

Ray, P.C., *Life and Experiences of a Bengali Chemist*, (Chukerverty, Chatterjee & Co., Calcutta, 1932).

Richey, J.A., *Selections from Educational Records*, vol. II, (Government of India, Calcutta, 1922).

Sen, S.N., *Scientific and Technical Education in India (1781-1900),* (Indian National Science Academy, New Delhi, 1991); *A Century for the Cultivation of Science: Early Period,* (Indian Association for the Cultivation of Science, Calcutta, 1976).

Sharp, H., *Selections from Educational Records*, vol. 1, (Government of India, 1920).

Singer, C., *A Short History of Science to the Nineteenth Century,* (Clarendon Press, Oxford, 1947).

Sircar, M.L., 'On the Desirability of a National Institute for the Cultivation of the Science by the Natives of India', (*Calcutta Journal of Medicine,* 2, 1869).

Subbarayappa, B.V., 'Western Science in India up to the End of the Nineteenth Century' in *A Concise History of Science in India*, revised edn. (Universities Press, Hyderabad, 2008).

Tandon, P.N. and Mohan Ram, H.Y. (eds), *Pursuit and Promotion of Science: Indian Experience,* (Indian National Science Academy, New Delhi, 2001).

Venkataraman, G., *Journey into Light: Life and Science of C.V.Raman,* (Indian Academy of Sciences, Bangalore, 1988).

Zaheer, B., *The Science of Empire: Scientific Knowledge, Civilization and Colonial Rule in India,* (State University of New York Press, Albany, 1996).

INDEX

www.ingramcontent.com/pod-product-compliance
Lightning Source LLC
Chambersburg PA
CBHW062132160726
48282CB00027B/796

* 9 7 8 9 3 8 6 2 4 0 4 1 5 *